NUMERICAL MATHEMATICS

NUMERICAL MATHEMATICS

Theory and Computer Applications

Carl-Erik Fröberg
Lund University

154 02 **1985**

The Benjamin/Cummings Publishing Company, Inc.

Menlo Park, California • Reading, Massachusetts
Don Mills, Ontario • Wokingham, U.K. • Amsterdam • Sydney
Singapore • Tokyo • Mexico City • Bogota • Santiago • San Juan

Sponsoring Editor: Richard W. Mixter
Cover Design: Leigh McLellan
Production Coordination: Editing, Design & Production, Inc.

Manufactured in the United States of America

ABCDEFGHIJ-HA-8987654

Library of Congress Cataloging in Publication Data

Fröberg, Carl Erik.
  Numerical mathematics.

  Rev. ed. of: Introduction to numerical analysis.
2nd ed. 1969.
  Bibliography: p.
  Includes index.
  1. Numerical analysis.   2. Numerical analysis—Data
processing.   I. Fröberg, Carl Erik.   Lärobok i
numerisk analys.   English.   II. Title.
QA297.F6813   1985   511   85-3880
ISBN 0-8053-2530-1

Preface

This textbook can be considered to be a third edition of *Introduction to Numerical Analysis,* 2nd ed., Addison-Wesley, 1969. However, it has been completely revised and restructured, and for several reasons it seemed natural to launch a new title. This title is actually in line with modern trends at most universities and with the material presented in the book.

The contents is now divided into five parts, the first presenting a fairly broad mathematical introduction. This arrangement needs some comments. Most numerical methods depend on some rather general mathematical ideas. For many of them, for example, the basic theory of calculus and algebra must be considered to be a reasonable prerequisite; while other topics such as vectors and matrices, series expansions, orthogonal functions, and linear operators might be less familiar. Hence, a rather thorough exposition emphasizing the most common applications has been considered to be sensible. Some other concepts such as special functions and Laplace transformation are not often found in numerical analysis texts; however, they are believed to add considerably to the usefulness and value of the book. Already the previous editions have had some reputation as reference books on a modest scale, and it seems desirable to preserve this property.

The other four parts cover more traditional areas. A typical elementary course could be selected as follows: Chapters 1, 2.1, 2.2, 3.1, 5.1, 5.2, 10.1, 10.2, 10.4, 11.1 to 11.4, 14, 15.1 to 15.4, 15.7, 17.1, 17.2, 18, 19.1 to 19.4. The reader will find numerous references to the mathematical introduction, which may be consulted whenever there is a need for more precise information on the mathematics behind the numerical methods. Concerning the subjects treated in the text, several modern important problems have also been discussed. We only mention pseudoinverses, orthogonal polynomials, fast Fourier transforms, spline approximation, stiff differential equations, the finite element method, and a great selection of methods for solving boundary value as well as eigenvalue problems.

It has become customary to present computer programs in some suitable language such as FORTRAN or Pascal. Still, the decision to refrain from such programs was not a difficult one. With the availability of large and modern program packages, there is hardly need for such programs in a modern textbook. In fact, this view is shared by many reviewers, and it seems safe to assume that such programs will be run only occasionally. Further, programs that are really good should soon become part of a suitable package.

Most of the exercises in this text have been included in examinations in numerical mathematics at Lund University. All of them are provided with answers, and in several cases hints for solution are given.

Quite a few of my colleagues have read all or part of the manuscript and have supplied valuable corrections, comments, and suggestions. In particular I want to thank professor Axel Ruhe, Gothenburg; Torgil Ekman, Lund; and Jan Bohman, Lund. I also express my sincere gratitude to two anonymous referees.

Carl-Erik Fröberg
Lund, February 1985

Contents

Part One

MATHEMATICAL INTRODUCTION

CHAPTER 1

Numerical Computation

Numerical analysis is a science—computation is an art.

§1.1. REPRESENTATION OF NUMBERS

In the history of mankind, numbers have always played an important role. For obvious reasons, our knowledge of all phases in the development of number representation in different cultures is rather incomplete. Manipulations with just one or a few units could probably be mastered without difficulty, but problems arose when it became necessary to define higher levels, where each unit on one level was supposed to be equivalent to a certain number of units on the level below. Different choices were made in different cultures. The Babylonians chose 60 as base, and we still have reminders in our time units and in our angular measures. In some ancient cultures, such as the Maya Indians, the number 20 played a similar role (also consider the French word *quatre-vingt* = 4 times 20 = 80). The number 10 represented a very natural choice, founded on the simple fact that we have 10 fingers. As a matter of fact, 10 strikes a good balance between the number of symbols in the system and the number of digits necessary to represent a given number. The Roman system was more like a currency system with special notations I, V, X, L, C, D, and M for 1, 5, 10, 50, 100, 500, and 1000. A number was represented in this system by use of just additions and subtractions, but the system was very clumsy and ill-suited for any kind of numerical computation.

A decisive step forward was taken by the Hindus about A.D. 500. Instead of introducing special symbols for higher units (10, 100, 1000 and so forth), they used the ordinary digits 0, 1, 2, . . . , 9 with the important convention that the position of a digit in a number carries information about its value. For instance, a position two steps to the left of the unit digit indicates that the digit must be multiplied by 100. Similarly, for fractions the value of a certain digit was one tenth, one hundredth, and so forth, depending on whether the position was one, two, or more steps to the right of the decimal point. This scheme turned out to be extremely practical, in particular for performing all kinds of numerical computations.

1

A *position system* is defined with respect to a certain *base*, which is always a positive integer $N > 1$; the best known examples are $N = 2$ (binary); $N = 8$ (octal); $N = 10$ (decimal); $N = 12$ (duodecimal); and $N = 16$ (sedecimal or hexadecimal, with a linguistic preference for the purely Latin word sedecimal against hexadecimal, which is half Greek and half Latin). It should be noted that $N = 2, 8,$ and 16 are closely related since they are all powers of 2; conversion between these systems is practically trivial.

A real positive number x has a unique representation

$$x = a_M N^M + a_{M-1} N^{M-1} + \cdots + a_0 + a_{-1} N^{-1} + a_{-2} N^{-2} + \cdots$$

with M a finite number and $0 \leq a_k < N$. As a rule, the number of coefficients with negative indices is infinite; only if $x = A/N^n$ with A and n integers do we have an exact representation.

Conversion to another number system with base Q is performed within the number system with base N by successive divisions with Q for the integer part and successive multiplications with Q for the fractional part. We must then have Q symbols for the digits $0, 1, 2, \ldots, Q - 1$.

Fixed and Floating Point Representation

The representation used so far is said to be fixed: all digits stand directly for their ordinary values. However, if numbers get very large or very small, this representation is rather impractical, and instead we can use a so-called floating point representation:

$$x = p \cdot N^q$$

where p is the *mantissa*, usually adjusted to a value close to 1, while N is the base and q the *exponent*, an integer that can be positive, negative, or zero. The notation E is often used as a delimiter between the two parts of a number with base 10, e.g., 1.732E3 with the meaning 1732, or 2.4168E − 8 with the meaning 0.000000024168.

When numerical computations are performed on numbers represented in a computer, special care must be exercised because the traditional rules (associative, commutative, and distributive) are no longer universally valid. We demonstrate here that there are cases when even a simple summation gives different results on forward and backward evaluation.

Suppose that we want to compute $\sum_{n=1}^{N} (1/n^2)$ on a computer that can store numbers with at most 23 binary digits. With summation done forward, all numbers greater than about 3000 will contribute nothing, since $3000^{-2} < 2^{-23}$ and the accumulated sum is already >1.6. We say that all terms from about $n = 3000$ are "out-shifted." If instead we do the summation backward, the sum of the terms with index greater than M can be estimated to be less than $\int_M^\infty dx/x^2 = 1/M$; for $M = 3000$ this will be approximately 0.000333, which is appreciably larger than $2^{-23} \simeq 0.000000119$. The results for $N = 4000$ and $10,000$ are given below; S_1 is obtained by forward summation and S_2 by backward summation.

N	S_1	S_2
4000	1.64464	1.64468
10000	1.64464	1.64483

The exact value of the infinite sum is $\pi^2/6 \simeq 1.64493$.

§1.2. ON THE ORIGIN AND GROWTH OF ERRORS

There are three main sources for errors, at least for what could be called legitimate errors. First, when a problem is to be treated there are usually initial data, which may originate in some kind of physical measurement and hence are only approximate. Second, as a rule we must apply an approximate method, e.g., replacing an infinite series by a finite one, a derivative by some difference expression, or an integral by a weighted sum. Third, all numerical computations must be carried out with a finite number of digits; i.e., although most numbers are irrational, we have to approximate them with rational numbers. These three error types are called *initial*, *truncation*, and *rounding* errors.

Obviously, we have not considered a fourth error type, namely the one that is due to malfunctioning of a human being, often in connection with a computer program, or of a computer. Such errors are quite common, but their treatment falls essentially outside the scope of the present discussion. However, we stress the importance of reliable checking possibilities; in many cases identities or alternative formulas can be used.

Initial errors may be associated with certain parameters entering an equation, but they may also appear in the starting values of a differential equation, for instance. What is of special importance is the long range influence on the computed solution, a feature that will be discussed to some extent in the next section.

Truncation errors, as a rule, appear when an infinite process is replaced by a finite one. If we desire an infinite sum $S = \sum_{k=0}^{\infty} a_k$ we instead compute a finite sum $S_n = \sum_{k=0}^{n} a_k$. In some cases we may be able to estimate the remainder term $R_n = S - S_n$, in some cases not; under all circumstances a truncation error will be created. Quite often the remainder term will behave as $C \cdot h^p$, where h is an interval length and p an integer ≥ 1. Denoting the truncation error with ε, this fact can be expressed as $\varepsilon = O(h^p)$.* Typical situations are:

$$y'_n = (y_{n+1} - y_{n-1})/2h + O(h^2)$$

$$\int_{x_n}^{x_{n+1}} y \, dx = \frac{h}{2} (y_n + y_{n+1}) + O(h^3).$$

* This notation is explained in Section 1.4.

As we will see later, there is no guarantee that a method with a high order truncation error will perform better than other methods. We also mention that a local error of size $O(h^p)$ will often lead to a total error of lower order, e.g. $O(h^{p-1})$ or $O(h^{p-2})$.

Rounding errors occur in many cases when two floating numbers are added or subtracted, and almost always in multiplication and division. Let a and b be two exact numbers approximated with x and y, and further let $|x - a| \leq \delta x$ and $|y - b| \leq \delta y$; then

$$|(x + y) - (a + b)| \leq \delta x + \delta y$$
$$|(x - y) - (a - b)| \leq \delta x + \delta y.$$

For multiplication and division we get

$$|xy - ab| \leq |y|\, \delta x + |x|\, \delta y$$
$$|x/y - a/b| \leq (|y|\, \delta x + |x|\, \delta y)/|y|^2.$$

Note that in a computer where all numbers must be represented with a fixed number of digits (N), one digit may be lost on addition or subtraction, N digits on multiplication, and an infinite number of digits on division. However, there are special problems in connection with very large or very small numbers. Assume that the largest and smallest positive numbers that can be stored in a certain computer are $2^N - 1$ and 2^{-N} respectively; N can be of the order a few hundreds. Then numbers greater than $2^N - 1$ will cause *overflow*, while numbers less than 2^{-N} will cause *underflow*. Both these occurrences may be disastrous in a practical computation, and reliable safety-valves must be provided, either in hardware or in software.

Let us now consider the computation of a function of one or several variables subject to errors. Denoting the function by $F = F(x_1, x_2, \ldots, x_n)$ and the individual errors by $\Delta x_1, \Delta x_2, \ldots, \Delta x_m$ we have:

$$\Delta F \simeq \frac{\partial F}{\partial x_1} \Delta x_1 + \frac{\partial F}{\partial x_2} \Delta x_2 + \cdots + \frac{\partial F}{\partial x_n} \Delta x_n.$$

Here we have used only the first-order terms in the Taylor expansion. The maximum error is approximately

$$(\Delta F)_{\max} \leq \sum_{k=1}^{n} |\partial F/\partial x_k|\,|\Delta x_k|.$$

If $F = x_1 + x_2 + \cdots + x_n$ we have simply

$$(\Delta F)_{\max} \leq |\Delta x_1| + |\Delta x_2| + \cdots + |\Delta x_n|,$$

and if $F = x_1^{m_1} x_2^{m_2} \cdots x_n^{m_n}$ then

$$\left(\frac{\Delta F}{F}\right)_{\max} \leq |m_1\, \Delta x_1/x_1| + |m_2\, \Delta x_2/x_2| + \cdots + |m_n\, \Delta x_n/x_n|.$$

Propagation of Errors

We return to our three main types of error with the intention to study their propagation. In a numerical process it is generally difficult to distinguish one type of error from another, but there are still clear differences. The initial errors are present from the very beginning, while the other types build up during the individual steps of the process. It is often possible to study the isolated effect of initial errors on the final solution by using special precautions to eliminate truncation and rounding errors. If the solution is very sensitive to initial errors, we have an *ill-conditioned* problem (see the chapter on linear equations). Quite often an ill-conditioned numerical problem indicates that the original problem is not well-posed. For example, one might have tried to combine conditions that from a physical point of view are not consistent, or are otherwise unnatural. This type of ill-conditioning is dependent only on the problem, not on the method used for its solution.

Rounding errors have a property that makes them somewhat easier to handle. In most cases they can be considered to be approximately random, and if we treat them statistically we obtain much smaller probabilistic bounds than would be the case if we used maximum values. However, when a rounding error has been introduced, it will add to the other errors and propagate in exactly the same way. This means that in every step we have an incoming error, which in general will be magnified through the operations performed during the step; meanwhile, a fresh error will be introduced. It might be helpful to compare this with a compound interest process.

It can be of some interest to illustrate the different types of error a little more explicitly. Suppose that we want to compute $f(x)$, where x is a real number and f is a real function which we so far do not specify any closer. In practical computations the number x must be approximated by a rational number x' since no computer can store numbers with an infinite number of decimals. The difference $x' - x$ constitutes the *initial error* while the difference $\varepsilon_1 = f(x') - f(x)$ is the corresponding *propagated error*. In many cases f is a function that must be replaced by a simpler function f_1 (often a truncated power series expansion of f). The difference $\varepsilon_2 = f_1(x') - f(x')$ is then the *truncation error*. The calculations performed by the computer, however, are not exact but pseudo-operations of a type that has just been discussed. The result is that instead of $f_1(x')$ we get another value $f_2(x')$, which is then a wrongly computed value of a wrong function of a wrong argument. The difference $\varepsilon_3 = f_2(x') - f_1(x')$ could be termed the *propagated error from the roundings*. The total error is

$$\varepsilon = f_2(x') - f(x) = \varepsilon_1 + \varepsilon_2 + \varepsilon_3.$$

We now choose the following specific example. Suppose that we want to determine $e^{1/3}$ and that all calculations are performed with four decimals. To start with, we try to compute $e^{0.3333}$ instead of $e^{1/3}$, and the propagated error becomes

$$\varepsilon_1 = e^{0.3333} - e^{1/3} = -0.00004\,65.$$

Next, we do not compute e^x but instead

$$1 + \frac{x}{1!} + \frac{x^2}{2!} + \frac{x^3}{3!} + \frac{x^4}{4!}$$

for $x = 0.3333$. Hence, the truncation error is

$$\varepsilon_2 = -\left(\frac{0.3333^5}{5!} + \frac{0.3333^6}{6!} + \cdots\right) = -0.00003\,63.$$

Finally, the summation of the truncated series is done with rounded values, giving the result

$$1 + 0.3333 + 0.0555 + 0.0062 + 0.0005 = 1.3955,$$

instead of $1.39552\,96$ obtained with seven decimals. Thus $\varepsilon_3 = -0.00002\,96$ and the total error is

$$\varepsilon = 1.3955 - e^{1/3} = \varepsilon_1 + \varepsilon_2 + \varepsilon_3 = -0.00011\,24.$$

Investigations of error propagation are, of course, particularly important in connection with iterative processes and computations where each value depends on its predecessors. Examples of such problems are in particular linear systems of equations, eigenvalue computations, and ordinary and partial differential equations. In the corresponding chapters we shall return to these problems in more explicit formulations.

In error estimations one can speak about *a-priori* estimations and *a-posteriori* estimations. As can be understood from the name, the first case is concerned with estimations performed *without* knowledge of the results to be computed. In the latter case the obtained results are used in the error analysis. Further the notions of *forward* and *backward analysis* should be mentioned. With forward analysis one follows the development of the errors from the initial values to the final result. In backward analysis one starts from a supposed error in the results, tracing it backward to see between which limits the initial values must lie to produce such an error. This technique was introduced by Wilkinson, who used it with great success for error analysis of linear systems of equations.

When different numerical methods are compared, one usually considers the truncation errors first. Then one investigates how the errors depend on some suitable parameter which in the ideal case tends toward 0 or ∞. Suppose that we consider the error ε as a function of h, where it is assumed that $h \to 0$. The error analysis can now be performed on several different levels. One might be content with showing that the method is convergent, i.e., $\varepsilon(h) \to 0$ when $h \to 0$. One might also derive results with respect to the convergence speed, e.g., $|\varepsilon(h)| \leq C\varphi(h)$ for some known function φ, where C is a constant whose value, however, is not known. It might also happen that one can prove $\varepsilon(h)/\varphi(h) \to 1$ when $h \to 0$, i.e., an asymptotic formula for the error. Finally, one may also derive actual error estimates of the type $|\varepsilon(h)| \leq \varphi(h)$ for all $h < h_0$. In this case $\varphi(h)$ is an upper limit for the error, giving us a possibility to guarantee a certain accuracy.

In many cases h stands for the step size in a grid. Consider, for example, what happens when an integral is replaced by a sum. Then the number of terms increases as h^{-1}, which implies a similar (for statistical reasons somewhat weaker) increase in the rounding errors. As a rule, one has to strike a balance between rounding and truncation errors in order to get the best possible accuracy.

Interval Analysis

It is well-known that error analysis as applied to complex operations is difficult and tedious, and in many cases gives rather unrealistic results. We will mention here a technique intended to get around this problem, namely *interval analysis*. The basic idea is that each number is represented by an *interval* instead of a numerical approximation. Suppose that a variable x is known to lie in an interval (a, b), and similarly, y in an interval (c, d). Then $a + c < x + y < b + d$. Hence the sum of two variables is again represented by an interval. If $a > 0$ and $c > 0$, we get analogously $ac < xy < bd$. If, further, $b < c$, then $c - b < y - x < d - a$ and $c/b < y/x < d/a$. It is obvious that difficulties will arise if 0 is contained in an interval. It has also turned out that special care must be exercised in more advanced applications in order to prevent unrealistic estimates.

---

§1.3. NUMERICAL CANCELLATION

In the previous section it has been shown how several fundamental arithmetical laws must be modified in numerical applications. Against this background it is not surprising that expressions which are completely equivalent from a mathematical point of view may turn out to be quite different numerically. We will restrict ourselves to a few examples on this matter.

The second-degree equation $x^2 - 2ax + \varepsilon = 0$ has the two solutions

$$x_1 = a + \sqrt{(a^2 - \varepsilon)} \qquad \text{and} \qquad x_2 = a - \sqrt{(a^2 - \varepsilon)}.$$

If $a > 0$ and ε is small compared with a, the root x_2 is expressed as the difference between two almost equal numbers, and a considerable amount of significance is lost. Instead, if we write

$$x_2 = \varepsilon/[a + (a^2 - \varepsilon)^{1/2}],$$

we obtain the root as approximately $\varepsilon/2a$ without loss of significance.

The following example demonstrates clearly the advantages and drawbacks of different numerical procedures. Consider the integral

$$I_n = \int_0^1 \frac{x^n}{x + 10}\, dx; \qquad n = 0, 1, 2, \ldots.$$

We find directly

$$I_0 = \int_0^1 dx/(x + 10) = \ln 1.1 \simeq 0.09531\,018$$

and further $I_n + 10I_{n-1} = 1/n$. Hence we obtain:

$$I_1 = 0.04689\,82 \qquad I_5 = 0.015$$
$$I_2 = 0.03101\,8 \qquad I_6 = 0.017$$
$$I_3 = 0.02315 \qquad I_7 = -0.027$$
$$I_4 = 0.0185 \qquad \vdots$$

Obviously, the values obtained by this method deteriorate quickly due to the fact that the error is multiplied by 10 in each step. If instead we expand and integrate, we get:

$$\frac{x^n}{10(1 + x/10)} = 0.1(x^n - x^{n+1}/10 + x^{n+2}/100 - \cdots)$$

and

$$I_n = 0.1/(n + 1) - 0.01/(n + 2) + 0.001/(n + 3) - \cdots$$

which converges quickly for all $n \geq 0$. For example, $n = 5$ gives $I_5 = 0.01535\,29$, and $n = 6$ gives $I_6 = 0.01313\,77$.

Another procedure suggests itself. Since $10 \leq x + 10 \leq 11$ we have

$$\frac{1}{11(n + 1)} < I_n < \frac{1}{10(n + 1)}.$$

Taking $n = 10$ we would get $1/121 < I_{10} < 1/110$ with the correct value closer to the lower limit, and we take approximately $I_{10} \simeq 1/120$. Then, using backward recursion we get:

n	I_n
10	0.0083
9	0.00917
8	0.01019 4
7	0.01148 06
6	0.01313 765
5	0.01535 2902
4	0.01846 47098
3	0.02315 35290 2
2	0.03101 79804 31
1	0.04689 82019 569
0	0.09531 01798 0431

The last value is in error by less than one unit in the 13th place. From this we conclude that backward recursion can be used in certain cases but is not a universal remedy.

Last, we also compute the exact value of the integral in the case $n = 6$. Performing the division in the integrand we get:

$$\frac{x^6}{x + 10} = x^5 - 10x^4 + 100x^3 - 1000x^2 + 10^4x - 10^5 + \frac{10^6}{x + 10}$$

and

$$
\begin{aligned}
I_6 &= 1/6 - 2 + 25 - 1000/3 + 5000 - 10^5 + 10^6 \ln 1.1 \\
&= -1/6 + \text{fractional part } (10^6 \ln 1.1) \\
&= -1/6 + 0.17980\,4324 = 0.01313\,7657.
\end{aligned}
$$

Here we have another example of numerical cancellation. The final answer is obtained as the difference between two large numbers, almost equal in size (about 10^5), with a corresponding loss of accuracy; we must give ln 1.1 to 14 significant digits in order to get eight digits in the final answer.

Another difficulty occurs when one or several roots of an algebraic equation are extremely sensitive to changes in the coefficients. Consider the polynomial

$$f(z) = z^n + a_1 z^{n-1} + a_2 z^{n-2} + \cdots + a_n,$$

and let r be a root of the equation $f(z) = 0$. Differentiating we get:

$$\left(\frac{\partial f}{\partial a_k}\right)_{z=r} = f'(r)\frac{\partial r}{\partial a_k} + r^{n-k} = 0.$$

Hence $\partial r/\partial a_k = -r^{n-k}/f'(r)$, and this relation is written in the following form:

$$\frac{\delta r}{r} = -\frac{a_k \cdot r^{n-k-1}}{f'(r)} \cdot \frac{\delta a_k}{a_k}.$$

Now put $A_k = |a_k \cdot r^{n-k-1}/f'(r)|$, and we can summarize as follows. Large values of A_k have the effect that small changes in the coefficient a_k cause large changes in the root r. Large values of A_k occur when r is large and $f'(r)$ is small; the latter is the case, for example, when some of the roots lie close together.

A well-known example has been given by Wilkinson:

$$(x + 1)(x + 2) \cdots (x + 20) = 0 \qquad \text{or} \qquad x^{20} + 210x^{19} + \cdots + 20! = 0.$$

We choose $k = 1$ and $\delta a_1 = 2^{-23}$, which means that the coefficient 210 is changed to 210.0000001192. Then the roots $-1, -2, \ldots, -8$ are shifted only slightly; among the remaining roots we find, for example, $-14 \pm 2.5i$, $-16.73 \pm 2.81i$, and -20.85. For $r = -16$ we obtain

$$A_1 = \frac{210 \cdot 16^{18}}{15!4!} = 3.2 \cdot 10^{10}.$$

This result indicates that we must have 10 guard digits, apart from those corresponding to the wanted accuracy. It should also be noted that the value of A_k cannot be used for computation of the exact root, since above we have given only the first order terms; here higher order terms are of decisive importance.

Equations where small changes in the coefficients cause large changes in the roots are said to be *ill-conditioned*. If one wants to determine the roots of such an equation, one has to work with a suitable number of guard digits. It may be difficult to tell offhand whether a given equation is ill-conditioned or not.

§1.4. ALGORITHMS AND COMPLEXITY

Notations

In general, we have good reasons to use the classical mathematical notations. Here we will mention just a few of them explicitly, and further we will add some which are of a more special nature.

If a variable x takes the values $a, a + h, a + 2h, \ldots, b - h, b$ where $b = a + nh$ (n integer), we write this $x = a(h)b$. In the same spirit, $x = 0(0.01)1(0.1)2(1)10$ means that x takes the values $0, 0.01, 0.02, \ldots, 1, 1.1, 1.2, \ldots, 2, 3, \ldots, 10$. This notation is very handy, particularly for the description of tables.

Often we have to use the integer part of a certain number, or the next higher integer. The following examples are self-explanatory:

$$\lfloor 7.2 \rfloor = 7; \quad \lfloor 5 \rfloor = 5; \quad \lfloor -3.6 \rfloor = -4.$$
$$\lceil 7.2 \rceil = 8; \quad \lceil 5 \rceil = 5; \quad \lceil -3.6 \rceil = -3.$$

Let $f(x)$ and $g(x)$ be given functions. If there exists a positive constant C such that $f(x) \leq Cg(x)$ for all x, we write this $f(x) = O(g(x))$ where the letter O stands for "order." If instead there is a positive constant D such that $f(x) \geq Dg(x)$ for all x, then we write this $f(x) = \Omega(g(x))$. If $f(x) = O(g(x))$ and $f(x) = \Omega(g(x))$ are both valid, then we write this $f(x) = \Theta(g(x))$.

Example 1

$$\tanh(x) = O(1), \quad x \geq 0.$$
$$\sin(x) = O(x), \quad x \geq 0.$$
$$\cosh(x) = \Omega(x), \quad x \geq 0.$$
$$(x^4 + x^2)/(x^2 + 2) = \Theta(x^2).$$

Finally, recall the very useful Kronecker symbol:

$$\delta_{ik} = \begin{cases} 0 & i \neq k \\ 1 & i = k. \end{cases}$$

Algorithms

An *algorithm* is defined as a step by step description of a procedure for solving a certain class of problems. Formally, the algorithm must consist of a finite number of steps, and when used it must always terminate. Well-known algorithms in numerical computations are Horner's rule for evaluation of a polynomial and its derivatives, and the usual rule for multiplication of two square matrices of the same size. With access to computers, many new algorithms have been proposed for classical problems. For example, it was generally believed that multiplication of two $n \times n$ matrices required n^3 multiplications and $n^2(n-1)$ additions, but in 1969 Strassen presented a method that took only $O(n^{2.81})$ operations; it has since been improved to better than $O(n^{5/2})$. The algorithm is described briefly below.

Let A and B be two matrices with dimension 2×2, and form the following expressions:

$$x_1 = (a_{11} + a_{22})(b_{11} + b_{22})$$
$$x_2 = (a_{21} + a_{22})b_{11}$$
$$x_3 = a_{11}(b_{12} - b_{22})$$
$$x_4 = a_{22}(b_{21} - b_{11})$$
$$x_5 = (a_{11} + a_{12})b_{22}$$
$$x_6 = (a_{21} - a_{11})(b_{11} + b_{22})$$
$$x_7 = (a_{12} - a_{22})(b_{21} + b_{22})$$

Then the matrix $C = AB$ has the elements:

$$c_{11} = x_1 + x_4 - x_5 + x_7$$
$$c_{12} = x_3 + x_5$$
$$c_{21} = x_2 + x_4$$
$$c_{22} = x_1 + x_3 - x_2 + x_6$$

Obviously only seven multiplications are required. The rules apply also when the elements are replaced by sub-matrices, and in this case the order of the factors becomes essential and must be as above. For matrices of orders other than powers of 2, the rules become somewhat more complicated, but the saving with respect to the number of multiplications still prevails. The number 2.81 comes from $\ln 7 / \ln 2$ or $\log_2 7$.

If a certain problem requires at most a time $O(n^s)$, where n is a parameter defining the size of an instance of a problem and s is a positive real number, we say that the problem can be solved in polynomial time. Other functions can also be included in the hierarchy of polynomials; since the function $\ln x$, for instance, increases more slowly than any power of x, a problem requiring the effort $O(n \ln n)$ can also be solved in polynomial time.

To be more specific, we have to introduce a theoretical model of computation. This can be achieved by a Turing machine, a theoretical device suggested in 1936.

It has a finite number of internal states and uses a potentially infinite tape as input, output, and memory. It can be shown that any digital computer can be simulated by a Turing machine; if a bound is placed on the amount of memory, a Turing machine can be simulated by a digital computer using a polynomial number of operations. Because of this polynomial equivalence we can just as well count the number of arithmetic operations when we deal with numerical algorithms. The corresponding order then defines the *complexity* of the problem.

Problems that can be solved in polynomial time by a deterministic algorithm (essentially a straightforward scheme without any guesses) are said to belong to class P. Almost all numerical problems belong to this class. However, there exist problems that require exponential time, and due to the rapid growth of the exponential function, such problems are called intractable. One example is provided by the so-called permanent, which for a square matrix is defined as the sum of all possible products of elements with exactly one element from each row and one element from each column.

There is also an intermediate class including a large number of problems of great practical importance. For these problems only exponential time algorithms are known, but on the other hand nobody has so far been able to demonstrate that they cannot be solved in polynomial time. Most of these problems can be included in a class called NP (N for non-deterministic, P for polynomial). This class contains problems that can be solved in polynomial time on a non-deterministic Turing machine, a device that is able to find the shortest way to the solution without having to search, or alternatively one that searches among all possible computational routes in parallel with no extra cost in time. It can be shown that certain of these problems are harder than others (in fact, several hundreds are known); they are said to be NP-complete (NPC). In fact, if any of these problems can be solved in polynomial time on a deterministic machine, then all the others can also be solved in polynomial time. Nobody actually believes this to be the case, but no proof has been found, and this question of whether $P = NP$ is one of the great open questions in mathematics.

In spite of the uncertainty that still prevails, research in this area has proved very successful. In order to deal with NP-problems without using exponential time, people have proposed a number of approximate or probabilistic algorithms. Such algorithms can give an approximate solution arbitrarily close to the real solution with a low probability of error. One example may be quoted concerning primality of very large numbers, where the probability that a certain number is composite may be $< 2^{-N}$ if N trials have been made at random. This problem belongs to NP but is believed not to belong to NPC. Whether in fact it belongs to P is not known either.

Another problem that for a long time was believed to belong to NP is linear programming: in the worst case the simplex algorithm uses exponential time. But in 1979 Khachian showed that linear programming in fact belongs to P.

NP-problems are mostly combinatorial in nature and often originate in graph theory or mathematical logic. In contrast, practical numerical problems are almost universally close to analytical problems, and the discreteness introduced does not in general result in combinatorial problems. As a consequence we get algorithms working in polynomial time $O(n^s)$ where s is often as low as 2 or 3. Here we just mention sorting and the fast Fourier transform, both requiring $O(n \ln n)$ operations.

---

§1.5.   USE OF COMPUTERS

Electronic computers appeared around 1950 and were primarily constructed for performing lengthy numerical calculations. One of the main difficulties during the early stages was that all programs had to be written in machine code. When FORTRAN was introduced around 1956–57, this was a major step forward. As a result of an international effort ALGOL entered the market around 1960. This was also an important event, since for the first time a grammar was defined containing precise and clear rules for how programs should be constructed. Since then a large number of programming languages have been developed. Among those generally used for numerical problems we mention FORTRAN (in a modernized version called FORTRAN 77), SIMULA and Pascal, as well as APL, which is a typical dialogue language.

Program Libraries

There now exist large collections of standard programs, usually written in FORTRAN, which can be called and used under different circumstances. It is often possible to call them also from a main program written in another language. Since such program packages are available at practically every computing center, we refrain from programming examples in the text.

With respect to the program packages we mention just two of the most important ones: IMSL (International Mathematical and Statistical Library) and NAG (Numerical Algorithms Group). In addition to this there also exist a number of special-purpose packages, e.g., for linear algebra, eigenvalue problems, and optimization.

A warning is appropriate here: the programs must be tested carefully before use. They are supposed to be fool-proof, but it is clearly impossible to foresee everything that can happen.

EXERCISES

1. Convert the sedecimal number $ABCDEF$ to decimal form, where A through F stand for 10 through 15.

2. Convert the decimal fraction 0.31416 to octal form.

3. Determine the maximum relative error when p_1 is calculated from the relation $p_1 v_1^k = p_2 v_2^k$ with $k = 1.4$. The maximum relative errors of v_1, v_2, and p_2 are 0.7, 0.75, and 1% respectively.

4. One wants to determine $1/(2 + \sqrt{3})^4$ having access to an approximate value of $\sqrt{3}$. Compare the relative errors on direct computation and on using the equivalent expression $97 - 56\sqrt{3}$.

5. Compute the smallest root of the equation $x^2 - 40x + 1 = 0$ as (a) $20 - \sqrt{399}$, (b) $(20 + \sqrt{399})^{-1}$ when the approximate value 19.97498 for $\sqrt{399}$ is used. Compare the errors in the two cases.

6. Let $J_n = \int_0^1 x^n (x^2 + 10x + 16)^{-1} dx$. Then $J_0 = (1/6) \ln(4/3) = 0.047947$ and $J_1 = \frac{1}{2} \ln(27/16) - 5J_0 = 0.021889$. Show that $J_{n+1} + 10J_n + 16J_{n-1} = 1/n$ and compute J_n for $n = 2, 3, \ldots, 7$. Are the results satisfactory?

CHAPTER 2

Vectors and Matrices

*"O servant of the Prophet," said the Sheik of the
Imperial Chibouk to the Mamoosh of the Invincible
Army, "how many unconquerable soldiers have we in
arms?"*

*"Upholder of the Faith," that dignitary replied
after examining his memoranda, "they are in numbers
as the leaves of the forest!"*

In the following we assume that the reader is familiar with the basic concepts of linear algebra. Nevertheless we will give a brief summary of different notions concerning vectors and matrices as they are used in connection with such problems as coordinate transformations. Eigenvalue problems will be treated in considerable detail, as well as matrix transformations and norms of vectors and matrices. We will also mention normal forms, pseudo-inverses and singular value decomposition.

In this and following chapters, small letters in boldface indicate vectors, while capital letters in boldface stand for matrices.

§2.1. VECTORS

A *vector* in n dimensions or of length n can be viewed as a quantity with n components in given order. If nothing else is said, it is supposed that the components are given *column*-wise:

$$x = \begin{pmatrix} x_1 \\ x_2 \\ \vdots \\ x_n \end{pmatrix}.$$

The corresponding row-vector x^T is obtained by transposition:

$$x^T = (x_1, x_2, \ldots, x_n).$$

A set of vectors $v_1, v_2, \ldots, v_p$ are said to be linearly independent if the only solution to the equation $c_1 v_1 + c_2 v_2 + \cdots + c_p v_p = 0$ is $c_1 = c_2 = \cdots = c_p = 0$.

The *scalar* or *inner product* of two vectors x and y is generally written (x, y) and defined through

$$(x, y) = x^T y = y^T x = \sum_{k=1}^{n} x_k y_k.$$

It is obviously a scalar. It is clear that the distributive law holds:

$$(ax + by, z) = a(x, z) + b(y, z)$$

where x, y, and z have the same dimension.

There is also an *outer* product, which is a matrix:

$$xy^T = \begin{pmatrix} x_1 y_1 & x_1 y_2 & \cdots & x_1 y_n \\ x_2 y_1 & x_2 y_2 & \cdots & x_2 y_n \\ \vdots & & & \\ x_m y_1 & x_m y_2 & \cdots & x_m y_n \end{pmatrix}.$$

Here x and y need not be of the same dimension.

Let us now suppose that two three-dimensional vectors x and y are given, and form $z = xy^T - yx^T$. Then:

$$z = \begin{pmatrix} 0 & x_1 y_2 - x_2 y_1 & x_1 y_3 - x_3 y_1 \\ x_2 y_1 - x_1 y_2 & 0 & x_2 y_3 - x_3 y_2 \\ x_3 y_1 - x_1 y_3 & x_3 y_2 - x_2 y_3 & 0 \end{pmatrix}$$

Putting

$$\begin{cases} x_2 y_3 - x_3 y_2 = v_1 \\ x_3 y_1 - x_1 y_3 = v_2 \\ x_1 y_2 - x_2 y_1 = v_3 \end{cases} \quad \text{we find} \quad z = \begin{pmatrix} 0 & v_3 & -v_2 \\ -v_3 & 0 & v_1 \\ v_2 & -v_1 & 0 \end{pmatrix}$$

and at least formally we can define a vector $v^T = (v_1, v_2, v_3)$. However, this quantity does not have the same properties as ordinary vectors. If we change x to $-x$ and y to $-y$, the quantity z as well as v remains the same. In older literature x and y were called *polar* vectors while v was called an *axial* vector. As a matter of fact, it is just a coincidence that the *bivector* z can be represented by three values in three dimensions. Already in four dimensions we need six values (in general $n(n-1)/2$ if n is the dimension). However, we prefer to characterize z as an antisymmetric tensor of second order, formed in a special way from two vectors. Polar and axial vectors are used extensively in classical mechanics.

A fundamental relation is the Cauchy–Schwarz inequality:

(2.1.1) $$(x, y)^2 \le (x, x)(y, y).$$

The proof follows directly since we have

$$(ax + y, ax + y) = a^2(x, x) + 2a(x, y) + (y, y) \ge 0$$

for all values of a. Supposing $x \neq 0$ and taking $a = -(x, y)/(x, x)$, we get the desired relation. It can be generalized to the complex case by a slight modification in the definition of the inner product.

Now suppose that we have a system of vectors $v_1, v_2, \ldots, v_p$ with the property $(v_i, v_k) = \delta_{ik}$ where δ_{ik} is the Kronecker tensor:

$$(2.1.2) \qquad \delta_{ik} = \begin{cases} 0, & i \neq k \\ 1, & i = k. \end{cases}$$

Then these vectors are said to form an *orthonormal* system. We assume that we have a set of linearly independent vectors $u_1, u_2, \ldots, u_p$ in n dimensions, and from these we want to construct an orthonormal system. We start with $v_1 = u_1/(u_1, u_1)^{1/2}$ and assume that we have already constructed k orthonormal vectors $v_1, v_2, \ldots, v_k$. Then we form

$$w_{k+1} = u_{k+1} - \sum_{r=1}^{k} c_r v_r$$

and we want to choose the coefficients c_r in such a way that for $s = 1(1)k$ we have $(w_{k+1}, v_s) = 0$. Hence:

$$0 = (u_{k+1}, v_s) - \sum_{r=1}^{k} c_r(v_r, v_s) = (u_{k+1}, v_s) - c_s$$

since $(v_r, v_s) = \delta_{rs}$. Thus $c_s = (u_{k+1}, v_s)$ and w_{k+1} is known and orthogonal to the previous vectors $v_1, \ldots, v_k$. It only remains to normalize w_{k+1}, and so we have finally $v_{k+1} = w_{k+1}/(w_{k+1}, w_{k+1})^{1/2}$. This algorithm is known as the Gram-Schmidt orthonormalization procedure.

In different applications we often meet the problem of assigning some kind of magnitude to a vector, and we will now define a quantity serving this purpose. We call such a value associated with a given vector x a *vector norm*, denoted $\|x\|$, if it satisfies the following conditions:

1. $\|x\| > 0$ if $x \neq 0$, and $\|x\| = 0$ if $x = 0$.
2. $\|cx\| = |c| \cdot \|x\|$, c an arbitrary complex number.
3. $\|x + y\| \leq \|x\| + \|y\|$.

The most common norms are

1. The maximum norm $\|x\| = \|x\|_\infty = \max_i |x_i|$.
2. The absolute norm $\|x\| = \|x\|_1 = \sum_i |x_i|$.
3. The Euclidean norm $\|x\| = \|x\|_2 = (x, x)^{1/2}$.

It is easy to verify that these three norms obey the axioms above; for example, the triangle inequality in case (3) follows from the Cauchy–Schwarz inequality.

The three norms mentioned above are special cases of a more general L_p-norm (sometimes called the Hölder norm), defined by

$$\|x\| = \|x\|_p = \left(\sum_{i=1}^{n} |x_i|^p \right)^{1/p}$$

for $p = \infty$, $p = 1$, and $p = 2$, respectively.

§2.2.   MATRICES

A *matrix* is defined as an ordered rectangular array of numbers. We denote the number of rows by m and the number of columns by n. A matrix of this kind will be characterized by the type symbol (m, n). The element in the ith row and the kth column of the matrix A is denoted by $(A)_{ik} = a_{ik}$. The whole matrix is usually written

$$A = \begin{bmatrix} a_{11} & a_{12} & \cdots & a_{1n} \\ a_{21} & a_{22} & \cdots & a_{2n} \\ \vdots & & & \\ a_{m1} & a_{m2} & \cdots & a_{mn} \end{bmatrix},$$

or sometimes in the more compact form

$$A = (a_{ik}).$$

In the latter case, however, it must be agreed upon what type of matrix it is. If $m = n$, the matrix is *square*; when $m = 1$, we have a *row vector*, and for $n = 1$, a *column vector*. If $m = n = 1$, the matrix is usually identified with the number represented by this single element.

Matrix Operations

Addition and subtraction are defined only for matrices of the same type, that is, for matrices with the same number of rows and the same number of columns. If A and B are matrices of type (m, n), then $C = A \pm B$ is defined by $c_{ik} = a_{ik} \pm b_{ik}$. Further, if λ is an ordinary number, λA is the matrix with the general element equal to λa_{ik}. If all $a_{ik} = 0$, the matrix is said to be a null matrix.

The definition of matrix multiplication is closely connected with the theory of coordinate transformations. We start with the linear transformation

$$z_i = \sum_{r=1}^{n} a_{ir} y_r \qquad (i = 1, 2, \dots, m).$$

If the quantities $y_1, y_2, \dots, y_n$ can be expressed by means of the quantities $x_1, x_2, \dots, x_p$ through a linear transformation

$$y_r = \sum_{k=1}^{p} b_{rk} x_k \qquad (r = 1, 2, \dots, n),$$

we find by substitution

$$z_i = \sum_{r=1}^{n} a_{ir} \sum_{k=1}^{p} b_{rk}x_k = \sum_{k=1}^{p} \left(\sum_{r=1}^{n} a_{ir}b_{rk} \right) x_k.$$

In this way we obtain z_i, expressed directly by means of x_k through a composed transformation, and we can write

$$z_i = \sum_{k=1}^{p} c_{ik}x_k \qquad \text{with} \qquad c_{ik} = \sum_{r=1}^{n} a_{ir}b_{rk}.$$

In view of this, matrix multiplication should be defined as follows. Let A be of type (m, n) and B of type (n, p). Then $C = AB$ if

(2.2.1) $$c_{ik} = \sum_{r=1}^{n} a_{ir}b_{rk},$$

where C will be of type (m, p).

It should be observed that the commutative law of multiplication is not satisfied. Even if BA is defined (which is the case if $m = p$), and even if BA is of the same type as AB (which is the case if $m = n = p$), we have in general

$$BA \neq AB.$$

The relation $AB = BA$ is valid only for some rather special matrices.

Transposition means exchange of rows and columns. We will denote the transposed matrix of A by A^T. Thus $(A^T)_{ik} = (A)_{ki}$. *Conjugation* of a matrix containing complex elements means that all elements are conjugated; the usual notation is A^*. If a matrix A is transposed and conjugated at the same time, the following symbol is often used:

$$(A^*)^T = (A^T)^* = A^H.$$

Let us now treat square matrices (n, n), row vectors $(1, n)$, and column vectors $(n, 1)$. First of all, we define the unit matrix or the identity matrix I:

$$(I)_{ik} = \delta_{ik}.$$

Thus we have

$$I = \begin{bmatrix} 1 & 0 & 0 & \cdots & 0 \\ 0 & 1 & 0 & \cdots & 0 \\ \vdots & & & & \\ 0 & 0 & 0 & \cdots & 1 \end{bmatrix}.$$

In a square matrix, the elements a_{ii} form a diagonal, the so-called *main diagonal*. Hence the unit matrix is characterized by ones in the main diagonal, zeros otherwise. Now form $AI = B$:

$$b_{ik} = \sum_{r=1}^{n} a_{ir}\delta_{rk} = a_{ik},$$

since $\delta_{rk} = 1$ only when $r = k$.

Analogously, we form $IA = C$ with

$$c_{ik} = \sum_{r=1}^{n} \delta_{ir} a_{rk} = a_{ik}.$$

Hence

$$AI = IA = A.$$

The sum of the diagonal elements is called the *trace* of the matrix:

$$\text{tr } A = \sum_{i=1}^{n} a_{ii}.$$

A real symmetric matrix A is said to be *positive definite* if $x^H A x > 0$ for all vectors $x \neq 0$. If, instead, $x^H A x \geq 0$, A is said to be *positive semidefinite*.

The determinant of a matrix A is a pure number, denoted by det A, which can be computed from A by use of certain rules. Generally, it is the sum of all possible products of elements where exactly one element has been taken from every row and every column, with a plus or a minus sign appended according as the permutation of indices is even or odd.

$$\det A = \sum \pm a_{1,i_1} a_{2,i_2} \cdots a_{n,i_n}.$$

The sign corresponding to the indices $(i_1, i_2, \ldots, i_n)$ is given by $(-1)^i$ where i is the number of inversions in the list, i.e., how many times it happens that a larger number precedes a smaller one. For example, the indices 3, 5, 2, 4, 1 have 7 inversions, and hence the term $a_{13} a_{25} a_{32} a_{44} a_{51}$ in a fifth order determinant should appear with a minus sign. This general rule for computing a determinant is impossible to use in practical work because of the large number of terms ($n!$); already for $n = 10$, we get more than 3 million terms. Instead, the following rules, which are completely sufficient, are used:

1. A determinant remains unchanged if one row (column) multiplied by a number λ is added to another row (column).

2. If in the ith row or kth column only the element $a_{ik} \neq 0$, then
$$\det A = (-1)^{i+k} a_{ik} \det A_{ik}$$
   where A_{ik} is the matrix obtained when the ith row and the kth column are deleted in A.

3. If A is triangular or diagonal, then
$$\det A = a_{11} a_{22} \cdots a_{nn}.$$

From these rules it follows, for example, that det $A = 0$ if two rows or two columns are proportional or, more generally, if one row (column) can be written as a linear combination of some of the other rows (columns).

We give a short proof that if $C = AB$, then det $C = $ det A det B. Let $(k_1, k_2, \ldots, k_n)$ be a permutation of $(1, 2, \ldots, n)$, and let $k = 0$ or 1, depending on whether the permutation is even or odd. Then we have from the definition of determinants (with summation over all permutations (K)):

$$\det C = \sum_{(K)} (-1)^k c_{1k_1} c_{2k_2} \cdots c_{nk_n}$$

$$= \sum_{(K)} (-1)^k \sum_{r_1=1}^{n} a_{1r_1} b_{r_1 k_1} \cdots \sum_{r_n=1}^{n} a_{nr_n} b_{r_n k_n}.$$

Here $r_1, r_2, \ldots, r_n$ take all values from 1 to n, but if we collect the a-factors we see that we get contributions only for indices $(r_1, r_2, \ldots, r_n)$ which are permutations of $(1, 2, \ldots, n)$; otherwise the result is zero corresponding to a determinant with some rows equal. Hence we find:

$$\det C = \sum_{(K)} (-1)^k \sum_{(R)} a_{1r_1} a_{2r_2} \cdots a_{nr_n} \times b_{r_1 k_1} \cdots b_{r_n k_n}$$

$$= \sum_{(R)} (-1)^r a_{1r_1} \cdots a_{nr_n} \sum_{(K)} (-1)^r (-1)^k b_{r_1 k_1} \cdots b_{r_n k_n}$$

$$= \det A \det B.$$

If $\det A = 0$, A is said to be singular; if $\det A \neq 0$, A is nonsingular. Now let A be given with $\det A \neq 0$. We will try to find a matrix X such that $AX = I$. When formulated algebraically, the problem leads to n systems of linear equations, all with the same coefficient matrix but with different right-hand sides. Since A is nonsingular, every system has exactly one solution (a column vector), and these n column vectors together form a uniquely determined matrix X. In a similar way, we see that the system $YA = I$ also has a unique solution Y. Premultiplication of $AX = I$ with Y and postmultiplication of $YA = I$ with X give the result

$$YAX = YI = Y; \qquad YAX = IX = X,$$

and hence $X = Y$. In this way we have proved that if A is non-singular, a right inverse and a left inverse exist and are equal. The inverse is denoted by A^{-1}:

$$AA^{-1} = A^{-1}A = I.$$

Fundamental Computing Laws for Matrices

As is understood directly, the associative and commutative laws are valid for addition:

$$(A + B) + C = A + (B + C),$$
$$A + B = B + A.$$

Further, the associative and distributive laws for multiplication are also obeyed:

$$(AB)C = A(BC),$$
$$A(B + C) = AB + AC,$$
$$(A + B)C = AC + BC.$$

For transposition, we have the laws

$$(A + B)^T = A^T + B^T,$$
$$(AB)^T = B^T A^T,$$

and for inversion,

$$(AB)^{-1} = B^{-1}A^{-1}.$$

First, we prove $(AB)^T = B^T A^T$. Put $A^T = C$, $B^T = D$, and $AB = G$. Then $c_{ik} = a_{ki}$, $d_{ik} = b_{ki}$, and further

$$(G^T)_{ik} = g_{ki} = \sum a_{kr} b_{ri} = \sum c_{rk} d_{ir} = \sum d_{ir} c_{rk} = (DC)_{ik}.$$

Hence $G^T = DC$ or $(AB)^T = B^T A^T$. It follows directly from the definition that $(AB)^{-1} = B^{-1}A^{-1}$, since

$$(AB)(B^{-1}A^{-1}) = (B^{-1}A^{-1})(AB) = I.$$

From the definition of the inverse, we have $AA^{-1} = I$ and $(A^T)^{-1}A^T = I$. Transposing the first of these equations, we get $(A^{-1})^T A^T = I$, and comparing, we find $(A^T)^{-1} = (A^{-1})^T$.

A matrix is said to have *rank r* if it has r linearly independent rows or columns. If a matrix is of rank r, an arbitrary row vector or column vector can be written as a linear combination of at most r vectors of the remaining ones, since otherwise we would have $r + 1$ linearly independent vectors and the rank would be at least $r + 1$.

§2.3. EIGENVALUES AND EIGENVECTORS

If for a given square matrix A one can find a number λ and a vector u such that the equation $Au = \lambda u$ is satisfied, λ is said to be an *eigenvalue* or *characteristic value* and u an *eigenvector* or *characteristic vector* of the matrix A. The equation can also be written

$$(A - \lambda I)u = 0.$$

This is a linear homogeneous system of equations. As is well known, nontrivial solutions exist only if $\det(A - \lambda I) = 0$. Thus we obtain an algebraic equation of degree n in λ. Explicitly, it has the following form:

(2.3.1)
$$\begin{vmatrix} a_{11} - \lambda & a_{12} & a_{13} & \cdots & a_{1n} \\ a_{21} & a_{22} - \lambda & a_{23} & \cdots & a_{2n} \\ \vdots & & & & \\ a_{n1} & a_{n2} & a_{n3} & \cdots & a_{nn} - \lambda \end{vmatrix} = 0.$$

We notice at once that one term in this equation is

$$(a_{11} - \lambda)(a_{22} - \lambda) \cdots (a_{nn} - \lambda)$$

while the other terms contain λ in at most power $(n - 2)$. Hence the equation can be written:

$$\lambda^n - (a_{11} + a_{22} + \cdots + a_{nn})\lambda^{n-1} + \cdots + (-1)^n \det A = 0.$$

Thus

(2.3.2)
$$\sum_{i=1}^{n} \lambda_i = \operatorname{tr} A; \qquad \prod_{i=1}^{n} \lambda_i = \det A.$$

Equation (2.3.1) is called the *characteristic equation*; the roots are the n eigenvalues of the matrix A. Thus a square matrix of order n has always n eigenvalues; in special cases some or all of them may coincide.

If the equation $Au = \lambda u$ is premultiplied by A, we get $A^2 u = \lambda A u = \lambda^2 u$ and, analogously, $A^m u = \lambda^m u$. From this we see that if A has the eigenvalues $\lambda_1, \lambda_2, \ldots, \lambda_n$, then A^m has the eigenvalues $\lambda_1^m, \lambda_2^m, \ldots, \lambda_n^m$, and the eigenvectors of A are also eigenvectors of A^m. We find directly that

$$\operatorname{tr} A^r = \sum_{i=1}^{n} \lambda_i^r.$$

Assume that all eigenvalues of A are different. Then the eigenvectors are linearly independent and form a basis, i.e., every n-dimensional vector can be written as a linear combination of the eigenvectors. To prove this, suppose to the contrary that we have the linear relation

$$\sum_{i=1}^{q} c_i u_i = 0, \qquad c_i \neq 0,$$

where u_i are the eigenvectors and $q \leq n$. It is further understood that there is no similar relation with a smaller value of q. Operating with the matrix A we find

$$\sum_{i=1}^{q} c_i A u_i = \sum_{i=1}^{q} c_i \lambda_i u_i = 0.$$

Eliminating any particular u_q between these two relations, we get another relation containing only $q - 1$ vectors, against our assumption.

We will now investigate what happens to the eigenvalues and eigenvectors when a coordinate transformation is performed. Let us choose a new set of basis vectors, naturally linearly independent, and form a matrix S with these vectors as columns. If the old coordinates are denoted by x and y and the new ones by x' and y', we have:

$$x = Sx'; \qquad y = Sy'.$$

Then the mapping $y = Ax$ goes over into:

$$Sy' = ASx' \quad \text{or} \quad y' = Bx' \quad \text{with} \quad B = S^{-1}AS$$

(note that S is nonsingular). If u is an eigenvector of A corresponding to the eigenvalue λ (or $Au = \lambda u$), then v is an eigenvector of B corresponding to the same eigenvalue if $u = Sv$, since

$$Bv = S^{-1}ASv = S^{-1}Au = \lambda S^{-1}u = \lambda v.$$

If rows and columns are interchanged in the matrix A, the determinant is unchanged, and hence the eigenvalues of A and A^T are the same, while the eigenvectors, as a rule, are different.

We will now discuss relations between the eigenvectors of the matrices A and A^H. For simplicity we assume that all eigenvalues of A are different. Further we denote the eigenvectors of A by $u_1, u_2, \ldots, u_n$ and those of A^H by $v_1, v_2, \ldots, v_n$. Thus we have:

$$\begin{cases} Au_i = \lambda_i u_i \\ A^H v_k = \lambda_k^* v_k. \end{cases}$$

Form the conjugate transpose of the first equation and postmultiply by v_k:

$$u_i^H A^H v_k = \lambda_i^* u_i^H v_k.$$

Premultiplication of the second equation by u_i^H gives:

$$u_i^H A^H v_k = \lambda_k^* u_i^H v_k$$

Subtracting and observing that $\lambda_i \neq \lambda_k$, we find $u_i^H v_k = 0$. The two groups of vectors u_i and v_i are said to be *biorthogonal*. Using the fact that an arbitrary vector $w \neq 0$ can be represented as $w = \sum c_i v_i$, we see that $u_i^H w = c_i u_i^H v_i \neq 0$, since otherwise u_i would be zero. By a simple normalization we can choose the vectors in such a way that $u_i^H v_i = 1$. From the column vectors u_i and v_i we form two matrices U and V, and from the eigenvalues λ_i we form a diagonal matrix D. Then we have:

$$U^H V = V^H U = I; \qquad AU = UD; \qquad A^H V = VD.$$

Consequently $U^{-1} A U = D$ and $V^{-1} A^H V = D^*$.

Now we specialize by assuming that A is *Hermitian*, i.e., $A^H = A$. Then $U = V$ and $U^H U = I$, i.e., $U^H = U^{-1}$. A matrix with this property is said to be *unitary*. Hence we have proved that an Hermitian matrix with all eigenvalues unequal can be diagonalized by a unitary matrix. If the matrix A is real and symmetric, then changing the notation from U to Q we get $Q^T Q = I$ or $Q^T = Q^{-1}$. A matrix with this property is said to be *orthogonal*. Hence we have the result.

A real symmetric matrix with all eigenvalues unequal can be diagonalized by an orthogonal matrix.

As is shown in connection with Schur's lemma (see Section 2.7), the condition that the eigenvalues must be different can be dropped.

A simultaneous orthogonal transformation of two vectors leaves the scalar product invariant: $x = Qu$ and $y = Qv$ implies

$$x^T y = (Qu)^T Qv = u^T Q^T Qv = u^T v.$$

We will now indicate how we can obtain specific results concerning the nature of the eigenvalues and the eigenvectors when the matrix has some special property. Let us first assume that A is Hermitian: $A^H = A$. Then from $Ax = \lambda x$ we get $x^H A^H = x^H A = \lambda^* x^H$. Premultiplying the first equation by x^H and postmultiplying the second by x, we get:

$$x^H A x = \lambda x^H x \quad \text{and} \quad x^H A x = \lambda^* x^H x.$$

Since $x^H x \neq 0$, we obtain $\lambda = \lambda^*$, i.e., λ is real.

Let us arrange the eigenvalues of an Hermitian matrix $A = A^H$ in decreasing order: $\lambda_1 \geq \lambda_2 \geq \cdots \geq \lambda_n$. Then λ_1 and λ_n can be characterized in the following way:

$$\lambda_1 = \max R(x), \qquad \lambda_n = \min R(x),$$

both taken over $x \neq 0$, where $R(x)$ is the *Rayleigh quotient*

(2.3.3) $$R(x) = x^H A x / x^H x.$$

For the proof we use the fact that A can be diagonalized by some unitary matrix U; that is, $U^{-1}AU = D$. Putting $x = Uy$ we obtain:

$$R(x) = \frac{x^H A x}{x^H x} = \frac{y^H U^H A U y}{y^H U^H U y} = \frac{y^H D y}{y^H y}$$

$$= \sum \lambda_i |y_i|^2 / \sum |y_i|^2 \leq \sum \lambda_1 |y_i|^2 / \sum |y_i|^2 = \lambda_1$$

and similarly $R(x) \geq \sum \lambda_n |y_i|^2 / \sum |y_i|^2 = \lambda_n$. The Rayleigh quotient is used extensively for numerical computation of eigenvalues by iterative methods.

If instead A is unitary, i.e., $A^H = A^{-1}$, then

$$Ax = \lambda x \quad \text{and} \quad x^H A^H = x^H A^{-1} = \lambda^* x^H.$$

Forming the scalar products of the left and right hand sides of these equations, we obtain $x^H A^{-1} A x = \lambda^* \lambda x^H x$, and since $x^H x \neq 0$ we can divide to get $|\lambda|^2 = 1$ or $|\lambda| = 1$.

In a similar way other results of this type can be deduced, and they are presented in the table below.

Type of matrix	Definition	Eigenvalues	Eigenvectors
Hermitian	$A^H = A$	λ real	u complex
Real, symmetric	$A^T = A^* = A$	λ real	u real
Anti-Hermitian	$A^H = -A$	λ purely imaginary	u complex
Antisymmetric	$A^T = -A$		If $\lambda \neq 0$, then $u^T u = 0$
Real, antisymmetric	$A^T = -A; A^* = A$	λ purely imaginary (or 0)	u complex (except when $\lambda = 0$)
Orthogonal	$A^T = A^{-1}$		If $\lambda \neq \pm 1$, then $u^T u = 0$
Real, orthogonal	$A^T = A^{-1}; A^* = A$	$|\lambda| = 1$	u complex
Unitary	$A^H = A^{-1}$	$|\lambda| = 1$	u complex

Finally, we point out that if U_1 and U_2 are unitary, then $U = U_1 U_2$ is also unitary:

$$U^H = U_2^H U_1^H = U_2^{-1} U_1^{-1} = (U_1 U_2)^{-1} = U^{-1}.$$

Similarly, we can show that if Q_1 and Q_2 are orthogonal, then the same is true for $Q = Q_1 Q_2$. For a real orthogonal matrix, the characteristic equation is real, and hence complex roots are conjugate imaginaries. Since the absolute value is 1, they must be of the form $e^{i\theta}$, $e^{-i\theta}$; further, a real orthogonal matrix of odd order must have at least one eigenvalue $+1$ or -1.

We have seen that Hermitian matrices can be diagonalized by unitary matrices. We mention here that the necessary and sufficient condition that A can be diagonalized with a unitary matrix is that A is *normal*, i.e., $A^H A = A A^H$. This condition is trivially satisfied if A is Hermitian.

We also mention another type of matrix product, the so-called *tensor product* or *Kronecker product*, denoted $A \otimes B$, which is used in certain applications. If A is of type $m \times n$, and B of type $p \times q$, then $A \otimes B$ is of type $mp \times nq$ and has the form of a block matrix:

$$A \otimes B = \begin{pmatrix} a_{11}B & a_{12}B & \cdots & a_{1n}B \\ \vdots & & & \\ a_{m1}B & a_{m2}B & \cdots & a_{mn}B \end{pmatrix}.$$

If in particular A is square $m \times m$ with eigenvalues $\lambda_1, \ldots, \lambda_m$, and B is square $n \times n$ with eigenvalues $\mu_1, \ldots, \mu_n$, then $C = A \otimes B$ has the eigenvalues $\lambda_i \mu_k$, $i = 1(1)m$, $k = 1(1)n$. Further, if A and B are nonsingular, then

$$(A \otimes B)^{-1} = A^{-1} \otimes B^{-1}.$$

---

§2.4. MATRIX NORMS

We define a matrix norm $\|A\|$ as a pure number associated with a matrix A, using the same axiom system as for vector norms:

1. $\|A\| > 0$ if $A \neq 0$, and $\|A\| = 0$ if $A = 0$.
2. $\|cA\| = |c|\,\|A\|$, c an arbitrary number.
3. $\|A + B\| \leq \|A\| + \|B\|$.

The most common matrix norms are:

1. The maximum norms

$$\|A\|_\infty = \max_i \sum_k |a_{ik}| \quad \text{and} \quad \|A\|_1 = \max_k \sum_i |a_{ik}|.$$

2. The Euclidean norm

$$\|A\|_E = [\operatorname{tr}(A^H A)]^{1/2} = [\operatorname{tr}(A A^H)]^{1/2} = \left(\sum_{i,k} |a_{ik}|^2 \right)^{1/2}.$$

3. The Hilbert or spectral norm

$$\|A\|_2 = \lambda_1^{1/2} \text{ where } \lambda_1 \text{ is the largest eigenvalue of } A^H A.$$

Since matrices and vectors often appear together, it is natural to relate the corresponding norms. Then it is easy to show that a given vector norm *induces* a matrix norm through the definition:

$$\|A\| = \sup_{x \neq 0} \|Ax\| / \|x\|$$

or equivalently $\|A\| = \sup\|Ax\|$, taken over $\|x\| = 1$. Here we have exclusively well-defined vector norms on the right hand sides. A matrix norm defined in this way has the obvious property:

$$\|Ax\| \leq \|A\| \cdot \|x\|.$$

A vector norm and a matrix norm that satisfy this condition are said to be *compatible*. Choosing $A = I$ and a vector x with norm 1 and such that $\|Ax\|$ is maximized, we obtain $\|I\| = 1$, valid for a matrix norm induced by some vector norm. Since $\|I\|_E = n^{1/2}$ we understand that the Euclidean norm is not induced by any vector norm. Nevertheless, we can show that it is compatible with the Euclidean vector norm. This follows because

$$\|Ax\|_2^2 = \sum_i \left| \sum_k a_{ik} x_k \right|^2 \leq \sum_i \left\{ \sum_k |a_{ik}|^2 \sum_k |x_k|^2 \right\}$$
$$= \sum_{i,k} |a_{ik}|^2 \sum_k |x_k|^2 = \|A\|_E^2 \|x\|_2^2$$

and hence $\|Ax\|_2 \leq \|A\|_E \|x\|_2$.

Suppose now that $\|A\|$ is induced by a certain vector norm. We will show that conditions 1–3 are satisfied together with one additional inequality. Obviously, 1 and 2 are trivial. The third condition follows if we choose a vector x with $\|x\| = 1$ such that $\|(A + B)x\|$ takes its maximum value. Then:

$$\|A + B\| = \|(A + B)x\| = \|Ax + Bx\| \leq \|Ax\| + \|Bx\|$$
$$\leq \|A\| \cdot \|x\| + \|B\| \cdot \|x\| = \|A\| + \|B\|.$$

Further, let y be a vector with $\|y\| = 1$ such that $\|ABy\|$ takes its maximum value. Then we have:

$$\|AB\| = \|ABy\| \leq \|A\| \cdot \|By\| \leq \|A\| \cdot \|B\| \cdot \|y\| = \|A\| \cdot \|B\|.$$

Even though the Euclidean norm is not induced by any vector norm, we can prove that $\|AB\|_E \leq \|A\|_E \|B\|_E$, and further we have:

$$\|A\|_2 \leq \|A\|_E \leq n^{1/2} \|A\|_2 \quad \text{and} \quad \|A\|_2^2 \leq \|A\|_1 \|A\|_\infty.$$

However, in many cases the bound $\|AB\|_E \leq \|A\|_2 \|B\|_E$ gives a better estimate, particularly if A is close to I, since $\|I\|_2 = 1$ while $\|I\|_E = \sqrt{n}$.

A matrix norm with the property $\|AB\| \leq \|A\| \cdot \|B\|$ is said to be *sub-multiplicative*. Using similar methods as above, we can prove that the matrix norms $\|A\|_1$ and $\|A\|_\infty$ are induced by the corresponding vector norms $\|x\|_1$ and $\|x\|_\infty$. We shall prove that the Hilbert or spectral norm is induced by the Euclidean vector norm. We choose a vector x such that $\|x\|_2 = 1$ as well as a matrix A. Then $\|Ax\|_2^2 = (Ax)^H Ax = x^H A^H Ax = x^H Hx$ where the matrix H is Hermitian. Its eigenvalues are real and non-negative: $\lambda_1 \geq \lambda_2 \geq \cdots \geq \lambda_n \geq 0$, and the corresponding eigenvectors $u_1, u_2, \ldots, u_n$ are supposed to be linearly independent and normalized. Then an arbitrary vector can be written $x = c_1 u_1 + c_2 u_2 + \cdots + c_n u_n = Uc$ if we denote by U the matrix formed by $u_1, u_2, \ldots, u_n$ and by c the column vector with elements c_k. As shown previously, U is unitary, and since $\|x\|_2 = 1$ we get:

$$x^H x = c^H U^H U c = c^H c = \sum |c_k|^2 = 1.$$

Further we have:

$$Hx = \sum c_k \lambda_k u_k = UDC$$

where $D = \text{diag}(\lambda_1, \lambda_2, \ldots, \lambda_n)$ and hence

$$\|Ax\|_2^2 = x^H H x = c^H U^H U D c = c^H D c = \sum \lambda_k |c_k|^2 \leq \lambda_1 \sum |c_k|^2 = \lambda_1.$$

On the other hand, taking $x = u_1$ we find $\|Au_1\|_2^2 = \lambda_1$; and since $\|A\|_2 = \sup \|Ax\|_2$ when $\|x\|_2 = 1$, we have directly $\|A\|_2 = \lambda_1^{1/2}$.

Here we also give a proof that for any matrix norm and λ an arbitrary eigenvalue we have $|\lambda| \leq \|A\|$. This follows because if x is an eigenvector, then $|\lambda| \cdot \|x\| = \|\lambda x\| = \|Ax\| \leq \|A\| \cdot \|x\|$.

In many applications another quantity is also useful, namely the *spectral radius* $\rho(A) = \max_i |\lambda_i|$ where λ_i are the eigenvalues of A. This is not a matrix norm, as can be seen from the fact that $A = \begin{pmatrix} 0 & 0 \\ 1 & 0 \end{pmatrix}$ and $B = \begin{pmatrix} 0 & 1 \\ 0 & 0 \end{pmatrix}$ violate the triangle inequality $(\rho(A) = \rho(B) = 0$, but $\rho(A + B) = 1)$. It has the property $\rho(A) \leq \|A\|$ for all norms, and further $\lim_{n \to \infty} \|A^n\|^{1/m} = \rho(A)$. In particular, if A is Hermitian we have $\rho(A) = \|A\|_2$.

In the theory of differential equations, a quantity $\mu(A)$ called the *logarithmic norm* is sometimes used. However, since it can assume also negative values, it is not a matrix norm. It is defined as follows:

$$\mu(A) = \lim_{\varepsilon \to +0} \frac{\|I + \varepsilon A\|_2 - 1}{\varepsilon}.$$

Since $\|A\|_2^2 = \|A^H\|_2^2 = \|A^H A\|_2$ we have, neglecting higher order terms in ε: $\|I + \varepsilon A\| = \|I + \varepsilon(A + A^H)\|^{1/2}$ and hence $\mu(A) = \lambda_1$, where λ_1 is the largest eigenvalue of the matrix $H = \frac{1}{2}(A + A^H)$. Further we mention the estimate $\alpha(A) < \mu(A)$, where $\alpha(A)$ is the *spectral abscissa* defined through $\alpha(A) = \max \text{Re}(\lambda(A))$.

In connection with error analysis for the solution of linear systems, we need estimates for the norms of $(I + E)^{-1}$ and $(I - E)^{-1}$, where E is a "small" matrix, i.e., $\|E\| < 1$. We start from the identity $(I - E)^{-1} = E(I - E)^{-1} + I$ and obtain:

$$\|(I - E)^{-1}\| \leq \|E(I - E)^{-1}\| + \|I\| \leq \|E\| \|(I - E)^{-1}\| + 1.$$

Hence $(1 - \|E\|) \|(I - E)^{-1}\| \leq 1$ and

$$\|(I - E)^{-1}\| \leq 1/(1 - \|E\|).$$

On the other hand, since $I = (I - E)^{-1} - E(I - E)^{-1}$ we find:

$$1 \leq \|(I - E)^{-1}\| + \|E(I - E)^{-1}\| \leq (1 + \|E\|) \|(I - E)^{-1}\|$$

which gives $1/(1 + \|E\|) \leq \|(I - E)^{-1}\|$. Replacing E by $-E$ and observing that $\|E\| = \|-E\|$, we get a similar estimate for $\|(I + E)^{-1}\|$:

(2.4.1) $$1/(1 + \|E\|) \leq \|(I \pm E)^{-1}\| \leq 1/(1 - \|E\|).$$

§2.5. MATRIX FUNCTIONS

Let us start with a polynomial $f(z) = z^n + a_1 z^{n-1} + \cdots + a_n$, where z is a variable and $a_1, a_2, \ldots, a_n$ are real or complex constants. Replacing z by A, where A is a square matrix, we get a matrix polynomial

$$f(A) = A^n + a_1 A^{n-1} + \cdots + a_{n-1} A + a_n I.$$

Here $f(A)$ is a square matrix of the same order as A. We will now investigate eigenvalues and eigenvectors of $f(A)$. Suppose that λ is an eigenvalue and u the corresponding eigenvector of A: $Au = \lambda u$. Premultiplication with A gives the result $A^2 u = \lambda A u = \lambda^2 u$, that is, λ^2 is an eigenvalue of A^2; the eigenvector is unchanged. In the same way we find that $f(\lambda)$ is an eigenvalue of the matrix $f(A)$.

It is natural to generalize to an infinite power series:

$$f(z) = a_0 + a_1 z + a_2 z^2 + \cdots.$$

We suppose that the series is convergent in the domain $|z| < R$. Then it can be shown that if A is a square matrix with all eigenvalues less than R in absolute value, then the series $a_0 I + a_1 A + a_2 A^2 + \cdots$ is convergent. The sum of the series will be denoted by $f(A)$. Some series of this kind converge for all finite matrices, for example,

$$e^A = I + A + \frac{A^2}{2!} + \frac{A^2}{3!} + \cdots,$$

while, for example, the series $(I - A)^{-1} = I + A + A^2 + \cdots$ converges only if all eigenvalues of A are less than 1 in absolute value.

If a matrix A undergoes a similarity transformation with a matrix S,

$$(2.5.1) \qquad\qquad S^{-1}AS = B,$$

then the same is true for A^2: $S^{-1}A^2 S = (S^{-1}AS)(S^{-1}AS) = B^2$. Obviously, this can be directly generalized to polynomials: $S^{-1}f(A)S = f(B)$. Generalization to the case when f also contains negative powers is immediate, since

$$S^{-1}A^{-1}S = (S^{-1}AS)^{-1} = B^{-1}.$$

We also conclude directly that

$$S^{-1}\left(\frac{f(A)}{g(A)}\right)S = \frac{f(B)}{g(B)},$$

where f and g are polynomials. The division has a good meaning, since we have full commutativity. Under certain conditions the results obtained can be further generalized, but we do not want to go into more detail right now.

Cayley–Hamilton's Theorem. *Any square matrix A satisfies its own characteristic equation.*

We give a proof in the special case when all eigenvalues are different and hence the eigenvectors form a nonsingular matrix S. Let $f(\lambda)$ be the characteristic polynomial. Thus

$$(-1)^n f(\lambda) = \det(\lambda I - A) = \lambda^n + c_1 \lambda^{n-1} + \cdots + c_n.$$

Let D be a diagonal matrix with $d_{ii} = \lambda_i$. Then $S^{-1}AS = D$, and further

$$S^{-1}f(A)S = f(D) = \begin{pmatrix} f(\lambda_1) & 0 & \cdots & 0 \\ 0 & f(\lambda_2) & \cdots & 0 \\ \vdots & & & \\ 0 & 0 & \cdots & f(\lambda_n) \end{pmatrix} = 0.$$

Premultiplication with S and postmultiplication with S^{-1} gives

(2.5.2) $$f(A) = 0.$$

A general proof can be obtained from this by observing that the coefficients of the characteristic polynomial are continuous functions of the matrix elements. Hence the case when some eigenvalues are equal can be considered as a limiting case, and in fact, with small modifications the given proof is still valid.

Cayley–Hamilton's theorem plays an important role in the theory of matrices as well as in many different applications. It should be mentioned here that for every square matrix A, there exists a unique polynomial $m(\lambda)$ of lowest degree with the leading coefficient 1 such that $m(A) = 0$. It is easy to prove that $m(\lambda)$ is a factor of $f(\lambda)$, since we can write $f(\lambda) = m(\lambda)q(\lambda) + r(\lambda)$, where $r(\lambda)$ is of lower degree than $m(\lambda)$. Replacing λ by A, we get

$$f(A) = m(A)q(A) + r(A) \quad \text{or} \quad r(A) = 0,$$

since $f(A) = 0$ and $m(A) = 0$. However, this is in contradiction to our supposition that $m(\lambda)$ is the *minimum polynomial*. In most cases we have $m(\lambda) = f(\lambda)$.

Now consider a matrix polynomial in A:

$$F(A) = A^n + a_1 A^{N-1} + \cdots + a_N I \qquad (N \geq n).$$

Dividing $F(x)$ by $f(x)$, we get

$$F(x) = f(x)Q(x) + R(x),$$

where $R(x)$ is of degree $\leq n - 1$. Replacing x by A, we find

(2.5.3) $$F(A) = f(A)Q(A) + R(A) = R(A).$$

Without proof we mention that this formula is also valid for functions other than polynomials.

Example 1.   Consider e^A where A is a two-by-two-matrix. From the preceding discussion, we have

$$e^A = aA + bI.$$

Suppose that the eigenvalues of A are λ_1 and λ_2 $(\lambda_1 \neq \lambda_2)$. Then e^A and $aA + bI$ have the eigenvalues e^{λ_1}, e^{λ_2}, and $a\lambda_1 + b$, $a\lambda_2 + b$, respectively. Hence

$$\begin{cases} e^{\lambda_1} = a\lambda_1 + b, \\ e^{\lambda_2} = a\lambda_2 + b. \end{cases}$$

From this system we solve for a and b:

$$a = \frac{e^{\lambda_2} - e^{\lambda_1}}{\lambda_2 - \lambda_1}; \qquad b = \frac{\lambda_2 e^{\lambda_1} - \lambda_1 e^{\lambda_2}}{\lambda_2 - \lambda_1}.$$

§2.6.   LOCALIZATION OF EIGENVALUES

As shown previously, we have $|\lambda| \leq \|A\|$ for any matrix norm. Hence, taking the maximum norms we get:

(2.6.1.)
$$\begin{cases} |\lambda| \leq \max_i \sum_k |a_{ik}| \\ |\lambda| \leq \max_k \sum_i |a_{ik}| \end{cases}$$

and naturally the better of these estimates is preferred.

If we denote $\mathrm{diag}(a_{11}, a_{22}, \ldots, a_{nn})$ by D, and assume that $\lambda \neq a_{ii}$, $i = 1(1)n$, is an eigenvalue of A, and x is the corresponding eigenvector, then we have:

$$(A - D)x = (\lambda I - D)x, \quad \text{or} \quad x = (\lambda I - D)^{-1}(A - D)x.$$

Taking norms on both sides, we obtain:

$$\|x\|_\infty = \|(\lambda I - D)^{-1}(A - D)x\|_\infty \leq \|(\lambda I - D)^{-1}(A - D)\|_\infty \cdot \|x\|_\infty$$

or

$$1 \leq \|(\lambda I - D)^{-1}(A - D)\|_\infty = \max_i |\lambda - a_{ii}|^{-1} \sum_{k \neq i} |a_{ik}|$$

where we have used only the definition of the norm. From this we obtain Gershgorin's famous theorem:

Gershgorin's Theorem.   *The union of all disks defined by*

(2.6.2)
$$|\lambda - a_{ii}| \leq \sum_{k \neq i} |a_{ik}|, \qquad i = 1(1)n,$$

contains all the eigenvalues of A.

Since A^T and A have the same eigenvalues, we also get another family of discs which may contribute more information concerning the eigenvalues. The circles are known as Gershgorin's circles. Using simple similarity transformations by suitable diagonal matrices, it is often possible to improve the estimates considerably.

Let us now consider the following family of matrices:

$$A(t) = D + t(A - D), \qquad 0 \le t \le 1.$$

Then $A(0) = D = \text{diag}(a_{ii})$ and $A(1) = A$. Applying Gershgorin's theorem on $A(0)$ we see that all circles have degenerated into points. If t is now gradually increased, the points will become circles which grow more and more. From the beginning the circles are disjoint and every circle contains exactly one eigenvalue. When two circles overlap partially, exactly two eigenvalues are contained, and so on. Using this idea one can often get more precise information on the distribution of the eigenvalues, at least as long as the domains are disjoint.

Here we mention briefly how to construct a matrix with a given characteristic polynomial $f(z) = z^n + a_1 z^{n-1} + \cdots + a_n$. It is easy to show that the matrix A below, called the *companion matrix*:

$$A = \begin{pmatrix} 0 & 0 & \cdots & 0 & -a_n \\ 1 & 0 & \cdots & 0 & -a_{n-1} \\ 0 & 1 & \cdots & 0 & -a_{n-2} \\ \vdots & & & & \\ 0 & 0 & \cdots & 1 & -a_1 \end{pmatrix}$$

has the desired property. Forming the determinant of $(A - \lambda I)$ we get:

$$\det(A - \lambda I) = \begin{vmatrix} -\lambda & 0 & \cdots & 0 & -a_n \\ 1 & -\lambda & \cdots & 0 & -a_{n-1} \\ \vdots & & & & \\ 0 & 0 & \cdots & -\lambda & -a_2 \\ 0 & 0 & \cdots & 1 & -\lambda - a_1 \end{vmatrix}.$$

We multiply the last line by λ and add it to the preceding one, multiply this line by λ and add it to the preceding one, and so on, and in this way we find $f(\lambda) = 0$.

§2.7. NORMAL FORMS

We have already shown that there are several situations when matrices can be transformed to diagonal form. For instance, this is the case if all eigenvalues are unequal. Then it is possible to construct a transformation matrix from the eigenvectors that are linearly independent. If the matrix is Hermitian, we have even proved that it can be diagonalized by a unitary matrix. We will now discuss what can be done in more general cases.

First we are going to prove a lemma that is due to Schur.

Schur's lemma. *Given an arbitrary matrix A, one can find a unitary matrix U such that $T = U^{-1}AU$ is triangular.*

Proof. Let the eigenvalues of A be $\lambda_1, \lambda_2, \ldots, \lambda_n$ and determine U_1 so that

$$U_1^{-1}AU_1 = A_1,$$

where A_1 has the following form:

$$A_1 = \begin{pmatrix} \lambda_1 & \alpha_{12} & \alpha_{13} & \cdots & \alpha_{1n} \\ 0 & \alpha_{22} & \alpha_{23} & \cdots & \alpha_{2n} \\ 0 & \alpha_{32} & \alpha_{33} & \cdots & \alpha_{3n} \\ \vdots & & & & \\ 0 & \alpha_{n2} & \alpha_{n3} & \cdots & \alpha_{nn} \end{pmatrix}.$$

As is easily found, the first column of U_1 must be a multiple of x_1, where $Ax_1 = \lambda_1 x_1$. Otherwise, the elements of U_1 can be chosen arbitrarily. The choice can be made in such a way that U_1 becomes unitary; essentially this corresponds to the well-known fact that in an n-dimensional space, it is always possible to choose n mutually orthogonal unit vectors. Analogously, we determine a unitary matrix U_2 which transforms the elements $\alpha_{32}, \alpha_{42}, \ldots, \alpha_{n2}$ to zero. This matrix U_2 has all elements in the first row and the first column equal to zero except the corner element, which is 1. The $(n-1)$-dimensional matrix obtained when the first row and the first column of A_1 are removed has, of course, the eigenvalues $\lambda_2, \lambda_3, \ldots, \lambda_n$, and the corresponding eigenvectors $x_2', x_3', \ldots, x_n'$, where x_i' is obtained from x_i by removing the first component. It is obvious that after $n-1$ rotations of this kind, a triangular matrix is formed and further that the total transformation matrix $U = U_1 U_2 \cdots U_{n-1}$ is unitary, since the product of two unitary matrices is also unitary.

In particular, if A is Hermitian, i.e., $A^H = A$, then it follows directly that $T^H = T$, and hence T is diagonal. This constitutes an alternative proof that every Hermitian matrix can be diagonalized by a unitary matrix.

As mentioned before, a normal matrix is characterized by the condition $A^H A = AA^H$. It can be shown that normal matrices are the most general family of matrices that can be diagonalized by unitary matrices. Obviously, Hermitian matrices are also normal.

We will now turn to the general case, in which some eigenvalues may be multiple. It is easy to give examples showing that there exist matrices that cannot be diagonalized. A very simple one is

$$A = \begin{pmatrix} 2 & 1 \\ 0 & 2 \end{pmatrix}.$$

Without proof we now state an important theorem by Jordan:

Every matrix A can be transformed with a regular matrix S as follows:

$$S^{-1}AS = J,$$

where

(2.7.1)
$$J = \begin{pmatrix} J_1 & 0 & \cdots & 0 \\ 0 & J_2 & \cdots & 0 \\ & & \ddots & \\ 0 & 0 & \cdots & J_m \end{pmatrix}.$$

The J_i are so-called Jordan boxes, that is, submatrices of the form

$$\lambda_i, \quad \begin{pmatrix} \lambda_i & 1 \\ 0 & \lambda_i \end{pmatrix}, \quad \begin{pmatrix} \lambda_i & 1 & 0 \\ 0 & \lambda_i & 1 \\ 0 & 0 & \lambda_i \end{pmatrix}, \quad \begin{pmatrix} \lambda_i & 1 & 0 & 0 \\ 0 & \lambda_i & 1 & 0 \\ 0 & 0 & \lambda_i & 1 \\ 0 & 0 & 0 & \lambda_i \end{pmatrix}, \dots$$

In one box we have the same eigenvalue λ_i in the main diagonal, but it can appear also in other boxes. The representation (2.7.1) is called *Jordan's normal form*. If, in particular, all λ_i are different, J becomes a diagonal matrix.

The occurrence of multiple eigenvalues causes degeneration of different kinds. If an eigenvalue is represented in a Jordan-box with at least 2 rows, we obtain a lower number of linearly independent eigenvectors than the order of the matrix. A matrix with this property is said to be *defective*. If one or more eigenvalues are multiple but in such a way that the corresponding Jordan-boxes contain just one element (that is, no ones are present above the main diagonal), we have another degenerate case characterized by the fact that the minimum polynomial is of lower degree than the characteristic polynomial. A matrix with this property is said to be *derogatory*. Both these degenerate cases can appear separately or combined. We give a simple example of a matrix that is degenerate in both respects. With

$$A = \begin{pmatrix} 2 & 1 & 0 & 0 \\ 0 & 2 & 0 & 0 \\ 0 & 0 & 2 & 1 \\ 0 & 0 & 0 & 3 \end{pmatrix}$$

we find only three linearly independent eigenvectors:

$$\begin{pmatrix} 1 \\ 0 \\ 0 \\ 0 \end{pmatrix}, \quad \begin{pmatrix} 0 \\ 0 \\ 1 \\ 0 \end{pmatrix}, \quad \text{and} \quad \begin{pmatrix} 0 \\ 0 \\ 1 \\ 1 \end{pmatrix}.$$

Hence A is defective. Further, the minimum polynomial turns out to be $m(\lambda) = (\lambda - 2)^2(\lambda - 3)$, which is of degree 3, while the characteristic polynomial $f(\lambda) = (\lambda - 2)^3(\lambda - 3)$ is of degree 4. Hence A is also derogatory.

---

§2.8. RECTANGULAR MATRICES

We have seen that for square matrices with all eigenvalues different, there exist partitions of the form $A = S^{-1}DS$ where D is diagonal. It is then natural to ask whether there is some similar decomposition for rectangular matrices. This question is answered by the following theorem.

Let A be an arbitrary $m \times n$ matrix. Then there is a unitary $m \times m$ matrix U and a unitary $n \times n$ matrix V such that $U^H A V = \Sigma$ is an $m \times n$ matrix with $\Sigma_{ii} = \sigma_i$, $i = 1, 2, \ldots, r$ and $\sigma_1 \geq \sigma_2 \geq \cdots \geq \sigma_r \geq 0$ where r is the rank of A and all other elements Σ_{ik} are zero. The same values $\sigma_1, \sigma_2, \ldots, \sigma_r$ appear in the partition of the $n \times m$ matrix A^H.

For a given matrix A the $n \times n$ matrix $A^H A$ is obviously positive semidefinite since $x^H A^H A x = \|Ax\|_2^2 \geq 0$ for all x. Hence the eigenvalues of $A^H A$ are non-negative, $\lambda_1 \geq \lambda_2 \geq \cdots \geq \lambda_n \geq 0$, and we can write $\lambda_k = \sigma_k^2$ with $\sigma_k \geq 0$. These numbers $\sigma_1 \geq \sigma_2 \geq \cdots \geq \sigma_n \geq 0$ are called *singular values* of A.

We have the following characterization:

$$\sigma_1 = \max_{x \neq 0} \|Ax\|_2 / \|x\|_2; \qquad \sigma_n = \min_{x \neq 0} \|Ax\|_2 / \|x\|_2.$$

The proof of the singular value decomposition theorem is carried with induction from $(m-1, n-1)$ to (m, n), much along the same lines as in the proof that an arbitrary square matrix can be triangularized with unitary matrices. However, we omit the details.

We can interpret A as a mapping from an n-dimensional to an m-dimensional space: $y = Ax$ where y, A, and x are of type $m \times 1$, $m \times n$, and $n \times 1$, respectively (we suppose $m > n$). Performing coordinate transformations with unitary matrices U and V, we have:

$$y = U\eta \quad \text{with} \quad U^{-1} = U^H \quad (m \times m)$$
$$x = V\xi \quad \text{with} \quad V^{-1} = V^H \quad (n \times n).$$

Hence $U_\eta = AV\xi$ and $\eta = U^H A V \xi$. As mentioned before, it is now possible to choose U and V so that $\Sigma = U^H A V$ is "pseudodiagonal," and then we get $A = U\Sigma V^H$. The $m \times n$ matrix Σ will have the form:

$$\Sigma = \begin{pmatrix} \sigma_1 & 0 & 0 & \cdots & 0 \\ 0 & \sigma_2 & 0 & \cdots & 0 \\ \vdots & & & \ddots & \\ 0 & 0 & 0 & \cdots & \sigma_n \\ 0 & 0 & 0 & \cdots & 0 \\ \vdots & & & & \\ 0 & 0 & 0 & \cdots & 0 \end{pmatrix}$$

where σ_i are real and non-negative. Forming

$$A^H A = V\Sigma^H U^H U\Sigma V^H = V\Sigma^H \Sigma V^H$$

we see that $S = \Sigma^H \Sigma = \text{diag}(\sigma_1^2, \sigma_2^2, \ldots, \sigma_n^2)$ (of type $n \times n$). Since S is obtained from $A^H A$ by a similarity transformation, the eigenvalues $\sigma_1^2, \ldots, \sigma_n^2$ are obtained from $\det(A^H A - \lambda I) = 0$.

If A is a rectangular matrix, the problems of solving the system $Ax = b$, of finding an inverse, and of solving the eigenvalue problem $Ax = \lambda x$ to some extent lose their original meaning. We will here try to generalize these notions in such a way that they become meaningful after reasonable modifications. It will then be useful to consider some questions that give rise to these problems. A system of equations with more equations than unknowns can be solved only to the extent that a suitably defined error quantity is minimized (see Chapter 3). It turns out to be convenient to define a *pseudo-inverse*, which plays a central role for all three problems mentioned above. It coincides with the usual inverse A^{-1} when A is a square, non-singular matrix, but otherwise it has the properties quoted below. It can be shown that the four equations

$$AXA = A, \qquad XAX = X, \qquad (AX)^H = AX, \qquad (XA)^H = XA$$

have a unique solution X when A is an arbitrary matrix. This solution is called the *pseudo-inverse* of A and is denoted $X = A^+$. If A is of type $m \times n$ we have:

(2.8.1)
$$\begin{aligned} &m > n\colon A^+ = (A^H A)^{-1} A^H &&\text{(type } n \times m) \\ &m < n\colon A^+ = A^H (A A^H)^{-1} &&\text{(type } n \times m) \\ &m = n\colon A^+ = A^{-1} &&\text{if } A \text{ is non-singular} \end{aligned}$$

As will be demonstrated in examples below, a pseudo-inverse exists even if A is square but singular.

The following relations are easily proved:

(2.8.2)
$$\begin{aligned} A^{++} &= A \\ (A^H)^+ &= (A^+)^H \\ (A^H A)^+ &= A^+ (A^+)^H. \end{aligned}$$

If U and V are unitary, then $(UAV)^+ = V^H A^+ U^H$. Further, A, $A^H A$, A^+, and $A^+ A$ are all of rank $r = \operatorname{tr}(A^+ A)$.

Again assuming $m > n$ and the decomposition $A = U\Sigma V^H$, we get $A^H A = V\Sigma^H \Sigma V^H$ and $(A^H A)^{-1} = V(\Sigma^H \Sigma)^{-1} V^H$. Hence

$$A^+ = (A^H A)^{-1} A^H = V(\Sigma^H \Sigma)^{-1} V^H V \Sigma^H U^H = V(\Sigma^H \Sigma)^{-1} \Sigma^H U^H = V\Sigma^+ U^H$$

where

$$\Sigma^+ = \begin{pmatrix} 1/\sigma_1 & \cdots & 0 & 0 & \cdots & 0 \\ \vdots & \ddots & & & & \\ 0 & \cdots & 1/\sigma_n & 0 & \cdots & 0 \end{pmatrix}.$$

We see that if the singular value decomposition is known, then the problem of determining A^+ is reduced to the simpler problem of determining Σ^+.

If A is a square matrix that is singular, we can use a method sketched below. Let $C = A^H A$, implying $C^H = C$; then obviously C is also singular. We will now try to determine a matrix B such that $BC^2 = C$. This can be done if we observe that according to Cayley–Hamilton's theorem, C satisfies the characteristic equation and, of course, also the minimal equation. If, for example, $C^4 - aC^3 + bC^2 - cC = 0$, then we choose $B = c^{-1}[C^2 - aC + bI]$ and verify directly that $BC^2 = c^{-1}[C^4 - aC^3 + bC^2] = C$. If the C-term is missing, we use the minimal equation instead.

Putting $BC = D$ and observing that $C = A^H A$ we get $DA^H A - A^H A = 0$. We multiply this from the right by $(D - I)^H$:

$$0 = (DA^H A - A^H A)(D^H - I) = DA^H AD^H - DA^H A - A^H AD^H + A^H A$$
$$= (DA^H - A^H)(DA^H - A^H)^H$$

implying that $DA^H = A^H$. Hence since $D = BC = BA^H A$ we have $BA^H AA^H = A^H$ or, taking the Hermitian conjugate, $AA^H AB^H = AA^H AB = A$ (note that B is a linear combination of I and powers of C, implying $B^H = B$). But $AA^H AB = ACB = ABC$ because B and C commute, and putting $X = BA^H$ we find:

$$A = AA^H AB = ABA^H A = AXA,$$

which is the first of the four equations defining the pseudo-inverse. The second equation can be verified using a similar technique, while the third and fourth equations are trivial. Hence our result is that $A^+ = BA^H$, where B is obtained from the characteristic equation as indicated above.

Example 1. $\quad A = \begin{pmatrix} 1 & 1 \\ 3 & 2 \\ 7 & 4 \end{pmatrix} \quad m = 3, n = 2.$

$$A^H A = \begin{pmatrix} 59 & 35 \\ 35 & 21 \end{pmatrix} \quad (A^H A)^{-1} = \frac{1}{14}\begin{pmatrix} 21 & -35 \\ -35 & 59 \end{pmatrix}$$

$$A^+ = \frac{1}{14}\begin{pmatrix} 21 & -35 \\ -35 & 59 \end{pmatrix}\begin{pmatrix} 1 & 3 & 7 \\ 1 & 2 & 4 \end{pmatrix} = \frac{1}{14}\begin{pmatrix} -14 & -7 & 7 \\ 24 & 13 & -9 \end{pmatrix}.$$

Example 2. $\quad A = \begin{pmatrix} 4 & -1 & 1 \\ 2 & 3 & 0 \\ 6 & 2 & 1 \end{pmatrix} \quad A^H A = C = \begin{pmatrix} 56 & 14 & 10 \\ 14 & 14 & 1 \\ 10 & 1 & 2 \end{pmatrix}$

$$\det(C - \lambda I) = \lambda^3 - 72\lambda^2 + 627\lambda = 0.$$

According to Cayley–Hamilton's theorem we have $C^3 - 72C^2 + 627C = 0$. We now form $B = (-1/627)(C - 72I)$ and obtain after easy calculations:

$$A^+ = \frac{1}{627}\begin{pmatrix} 68 & -10 & 58 \\ -115 & 146 & 31 \\ 31 & -23 & 8 \end{pmatrix}.$$

Example 3. If A is 0 and of type $m \times n$, then A^+ is also 0 but of type $n \times m$. Note that we cannot obtain A^+ by continuity arguments when A is square and singular.

Example 4. $\quad A = \begin{pmatrix} 1 & 2 \\ 2 & 1 \\ -3 & 1 \\ -1 & -3 \end{pmatrix} \quad$ and hence $\quad A^H = \begin{pmatrix} 1 & 2 & -3 & -1 \\ 2 & 1 & 1 & -3 \end{pmatrix}.$

Thus we find $A^H A = \begin{pmatrix} 15 & 4 \\ 4 & 15 \end{pmatrix}$ with the characteristic equation $\lambda^2 - 30\lambda + 209 = 0$ and the eigenvalues $\lambda_1 = 19$, $\lambda_2 = 11$. The corresponding eigenvectors are $v_1 = \begin{pmatrix} 1 \\ 1 \end{pmatrix}$, $v_2 = \begin{pmatrix} -1 \\ 1 \end{pmatrix}$. Further we have the inverse

$$(A^H A)^{-1} = \frac{1}{209} \begin{pmatrix} 15 & -4 \\ -4 & 15 \end{pmatrix}$$

and the pseudo-inverse of A:

$$A^+ = (A^H A)^{-1} A^H = \frac{1}{209} \begin{pmatrix} 7 & 26 & -49 & -3 \\ 26 & 7 & 27 & -41 \end{pmatrix}.$$

Obviously $A^+ A = \begin{pmatrix} 1 & 0 \\ 0 & 1 \end{pmatrix}$ as it should be.

We will now try to find two more vectors, orthogonal to the original column vectors of A and orthogonal to each other. After some manipulation we find:

$$w_1 = \begin{pmatrix} 0 \\ 2 \\ 1 \\ 1 \end{pmatrix}, \qquad w_2 = \begin{pmatrix} -30 \\ 8 \\ 1 \\ -17 \end{pmatrix}$$

and hence:

$$\frac{1}{209} \begin{pmatrix} 7 & 26 & -49 & -3 \\ 26 & 7 & 27 & -41 \end{pmatrix} \underbrace{\begin{pmatrix} 1 & 2 \\ 2 & 1 \\ -3 & 1 \\ -1 & -3 \end{pmatrix}}_{A} \underbrace{\begin{pmatrix} 0 & -30 \\ 2 & 8 \\ 1 & 1 \\ 1 & -17 \end{pmatrix}}_{N} = \begin{pmatrix} 1 & 0 & 0 & 0 \\ 0 & 1 & 0 & 0 \end{pmatrix}.$$

We can say that A represents a two-dimensional space corresponding to the original problem, while N represents a null-space. In general they have rank r and $n - r$ respectively, where r is defined as before. We now form the matrix AA^H:

$$AA^H = \begin{pmatrix} 5 & 4 & -1 & -7 \\ 4 & 5 & -5 & -5 \\ -1 & -5 & 10 & 0 \\ -7 & -5 & 0 & 10 \end{pmatrix}.$$

It has rank 2 and the characteristic equation $\lambda^4 - 30\lambda^3 + 209\lambda^2 = 0$ with the eigenvalues 19, 11, 0, and 0. We note that the same eigenvalues as for $A^H A$ appear, but in addition we have also two zero eigenvalues. The corresponding eigenvectors are:

$$\boldsymbol{u}_1 = \begin{pmatrix} 3 \\ 3 \\ -2 \\ -4 \end{pmatrix}, \quad \boldsymbol{u}_2 = \begin{pmatrix} 1 \\ -1 \\ 4 \\ -2 \end{pmatrix}, \quad \boldsymbol{u}_3 = \begin{pmatrix} 0 \\ 2 \\ 1 \\ 1 \end{pmatrix}, \quad \boldsymbol{u}_4 = \begin{pmatrix} -30 \\ 8 \\ 1 \\ -17 \end{pmatrix}.$$

Note that $\boldsymbol{u}_3$ and $\boldsymbol{u}_4$ are equal to $\boldsymbol{w}_1$ and $\boldsymbol{w}_2$, and of course all $\boldsymbol{u}_i$ are mutually orthogonal. The original two column vectors of A are linear combinations of $\boldsymbol{u}_1$ and $\boldsymbol{u}_2$; in fact we have

$$A(1) = \frac{1}{2}(\boldsymbol{u}_1 - \boldsymbol{u}_2); \qquad A(2) = \frac{1}{2}(\boldsymbol{u}_1 + \boldsymbol{u}_2).$$

When dealing with ordinary Hermitian square matrices, we know that they can be diagonalized with unitary matrices formed from the eigenvectors. A similar argument shows that U is built from the eigenvectors of AA^H, while V consists of the eigenvectors of $A^H A$. In the numerical example we obtain after normalization, for brevity putting $a = 38^{-1/2}$, $b = 22^{-1/2}$, $c = 6^{-1/2}$, and $d = 1254^{-1/2}$:

$$U = \begin{pmatrix} 3a & b & 0 & -30d \\ 3a & -b & 2c & 8d \\ -2a & 4b & c & d \\ -4a & -2b & c & -17d \end{pmatrix}; \quad V = \begin{pmatrix} 1/\sqrt{2} & -1/\sqrt{2} \\ 1/\sqrt{2} & 1/\sqrt{2} \end{pmatrix}.$$

Further, inserting

$$\Sigma = \begin{pmatrix} \sqrt{19} & 0 \\ 0 & \sqrt{11} \\ 0 & 0 \\ 0 & 0 \end{pmatrix}$$

we compute

$$U\Sigma V^H = \begin{pmatrix} 1 & 2 \\ 2 & 1 \\ -3 & 1 \\ -1 & -3 \end{pmatrix},$$

which is equal to A in accordance with the theory. Further, since

$$\Sigma^+ = \begin{pmatrix} 1/\sqrt{19} & 0 & 0 & 0 \\ 0 & 1/\sqrt{11} & 0 & 0 \end{pmatrix},$$

we get

$$\Sigma^+ U^H = \frac{1}{\sqrt{2}} \begin{pmatrix} 3/19 & 3/19 & -2/19 & -4/19 \\ 1/11 & -1/11 & 4/11 & -2/11 \end{pmatrix}$$

and

$$A^+ = V\Sigma^+ U^H = \frac{1}{209}\begin{pmatrix} 7 & 26 & 49 & -3 \\ 26 & 7 & 27 & -41 \end{pmatrix}$$

in accordance with our previous result.

EXERCISES

1. Find a vector w that is orthogonal to $u = (1, 2, 3)$ and $v = (4, 2, 1)$.

2. Show that the vectors $x = (1, 1, 0, 1)$, $y = (2, 5, -3, 4)$, $z = (3, -1, 2, 0)$ and $u = (1, 8, -5, 6)$ are linearly dependent. Determine constants a, b, and c such that $u = ax + by + cz$.

3. The matrix A is defined through $A = xy^T$ where x and y are vectors of length n. Find the eigenvalues and the corresponding eigenvectors, together with the characteristic equation and the minimal equation.

4. A is a matrix of order $n \times n$ where all a_{ik} are zero for $i \leq k$. Show that $(I - A)^{-1} = I + A + A^2 + \cdots + A^{n-1}$, where I is the unit matrix.

5. A and B are matrices of order $n \times n$. Prove that the relation $AB - BA = I$ cannot hold.

6. Show that an antisymmetric matrix of odd order is singular.

7. Show that $B = e^A$ is an orthogonal matrix if A is antisymmetric. Find B if $A = \begin{pmatrix} 0 & a \\ -a & 0 \end{pmatrix}$.

8. The matrix $A = \begin{pmatrix} 6 & 2 \\ 1 & 5 \end{pmatrix}$ is given. Then $\cos A$ can be written in the form $aA + bI$. Find the constants a and b.

9. The matrix $A = \begin{pmatrix} 0.9 & 0.2 \\ 0.3 & 0.4 \end{pmatrix}$ is given. Compute $\lim_{n \to \infty} A^n$.

10. Show by using Gershgorin's theorem that $x^T A x > 0$ for all real $x \neq 0$ when

$$A = \begin{pmatrix} 10 & 3 & -1 & 5 \\ 3 & 8 & 1 & -2 \\ -1 & 1 & 6 & 0 \\ 5 & -2 & 0 & 9 \end{pmatrix}.$$

11. The matrix $A = \begin{pmatrix} 4 & 2 \\ 1 & 3 \end{pmatrix}$ is given. Compute $\|A\|_1$, $\|A\|_2$, $\|A\|_\infty$, $\|A\|_E$, and $\rho(A)$.

12. Find a 3×3 matrix with the eigenvalues 6, 2, and -1 and the corresponding eigenvectors $\begin{pmatrix} 2 \\ 3 \\ -2 \end{pmatrix}$, $\begin{pmatrix} 9 \\ 5 \\ 4 \end{pmatrix}$, and $\begin{pmatrix} 4 \\ 4 \\ -1 \end{pmatrix}$.

13. A square matrix that has exactly one element equal to 1 in every row and in every column, and all other elements zero, is called a permutation matrix. Show that a permutation matrix is orthogonal, and that all eigenvalues are 1 in absolute value.

14. The normal matrix A is transformed to diagonal form D by a unitary matrix U: $D = U^{-1}AU$. Find U and D when $A = \begin{pmatrix} a & b \\ -b & a \end{pmatrix}$, a and b real.

15. A matrix $A = \begin{pmatrix} a & b \\ c & d \end{pmatrix}$ with complex elements is given. A is transformed with a unitary matrix U:

$$U^{-1}AU = A' = \begin{pmatrix} a' & b' \\ c' & d' \end{pmatrix},$$

where we choose

$$U = \begin{pmatrix} p & -p\alpha^* \\ p\alpha & p \end{pmatrix}$$

with p real. The p and α are determined from the conditions that U is unitary and that A' is an upper triangular matrix (that is, $c' = 0$). Find p and α expressed in a, b, c, and d, and use the result for computing a' and d' if

$$A = \begin{pmatrix} 9 & 10 \\ -2 & 5 \end{pmatrix}.$$

16. One wants to construct an $n \times n$-matrix P_r such that $P_r a_1 = P_r a_2 = \cdots = P_r a_r = 0$ where $a_1, a_2, \ldots, a_r$ are given vectors with n components $(r < n)$ that are linearly independent. A matrix A_r of type $r \times n$ is formed by writing the row vectors $a_1^T, a_2^T, \ldots, a_r^T$ downward in this order. Show that $P_r = I - A_r^T(A_r A_r^T)^{-1}A_r$ has the desired property and that P_r is singular when $r \geq 1$ (P_r is called a projection matrix). Finally, compute P_r for $r = 2$ when $a_1^T = (1, 3, 0, -2)$ and $a_2^T = (0, 4, -1, 3)$.

17. The matrix A is given by $A = \begin{pmatrix} 1.2 & -0.1 \\ 0.4 & 0.8 \end{pmatrix}$. Show that $\lim_{n \to \infty} A^n$ does not exist. Find A^n explicitly as a function of n.

18. The matrix

$$A = \begin{pmatrix} 3 & 1 & 1 & 2 \\ 1 & 3 & 2 & 1 \\ 1 & 2 & 3 & 1 \\ 2 & 1 & 1 & 3 \end{pmatrix}$$

is given. Find all eigenvalues and eigenvectors together with the minimal equation. Decide whether A is defective or derogatory, or possibly both.

19. Solve the same problem as in Exercise 18 when

$$A = \begin{pmatrix} -7 & 3 & 3 \\ -21 & 9 & 7 \\ -6 & 2 & 4 \end{pmatrix}.$$

20. *A* and *B* are square matrices of order *n*. Show that the eigenvalues of the block matrix $C = \begin{pmatrix} A & B \\ B & A \end{pmatrix}$ are eigenvalues of $A + B$ and of $A - B$. Numerical example: $A = \begin{pmatrix} 11 & 4 \\ 3 & 7 \end{pmatrix}$, $B = \begin{pmatrix} 3 & 0 \\ 1 & 1 \end{pmatrix}$. Eigenvalues of *C*: 16, 10, 6, 4.

21. The matrix

$$A = \begin{pmatrix} 1 & 0 & 10 & 0 \\ 2 & 3 & 1 & 3 \\ 1 & 1 & 1 & 2 \\ 1 & 2 & 2 & 1 \end{pmatrix}$$

is given. Put $a = \max_i |\lambda_i|$ where λ_i are the eigenvalues, and estimate *a* by considering A, A^2, A^3, and A^4. Also compute *a* exactly.

22. Show that

$$A = \begin{pmatrix} 1 & 1+i & -1 \\ 1-i & 6 & -3+i \\ -1 & -3-i & 11 \end{pmatrix}$$

is positive definite.

23. The matrix $A = \begin{pmatrix} 4 & 2 \\ 1 & 3 \end{pmatrix}$ is given. Compute ln *A*, i.e., the matrix *X* which satisfies the equation $e^X = A$.

24. *A* and *B* are square matrices of the same size. Prove that *AB* and *BA* have the same eigenvalues.

25. Find the pseudo-inverse of

$$A = \begin{pmatrix} 2 & 0 & -1 \\ 1 & 3 & -3 \\ 0 & 1 & -1 \\ -1 & 1 & 0 \end{pmatrix}.$$

26. Find the pseudo-inverse of

$$A = \begin{pmatrix} 1 & 2 & 3 \\ 2 & 4 & 6 \\ 4 & 8 & 12 \end{pmatrix}.$$

CHAPTER 3

Series Expansions

Let n be a large number, say 3!

E. Landau

Many problems in numerical analysis are concerned with functions that are regular and can be expressed as Taylor series expansions in some interval. Singularities may occur, but as a rule they can be handled separately. Quite often power series expansions are extremely useful tools: they can be differentiated and integrated under very general conditions, and the function values can usually be computed easily. In practical applications there are hardly any difficulties in estimating the error when the series is truncated, even if the exact remainder term is not easily found.

Power series expansions are often used in the solution of differential and integral equations; the coefficients are then mostly computed successively. Many numerical rules also depend heavily on series expansions, particularly when the formulas are derived by operator techniques.

§3.1. POWER SERIES

A power series has the general form

$$f(z) = \sum_{k=0}^{\infty} a_k z^k$$

where a_k are given (real or complex) coefficients and z is a real or complex variable. If ρ is the convergence radius, this formula can be used for computing $f(z)$ in the domain $|z| < \rho$. In most applications only real values appear, i.e., both a_k and $z = x$ are real. Then we have convergence in the interval $-\rho < x < \rho$. The slightly more general case when the origin is shifted to another point

$$g(z) = \sum_{k=0}^{\infty} a_k (z - \alpha)^k$$

does not need to be treated separately.

In practical applications the series must be truncated, and we write instead

$$f(x) = \sum_{k=0}^{n-1} a_k x^k + R_n$$

where R_n is the remainder term. If the function $f(x)$ is given, there are well-known expressions for R_n, often of the form

$$R_n = (x^n/n!) f^{(n)}(\theta x), \qquad 0 \le \theta \le 1.$$

However, if only the coefficients a_k are known it may be more difficult to find a realistic expression for R_n. In practice one usually keeps track of the successive terms in the series, and if they decrease at a reasonable speed it will in most cases be possible to obtain an estimate of the remainder term. It should be stressed that the situation is quite different with asymptotic series, since they are generally divergent.

We suppose that the Maclaurin series for the elementary functions $\exp(x)$, $\cos x$, $\sin x$, $\arccos x$, $\arcsin x$, $\arctan x$, $\cosh x$, $\sinh x$, $\ln(1 + x)$, and $(1 + x)^n$ are known. However, the function $(1 + x)^n$ appears so often in different contexts that we find it reasonable to treat a few special cases explicitly, namely $n = \pm 1/2$. We have

$$(1 + x)^{1/2} = \sum_{k=0}^{\infty} \binom{1/2}{k} x^k \quad \text{where} \quad \binom{1/2}{k} = \frac{(1/2)(-1/2)(-3/2) \cdots (1/2 - k + 1)}{k!}$$

$$= \frac{(-1)^{k-1} \cdot 1 \cdot 3 \cdot 5 \cdots (2k - 3)(2k - 1)}{k! 2^k (2k - 1)} \cdot \underbrace{\frac{2 \cdot 4 \cdot 6 \cdots (2k)}{k! 2^k}}_{= 1}$$

$$= \frac{(-1)^{k-1}(2k)!}{2^{2k}(k!)^2(2k - 1)} \quad (k \ge 1); \qquad \binom{1/2}{0} = 1.$$

In a similar way we get $\binom{-1/2}{k} = \frac{(-1)^k (2k)!}{2^{2k}(k!)^2}$ and hence

(3.1.1) $$(1 + x)^{1/2} = 1 + \frac{x}{2} - \frac{x^2}{8} + \frac{x^3}{16} - \frac{5x^4}{128} + \frac{7x^5}{256} - \frac{21x^6}{1024} + \cdots$$

(3.1.2) $$(1 + x)^{-1/2} = 1 - \frac{x}{2} + \frac{3x^2}{8} - \frac{5x^3}{16} + \frac{35x^4}{128} - \frac{63x^5}{256} + \frac{231x^6}{1024} - \cdots$$

Further, through repeated differentiation from $(1 - x)^{-1} = 1 + x + x^2 + x^3 + \cdots$ we get:

$$(1 - x)^{-2} = 1 + 2x + 3x^2 + 4x^3 + 5x^4 + \cdots$$
$$(1 - x)^{-3} = 1 + 3x + 6x^2 + 10x^3 + 15x^4 + \cdots$$

and in general:

$$(1 - x)^{-n-1} = \sum_{k=0}^{\infty} \binom{n+k}{k} x^k$$

with convergence in the interval $-1 < x < 1$.

Addition and subtraction of two power series is immediate. Multiplication is also rather simple: putting

$$f(x) = \sum_{k=0}^{\infty} a_k x^k, \quad g(x) = \sum_{k=0}^{\infty} b_k x^k$$

we get

$$f(x)g(x) = \sum_{n=0}^{\infty} c_n x^n \quad \text{where} \quad c_n = \sum_{k=0}^{n} a_k b_{n-k}.$$

Division is accomplished by the indeterminate coefficients method. If we want to compute $h(x) = f(x)/g(x)$ with $f(x)$ and $g(x)$ as defined above, we put $h(x) = \sum_{k=0}^{\infty} c_k x^k$ and obtain after identification:

$$\begin{aligned}
b_0 c_0 &= a_0 \\
b_1 c_0 + b_0 c_1 &= a_1 \\
b_2 c_0 + b_1 c_1 + b_0 c_2 &= a_2 \\
b_3 c_0 + b_2 c_1 + b_1 c_2 + b_0 c_3 &= a_3 \\
&\;\;\vdots
\end{aligned}$$

This is a triangular system that can easily be solved by successive substitution.

We also mention an important inversion theorem. Suppose that x is given as a power series of y starting with a nonzero first order term:

$$x = P(y) = \sum_{r=1}^{\infty} c_r y^r.$$

Now we want y expressed as a power series in x:

$$y = \sum_{n=1}^{\infty} k_n x^n.$$

Then, according to a theorem by Bürmann–Lagrange, the coefficients k_n are given by the expression

(3.13)
$$k_n = (1/n!)\{D_t^{(n-1)}[t/P(t)]^n\}_{t=0}$$

where D_t means differentiation with respect to t. The proof relies on complex integration; however, we refrain from further details.

Example 1. Suppose

$$x = P(y) = y e^{-y} = y - y^2 + y^3/2 - y^4/6 + \cdots$$

Then $t/P(t) = e^t$ and

$$k_n = (1/n!)\{D^{(n-1)}e^{nt}\}_{t=0} = (1/n!)n^{n-1}\{e^{nt}\}_{t=0} = n^{n-1}/n!$$

Hence

$$y = x + x^2 + 3x^3/2 + 8x^4/3 + 125x^5/24 + \cdots$$

---

§3.2. GENERATING FUNCTIONS

In many applications, particularly in combinatorics, number theory, graph theory, and computer science in general, certain functions $g(n)$ appear that have a meaning only when n is a nonnegative integer. Often these values count the number of occurrences of a special event for different values of n. It is often convenient to construct a polynomial

$$f(x) = \sum_{n=0}^{\infty} a_n x^n$$

where $a_n = g(n)$ and x is a purely formal variable. Such a function $f(x)$ is called a *generating function*; in many cases it is possible to express $f(x)$ in closed form. Normally the coefficients are integers, but this is by no means a necessary condition. Below we give some examples to demonstrate the power of this technique.

Example 1.   Given a positive integer n. Then a *partition* is a representation of n as the sum of positive integers where different permutations are not counted; e.g., $1 + 1 + 2 + 5$ and $2 + 1 + 5 + 1$ are considered to be identical partitions of the number 9. We now want to determine the number of partitions of n for different values of n. We introduce a variable x and let x represent 1; let x^2 represent $1 + 1$ or 2; let x^3 represent $1 + 1 + 1$ or 3; let x^4 represent $1 + 1 + 1 + 1$ or $2 + 2$ or 4; and so on. A proper choice is made as follows: the first polynomial in parentheses below refers to ones, the second to twos, the third to threes, and so forth. Hence we consider:

$$\underbrace{(1 + x + x^2 + x^3 + \cdots)}_{\text{powers of } x}\underbrace{(1 + x^2 + x^4 + x^6 + \cdots)}_{\text{powers of } x^2}\underbrace{(1 + x^3 + x^6 + x^9 + \cdots)}_{\text{powers of } x^3}$$

$$\times (1 + x^4 + x^8 + x^{12} + \cdots)(1 + x^5 + x^{10} + x^{15} + \cdots)\cdots = \sum_{n=0}^{\infty} p(n)x^n.$$

Taking in turn x^3 from the first parentheses, x^4 from the second, x^3 from the third, and 1 from all the following ones, we get a term x^{10} corresponding to the partition

$10 = (1 + 1 + 1) + (2 + 2) + 3$. It is obvious that $p(n) = p_n$ is the number of partitions of n. An easy computation gives $p_0 = p_1 = 1$, $p_2 = 2$, $p_3 = 3$, $p_4 = 5$, $p_5 = 7$ and so on; these values can also be verified directly without difficulty. However, using another line of approach, supposing $|x| < 1$, we get

$$(1 - x)^{-1}(1 - x^2)^{-1}(1 - x^3)^{-1} \cdots = \sum p(n)x^n \quad \text{or}$$
$$(1 - x)(1 - x^2)(1 - x^3) \cdots (1 + p_1 x + p_2 x^2 + \cdots) = 1.$$

According to a famous theorem by Euler we have:

$$\prod_{k=1}^{\infty} (1 - x^k) = 1 + \sum_{n=1}^{\infty} (-1)^n [x^{n(3n-1)/2} + x^{n(3n+1)/2}]$$
$$= 1 - x - x^2 + x^5 + x^7 - x^{12} - x^{15} + x^{22} + x^{26} - \cdots$$

Hence, multiplying and identifying:

$$p(n) = p(n - 1) + p(n - 2) - p(n - 5) - p(n - 7) + p(n - 12) + p(n - 15) - \cdots$$

The number of terms to the right is only of the order $2\sqrt{(2n/3)}$.

Example 2. The Fibonacci* numbers. These numbers are usually defined through $a_0 = a_1 = 1$; $a_{n+2} = a_{n+1} + a_n$. The generating function is obviously $(1 - x - x^2)^{-1}$, since multiplying we get:

$$(1 - x - x^2)(a_0 + a_1 x + a_2 x^2 + \cdots) = 1$$

which leads to the defining relation above. Naturally, the generating function is closely connected with the characteristic equation of the difference equation $a_{n+2} - a_{n+1} - a_n = 0$. Expanding $f(x) = (1 - x - x^2)^{-1}$ in a geometric series, we can write

$$f(x) = 1 + \sum_{n=1}^{\infty} x^n (1 + x)^n = 1 + \sum_{n=1}^{\infty} x^n \sum_{k=0}^{n} \binom{n}{k} x^k = 1 + \sum_{r=1}^{\infty} a_r x^r$$

where

$$a_r = \sum_{n=\lceil r/2 \rceil}^{r} \binom{n}{r - n} \qquad \text{(cf. Section 1.4 for the notation } \lceil x \rceil)$$

Taking $r = 7$, we would get:

$$a_7 = \sum_{n=4}^{7} \binom{n}{7 - n} = \binom{4}{3} + \binom{5}{2} + \binom{6}{1} + \binom{7}{0} = 4 + 10 + 6 + 1 = 21.$$

This is the eighth term in the series $1, 1, 2, 3, 5, 8, 13, 21, 34, 55, \ldots$ as it should be.

Example 3. Binary trees. This notion is best explained by the figure below, where n is the number of nodes and b_n the number of binary trees.

* Pronounced Feebonatchi.

$$n = b_1 = 1 \qquad n = b_2 = 2 \qquad n = 3; \quad b_3 = 5$$

$$n = 4; \quad b_4 = 14$$

Let $B(x) = b_0 + b_1 x + b_2 x^2 + \cdots$ and define $b_0 = 1$. Then a tree with n nodes can be formed from a left tree with k nodes ($k = 0, 1, \ldots, n-1$) and a right tree with $n - k - 1$ nodes (either of them may be empty). Hence

$$b_n = b_0 b_{n-1} + b_1 b_{n-2} + \cdots + b_{n-1} b_0, \qquad n \ge 1.$$

Obviously,

$$B(x)^2 = b_0^2 + (b_0 b_1 + b_1 b_0)x + (b_0 b_2 + b_1^2 + b_2 b_0)x^2 + \cdots$$
$$= b_1 + b_2 x + b_3 x^2 + \cdots$$

and so we get $xB(x)^2 = B(x) - 1$ with the solution

$$B(x) = \frac{1 - \sqrt{(1 - 4x)}}{2x} = (2x)^{-1}\left(1 - \sum_{n=0}^{\infty} \binom{1/2}{n}(-4x)^n\right)$$

or, using (3.1.1):

$$B(x) = \sum_{n=0}^{\infty} \frac{(2n)!}{n!(n + 1)!} x^n$$
$$= 1 + x + 2x^2 + 5x^3 + 14x^4 + 42x^5 + 132x^6 + 429x^7$$
$$+ 1430x^8 + 4862x^9 + 16796x^{10} + \cdots$$

By the aid of Stirling's formula (7.2.9) we find the following asymptotic expression:

$$b_n = \frac{(2n)!}{n!(n + 1)!} \sim \pi^{-1/2} 4^n n^{-3/2}.$$

§3.3. ASYMPTOTIC SERIES

Important tools for computation of functions are the asymptotic series. They are defined in the following way. Suppose that $s_n(z) = \sum_{k=0}^{n} a_k z^k$ and further that

$$\lim_{z \to 0} \frac{|f(z) - s_n(z)|}{|z|^n} = 0$$

for an arbitrary fixed value of n; then the following notation is used: $f(z) \sim \sum_{k=0}^{\infty} a_k z^k$ when $z \to 0$. In a certain sense, the series expansion represents the function $f(z)$ in such a way that the nth partial sum approximates $f(z)$ better than $|z|^n$ approximates 0. The remarkable fact is that we do not claim that the series is convergent, and as a matter of fact it is in general *divergent*. In this case we say that the series is a *semiconvergent* expansion of the function $f(z)$.

An asymptotic series expansion around the point z_0 is written in the form $f(z) \sim \sum_{k=0}^{\infty} a_k (z - z_0)^k$, with the meaning

$$\lim_{z \to z_0} \frac{|f(z) - s_n(z)|}{|z - z_0|^n} = 0$$

for a fixed value of n when $s_n(z) = \sum_{k=0}^{n} a_k (z - z_0)^k$.

Asymptotic expansions at infinity are of special interest. The following notation is used: $f(z) \sim \sum_{k=0}^{\infty} a_k z^{-k}$ when $z \to \infty$ if

$$\lim_{z \to \infty} |z|^n |f(z) - s_n(z)| = 0$$

for a fixed value of n when $s_n(z) = \sum_{k=0}^{n} a_k z^{-k}$. In many cases $f(z)$ has no asymptotic series expansion, but it might be possible to obtain such an expansion if a suitable function is subtracted or if we divide by an appropriate function. The following notation should present no difficulties:

$$f(z) \sim g(z) + h(z) \cdot \sum_{k=0}^{\infty} a_k z^{-k}.$$

It is obvious that $g(z)$ and $h(z)$ must be independent of n.

Example 1

$$f(x) = \int_x^\infty \frac{e^{-t}}{t} \, dt = \frac{e^{-x}}{x} - \int_x^\infty \frac{e^{-t}}{t^2} \, dt = \frac{e^{-x}}{x} - \frac{e^{-x}}{x^2} + 2! \int_x^\infty \frac{e^{-t}}{t^3} \, dt = \cdots$$

$$= \frac{e^{-x}}{x} \left[1 - \frac{1}{x} + \frac{2!}{x^2} - \frac{3!}{x^3} + \cdots + (-1)^{n-1} \cdot \frac{(n-1)!}{x^{n-1}} \right] + R_n,$$

with

$$R_n = (-1)^n \cdot n! \cdot \int_x^\infty \frac{e^{-t}}{t^{n+1}} \, dt.$$

Thus we have $|R_n| < e^{-x} n! / x^{n+1}$, and by definition,

$$f(x) \sim \frac{e^{-x}}{x} \left[1 - \frac{1}{x} + \frac{2!}{x^2} - \cdots + (-1)^n \frac{n!}{x^n} + \cdots \right].$$

For large values of x the terms first decrease in absolute value, but diverge when $n \to \infty$, since x is fixed.

Thus $f(x)$ is expressed as a series expansion with a remainder term, whose absolute value first decreases but later on increases to infinity from a certain value of n. Hence the series is divergent, but if it is truncated after n terms where n is chosen so that $|R_n| < \varepsilon$, the truncated series can nevertheless be used, giving a maximum error less than ε.

In our special example we denote the partial sums with S_n, and for $x = 10$ we obtain the following table:

n	$10^7 \cdot e^x \cdot S_n$	$10^7 \cdot e^x \cdot R^n$	n	$10^7 \cdot e^x \cdot S_n$	$10^7 \cdot e^x \cdot R_n$
1	1000000	-84367	11	915891	-186
2	900000	$+15633$	12	915420	$+213$
3	920000	-4367	13	915899	-266
4	914000	$+1633$	14	915276	$+357$
5	916400	-767	15	916148	-515
6	915200	$+433$	16	914840	$+793$
7	915920	-287	17	916933	-1300
8	915416	$+217$	18	913376	$+2257$
9	915819	-186	19	919778	-4145
10	915456	$+177$	20	907614	$+8019$

The correct value is

$$e^{10} \int_{10}^{\infty} \frac{e^{-t}}{t}\, dt = 0.0915633.$$

In general, the error is of the same order of magnitude as the first neglected term.

§3.4. FOURIER SERIES

Sums and integrals involving complex exponential functions or trigonometric functions play an important role in theoretical and applied mathematics. We will here give an exposition of the basic facts even if full proofs cannot always be given.

We start with some fundamental summation formulas. Let n be an integer, $n > 1$. Then

(3.4.1) $$\sum_{r=0}^{n-1} \exp(2\pi i r j/n) \exp(-2\pi i r k/n) = n\delta_{jk}$$

where $0 \le j, k \le n - 1$, and as usual $\delta_{jk} = 1$ if $j = k$ and 0 otherwise. The proof is immediate, because putting $w = \exp(2\pi i/n)$ we can write the sum

$$\sum_{r=0}^{n-1} w^{(j-k)r} = (w^{(j-k)n} - 1)/(w^{j-k} - 1)$$

and since $w^n = 1$ this is equal to 0 if $j \neq k$ (note that $w^{j-k} \neq 1$ in this case). If $j = k$, all n terms in the sum equal unity, giving a total of n.

The relation above can be slightly generalized. Removing the restrictions on j and k and putting

$$\delta(m, n) = \begin{cases} 1 & \text{if } n \text{ divides } m, \\ 0 & \text{otherwise} \end{cases}$$

we find:

(3.4.2)
$$\sum_{r=0}^{n-1} \exp(2\pi i r j/n) \exp(2\pi i r k/n) = n\delta(j + k, n).$$

We can now use these results for establishing similar trigonometric formulas:

$$\sum_{r=0}^{n-1} \cos(2\pi r j/n) \cos(2\pi r k/n) = \frac{1}{2} \sum_{r=0}^{n-1} \left[\cos(2\pi r(j + k)/n) + \cos(2\pi r(j - k)/n) \right]$$

$$= \frac{1}{2} \operatorname{Re} \sum_{r=0}^{n-1} \left[\exp(2\pi i r(j + k)/n) + \exp(2\pi i r(j - k)/n) \right]$$

$$= (n/2)[\delta(j + k, n) + \delta(j - k, n)].$$

Similarly,

$$\sum_{r=0}^{n} \sin(2\pi r j/n) \sin(2\pi r k/n) = (n/2)[\delta(j - k, n) - \delta(j + k, n)]$$

and

$$\sum_{r=0}^{n} \sin(2\pi r j/n) \cos(2\pi r k/n) = 0.$$

For integrals of the same structure we find directly (j, k integers, ≥ 0):

(3.4.3)
$$\int_0^{2\pi} \exp(ijx) \exp(-ikx) \, dx = 2\pi \delta_{jk}.$$

We also have:

$$\int_0^{2\pi} \exp(ijx) \exp(ikx) \, dx = 2\pi \delta_{j0} \delta_{k0}.$$

Adding or subtracting these relations and splitting into real and imaginary parts, we obtain three types of integrals:

$$I_1 = \int_0^{2\pi} \cos rx \cos sx \, dx; \qquad I_2 = \int_0^{2\pi} \sin rx \sin sx \, dx;$$

and

$$I_3 = \int_0^{2\pi} \cos rx \sin sx \, dx.$$

The following values are obtained:

	I_1	I_2	I_3
$r \neq s$	0	0	0
$r = s \neq 0$	π	π	0
$r = s = 0$	2π	0	0

The summation and integration formulas above are special cases of orthogonality relations. The more general aspects of this topic will be treated in a subsequent section. However, we mention here that the summation formulas are the basis for harmonic analysis, while the integral relations are used in the computation of Fourier coefficients.

Expansion of a Function in a Fourier Series

A function $f(x)$ with the property $f(x + 2\pi) = f(x)$ is said to be periodic with period 2π. In the following we will treat such functions in an interval of length 2π, and we then observe that this interval can be chosen deliberately. Usually we will take either $(0, 2\pi)$ or $(-\pi, \pi)$. It can now be proved that under certain rather mild conditions on $f(x)$ there exists a representation

(3.4.4)
$$f(x) = \frac{1}{2} a_0 + \sum_{r=1}^{\infty} (a_r \cos rx + b_r \sin rx).$$

If we multiply this relation with $\cos sx$ and $\sin sx$ respectively and integrate, using the orthogonality expressions above, we find:

(3.4.5)
$$a_s = \pi^{-1} \int_0^{2\pi} f(x) \cos sx \, dx$$

$$b_s = \pi^{-1} \int_0^{2\pi} f(x) \sin sx \, dx.$$

Shifting the interval to $(-\pi, \pi)$, we see that if $f(x)$ is an even function then all b_s are zero, while if $f(x)$ is odd then all a_s are zero.

Example 1. Expand the function $y = f(x)$ in a Fourier series, where

$$f(x) = \begin{cases} +1, & 0 < x < \pi, \\ -1, & \pi < x < 2\pi. \end{cases}$$

We find

$$a_n = \frac{1}{\pi} \int_0^{\pi} \cos nx \, dx - \frac{1}{\pi} \int_{\pi}^{2\pi} \cos nx \, dx = 0,$$

$$b_n = \frac{1}{\pi} \int_0^{\pi} \sin nx \, dx - \frac{1}{\pi} \int_{\pi}^{2\pi} \sin nx \, dx = \frac{2}{\pi n} [1 - (-1)^n],$$

and hence

$$b_n = \begin{cases} 0, & n \text{ even,} \\ 4/\pi n, & n \text{ odd.} \end{cases}$$

Thus we obtain

$$\frac{\pi}{4} = \sin x + \frac{\sin 3x}{3} + \frac{\sin 5x}{5} + \cdots, \qquad 0 < x < \pi.$$

The function $f(x)$ is represented by a rectangular wave (see Figure 3.4.2). Since there are sudden jumps between -1 and $+1$, it is of some interest to examine the behavior of the series expansion at these points. We put

$$S_{2n+1}(x) = \frac{4}{\pi} \sum_{k=0}^{n} \frac{\sin(2k+1)x}{2k+1}$$

and we are interested in what happens when $n \to \infty$ and $x \to 0$. We get

$$S_{2n+1} = \frac{4}{\pi} \sum_{k=0}^{n} \int_0^x \cos(2k+1)t \, dt$$

$$= \frac{2}{\pi} \int_0^x \sum_{k=0}^{n} \left[\exp((2k+1)it) + \exp(-(2k+1)it) \right] dt$$

$$= \frac{2}{\pi} \int_0^x \frac{\exp(2i(n+1)t) - \exp(-2i(n+1)t)}{\exp(it) - \exp(-it)} \, dt$$

$$= \frac{2}{\pi} \int_0^x \frac{\sin 2(n+1)t}{\sin t} \, dt.$$

Now we are particularly interested in the case when x is small, so with sufficient accuracy we can replace $\sin t$ in the denominator by t. Putting $u = 2(n+1)t$ and $v = 2(n+1)x$, we get:

$$S_{2n+1} \simeq (2/\pi) \int_0^v (\sin u/u) \, du.$$

We can now investigate what happens when $x \to 0$ and when $n \to \infty$. If we put $x = 0$ exactly, then v becomes 0 and $S_{2n+1} = 0$ irrespective of n. However, if x has a small but fixed value, then as $n \to \infty$ also $v \to \infty$, and S_{2n+1} will tend to $(2/\pi) \int_0^\infty (\sin u/u) \, du = 1$.

Let us now discuss the case when x and n are not independent of each other. It is then natural to consider the function

$$\text{Si}(v) = \int_0^v (\sin u/u) \, du,$$

known by the classical name "Sinus integralis" (also see equation (7.4.5)). It is easy to see that the function takes its maximum value when $v = \pi$, followed by a series of smaller maxima for $v = 3\pi, 5\pi, \ldots$ approaching the limiting value $\pi/2$ (see Figure 3.4.1). By numerical computation the first maximum is found to be 1.851937, and hence if we choose $x_n = \pi/2(n+1)$, $x_{n+1} = \pi/2(n+2), \ldots$ we get the fixed maximum

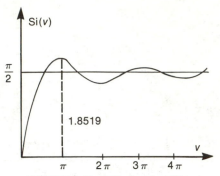

FIGURE 3.4.1.   The function Si(v).

value for S_{2n+1} of $(2/\pi) \cdot 1.851937 = 1.17898$. In fact, we can get any value between -1.18 and $+1.18$ by making a suitable passage to the limit. This phenomenon was first discovered by Willard Gibbs and is usually called Gibb's phenomenon (see Figure 3.4.2).

Example 2

$$f(x) = \begin{cases} \pi/4 - x/2, & 0 \le x \le \pi, \\ -3\pi/4 + x/2, & \pi \le x \le 2\pi. \end{cases}$$

We have

$$a_n = \frac{1}{\pi} \int_0^\pi \left(\frac{\pi}{4} - \frac{x}{2}\right) \cos nx \, dx + \frac{1}{\pi} \int_\pi^{2\pi} \left(-\frac{3\pi}{4} + \frac{x}{2}\right) \cos nx \, dx$$

$$= \begin{cases} 0, & n \text{ even}, \\ 2/\pi n^2, & n \text{ odd}. \end{cases}$$

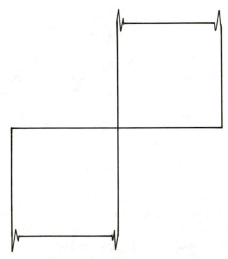

FIGURE 3.4.2.   Gibb's phenomenon.

Further, we easily find $b_n = 0$, and hence

$$\frac{\pi}{4} - \frac{x}{2} = \frac{2}{\pi}\left(\cos x + \frac{\cos 3x}{3^2} + \frac{\cos 5x}{5^2} + \cdots\right), \qquad 0 \le x \le \pi.$$

For $x = 0$ we obtain

$$\frac{\pi^2}{8} = 1 + \frac{1}{3^2} + \frac{1}{5^2} + \cdots$$

When a function is represented as a trigonometric series, this means that only certain discrete frequencies have been used. Passing instead to the integral form, we actually use all frequencies. Here Fourier's integral theorem plays a central role. It can be expressed in the following form.

Let $f(x)$ be a continuous, integrable function. Then the following relations hold:

$$\begin{cases} F(t) = (2\pi)^{-1/2} \displaystyle\int_{-\infty}^{\infty} f(x)\exp(itx)\,dx \\ f(x) = (2\pi)^{-1/2} \displaystyle\int_{-\infty}^{\infty} F(t)\exp(-itx)\,dt. \end{cases}$$

We must refrain from a proof; we only mention that the following idea is important here. It can be understood that the integral

$$\lim_{w \to \infty} \int_a^b f(x)(\sin wx/x)\,dx = \begin{cases} 0, & a > 0 \\ (\pi/2)f(+0), & a = 0 \end{cases}$$

where $b > a$, and where $f(x)$ has no infinite discontinuities, only a finite number of finite discontinuities, and only a finite number of maxima and minima in the interval (a, b).

There exists a similar formula pair involving the cosine function:

(3.4.7)
$$\begin{cases} F(t) = (\pi/2)^{-1/2} \displaystyle\int_0^{\infty} f(x)\cos xt\,dx \\ f(x) = (\pi/2)^{-1/2} \displaystyle\int_0^{\infty} F(t)\cos xt\,dt. \end{cases}$$

As we shall see in a forthcoming section on Laplace transforms, the Fourier transforms provide the basis for this theory.

---

§3.5. BERNOULLI NUMBERS AND BERNOULLI POLYNOMIALS

The Bernoulli numbers of order n are defined through

(3.5.1)
$$[t/(e^t - 1)]^n = \sum_{k=0}^{\infty} B_k^{(n)} t^k/k!$$

The Bernoulli polynomials of order n and degree k are defined through

(3.5.2) $$e^{xt}[t/(e^t - 1)]^n = \sum_{k=0}^{\infty} B_k^{(n)}(x)t^k/k!$$

In the following we shall exclusively treat the case $n = 1$, and the corresponding numbers and polynomials will be called just Bernoulli numbers and Bernoulli polynomials:

(3.5.3) $$t/(e^t - 1) = \sum_{k=0}^{\infty} A_k t^k/k!$$

(3.5.4) $$e^{xt}t/(e^t - 1) = \sum_{k=0}^{\infty} B_k(x)t^k/k!$$

We have chosen the notation A_k because we want to reserve B_k for other numbers closely associated with the A's.

From relation (3.5.3), multiplying with the factor $e^t - 1$, we get:

$$t = \sum_{k=0}^{\infty} A_k t^k/k! \sum_{r=1}^{\infty} t^r/r! = \sum_{k=1}^{\infty} t^k \sum_{i=0}^{k-1} A_i/i!(k-i)!$$

Hence we find:

$$A_0 = 1; \quad \sum_{i=0}^{k-1} \binom{k}{i} A_i = 0, \quad k = 2, 3, 4, \ldots$$

Written out explicitly, this is a linear system of equations that can be solved successively without difficulty:

(3.5.5) $$\begin{cases} A_0 = 1, \\ A_0 + 2A_1 = 0, \\ A_0 + 3A_2 = 0, \\ A_0 + 4A_1 + 6A_2 + 4A_3 = 0, \\ \vdots \end{cases}$$

which gives $A_1 = -\frac{1}{2}$; $A_2 = \frac{1}{6}$; $A_3 = 0$; $A_4 = -\frac{1}{30}$; $A_5 = 0$; $A_6 = \frac{1}{42}$; $\ldots$ Using a symbolic notation, we can write the system in a very elegant way, namely,

$$(A + 1)^s - A^s = 0, \quad s = 2, 3, 4, \ldots$$

with the convention that all powers A^i should be replaced by A_i.

Now we treat (3.5.4) in a similar way and obtain

$$\sum_{r=0}^{\infty} \frac{x^r t^r}{r!} \cdot \sum_{k=0}^{\infty} A_k \frac{t^k}{k!} = \sum_{n=0}^{\infty} B_n(x) \frac{t^n}{n!}.$$

Identifying we get

$$B_n(x) = \sum_{r=0}^{n} \binom{n}{r} A_{n-r} x^r.$$

Symbolically, we can write $B_n(x) = (A + x)^n$, with the same convention $A^i \to A_i$ as before. Summing up, we have the following simple rules for the formation of Bernoulli numbers and polynomials:

$$(3.5.6) \qquad \begin{cases} A_0 = 1, \\ (A + 1)^n - A^n = 0 & \text{with} \qquad A^k \to A_k, \\ B_n(x) = (A + x)^n & \text{with} \qquad A^k \to A_k. \end{cases}$$

The first Bernoulli polynomials are

$$B_0(x) = 1,$$
$$B_1(x) = x - 1/2,$$
$$B_2(x) = x^2 - x + 1/6,$$
$$B_3(x) = x^3 - 3x^2/2 + x/2,$$
$$B_4(x) = x^4 - 2x^3 + x^2 - 1/30,$$
$$B_5(x) = x^5 - 5x^4/2 + 5x^3/3 - x/6,$$
$$B_6(x) = x^6 - 3x^5 + 5x^4/2 - x^2/2 + 1/42.$$

Equation (3.5.4) can be written

$$e^{xt} \frac{t}{e^t - 1} = 1 + \sum_{n=1}^{\infty} B_n(x) \frac{t^n}{n!},$$

and differentiating with respect to x, we obtain

$$e^{xt} \frac{t}{e^t - 1} = \sum_{n=1}^{\infty} \frac{B'_n(x)}{n} \frac{t^{n-1}}{(n-1)!} = \sum_{n=1}^{\infty} B_{n-1}(x) \frac{t^{n-1}}{(n-1)!}.$$

Comparing the coefficients, we get

$$B'_n(x) = n B_{n-1}(x).$$

The same relation is obtained by differentiating the symbolic equation

$$B_n(x) = (A + x)^n.$$

Now let x be an arbitrary complex number. From $(A + 1)^i = A^i$ we also have

$$\sum_{i=2}^{n} \binom{n}{i} (A + 1)^i x^{n-i} = \sum_{i=2}^{n} \binom{n}{i} A^i x^{n-i},$$

or

$$(A + 1 + x)^n - n(A + 1)x^{n-1} = (A + x)^n - nAx^{n-1}.$$

Hence we find the following formula, which we shall use later:

$$(3.5.7) \qquad (A + 1 + x)^n - (A + x)^n = nx^{n-1}.$$

It can also be written $\Delta(A + x)^n = Dx^n$, or, since $(A + x)^n = B_n(x)$,

$$\Delta B_n(x) = Dx^n.$$

Conversely, if we want to find a polynomial $B_n(x)$ fulfilling this equation, we have:

$$B_n(x) = \frac{D}{e^D - 1} x^n = \left(\sum_{k=0}^{\infty} A_k \frac{D^k}{k!} \right) x^n = \sum_{k=0}^{n} \binom{n}{k} A_k x^{n-k} = (A + x)^n$$

in agreement with (3.5.6).

Integrating $B_n(t) = B'_{n+1}(t)/(n + 1)$, we obtain

$$\int_x^{x+1} B_n(t) \, dt = \frac{B_{n+1}(x + 1) - B_{n+1}(x)}{n + 1} = \frac{\Delta B_{n+1}(x)}{n + 1} = \frac{Dx^{n+1}}{n + 1} = x^n.$$

Putting $x = 0$, we find for $n = 1, 2, 3, \ldots$,

$$\int_0^1 B_n(t) \, dt = 0.$$

The relations

$$\begin{cases} B_0(t) = 1, \\ B'_n(t) = nB_{n-1}(t), \\ \int_0^1 B_n(t) \, dt = 0, \qquad n = 1, 2, 3, \ldots, \end{cases}$$

give a simple and direct method for determination of $B_n(t)$, and hence also of $A_n = B_n(0)$.

From (3.5.5) we obtained $A_3 = A_5 = 0$, and we shall now prove that $A_{2k+1} = 0$ for $k = 1, 2, 3, \ldots$ Forming

$$f(t) = \frac{t}{e^t - 1} + \frac{t}{2} = \frac{t}{2} \cdot \frac{e^t + 1}{e^t - 1} = \frac{t}{2} \coth \frac{t}{2},$$

we see at once that $f(-t) = f(t)$, which concludes the proof. Hence, there are only even powers of t left, and we introduce the common notation

(3.5.8) $$B_k = (-1)^{k+1} A_{2k}, \qquad k = 1, 2, 3, \ldots.$$

The first Bernoulli numbers B_k are

$B_1 = 1/6$	$B_7 = 7/6$	$B_{13} = 8553103/6$
$B_2 = 1/30$	$B_8 = 3617/510$	$B_{14} = 23749461029/870$
$B_3 = 1/42$	$B_9 = 43867/798$	$B_{15} = 8615841276005/14322$
$B_4 = 1/30$	$B_{10} = 174611/330$	$B_{16} = 7709321041217/510$
$B_5 = 5/66$	$B_{11} = 854513/138$	$B_{17} = 2577687858367/6$
$B_6 = 691/2730$	$B_{12} = 236364091/2730$	

We now get

(3.5.9)
$$\frac{t}{e^t - 1} = 1 - \frac{t}{2} + B_1 \frac{t^2}{2!} - B_2 \frac{t^4}{4!} + B_3 \frac{t^6}{6!} - \cdots$$

Above we discussed the even function

$$\frac{t}{e^t - 1} + \frac{t}{2} = \frac{t}{2} \coth \frac{t}{2}.$$

Putting $t/2 = x$, we obtain the expansion

(3.5.10)
$$x \coth x = 1 + B_1 \frac{(2x)^2}{2!} - B_2 \frac{(2x)^4}{4!} + B_3 \frac{(2x)^6}{6!} - \cdots$$

Replacing x by ix, we get

$$x \cot x = 1 - B_1 \frac{(2x)^2}{2!} - B_2 \frac{(2x)^4}{4!} - B_3 \frac{(2x)^6}{6!} - \cdots,$$

or

(3.5.11)
$$1 - x \cot x = \sum_{p=1}^{\infty} \frac{2^{2p}}{(2p)!} B_p x^{2p}.$$

Using the identity $\tan x = \cot x - 2 \cot 2x$, we find

(3.5.12)
$$\tan x = \sum_{p=1}^{\infty} \frac{2^{2p}(2^{2p} - 1)}{(2p)!} B_p x^{2p-1}.$$

In explicit form we finally get the following set of formulas:

$$x/(e^x - 1) = 1 - \frac{x}{2} + \frac{x^2}{12} - \frac{x^4}{720} + \frac{x^6}{30240} - \frac{x^8}{1209600} + \frac{x^{10}}{47900160} - \cdots$$

$$\tan x = x + \frac{x^3}{3} + \frac{2x^5}{15} + \frac{17x^7}{315} + \frac{62x^9}{2835} + \frac{1382x^{11}}{155925} + \cdots$$

$$x \cot x = 1 - \frac{x^2}{3} - \frac{x^4}{45} - \frac{2x^6}{945} - \frac{x^8}{4725} - \frac{2x^{10}}{93555} - \cdots$$

$$\tanh x = x - \frac{x^3}{3} + \frac{2x^5}{15} - \frac{17x^7}{315} + \frac{62x^9}{2835} - \frac{1382x^{11}}{155925} + \cdots$$

$$x \coth x = 1 + \frac{x^2}{3} - \frac{x^4}{45} + \frac{2x^6}{945} - \frac{x^8}{4725} + \frac{2x^{10}}{93555} - \cdots$$

The first series converges for $|x| < 2\pi$, the second and the fourth for $|x| < \pi/2$, and the third and the fifth for $|x| < \pi$. These results can be obtained from the formula

(3.5.13)
$$\sum_{n=1}^{\infty} n^{-2p} = \frac{(2\pi)^{2p} B_p}{2(2p)!}$$

which is valid for $p = 1, 2, 3, \ldots$ (proved in (7.5.7)). For increasing values of p the left hand side gets very close to 1, and we find the asymptotic behavior of the Bernoulli numbers:

(3.5.14)
$$B_p \sim \frac{2(2p)!}{(2\pi)^{2p}}.$$

§3.6.   CONTINUED FRACTIONS

A finite continued fraction is an expression of the form

$$b_0 + \cfrac{a_1}{b_1 + \cfrac{a_2}{b_2 + \cfrac{a_3}{b_3 + \cfrac{\ddots}{b_{n-1} + \cfrac{a_n}{b_n}}}}}$$

A convenient notation for this expression is the following:

$$b_0 + \frac{a_1}{b_1} + \frac{a_2}{b_2} + \cdots \frac{a_n}{b_n} = \frac{A_n}{B_n}.$$

A_r and B_r are polynomials in the elements a_s and b_s, $s = 0, 1, \ldots, r$. As is easily found, the following recursion formulas are valid:

(3.6.1)
$$\begin{cases} A_r = b_r A_{r-1} + a_r A_{r-2}, \\ B_r = b_r B_{r-1} + a_r B_{r-2}. \end{cases}$$

Here $A_{-1} = 1$, $B_{-1} = 0$, $A_0 = b_0$, and $B_0 = 1$. The formula is proved by induction, taking into account that A_{r+1}/B_{r+1} is obtained from A_r/B_r if b_r is replaced by $b_r + a_{r+1}/b_{r+1}$.

Generalization to infinite continued fractions is obvious; such a fraction is said to be convergent if $\lim_{n \to \infty} A_n/B_n$ exists.

Continued fractions containing a variable z are often quite useful for representation of functions; in particular, such expansions often have much better convergence properties than, for example, the power-series expansions. Space limitations make it impossible to give a description of the relevant theory, and so we restrict ourselves to a few examples.

$$\arctan x = \frac{x}{1} + \frac{x^2}{3} + \frac{4x^2}{5} + \frac{9x^2}{7} + \frac{16x^2}{9} + \frac{25x^2}{11} + \cdots$$

$$\tan x = \frac{x}{1} - \frac{x^2}{3} - \frac{x^2}{5} - \frac{x^2}{7} - \cdots$$

$$\tanh x = \frac{x}{1} + \frac{x^2}{3} + \frac{x^2}{5} + \frac{x^2}{7} + \cdots$$

$$e^x = 1 + \frac{x}{1} - \frac{x}{2} + \frac{x}{3} - \frac{x}{2} + \frac{x}{5} - \frac{x}{2} + \frac{x}{7} - \frac{x}{2} + \frac{x}{9} - \cdots$$

$$\int_0^\infty \frac{e^{-u}\,du}{x+u} = \frac{1}{x} + \frac{1}{1} + \frac{1}{x} + \frac{2}{1} + \frac{2}{x} + \frac{3}{1} + \frac{3}{x} + \cdots$$

$$e^x \int_x^\infty \frac{e^{-u}}{u}\,du = \frac{1}{x+1} - \frac{1}{x+3} - \frac{4}{x+5} - \frac{9}{x+7} - \frac{16}{x+9} - \cdots$$

EXERCISES

1. Give the first six terms in the power series for $(1+x)^{1/3}$ and for $(1+x)^{-1/3}$.

2. The function $y = 4(1 + x + 3x^2/16)^{1/2} - (1+x)^{1/2} - 3 - 3x/2$ has a very flat minimum at $x = 0$. Compute the first two terms in the power series expansion.

3. Two power series with coefficients a_r and A_r are related through

$$\exp\left\{\sum_{r=1}^\infty a_r x^r\right\} = 1 + \sum_{r=1}^\infty A_r x^r.$$

Find a formula from which A_n can be determined by use of the previous values $A_r, r = 1(1)n - 1$, and $a_r, r = 1(1)n$. Applying this formula, find the first terms in the expansion $e^{\sin x}$.

4. The number of nonnegative integer solutions of the equation $x + y + 2z = n$ (n integer, ≥ 0) is denoted by a_n. Find the generating function and a recursive formula for a_n. Compute the first 8 values of a_n.

5. Obtain an asymptotic series expansion for the function $y = e^{-x^2} \cdot \int_0^x e^{t^2}\,dt$ by first proving that y satisfies the differential equation $y' = 1 - 2xy$.

6. Find the coefficients in the asymptotic formula

$$\int_z^\infty \left(\frac{\sin t}{t}\right) dt = \cos x \left(\frac{a_1}{x} + \frac{a_2}{x^2} + \cdots\right) + \sin x \left(\frac{b_1}{x} + \frac{b_2}{x^2} + \cdots\right).$$

7. The equation $\tan x = x$ has an infinite number of solutions drawing closer and closer to $a = (n + \frac{1}{2})\pi$, $n = 1, 2, 3, \ldots$ Put $x = a - z$, where z is supposed to be a small number, expand $\tan z$ in powers of z, and put $z = A/a + B/a^3 + C/a^5 + \cdots$. Determine the constants A, B, and C, and with this approximation compute the root corresponding to $n = 6$ to six decimals.

8. The equation $\sin x = 1/x$ has infinitely many solutions in the neighborhood of the points $x = n\pi$. For sufficiently large even values of n, the roots ξ_n can be written as $n\pi + \alpha + A\alpha^3 + B\alpha^5 + C\alpha^7 + \cdots$, where $A, B, C, \ldots$ are constants and $\alpha = 1/n\pi$. Find A and B.

9. For values of k slightly less than 1, the equation $\sin x = kx$ has two roots $x = \pm x_0$ close to 0. These roots can be computed by expanding $\sin x$ in a power series and putting $6(1 - k) = s$; $x^2 = y$. Then s is obtained as a power series in y. Next, a new series expansion is attempted: $y = s + \alpha_1 s^2 + \alpha_2 s^3 + \alpha_3 s^4 + \cdots$, with unknown coefficients $\alpha_1, \alpha_2, \alpha_3, \ldots$ Find the first three coefficients, and use the result for solving the equation $\sin x = 0.95x$ (six decimals).

10. The number of structurally isomeric and stereoisomeric alcohols as a function of the number of carbon atoms n is defined through the generating function $f(x) = 1 + a_1 x + a_2 x^2 + \cdots$ where we have:

$$f(x) = 1 + \frac{x}{3}[f(x)^3 + 2f(x^3)].$$

Find a_n when $n = 1(1)6$.

11. Find $\cos pz$ as a power series of x when $x = 2 \sin(z/2)$.

12. Develop the functions $|\sin x|$ and $|\cos x|$ in Fourier series.

13. Develop the function $f(x)$ in a Fourier series when

$$f(x) = \begin{cases} x + \pi, & -\pi \leq x \leq -\pi/2 \\ \pi/2, & -\pi/2 < x < \pi/2 \\ -x + \pi, & \pi/2 \leq x \leq \pi \end{cases}.$$

14. It is known that

$$\int_0^{\pi/2} \sin^{2k} x \, dx = \pi 2^{-2k-1}(2k)!/(k!)^2$$

and

$$\int_0^{\pi/2} \sin^{2k+1} x \, dx = 2^{2k}(k!)^2/(2k + 1)!$$

Try to determine numerically the constants a and b in an asymptotic formula:

$$\int_0^{\pi/2} \sin^n x \, dx \sim an^{-1/2} \exp(-b/n).$$

15. Find power series expansions for the functions $y = \exp(\exp x - 1)$ and $y = \exp(1 - \exp(-x))$.

16. For $|x| < 1$, there exists an expansion of the form

$$e^{-x} = (1 + a_1 x)(1 + a_2 x^2)(1 + a_3 x^3) \cdots$$

Determine the constants $a_1, a_2, \ldots, a_6$.

17. Find an expansion $\sum_{j,k=1}^{\infty} a_{jk} \sin j\pi x \sin k\pi y = 1$ inside the unit square.

CHAPTER 4

Orthogonal Functions

*To know the vintage and quality of a wine one
need not drink the whole cask.*

Oscar Wilde

§4.1. GENERAL CONCEPTS

We first consider a *normed linear space* consisting of a set of elements $x_1, x_2,$ $x_3, \ldots$. For each element x in the set we have defined a nonnegative number $\|x\|$, which we call the *norm* of x, with the following properties (y also belongs to the set):

$$\|x\| \geq 0.$$
$$\|\alpha x\| = |\alpha| \|x\|, \quad \alpha \text{ a complex number.}$$
$$\|x + y\| \leq \|x\| + \|y\|.$$
$$\|x\| = 0 \quad \text{implies} \quad x = 0.$$

These conditions are exactly the same as those defined for vector and matrix norms. Now suppose that the elements have the following property: In every sequence $\{x_n\}_{n=1}^{\infty}$ such that $\|x_n - x_m\| \to 0$ if n and m tend to ∞ independently, there exists a limit element x such that $\|x_n - x\| \to 0$ when $n \to \infty$. The space is then said to be *complete* and is called a *Banach space*.

If, in particular, the norm is given by a *scalar product* according to

$$\|x\| = (x, x)^{1/2}$$

then the space is called a *Hilbert space*. The scalar product can be defined in different ways, but it must obey these rules:

$$(ax + by, z) = a(x, z) + b(y, z) \quad \text{and}$$
$$(x, y) = (y, x)^* \quad (\text{or } (x, y) = (y, x) \text{ in the real case}).$$

For an ordinary vector space we have the usual definition $(x, y) = \sum x_k y_k$. However, for a space consisting of functions as elements, another definition is used, namely

(4.1.1)
$$(f, g) = \int_a^b f(x) g(x) w(x) \, dx.$$

Here we have assumed the functions to be real; in the complex case $g(x)$ should be replaced by $g(x)^*$ in the integral. As usual, the function $w(x)$ is a weight function with the property $w(x) \geq 0$.

In the discrete case we define

$$(f, g) = \sum_{i=0}^{\infty} f(x_i)g(x_i)w_i$$

where $w_i \geq 0$ are given weights.

If the scalar product $(f, g) = 0$, we say that the functions are *orthogonal*, exactly as if f and g were vectors instead.

We also mention that there are other ways to define norms for functions:

$$\text{the maximum norm} \qquad \|f\|_\infty = \max_{a \leq x \leq b} |f(x)|$$

$$\text{the } L_p\text{-norm} \qquad \|f\|_p = \left[\int_a^b |f(x)|^p \, dx \right]^{1/p}.$$

In particular, for $p = 2$ we get the Euclidean norm. The norm we have just defined using a scalar product can be considered as a weighted Euclidean norm:

$$\|f\|_{2,w} = \left[\int_a^b f(x)^2 w(x) \, dx \right]^{1/2}.$$

Similarly, in the discrete case we can define what we call a *seminorm*, the only difference being that the last condition [that $\|x\| = 0$ implies $x = 0$] must be removed.

We can now prove some important theorems.

Theorem (Cauchy–Schwarz). *Let f and g be given real functions. Then $(f, g)^2 \leq (f, f)\, (g, g)$.*

Proof. We assume $g \neq 0$, which implies $(g, g) > 0$ (if $g = 0$ the result is trivial). We then have for all real values λ

$$0 \leq (f + \lambda g, f + \lambda g) = (f, f) + 2\lambda(f, g) + \lambda^2(g, g).$$

Choosing in particular $\lambda = -(f, g)/(g, g)$, we obtain

$$(f, f) - 2(f, g)^2/(g, g) + (f, g)^2(g, g) \geq 0$$

and the theorem follows. $\qquad\qquad\qquad\qquad\qquad\qquad\qquad\qquad\qquad\qquad\qquad \square$

Essentially the same proof was given for vectors.

We now assume that we have a family of functions $g_0, g_1, g_2, \ldots$ satisfying the orthonormality conditions $(g_i, g_k) = \delta_{ik}$. Further suppose that there exists a representation of a certain function f as a linear combination of the functions g_k:

$$f = \sum_{k=0}^{\infty} c_k g_k.$$

The coefficients c_k are determined by forming scalar products, and we find

$$(f, g_i) = \sum_{k=0}^{\infty} c_k(g_k, g_i) = \sum_{k=0}^{\infty} c_k \delta_{ik} = c_i.$$

Hence we have the result $c_k = (f, g_k)$.

If no such expansion exists, we try instead to find a best possible solution. We do this by choosing the coefficients c_k in such a way that the norm of $f - \sum c_k g_k$ is minimized. Hence

$$S = (f - \sum c_k g_k, f - \sum c_k g_k) = (f, f) - 2 \sum (f, g_k)c_k + \sum c_k^2$$

should be a minimum. The condition $\partial S/\partial c_k = 0$ gives the result $c_k = (f, g_k)$ exactly as before.

We can now prove an important theorem. With the choice $c_k = (f, g_k)$ we find:

$$0 \le (f - \sum c_k g_k, f - \sum c_k g_k) = (f, f) - 2 \sum c_k(f, g_k) + \sum c_i c_k(g_i, g_k)$$
$$= (f, f) - \sum c_k^2.$$

Hence we have proved the *Bessel inequality*

(4.1.2)
$$(f, f) - \sum_{k=0}^{\infty} c_k^2 \ge 0.$$

If we have equality:

$$(f, f) = \sum_{k=0}^{\infty} c_k^2,$$

this is called the *completeness relation* or *Parseval's relation*; it implies $f = \sum_{k=0}^{\infty} c_k g_k$.

The coefficients $c_k = (f, g_k)$ are often called *Fourier coefficients*, and we see that the Fourier series constitutes a special case of expansion when the trigonometric functions are used for this purpose.

Finally we state Pythagoras' theorem for functions: Provided $(f, g) = 0$, we have

$$\|f + g\|^2 = \|f\|^2 + \|g\|^2.$$

Proof. $(f + g, f + g) = (f, f) + 2(f, g) + (g, g) = (f, f) + (g, g).$ □

---

§4.2. ORTHOGONAL POLYNOMIALS

Different kinds of polynomials play an important role in applied and numerical mathematics. Chebyshev polynomials are essential in approximation theory, while Legendre, Laguerre, and Hermite polynomials appear in the theory of numerical integration. All these polynomials form orthogonal families, and they also satisfy certain second order differential equations. Further, there are general recursive relations between any three polynomials of consecutive degrees. In the following we will show how these properties can be derived in general when the interval and the weight function are given. Finally we will treat in some detail the polynomials mentioned above.

As before we assume an inner product of the form:

$$(p(x), q(x)) = \int_a^b w(x)p(x)q(x)\, dx$$

with $a < b$ and the weight function $w(x) \geq 0$. Further, we assume that all *moments* m_k, defined through

(4.2.1) $$m_k = \int_a^b x^k w(x)\, dx, \qquad k = 0, 1, 2, \ldots$$

exist and are finite, and last that $m_0 > 0$. We will in fact use these values to determine the coefficients of the polynomials.

All real polynomials constitute a vector space, and we easily understand that the vectors $1, x, x^2, x^3, \ldots$ are linearly independent. Therefore it must be possible to perform an orthonormalization by using the Gram–Schmidt method, and in this way we obtain a series of polynomials $P_n(x)$ of degree n, where $n = 0, 1, 2, \ldots$ Hence we have

$$(P_i(x), P_k(x)) = \int_a^b P_i(x)P_k(x)w(x)\, dx = \delta_{ik}.$$

Now suppose that $P_n(x)$ can be written

$$P_n(x) = \sum_{k=0}^n s_k x^k \quad \text{where} \quad s_n \neq 0.$$

Since P_n is orthogonal to all polynomials P_k with $k < n$, it must also be orthogonal to all powers x^k, $k = 0, 1, 2, \ldots, n - 1$. In this way we find

$$\left(\sum_{i=0}^n s_i x^i, x^k \right) = 0$$

or

$$\sum_{i=0}^n s_i \int_a^b x^{i+k} w(x)\, dx = \sum_{i=0}^n s_i m_{i+k} = 0.$$

We introduce $t_i = s_i/s_n$ and obtain the system:

(4.2.2)
$$\begin{cases} m_0 t_0 + m_1 t_1 + \cdots + m_{n-1} t_{n-1} = -m_n \\ m_1 t_0 + m_2 t_1 + \cdots + m_n t_{n-1} = -m_{n+1} \\ \vdots \\ m_{n-1} t_0 + m_n t_1 + \cdots + m_{2n-2} t_{n-1} = -m_{2n-1}. \end{cases}$$

We denote the coefficient matrix by M, where obviously $M_{ik} = m_{i+k}$ (where $i, k = 0(1)n - 1$). We prove that M is positive definite and hence also nonsingular. Assume

u to be an n-dimensional vector; then we have

$$(u, Mu) = \sum_{i,k} u_i M_{ik} u_k = \sum u_i m_{i+k} u_k$$

$$= \left(\sum_{i=0}^{n-1} u_i x^i, \sum_{k=0}^{n-1} u_k x^k \right) > 0$$

provided that u is not the null vector. From this we conclude that the system of equations above has a unique solution $t_0, t_1, \ldots, t_{n-1}$. We then form the monic polynomial*

$$p(x) = t_0 + t_1 x + \cdots + t_{n-1} x^{n-1} + x^n,$$

and clearly $P_n(x) = Np(x)$ where N is a normalization constant to be determined from the condition $(P_n(x), P_n(x)) = 1$. Since $t_n = 1$, we find

$$N^2(p(x), p(x)) = N^2 \sum_{i,k=0}^{n} t_i M_{ik} t_k = N^2 \sum_{i=0}^{n} t_i (Mt)_i = 1$$

where M now is the augmented matrix of size $(n + 1, n + 1)$. The vector Mt has the first n components equal to zero, as expressed in the system above. The last component has the value

$$m = \sum_{i=0}^{n} m_{n+i} t_i$$

and hence we have $N^2 m = 1$, giving $N = m^{-1/2}$.

Example 1.   We take $a = -1$, $b = 1$, $w(x) = 1$, and $n = 4$. The moments are:

$$m_r = \int_{-1}^{1} x^r \, dx = \begin{cases} 2/(r + 1), & r \text{ even} \\ 0, & r \text{ odd.} \end{cases}$$

We write the system of equations $Mt = d$ where

$$M = \begin{pmatrix} 2 & 0 & 2/3 & 0 \\ 0 & 2/3 & 0 & 2/5 \\ 2/3 & 0 & 2/5 & 0 \\ 0 & 2/5 & 0 & 2/7 \end{pmatrix}, \quad t = \begin{pmatrix} t_0 \\ t_1 \\ t_2 \\ t_3 \end{pmatrix}, \quad d = \begin{pmatrix} -2/5 \\ 0 \\ -2/7 \\ 0 \end{pmatrix}.$$

Solving this system, we get $t_0 = 3/35$, $t_2 = -6/7$, and $t_1 = t_3 = 0$. Hence

$$p(x) = x^4 - \frac{6}{7} x^2 + \frac{3}{35} \quad \text{and} \quad m = \frac{2}{5} t_0 + \frac{2}{7} t_2 + \frac{2}{9} = \frac{512}{44100}.$$

This gives $N = 105 \sqrt{2}/16$ and

$$P_4(x) = \frac{3\sqrt{2}}{16} (35x^4 - 30x^2 + 3).$$

---

* Monic means that the coefficient of the highest term is 1.

Apart from a numerical factor this is one of the Legendre polynomials, which will be treated in more detail later.

We shall now deduce a three-term recursive formula. We start by analyzing the expression $xP_n(x)$. Since it is a polynomial of degree $n + 1$ it must be a linear combination of $P_0, P_1, \ldots, P_{n+1}$:

$$xP_n = \sum_{i=0}^{n+1} c_i P_i \quad \text{where} \quad c_i = (xP_n, P_i).$$

But $(xP_n, P_i) = (P_n, xP_i) = 0$ if $i < n - 1$, because xP_i is then a linear combination of P_k, $k = 0(1)i + 1$. Thus we conclude that

$$xP_n = c_{n+1}P_{n+1} + c_n P_n + c_{n-1}P_{n-1}$$

with $c_{n+1} \neq 0$, so xP_n is a polynomial of degree $n + 1$. Introducing other notations, we can write this relation in the form

(4.2.3) $$P_{n+1} = (A_n x + B_n)P_n - C_n P_{n-1}$$

where A_n must be different from 0. Now suppose that

$$P_r(x) = a_r x^r + b_r x^{r-1} + \cdots$$

where we are particularly concerned with $r = n - 1$, n, and $n + 1$. Comparing coefficients for x^{n+1} and x^n in the recursion formula, we find directly:

$$\begin{cases} a_{n+1} = A_n a_n \\ b_{n+1} = A_n b_n + B_n a_n. \end{cases}$$

Finally, taking inner products with P_{n-1} we get

$$\begin{aligned} C_n &= A_n(xP_n, P_{n-1}) = A_n(P_n, xP_{n-1}) = A_n(P_n, (a_{n-1}/a_n)P_n + \cdots) \\ &= A_n a_{n-1}/a_n = a_{n+1}a_{n-1}/a_n^2. \end{aligned}$$

Hence we have:

(4.2.4) $$\begin{cases} A_n = a_{n+1}/a_n \\ B_n = A_n(b_{n+1}/a_{n+1} - b_n/a_n) \\ C_n = a_{n+1}a_{n-1}/a_n^2. \end{cases}$$

If we choose a certain value of x, we can form the vector P with components P_0, $P_1, \ldots, P_n$. Since $A_r \neq 0$, we can then write the system of recursion formulas in the following form ($P_{-1} = 0$):

$$\begin{cases} xP_0 = (-B_0/A_0)P_0 + (1/A_0)P_1 \\ xP_1 = (C_1/A_1)P_0 + (-B_1/A_1)P_1 + (1/A_1)P_2 \\ \quad \vdots \\ xP_n = (C_n/A_n)P_{n-1} + (-B_n[/A_n)P_n + (1/A_n)P_{n+1}. \end{cases}$$

We verify directly that $C_r/A_r = 1/A_{r-1} = a_{r-1}/a_r$, and we call this quantity β_r. Further, putting $b_r/a_r = c_r$ we have

$$-B_r/A_r = c_r - c_{r-1}$$

and we call this α_r. Introducing the matrix

(4.2.5)
$$T = \begin{pmatrix} \alpha_0 & \beta_1 & 0 & 0 & \cdots & 0 \\ \beta_1 & \alpha_1 & \beta_2 & 0 & \cdots & 0 \\ 0 & \beta_2 & \alpha_2 & \beta_3 & \cdots & 0 \\ \vdots & & & \ddots & \ddots & \vdots \\ 0 & 0 & 0 & \cdots & \beta_n & \alpha_n \end{pmatrix}$$

we obtain $TP = xP - (0 \; 0 \; \cdots \; 0 \; P_{n+1}/A_n)^T$. Now suppose that $x_j, \; j = 1(1)n + 1$, are the roots of $P_{n+1}(x) = 0$. Choose x equal to one of these roots and we obtain:

$$TP = x_j P$$

which means that x_j are the eigenvalues of the matrix T. Since T is real and symmetric, we know that all eigenvalues (i.e., all roots of $P_{n+1}(x)$), are real.

Let $x_1, x_2, \ldots, x_k$ be those roots of $P_n(x)$ that are contained in the interval $a \le x \le b$. Then obviously

$$(x - x_1)(x - x_2) \cdots (x - x_k)P_n(x)$$

is of constant sign in $[a, b]$. Hence

$$\int_a^b (x - x_1)(x - x_2) \cdots (x - x_k)P_n(x)w(x) \, dx \ne 0.$$

However, due to orthogonality this is possible only if $k = n$. We shall also prove that all zeros are simple. To show this, suppose that $x = x_1$ is a multiple root; then we can write $P_n(x) = (x - x_1)^2 p_{n-2}(x)$ and

$$P_n(x)p_{n-2}(x) = (P_n(x)/(x - x_1))^2 > 0.$$

This implies $\int_a^b P_n(x)p_{n-2}(x)w(x) \, dx > 0$, which is a contradiction since the integral is zero, again due to orthogonality. By this we have proved that all zeros are simple and lie in the interval $[a, b]$.

Finally, we mention that quite often we work with families of polynomials that are only mutually orthogonal; we do not require that they are normalized. If instead $(f_m, f_n) = h_n \delta_{mn}$, an easy calculation shows that A_n and B_n are not changed while C_n is multiplied by the factor h_n/h_{n-1}.

Christoffel–Darboux's Theorem

Theorem. *A system of orthonormal polynomials $P_n(x)$ satisfies the relation*

(4.2.6)
$$P_{n+1}(x)P_n(y) - P_n(x)P_{n+1}(y) = A_n(x - y) \sum_{k=0}^{n} P_k(x)P_k(y).$$

Proof.   We multiply the three-term recursion formula

$$P_{k+1}(x) = (A_k x + B_k) P_k(x) - C_k P_{k-1}(x)$$

by $P_k(y)$, writing afterwards the same relation with x and y interchanged:

$$\begin{cases} P_{k+1}(x)P_k(y) = (A_k x + B_k)P_k(x)P_k(y) - C_k P_{k-1}(x)P_k(y) \\ P_k(x)P_{k+1}(y) = (A_k y + B_k)P_k(y)P_k(x) - C_k P_{k-1}(y)P_k(x). \end{cases}$$

Subtracting and observing that $C_k = A_k/A_{k-1}$ we get, after division by A_k:

$$(x - y)P_k(x)P_k(y) = A_k^{-1}[P_{k+1}(x)P_k(y) - P_k(x)P_{k+1}(y)] - A_{k-1}^{-1}[P_k(x)P_{k-1}(y)$$
$$- P_{k-1}(x)P_k(y)] = G_k - G_{k-1}$$

with an obvious meaning for G_k. We note that G_{-1} vanishes since we have $1/A_{-1} = a_{-1}/a_0 = 0$ where, as before, a_k is the leading coefficient in $P_k(x)$. Adding all relations from $k = 0$ to $k = n$ we then obtain:

$$(x - y) \sum_{k=0}^{n} P_k(x)P_k(y) = G_n = A_n^{-1}[P_{n+1}(x)P_n(y) - P_n(x)P_{n+1}(y)]$$

which is exactly the relation to be proved.   □

Rodrigues' Formula

Suppose that we have the weight function $w(x)$ and further that a and b are finite with $a < b$. We then choose the function $z(x)$ in such a way that

(4.2.7) $$f(x) = \frac{1}{w(x)} \frac{d^n}{dx^n} \{(x - a)^n (b - x)^n z(x)\}$$

becomes a polynomial of exactly degree n. Under certain conditions for $z(x)$, the function $f(x)$ (apart from a trivial numerical factor) is equal to $p_n(x)$ belonging to the corresponding family of orthogonal polynomials.

We observe that the following condition must be satisfied:

$$I = \int_a^b f(x)r(x)w(x)\,dx = 0$$

where $r(x)$ is a polynomial of degree $n - 1$. The proof that this equation is valid is conducted by partial integration, and then we see that the function $z(x)$ must be such that $D^k\{(x - a)^n(b - x)^n z(x)\}$ disappears for $x = a$ and $x = b$ where $k = 0(1)n - 1$. We then obtain:

$$I = (-1)^n \int_a^b (x - a)^n (b - x)^n z(x) \frac{d^n r}{dx^n}\,dx = 0$$

since $d^n r/dx^n = 0$.

If one or both limits are infinite, we take instead:

$$f(x) = [w(x)]^{-1} D^n\{(b - x)^n z(x)\},$$
$$f(x) = [w(x)]^{-1} D^n\{(x - a)^n z(x)\}, \quad \text{or}$$
$$f(x) = [w(x)]^{-1} D^n z(x).$$

Well-known examples are $a = -1$, $b = 1$, $w(x) = 1$ (Legendre polynomials), $a = -1$, $b = 1$, $w(x) = (1 - x^2)^{-1/2}$ (Chebyshev polynomials), $a = 0$, $b = \infty$, $w(x) = e^{-x}$ (Laguerre polynomials), and $a = -\infty$, $b = \infty$, $w(x) = \exp(-x^2)$ (Hermite polynomials).

§4.3. LEGENDRE POLYNOMIALS

The Legendre polynomials are associated with the interval $(-1, 1)$ and the weight function $w(x) = 1$. In their nonnormalized form they are conveniently defined through

(4.3.1) $$P_0(x) = 1; \qquad P_n(x) = \frac{1}{2^n n!} \frac{d^n}{dx^n} (x^2 - 1)^n$$

(Rodrigues' formula). Writing

$$P_n(x) = \frac{2^n}{n!} D^n \left\{ \left(\frac{x - 1}{2} \right)^n \left(\frac{x + 1}{2} \right)^n \right\}$$

and applying the well-known formula for the nth derivative of a product, we obtain

$$P_n(x) = \sum_{k=0}^{n} \binom{n}{k}^2 \left(\frac{x - 1}{2} \right)^{n-k} \left(\frac{x + 1}{2} \right)^k.$$

We shall now prove that these polynomials obey a certain second order differential equation. Putting $y = (x^2 - 1)^n$ we have

$$y' = 2nx(x^2 - 1)^{n-1}$$

and hence, multiplying with $(x^2 - 1)$ and differentiating:

$$(x^2 - 1)y' = 2nxy$$
$$(x^2 - 1)y'' = 2(n - 1)xy' + 2ny$$
$$(x^2 - 1)y''' = 2(n - 2)xy'' + 2[n + (n - 1)]y'$$
$$(x^2 - 1)y^{(iv)} = 2(n - 3)xy''' + 2[n + (n - 1) + (n - 2)]y''$$
$$\vdots$$
$$(x^2 - 1)y^{(n+1)} = 2(n - n)xy^{(n)} + 2[n + (n - 1) + (n - 2) + \cdots + 2 + 1]y^{(n-1)}$$

Thus we obtain $(x^2 - 1)y^{(n+1)} = n(n + 1)y^{(n-1)}$ and differentiating once more:

(4.3.2) $$\frac{d}{dx}\left[(1 - x^2) \frac{dP_n}{dx} \right] + n(n + 1)P_n = 0.$$

We now proceed to prove in a more direct way that

$$\int_{-1}^{+1} x^r P_n(x)\, dx = 0, \qquad r = 0(1)n - 1.$$

Apart from a constant factor, this can be written:

$$\int_{-1}^{+1} x^r \frac{d^n}{dx^n}(x^2-1)^n \, dx = \left[x^r \frac{d^{n-1}}{dx^{n-1}}(x^2-1)^n \right]_{-1}^{+1} - \int_{-1}^{+1} r x^{r-1} \frac{d^{n-1}}{dx^{n-1}}(x^2-1)^n \, dx$$

$$\underbrace{}_{=0}$$

$$= \cdots = (-1)^r \cdot r! \int_{-1}^{+1} \frac{d^{n-r}}{dx^{n-r}}(x^2-1)^n \, dx = 0.$$

If $r = n$, we obtain:

$$(-1)^n \cdot n! \int_{-1}^{+1} (x^2-1)^n \, dx = 2 \cdot n! \int_0^1 (1-x^2)^n \, dx$$

$$= 2n! \int_0^{\pi/2} \cos^{2n+1} \varphi \, d\varphi$$

$$= 2n! \frac{2n(2n-2)\cdots 2}{(2n+1)(2n-1)\cdots 3}$$

$$= \frac{2^{2n+1}(n!)^3}{(2n+1)!}.$$

If $m \neq n$ we have $\int_{-1}^{+1} P_m(x)P_n(x) \, dx = 0$, since if $m < n$, then $P_m(x)$ is a polynomial of degree less than n and the integral vanishes in view of the relation above. Finally, we also compute

$$\int_{-1}^{+1} [P_n(x)]^2 \, dx = \int_{-1}^{+1} \left(\frac{1}{2^n n!} \frac{(2n)!}{n!} x^n + \cdots \right) \frac{1}{2^n n!} \frac{d^n}{dx^n}(x^2-1)^n \, dx$$

$$= \frac{(2n)!}{2^{2n}(n!)^3} \frac{2^{2n+1}(n!)^3}{(2n+1)!} = \frac{2}{2n+1},$$

since powers below x^n give no contributions to the integral. Hence we have the following relations:

(4.3.3)
$$\int_{-1}^{+1} x^r P_n(x) \, dx = 0, \qquad r = 0(1)n-1$$

$$\int_{-1}^{+1} P_m(x)P_n(x) \, dx = \frac{2}{2n+1} \cdot \delta_{m,n}.$$

When $m \neq n$, the last relation can also be proved easily directly from the differential equation:

$$(1-x^2)P_m'' - 2xP_m' + m(m+1)P_m = 0$$
$$(1-x^2)P_n'' - 2xP_n' + n(n+1)P_n = 0.$$

Multiplying the first equation by P_n and the second by $-P_m$ and adding, we obtain:

$$(1-x^2)(P_n P_m'' - P_m P_n'') - 2x(P_n P_m' - P_m P_n') + (m-n)(m+n+1)P_m P_n = 0.$$

Observing that the first two terms can be written

$$\frac{d}{dx}\left[(1-x^2)(P_nP'_m - P_mP'_n)\right]$$

and integrating from -1 to $+1$, provided that $m \neq n$, we get:

$$\int_{-1}^{+1} P_m(x)P_n(x)\,dx = 0.$$

We shall finally apply our three-term recursion formula (4.2.3). We find directly:

$$p_n = \frac{(2n)!}{2^n(n!)^2}, \qquad q_n = 0 \qquad \text{(defined through } P_n = p_n x^n + q_n x^{n-1} + \cdots\text{),}$$

giving $A_n = (2n+1)/(n+1)$; $B_n = 0$. Further, since $h_n = (P_n, P_n) = 2/(2n+1)$, we obtain $C_n = n/(n+1)$ and hence

(4.3.4) $$(n+1)P_{n+1} = (2n+1)xP_n - nP_{n-1}.$$

We will now prove that the Legendre polynomials can be defined by use of a generating function:

(4.3.5) $$(1 - 2xh + h^2)^{-1/2} = \sum_{n=0}^{\infty} P_n(x)h^n.$$

Differentiating with respect to h we obtain:

$$(x - h)(1 - 2xh + h^2)^{-3/2} = \sum nh^{n-1}P_n(x)$$

which can also be written

$$(x - h)\sum h^n P_n(x) = (1 - 2xh + h^2)\sum nh^{n-1}P_n(x).$$

Comparing coefficients of h^n, we find after simplification

$$(n+1)P_{n+1} - (2n+1)xP_n + nP_{n-1} = 0,$$

i.e., precisely the three-term recursion formula proved previously. By direct series expansion we have:

$$1 - \tfrac{1}{2}(-2xh + h^2) + \tfrac{3}{8}(-2xh + h^2)^2 + \cdots = \sum P_n(x)h^n$$

giving $P_0 = 1$ and $P_1 = x$ as it should be. Hence the coefficients P_n must become Legendre polynomials also for $n > 1$, since the recursion formula is the same.

Differentiating (4.3.5) with respect to x instead, after some trivial manipulations we get:

$$h\sum P_n h^n = (1 - 2xh + h^2)\sum P'_n h^n$$

and comparing coefficients for h^{n+1} we find:

$$P_n = P'_{n+1} - 2xP'_n + P'_{n-1}.$$

Differentiating the three-term recursion formula with respect to x, we get

$$(n + 1)P'_{n+1} = (2n + 1)(xP'_n + P_n) - nP'_{n-1}$$

and eliminating in an obvious manner yields

(4.3.6) $$P'_{n+1} = xP'_n + (n + 1)P_n.$$

Two more trivial eliminations give:

(4.3.7)
$$P'_{n-1} = xP'_n - nP_n$$
$$P'_{n+1} - P'_{n-1} = (2n + 1)P_n.$$

The first Legendre polynomials are:

$$P_0 = 1; \quad P_1 = x; \quad P_2 = (3x^2 - 1)/2; \quad P_3 = (5x^3 - 3x)/2;$$
$$P_4 = (35x^4 - 30x^2 + 3)/8; \quad P_5 = (63x^5 - 70x^3 + 15x)/8.$$

For the normalized polynomials with $P_n(x) = a_n x^n + b_n x^{n-1} + \cdots$ we have

$$a_n = \frac{(2n)!}{2^n (n!)^2} (n + \tfrac{1}{2})^{1/2} \quad \text{and} \quad b_n = 0.$$

The numbers α and β determining the tridiagonal matrix T in (4.2.5) are:

$$\alpha = (0, 0, 0, \ldots)$$
$$\beta = (1/\sqrt{(1 \cdot 3)}, 2/\sqrt{(3 \cdot 5)}, 3/\sqrt{(5 \cdot 7)}, 4/\sqrt{(7 \cdot 9)}, \ldots).$$

Hence $\alpha_r = 0$ and $\beta_r = r(4r^2 - 1)^{-1/2}$.

§4.4.   LAGUERRE POLYNOMIALS

The Laguerre polynomials are defined by

(4.4.1) $$L_n(x) = e^x \frac{d^n}{dx^n} (e^{-x} x^n)$$

corresponding to the interval $(0, \infty)$ and the weight function $w(x) = e^{-x}$. Using a technique similar to that for the Legendre polynomials, we find

(4.4.2)
$$\int_0^\infty x^r e^{-x} L_n(x)\, dx = 0, \qquad r = 0(1)n - 1,$$
$$\int_0^\infty x^n e^{-x} L_n(x)\, dx = (-1)^n (n!)^2.$$

From this we also conclude that

(4.4.3) $$\int_0^\infty e^{-x} L_m(x) L_n(x)\, dx = (n!)^2 \delta_{m,n}.$$

Putting $y = x^n e^{-x}$, we get $y' = (nx^{n-1} - x^n)e^{-x} = ny/x - y$ or $xy' + xy - ny = 0$. Successive differentiation gives

$$xy'' + [x - (n - 1)]y' + y = 0$$
$$xy''' + [x - (n - 2)]y'' + 2y' = 0$$
$$\vdots$$
$$xy^{(n+1)} + xy^{(n)} + ny^{(n-1)} = 0$$
$$xy^{(n+2)} + (x + 1)y^{(n+1)} + (n + 1)y^{(n)} = 0.$$

Observing that $y^{(n)} = e^{-x}L_n$, we get:

$$y^{(n+1)} = e^{-x}(L_n' - L_n)$$
$$y^{(n+2)} = e^{-x}(L_n'' - 2L_n' + L_n)$$

and inserting these expressions in the last equation above we find:

(4.4.4) $$xL_n'' + (1 - x)L_n' + nL_n = 0.$$

Further, with $L_n(x) = p_n x^n + q_n x^{n-1} + \cdots$ we find that $p_n = (-1)^n$ and $q_n = (-1)^{n-1}n^2$ and hence $A_n = -1$, $B_n = 2n + 1$. Since $h_n = (n!)^2$, we get $C_n = n^2$ and the recursion formula:

(4.4.5) $$L_{n+1} = (2n + 1 - x)L_n - n^2 L_{n-1}.$$

The Laguerre polynomials can also be defined by a generating function:

(4.4.6) $$(1 - t)^{-1} \exp(-xt/(1 - t)) = \sum_{n=0}^{\infty} L_n(x)t^n/n!$$

This is proved by differentiating with respect to t, which leads to the three-term recursion formula. Further, taking the first terms in the formula above we find $L_0 = 1$, $L_1 = 1 - x$, and consequently all other L_n must be Laguerre polynomials. Differentiating with respect to x we get

(4.4.7) $$L_n' = n(L_{n-1}' - L_{n-1}).$$

Differentiating again and eliminating the second derivatives using the differential equation, we obtain:

(4.4.8) $$xL_n' = n(L_n - nL_{n-1}).$$

The first polynomials are:

$$L_0 = 1; \quad L_1 = 1 - x; \quad L_2 = 2 - 4x + x^2;$$
$$L_3 = 6 - 18x + 9x^2 - x^3;$$
$$L_4 = 24 - 96x + 72x^2 - 16x^3 + x^4;$$
$$L_5 = 120 - 600x + 600x^2 - 200x^3 + 25x^4 - x^5.$$

Finally, we have $a_n = 1/n!$ and $b_n = -n^2/n!$. The numbers α and β defining the matrix T in (4.2.5) are:

$$\alpha = (1, 3, 5, 7, 9, \ldots)$$
$$\beta = (1, 2, 3, 4, 5, \ldots).$$

or $\alpha_r = 2r + 1$, $\beta_r = r$.

§4.5. HERMITE POLYNOMIALS

The Hermite polynomials are defined through

(4.5.1) $$H_n(x) = (-1)^n \exp(x^2) \frac{d^n}{dx^n} \exp(-x^2).$$

Using the same technique as before, we prove that they satisfy the differential equation:

(4.5.2) $$H_n'' - 2xH_n' + 2nH_n = 0.$$

Further, by successive partial integration we get

(4.5.3)
$$\int_{-\infty}^{\infty} \exp(-x^2) H_n(x) x^r \, dx = 0, \qquad r = 0(1)n - 1, \qquad \text{and}$$
$$\int_{-\infty}^{\infty} \exp(-x^2) H_n(x) x^n \, dx = \sqrt{\pi} n!.$$

From these results we obtain the orthogonality relation:

(4.5.4) $$\int_{-\infty}^{\infty} \exp(-x^2) H_m(x) H_n(x) \, dx = 2^n n! \sqrt{\pi} \, \delta_{m,n}.$$

Since $p_n = 2^n$ and $q_n = 0$, we get $A_n = 2$, $B_n = 0$, and $C_n = 2n$. Hence the recursion formula takes the form:

(4.5.5) $$H_{n+1} = 2xH_n - 2nH_{n-1}$$

and further we can easily prove that

(4.5.6) $$H_n' = 2nH_{n-1}.$$

The Hermite polynomials are related to the interval $(-\infty, \infty)$ and the weight function $w(x) = \exp(-x^2)$, and the first of them are:

$$H_0 = 1; \quad H_1 = 2x; \quad H_2 = 4x^2 - 2; \quad H_3 = 8x^3 - 12x;$$
$$H_4 = 16x^4 - 48x^2 + 12; \quad H_5 = 32x^5 - 16x^3 + 120x;$$
$$H_6 = 64x^6 - 480x^4 + 720x^2 - 120.$$

The Hermite polynomials can be defined from a generating function:

$$(4.5.7) \qquad \exp(2tx - t^2) = \sum_{n=0}^{\infty} H_n(x)t^n/n!$$

Finally, we have $a_n = \pi^{-1/4}(2^n/n!)^{1/2}$ and $b_n = 0$. The numbers α and β are $\alpha_n = 0$; $\beta_n = (n/2)^{1/2}$.

---

§4.6. CHEBYSHEV POLYNOMIALS

The Chebyshev polynomials are defined by the relation

$$(4.6.1) \qquad T_n(x) = \cos(n \arccos x),$$

where it is natural to put $T_{-n}(x) = T_n(x)$. Well-known trigonometric formulas give at once $T_{n+m}(x) + T_{n-m}(x) = 2T_n(x)T_m(x)$. For $m = 1$ we get in particular

$$(4.6.2) \qquad T_{n+1}(x) = 2xT_n(x) - T_{n-1}(x).$$

Since $T_0(x) = 1$ and $T_1(x) = x$, we can successively compute all $T_n(x)$. Putting, for a moment, $x = \cos \theta$, we get $y = T_n(x) = \cos n\theta$ and, further,

$$\frac{dy}{dx} = y' = \frac{n \sin n\theta}{\sin \theta}$$

and

$$y'' = \frac{-n^2 \cos n\theta + n \sin n\theta \cot \theta}{\sin^2 \theta} = -\frac{n^2 y}{1 - x^2} + \frac{xy'}{1 - x^2}.$$

Thus the polynomials $T_n(x)$ satisfy the differential equation

$$(4.6.3) \qquad (1 - x^2)y'' - xy' + n^2 y = 0.$$

As is inferred directly, the other fundamental solution is the function $S_n(x) = \sin(n \arccos x)$. In particular we have $S_0(x) = 0$ and $S_1(x) = \sqrt{(1 - x^2)}$. With $U_n(x) = S_n(x)/\sqrt{(1 - x^2)}$, we have in the same way as for (4.6.2):

$$(4.6.4) \qquad U_{n+1}(x) = 2xU_n(x) - U_{n-1}(x).$$

The first polynomials are

$T_0(x) = 1$	$U_0(x) = 0$
$T_1(x) = x$	$U_1(x) = 1$
$T_2(x) = 2x^2 - 1$	$U_2(x) = 2x$
$T_3(x) = 4x^3 - 3x$	$U_3(x) = 4x^2 - 1$
$T_4(x) = 8x^4 - 8x^2 + 1$	$U_4(x) = 8x^3 - 4x$
$T_5(x) = 16x^5 - 20x^3 + 5x$	$U_5(x) = 16x^4 - 12x^2 + 1$

Using the orthogonal relations for the cosine function, we have directly

$$(4.6.5) \qquad \int_{-1}^{+1} T_m(x)T_n(x) \cdot \frac{dx}{\sqrt{(1-x^2)}} = \begin{cases} 0, & m \neq n \\ \pi/2, & m = n \neq 0, \\ \pi, & m = n = 0, \end{cases}$$

that is, the polynomials $T_n(x)$ are orthogonal with the weight function $1/\sqrt{(1-x^2)}$.
   In this special case Rodrigues' formula has the following shape:

$$(4.6.6) \qquad T_n(x) = (-1)^n \frac{(2n)!}{2^n n!} (1-x^2)^{1/2} \frac{d^n}{dx^n} (1-x^2)^{n-1/2}.$$

Similarly we have:

$$(4.6.7) \qquad \int_{-1}^{+1} U_m(x)U_n(x)(1-x^2)^{1/2} \, dx = \begin{cases} 0, & m \neq n, \\ \pi/2, & m = n \neq 0, \\ 0, & m = n = 0. \end{cases}$$

The successive powers of x can also be expressed in terms of Chebyshev polynomials, and we find:

$$(4.6.8) \qquad \begin{aligned} 1 &= T_0, & x^4 &= \tfrac{1}{8}(3T_0 + 4T_2 + T_4), \\ x &= T_1, & x^5 &= \tfrac{1}{16}(10T_1 + 5T_3 + T_5), \\ x^2 &= \tfrac{1}{2}(T_0 + T_2), & x^6 &= \tfrac{1}{32}(10T_0 + 15T_2 + 6T_4 + T_6), \\ x^3 &= \tfrac{1}{4}(3T_1 + T_3), & x^7 &= \tfrac{1}{64}(35T_1 + 21T_3 + 7T_5 + T_7), \\ & & & \quad\vdots \end{aligned}$$

The general coefficient in the expansion $x^k = \sum c_n T_n$ becomes

$$c_n = \frac{2}{\pi} \cdot \int_{-1}^{+1} x^k T_n(x) \frac{dx}{\sqrt{(1-x^2)}} = \frac{2}{\pi} \int_0^{\pi} \cos^k \varphi \cos n\varphi \, d\varphi.$$

If k is odd, n takes the values $1, 3, 5, \ldots, k$, and if k is even, n takes the values $0, 2, 4, \ldots, k$. The integrand can be written

$$\frac{(e^{i\varphi} + e^{-i\varphi})^k}{2^k} \cdot \frac{e^{ni\varphi} + e^{-ni\varphi}}{2}$$

$$= \frac{e^{ki\varphi} + \binom{k}{1} e^{(k-2)i\varphi} + \binom{k}{2} e^{(k-4)i\varphi} + \cdots + e^{-ki\varphi}}{2^k} \cdot \frac{e^{ni\varphi} + e^{-ni\varphi}}{2}.$$

   The term $e^{(k-2r)i\varphi} \cdot e^{ni\varphi}$ becomes 1 if $r = (n+k)/2$, and analogously, we get 1 also for the term $e^{(2r-k)i\varphi} \cdot e^{-ni\varphi}$ with the same r. When we integrate, all other terms vanish, and we are left with

$$c_n = 2^{-k+1} \binom{k}{(n+k)/2}.$$

If $k = 0$ we have $c_n = 0$ except when $n = 0$, since $c_0 = 1$. Hence we can write symbolically $x^n \sim [\frac{1}{2}(T + T^{-1})]^n$ with $T^n \to T_n$, $T^{-n} \to T_n$. Finally, with $T_n(x)$ normalized, we have $T_n(x) = a_n x^n + b_n x^{n-1} + \cdots$ where $a_0 = 1/\sqrt{\pi}$ and $a_n = 2^n/(2\pi)^{1/2}$, $n \geq 1$; $b_n = 0$. The numbers α and β defining the matrix T are:

$$\alpha = (0, 0, 0, \ldots)$$
$$\beta = (1/\sqrt{2}, \tfrac{1}{2}, \tfrac{1}{2}, \tfrac{1}{2}, \ldots).$$

EXERCISES

1. Compute the polynomials Q_1, Q_2, and Q_3 from $Q_n = (x - a_n)Q_{n-1} - b_n Q_{n-2}$ with $Q_{-1} = 0$, $Q_0 = 1$, and with the orthogonality property
$$\int_0^1 Q_i(x)Q_k(x)\, dx = 0, \qquad i \neq k.$$
Also find a_1, a_2, a_3, b_2, and b_3.

2. A family of orthonormal polynomials is given over the interval (a, b) with the weight function $w(x) = 1$. Prove that $P_k(a)^2 = P_k(b)^2$.

3. The polynomials P_n are defined through
$$\int_0^1 w(x)P_i(x)P_k(x)\, dx = \delta_{ik}$$
where $w(x) = \arccos x$. Find P_0, P_1, and P_2.

4. The Chebyshev polynomials of the second kind are defined through
$$U_n(x) = (1 - x^2)^{-1/2} \sin(n \arccos x).$$
Show that $U_{n+1} = 2xU_n - U_{n-1}$ and further that the polynomials are orthogonal over the interval $(-1, 1)$ with the weight function $(1 - x^2)^{1/2}$.

5. A family of orthonormal polynomials P_n is given. Prove the relation
$$\sum_{k=0}^n P_k(x)^2 = A_n^{-1}(P'_{n+1}(x)P_n(x) - P_{n+1}(x)P'_n(x))$$
using Christoffel–Darboux's theorem.

6. Let P_n be the Legendre polynomials and $p_n = (n + \tfrac{1}{2})^{1/2} P_n$ the normalized polynomials. Differentiating the previous formula and using the differential equation, show that
$$(1 - x^2) \sum_{k=0}^n (2k + 1)P_k P'_k - x \sum_{k=0}^n (2k + 1)P_k^2 + (n + 1)^2 P_{n+1}P_n = 0.$$
We have $A_n = (2n + 1)^{1/2}(2n + 3)^{1/2}/(n + 1)$.

7. The polynomials P_n are orthonormal over the interval $(-1, 1)$ with the weight function $w(x) = |x|$. Compute these polynomials up to $n = 6$.

8. Develop the function $f(x) = \tfrac{1}{2} \ln[(1 + x)/(1 - x)]$ in a series of Chebyshev polynomials.

CHAPTER 5

Linear Operators

"I could have done it in a much more complicated
way," said the red Queen immensely proud.

Lewis Carroll

§5.1. GENERAL PROPERTIES

When one wishes to construct formulas for interpolation, numerical differentiation, quadrature, and summation, the operator technique proves to be a most useful tool. One of the greatest advantages is that one can sketch the type of formula desired in advance and then proceed directly toward the goal. Usually the deduction is also considerably simplified. It must be understood, however, that complete formulas, including remainder terms, are in general not obtained in this way.

The operators are assumed to operate on functions belonging to a linear function space. Such a space is defined as a set F of functions, $f, g, \ldots$, having the properties that $f \in F$, $g \in F$ implies $\alpha f + \beta g \in F$, where α and β are arbitrary constants. A linear operator is then a mapping of F on a linear function space F^*; usually one chooses $F^* = F$. An operator P being linear means that $P(\alpha f + \beta g) = \alpha P f + \beta P g$ for all f and $g \in F$ with α and β arbitrary constants. The operators that we are going to discuss fulfill the associative and distributive laws, but in general they do not commutate. If we let $P_1, P_2, P_3, \ldots$ denote operators, then the following laws are supposed to be valid:

$$(5.1.1) \quad \begin{cases} P_1 + (P_2 + P_3) = (P_1 + P_2) + P_3, \\ \quad P_1(P_2 P_3) = (P_1 P_2) P_3, \\ \quad P_1(P_2 + P_3) = P_1 P_2 + P_1 P_3. \end{cases}$$

Here we define the sum P of two operators P_1 and P_2 as the operator that transforms the function f into $P_1 f + P_2 f$; the product $P_1 P_2$ is defined as the operator that transforms f into $P_1(P_2 f)$.

Two operators P_1 and P_2 are said to be equal if $P_1 f = P_2 f$ for all functions $f \in F$. We also define the inverse P^{-1} of an operator P. If Q is such an operator that for all f we have $Qg = f$ if $Pf = g$, then Q is said to be an inverse of P. Here we must be careful, however, since it can occur that P^{-1} is not uniquely determined. In order to demonstrate this we suppose that $w(x)$ is a function which is annihilated by P, that is, $Pw(x) = 0$, and further that $f(x)$ represents a possible result of $P^{-1}g(x)$. Then $f(x) + w(x)$ is another possible result. In fact, we have $P(f(x) + w(x)) = Pf(x) = g(x)$. Hence we can write $P^{-1}Pf = f + w$, where w is an arbitrary function annihilated by P. Thus P^{-1} is a right inverse but not a left inverse. If the equation $Pw(x) = 0$ has only the trivial solution $w(x) = 0$, then P^{-1} also becomes a left inverse:

$$PP^{-1} = P^{-1}P = 1.$$

Then we say briefly that P^{-1} is the inverse of P.

§5.2.   SPECIAL OPERATORS

We now introduce the following six operators together with their defining equations.

$Ef(x) = f(x + h)$	The shift operator
$\Delta f(x) = f(x + h) - f(x)$	The forward-difference operator
$\nabla f(x) = f(x) - f(x - h)$	The backward-difference operator
$\delta f(x) = f(x + h/2) - f(x - h/2)$	The central-difference operator
$\mu f(x) = \frac{1}{2}[f(x + h/2) + f(x - h/2)]$	The mean-value operator
$Df(x) = f'(x)$	The differentiation operator

The first five operators depend on the interval length h and have a discrete character, while the last one is associated with a limit operation. An important task will be to express various operators in terms of other ones, particularly the central-difference operator. However, we will first discuss a few explicit formulas of great importance.

To begin with, we observe that the operator E^{-1} should be interpreted by $E^{-1}f(x) = f(x - h)$, and hence we have $EE^{-1} = E^{-1}E = 1 =$ the unit operator. Further we notice that when some of the operators act upon a polynomial, the result will be a polynomial of lower degree. This is the case for the difference and differentiation operators. We express this by saying that they have the property of *degrading*, and in such cases the inverse operation must create an indeterminacy. This is exemplified by the fact that the equations $Df(x) = 0$ and $\Delta f(x) = 0$ possess nontrivial solutions ($\neq 0$).

In order to find a relation between the operators E and D, we consider Taylor's formula:

$$f(x + h) = f(x) + hf'(x) + \frac{h^2}{2!} f''(x) + \cdots$$

In operator form we can write this:

$$Ef(x) = \left(1 + hD + \frac{h^2 D^2}{2!} + \cdots\right) f(x)$$
$$= e^{hD} f(x).$$

This formula can be used only for analytical functions. Note that e^{hD} is defined by its power series expansion, a technique which we will use frequently in the following sections. We can now write Taylor's formula in the compact form $E = e^{hD} = e^U$, where we have used the notation $U = hD$. Since $\delta = E^{1/2} - E^{-1/2}$, we get the important formula

$$\delta = 2 \sinh(U/2).$$

Below we present a table for converting the operators into each other.

	E	Δ	∇	δ	U
E	E	$\Delta + 1$	$(1 - \nabla)^{-1}$	$1 + \frac{1}{2}\delta^2 + \delta\sqrt{(1 + \delta^2/4)}$	e^U
Δ	$E - 1$	Δ	$(1 - \nabla)^{-1} - 1$	$\frac{1}{2}\delta^2 + \delta\sqrt{(1 + \delta^2/4)}$	$e^U - 1$
∇	$1 - E^{-1}$	$1 - (1 + \Delta)^{-1}$	∇	$-\frac{1}{2}\delta^2 + \delta\sqrt{(1 + \delta^2/4)}$	$1 - e^{-U}$
δ	$E^{1/2} - E^{-1/2}$	$\Delta(1 + \Delta)^{-1/2}$	$\nabla(1 - \nabla)^{-1/2}$	δ	$2\sinh(U/2)$
μ	$\frac{1}{2}(E^{1/2} + E^{-1/2})$	$(1 + \Delta/2)(1 + \Delta)^{-1/2}$	$(1 - \nabla/2)(1 - \nabla)^{-1/2}$	$\sqrt{(1 + \delta^2/4)}$	$\cosh(U/2)$
U	$\ln E$	$\ln(1 + \Delta)$	$\ln(1 - \nabla)^{-1}$	$2\sinh^{-1}(\delta/2)$	U

§5.3.   REPRESENTATION WITH DIFFERENCES

In numerical problems we often have to compute quantities involving derivatives or integrals, although in general we know the value of the function only at certain equidistant points. What can be computed directly is a difference table, and the problem then becomes one of constructing formulas that express differentiation and integration operators (as well as functions of these operators) in terms of differences. Suppose as before $U = hD$. Then we want to express $F(U)$ in terms of δ, using $\delta = 2\sinh(U/2)$. Putting $\delta := x$, $F := y$, and $U := z$, we get:

$$y = y(z); \qquad x = 2\sinh(z/2).$$

Differentiating we obtain:

$$dx = \cosh(z/2)\, dz = (1 + x^2/4)^{1/2}\, dz$$

or

$$dz/dx = (1 + x^2/4)^{-1/2}.$$

The general idea is now to differentiate y with respect to x repeatedly and then to compute the coefficients of the Maclaurin series. We treat only a few cases explicitly.

Example 1

$$\begin{cases} y = \dfrac{z}{\cosh(z/2)}; \\[2mm] \dfrac{dy}{dx} = \dfrac{\cosh(z/2) - (z/2)\sinh(z/2)}{\cosh^2(z/2)}\,\dfrac{dz}{dx} = \dfrac{1 - xy/4}{1 + x^2/4}. \end{cases}$$

Hence $(x^2 + 4)y' + xy = 4$. For $x = 0$ we get $y = 0$, and from the equation we get $y' = 1$.

Next we differentiate m times and obtain

$$\begin{cases} (x^2 + 4)y'' & + & 3xy' & + & y & = 0, \\ (x^2 + 4)y''' & + & 5xy'' & + & 4y' & = 0, \\ (x^2 + 4)y^{(iv)} & + & 7xy''' & + & 9y'' & = 0, \\ & & & & & \vdots \\ (x^2 + 4)y^{(m+1)} & + & (2m + 1)xy^{(m)} & + & m^2 y^{(m-1)} & = 0. \end{cases}$$

Putting $x = 0$ and $m = 2n$, we obtain $y^{(2n+1)}(0) = -n^2 y^{(2n-1)}(0)$, and hence $y^{(2n+1)}(0) = (-1)^n (n!)^2$. Thus we get

(5.3.1)
$$y = \sum_{n=0}^{\infty} (-1)^n \frac{(n!)^2}{(2n+1)!} x^{2n+1}$$

$$= x - \frac{x^3}{6} + \frac{x^5}{30} - \frac{x^7}{140} + \frac{x^9}{630} - \frac{x^{11}}{2772} + \frac{x^{13}}{12012} - \cdots.$$

Example 2

$$\begin{cases} y = z^2; \\[2mm] \dfrac{dy}{dx} = 2z\,\dfrac{dz}{dx} = 2\,\dfrac{z}{\cosh(z/2)} = 2\sum_{n=0}^{\infty} (-1)^n \dfrac{(n!)^2}{(2n+1)!} x^{2n+1}, \end{cases}$$

and after integration,

(5.3.2)
$$y = 2\sum_{n=1}^{\infty} (-1)^{n-1} \frac{[(n-1)!]^2}{(2n)!} x^{2n}$$

$$= x^2 - \frac{x^4}{12} + \frac{x^6}{90} - \frac{x^8}{560} + \frac{x^{10}}{3150} - \frac{x^{12}}{16632} + \cdots.$$

Example 3

$$\begin{cases} y = \cosh pz; \\[2mm] y' = \dfrac{dy}{dx} = \dfrac{p \sinh pz}{\cosh(z/2)}; \\[3mm] y'' = \dfrac{p^2 \cosh pz}{\cosh^2(z/2)} - \dfrac{p \sinh pz}{\cosh(z/2)}\dfrac{(1/2)\sinh(z/2)}{\cosh^2(z/2)} = \dfrac{p^2 y - xy'/4}{1 + x^2/4}. \end{cases}$$

Hence $(x^2 + 4)y'' + xy' - 4p^2y = 0$ with the initial conditions $x = 0$, $y = 1$, and $y' = 0$. Using the same technique as above, we obtain without difficulty

$$y^{(2n+2)}(0) = (p^2 - n^2)y^{(2n)}(0),$$

and hence

$$y = 1 + \frac{p^2x^2}{2!} + \frac{p^2(p^2 - 1)x^4}{4!} + \frac{p^2(p^2 - 1)(p^2 - 4)x^6}{6!} + \cdots .$$

In the preceding formula, we have computed $y = \cosh pz$. Differentiating with respect to x, we find

$$p \sinh pz \frac{1}{\cosh(z/2)} = p^2x + \frac{p^2(p^2 - 1)x^3}{3!} + \frac{p^2(p^2 - 1)(p^2 - 4)x^5}{5!} + \cdots ,$$

that is,

$$\frac{\sinh pz}{\sinh z} = p \left[1 + \frac{(p^2 - 1)x^2}{3!} + \frac{(p^2 - 1)(p^2 - 4)x^4}{5!} + \cdots \right]$$

which is more conveniently written

(5.3.3) $$\frac{\sinh pz}{\sinh z} = p + \binom{p + 1}{3}x^2 + \binom{p + 2}{5}x^4 + \cdots$$

The following list contains additional expansions, which are obtained with essentially the same methods as above.

$$U = \delta - \frac{\delta^3}{24} + \frac{3\delta^5}{640} - \frac{5\delta^7}{7168} + \frac{35\delta^9}{294912} - \frac{63\delta^{11}}{2883584} + \cdots ,$$

$$U = \mu\delta \left(1 - \frac{\delta^2}{6} + \frac{\delta^4}{30} - \frac{\delta^6}{140} + \frac{\delta^8}{630} - \frac{\delta^{10}}{2772} + \frac{\delta^{12}}{12012} - \cdots \right),$$

$$U^2 = \delta^2 - \frac{\delta^4}{12} + \frac{\delta^6}{90} - \frac{\delta^8}{560} + \frac{\delta^{10}}{3150} - \frac{\delta^{12}}{16632} + \cdots ,$$

$$\cosh pU = 1 + \frac{p^2\delta^2}{2!} + \frac{p^2(p^2 - 1)\delta^4}{4!} + \frac{p^2(p^2 - 1)(p^2 - 4)\delta^6}{6!} + \cdots ,$$

$$\frac{\sin pU}{\sinh U} = p + \binom{p + 1}{3}\delta^2 + \binom{p + 2}{5}\delta^4 + \binom{p + 3}{7}\delta^6 + \cdots ,$$

(5.3.4) $$\frac{\cosh pU}{\cosh(U/2)} = 1 + \binom{p + \frac{1}{2}}{2}\delta^2 + \binom{p + \frac{3}{2}}{4}\delta^4 + \binom{p + \frac{5}{2}}{6}\delta^6 + \cdots ,$$

$$\mu^{-1} = 1 - \frac{\delta^2}{8} + \frac{3\delta^4}{128} - \frac{5\delta^6}{1024} + \frac{35\delta^8}{32768} - \frac{63\delta^{10}}{262144} + \frac{231\delta^{12}}{4194304} - \cdots ,$$

$$\frac{\mu\delta}{U} = 1 + \frac{\delta^2}{6} - \frac{\delta^4}{180} + \frac{\delta^6}{1512} - \frac{23\delta^8}{226800} + \frac{263\delta^{10}}{14968800} - \cdots ,$$

$$\frac{\delta}{\mu U} = 1 - \frac{\delta^2}{12} + \frac{11\delta^4}{720} - \frac{191\delta^6}{60480} + \frac{2497\delta^8}{3628800} - \frac{14797\delta^{10}}{95800320} + \cdots ,$$

$$\frac{\delta^2}{U^2} = 1 + \frac{\delta^2}{12} - \frac{\delta^4}{240} + \frac{31\delta^6}{60480} - \frac{289\delta^8}{3628800} + \frac{317\delta^{10}}{22809600} - \cdots .$$

§5.4.   FACTORIALS AND STIRLING NUMBERS

The factorial polynomial of degree n, where n is a positive integer, is defined through

$$(5.4.1) \qquad p^{(n)} = p(p-1)(p-2)\cdots(p-n+1) = p!/(p-n)!$$

In the last expression we must assume that the factorial function $x!$ has been defined also for noninteger values of x. This is done by the formula $x! = \Gamma(x+1)$; for definition of the gamma function $\Gamma(x)$, see Section 7.2.

Among the important properties of such expressions, we note the following:

$$
\begin{aligned}
(p+1)^{(n)} - p^{(n)} &= (p+1)p(p-1)\cdots(p-n+2) - p(p-1)\cdots(p-n+1) \\
&= p(p-1)(p-2)\cdots(p-n+2)[p+1-(p-n+1)] \\
&= np(p-1)\cdots(p-n+2)
\end{aligned}
$$

or

$$(5.4.2) \qquad \Delta p^{(n)} = np^{(n-1)}.$$

Here the Δ-symbol operates on the variable p with $h = 1$.

We generalize directly to

$$\Delta^2 p^{(n)} = n(n-1)p^{(n-2)} = n^{(2)}p^{(n-2)}$$

and

$$(5.4.3) \qquad \Delta^k p^{(n)} = n(n-1)(n-2)\cdots(n-k+1)p^{(n-k)} = n^{(k)}p^{(n-k)}.$$

If $k = n$, we get $\Delta^n p^{(n)} = n!$

Until now we have defined factorials only for positive values of n. When $n > 1$, we have $p^{(n)} = (p-n+1)p^{(n-1)}$; and requiring this formula to hold also for $n = 1$ and $n = 0$, we get $p^{(0)} = 1$; $p^{(-1)} = 1/(p+1)$. Using the formula repeatedly for $n = -1, -2, \ldots$, we obtain:

$$(5.4.4) \qquad p^{(-n)} = \frac{1}{(p+1)(p+2)\cdots(p+n)} = \frac{1}{(p+n)^{(n)}}.$$

With this definition, the formula $p^{(n)} = (p-n+1)p^{(n-1)}$, as well as (5.4.2), holds also for negative values of n.

We shall now derive formulas for expressing a factorial as a sum of powers, and a power as a sum of factorials. Putting

$$(5.4.5) \qquad z^{(n)} = \sum_{k=1}^{n} \alpha_k^{(n)} z^k,$$

and using the identity $z^{(n+1)} = (z-n)z^{(n)}$, we obtain, on comparing the coefficients of z^k,

$$(5.4.6) \qquad \alpha_k^{(n+1)} = \alpha_{k-1}^{(n)} - n\alpha_k^{(n)},$$

with $\alpha_1^{(1)} = 1$ and $\alpha_0^{(n)} = 0$. These numbers are usually called *Stirling's numbers of the first kind*; they are presented in the table on next page.

k / n	1	2	3	4	5	6	7	8	9	10
1	1									
2	−1	1								
3	2	−3	1							
4	−6	11	−6	1						
5	24	−50	35	−10	1					
6	−120	274	−225	85	−15	1				
7	720	−1764	1624	−735	175	−21	1			
8	−5040	13068	−13132	6769	−1960	322	−28	1		
9	40320	−109584	118124	−67284	22449	−4536	546	−36	1	
10	−362880	1026576	−1172700	723680	−269325	63273	−9450	870	−45	1

For example, $z^{(5)} = 24z - 50z^2 + 35z^3 - 10z^4 + z^5$.

To obtain a power in terms of factorials, we observe that

$$z \cdot z^{(k)} = (z - k + k)z^{(k)}$$
$$= z^{(k+1)} + kz^{(k)}.$$

Putting

(5.4.7)
$$z^n = \sum_{k=1}^{n} \beta_k^{(n)} z^{(k)},$$

we obtain

$$z^n = z \cdot z^{n-1} = z \sum_{k=1}^{n-1} \beta_k^{(n-1)} z^{(k)}$$

$$= \sum_{k=1}^{n-1} \beta_k^{(n-1)} (z^{(k+1)} + kz^{(k)}).$$

Identifying,

(5.4.8)
$$\begin{cases} \beta_k^{(n)} = \beta_{k-1}^{(n-1)} + k\beta_k^{(n-1)}, & k = 1, 2, \ldots, n - 1 \\ \beta_n^{(n)} = \beta_{n-1}^{(n-1)} = 1. \end{cases}$$

The numbers β are called *Stirling's numbers of the second kind*; they are displayed in the table below:

k / n	1	2	3	4	5	6	7	8	9	10
1	1									
2	1	1								
3	1	3	1							
4	1	7	6	1						
5	1	15	25	10	1					
6	1	31	90	65	15	1				
7	1	63	301	350	140	21	1			
8	1	127	966	1701	1050	266	28	1		
9	1	255	3025	7770	6951	2646	462	36	1	
10	1	511	9330	34105	42525	22827	5880	750	45	1

For example, $z^5 = z^{(1)} + 15z^{(2)} + 25z^{(3)} + 10z^{(4)} + z^{(5)}$.

Both these tables, counted with n rows and n columns, form triangular matrices which we denote by A and B, respectively. Further, let u and v be the following column vectors:

$$u = \begin{pmatrix} z \\ z^2 \\ \vdots \\ z^n \end{pmatrix} \quad \text{and} \quad v = \begin{pmatrix} z^{(1)} \\ z^{(2)} \\ \vdots \\ z^{(n)} \end{pmatrix}.$$

Then we have

(5.4.9) $$v = Au; \quad u = Bv; \quad AB = BA = I.$$

The Stirling numbers $\alpha_k^{(n)}$ and $\beta_k^{(n)}$ are special cases of the so-called Bernoulli numbers $B_k^{(n)}$ of order n (see Section 3.5):

(5.4.10)
$$\begin{cases} \alpha_k^{(n)} = \binom{n-1}{k-1} B_{n-k}^{(n)}, \\ \beta_k^{(n)} = \binom{n}{k} B_{n-k}^{(-n)}. \end{cases}$$

We have seen that the factorials have very attractive properties with respect to the operator Δ. This ensures that sums of factorials are easily computed. In order to demonstrate this, we consider

$$\sum_{i=m}^{n} f(x_0 + ih) = \sum_{i=m}^{n} f_i,$$

where f is a function such that $\Delta F = f$, and obtain

$$\sum_{i=m}^{n} f_i = (F_{m+1} - F_m) + (F_{m+2} - F_{m+1}) + \cdots + (F_{n+1} - F_n)$$
$$= F_{n+1} - F_m.$$

Since $\Delta(p^{(n+1)}/(n+1)) = p^{(n)}$, we get

(5.4.11) $$\sum_{p=P}^{Q} p^{(n)} = \frac{(Q+1)^{(n+1)} - P^{(n+1)}}{n+1}.$$

Here P and Q need not be integers; on the other hand, the interval is supposed to be 1.

If, instead, the factorial is defined by

$$p^{(n)} = p(p-h) \cdots (p - nh + h),$$

the formula is slightly modified:

(5.4.12) $$\sum_{p=P}^{Q} p^{(n)} = \frac{(Q+h)^{(n+1)} - P^{(n+1)}}{(n+1)h}.$$

EXERCISES

1. Show that $\Delta - \nabla = \Delta\nabla$ and that $\Delta + \nabla = \Delta/\nabla - \nabla/\Delta$.

2. Prove the relations.

   (a) $\mu(f_k g_k) = \mu f_k \mu g_k + \frac{1}{4}\delta f_k\,\delta g_k$.

   (b) $\mu\left(\dfrac{f_k}{g_k}\right) = \dfrac{\mu f_k \mu g_k - \frac{1}{4}\delta f_k\,\delta g_k}{g_{k-1/2}g_{k+1/2}}$.

   (c) $\delta(f_k g_k) = \mu f_k\,\delta g_k + \mu g_k\,\delta f_k$.

   (d) $\delta\left(\dfrac{f_k}{g_k}\right) = \dfrac{\mu g_k\,\delta f_k - \mu f_k\,\delta g_k}{g_{k-1/2}g_{k+1/2}}$.

3. Prove that:

   (a) $\displaystyle\sum_{k=0}^{n-1}\Delta^2 f_k = \Delta f_n - \Delta f_0$.

   (b) $\displaystyle\sum_{k=0}^{n-1}\delta^2 f_{2k+1} = \tanh(U/2)(f_{2n} - f_0)$.

4. Prove that:

   (a) $\Delta\sqrt{f_k} = \dfrac{\Delta f_k}{\sqrt{f_k} + \sqrt{f_{k+1}}}$.

   (b) $p^{(2n)} = 2^{2n}\left(\dfrac{p}{2}\right)^{(n)}\left(\dfrac{p-1}{2}\right)^{(n)}$ (n integer, >0).

5. Find $\delta^2(x^2)$, $\delta^2(x^3)$, and $\delta^2(x^4)$ when $h = 1$. Use the result for determining a particular solution of the equation $\mu f(x) = 2x^4 + 4x^3 + 3x^2 + 3x + \frac{1}{8}$.

6. The sequence y_n is formed according to the rule

$$y_n = (-1)^{n-1}\left\{\binom{1/2}{n+1} + \binom{-1/2}{n+1}\right\}.$$

   Prove that $\Delta^n y_n = 0$.

7. The expression δy_0 cannot usually be computed directly from a difference scheme. Find its value expressed in known central differences.

8. (a) Show that

$$\Delta\binom{n}{i+1} = \binom{n}{i},$$

   where Δ operates on n, and hence that

$$\sum_{n=1}^{N}\binom{n}{i} = \binom{N+1}{i+1} - \binom{1}{i+1}.$$

   (b) Find the coefficients c_i in the expansion

$$n^5 = \sum_{i=0}^{5} c_i\binom{n}{i}$$

   and compute the smallest value N such that $\sum_{n=1}^{N} n^5 > 10^{10}$.

9. Express

$$J_0 y_0 = \int_{x_0 - h/2}^{x_0 + h/2} y(t)\,dt$$

   in central differences of y.

10. We define

$$\delta^2 f = f(x - h) - 2f(x) + f(x + h)$$

and

$$\delta'^2 f = f(x - nh) - 2f(x) + f(x + nh).$$

Obtain δ'^2 as a power series in δ^2 (the three first terms).

CHAPTER 6

Difference Equations

*My goodness, I'm feeling fine said Granny, waving
with her hand. Next birthday I'll be ninety-nine with
just one year left to thousand.*

Elias Sehlstedt*

§6.1. GENERAL CONCEPTS

We will now work with a function $y = y(x)$, which can be defined for all real
values of x, or only at certain points. In the latter case we will assume that the
points are equidistant (grid points), and further we suppose that the variable x has
been transformed in such a way that the interval length is 1, and the grid points
correspond to integer values of x. Then we shall also use the notation y_n for $y(n)$.

We define an ordinary *difference equation* as an equation which contains an independent variable x, a dependent variable y, and one or several differences Δy,
$\Delta^2 y, \ldots, \Delta^n y$. Of course, it is not an essential specialization to assume *forward*
differences.

If we use the formula $\Delta = E - 1$, we infer that the difference equation can be
trivially transformed to a relation of the type

(6.1.1) $$F(x, y(x), y(x + 1), \ldots, y(x + n)) = 0,$$

where n is called the order of the equation. As a matter of fact, an equation of this
kind should be called a *recurrence equation*, but we shall continue with the name
difference equation.

The treatment of general difference equations is extremely complicated, and here
we can discuss only some simple special cases. First of all we assume that the
equation is *linear*, that is, that F has the form

(6.1.2) $$p_0(x)y(x + n) + p_1(x)y(x + n - 1) + \cdots + p_n(x)y(x) = g(x).$$

If $g(x) = 0$, the equation is said to be homogeneous, and further we can specialize
to the case when all $p_i(x)$ are constant. We will restrict ourselves mainly to this

---

* This is a free, though not very poetic, translation of a humorous Swedish verse.

90

latter case. After a suitable translation, which moves the origin to the point x, such an equation can always be written

$$(6.1.3) \qquad y_n + c_1 y_{n-1} + \cdots + c_n y_0 = (E^n + c_1 E^{n-1} + \cdots + c_n) y_0$$
$$\equiv \varphi(E) y_0 = 0.$$

At first one might expect that several details from the theory of differential equations could be transferred directly to the theory of difference equations. However, in the latter case we meet difficulties of quite unexpected character and without correspondence in the former case. A fundamental question is, what is meant by a "solution of a difference equation"? In order to indicate that the question is legitimate, we consider the following example.

The equation $y_{n+1} = 3y_n + 2n - 1$ is a first-order inhomogeneous difference equation. Assuming that $y_0 = 1$, we see that y_n is uniquely determined for all *integer* values n. We find

n	0	1	2	3	4	5	6	$\cdots$	-1	-2	-3	$\cdots$
y_n	1	2	7	24	77	238	723	$\cdots$	$\frac{4}{3}$	$\frac{19}{9}$	$\frac{82}{27}$	$\cdots$

Since the solution appears only in certain points, it is called *discrete*; further, it is a *particular solution*, since it depends on a special value of y_0. Now we easily find that $y(n) = c3^n - n$, where c is a constant, satisfies the equation. The discrete solution, just discussed, is obtained if we take $c = 1$. The solution $y = 3^x - x$ is defined for all values of x and is called a *continuous particular solution*.

One might believe $y = c3^x - x$ to be the general solution, but it is easy to show that this is not the case. Consider, for example,

$$y(x) = 3^x(c + \tfrac{1}{3}\cos 2\pi x + \tfrac{2}{5}\cos 6\pi x - \tfrac{1}{4}\sin^2 3\pi x) - x,$$

and we see at once that the difference equation is satisfied again. In order to clarify this point, we put $y(x) = w(x)3^x - x$ and find $w(x + 1) = w(x)$. This is a new difference equation which, however, can be interpreted directly by saying that w must be a *periodic function* with period 1. We can choose $w = w(x)$ completely arbitrarily for $0 \le x < 1$, but the corresponding solution y does not, in general, obtain any of the properties of continuity, differentiability, and so on. If we claim such properties, suitable conditions must be imposed on $w(x)$. For continuity, $w(x)$ must be continuous, and further, $\lim_{x \to 1-0} w(x) = w(0)$.

In the remainder of this chapter, $w(x), w_1(x), w_2(x), \ldots$ will stand for periodic functions of x with period 1. Hence the general solution of the equation which we have discussed can be written

$$y(x) = w(x) \cdot 3^x - x.$$

On the other hand, consider a relation containing the independent variable x, the dependent variable y, and a periodic function $w(x)$:

$$F(x, y(x), w(x)) = 0.$$

Operating with E, we get

$$F(x + 1, y(x + 1), w(x + 1)) = F_1(x, y(x + 1), w(x)) = 0.$$

Eliminating $w(x)$ from these two equations, we obtain

$$G(x, y(x), y(x + 1)) = 0,$$

which is a first-order difference equation.

§6.2. LINEAR, HOMOGENEOUS DIFFERENCE EQUATIONS

We shall now discuss equation (6.1.3) in somewhat greater detail. We assume $c_n \neq 0$; otherwise, the equation would be of lower order. First we consider only the discrete particular solutions appearing when the function values $y_0, y_1, \ldots, y_{n-1}$ are given. Thus y_n can be determined from the difference equation, and then the whole procedure can be repeated with n replaced by $n + 1$. Clearly, all discrete solutions are uniquely determined from the n given parameter values.

On the other hand, putting $y = \lambda^x$, we get $E^r y = \lambda^r \cdot \lambda^x$ and

$$(\lambda^n + c_1 \lambda^{n-1} + \cdots + c_n)\lambda^x \equiv \varphi(\lambda) \cdot \lambda^x = 0.$$

Hence $y = \lambda^x$ is a solution of the difference equation if λ satisfies the algebraic equation

$$(6.2.1) \qquad \lambda^n + c_1 \lambda^{n-1} + \cdots + c_n = \varphi(\lambda) = 0.$$

This equation is called the *characteristic equation* of the difference equation.

Since our equation is linear and homogeneous, we infer that, provided u and v are solutions and a and b constants, then $au + bv$ is also a solution. Hence

$$y = A_1 \lambda_1^x + A_2 \lambda_2^x + \cdots + A_n \lambda_n^x$$

is a solution of (6.1.3) if $\lambda_1, \lambda_2, \ldots, \lambda_n$ (all assumed to be different from one another), are the roots of (6.2.1). The n parameters $A_1, A_2, \ldots, A_n$ can be determined in such a way that for $x = 0, 1, 2, \ldots, n - 1$, the variable y takes the assigned values $y_0, y_1, y_2, \ldots, y_{n-1}$. Thus we have constructed a *continuous* particular solution which coincides with the discrete solution at the grid points. As in a previous special example, we conclude that the general solution can be written

$$(6.2.2) \qquad y_0 = w_1(x) \cdot \lambda_1^x + w_2(x) \cdot \lambda_2^x + \cdots + w_n(x) \cdot \lambda_n^x.$$

If the characteristic equation has complex roots, this does not give rise to any special difficulties. For example, the equation $y(x + 2) + y(x) = 0$ has the solution

$$y = a \cdot i^x + b(-i)^x = a \cdot e^{i\pi x/2} + b \cdot e^{-i\pi x/2} = A \cos \frac{\pi x}{2} + B \sin \frac{\pi x}{2}.$$

So far, we have considered only the case when the characteristic equation has simple roots. For multiple roots we can use a limit procedure, analogous to the well-known process for differential equations. Thus we suppose that r and $r + \varepsilon$ are roots of the characteristic equation, that is, that r^x and $(r + \varepsilon)^x$ are solutions of the difference equation. Due to the linearity,

$$\lim_{\varepsilon \to 0} \frac{(r + \varepsilon)^x - r^x}{\varepsilon} = xr^{x-1}$$

is also a solution. Since multiplication with r is trivially allowed, we conclude that we have the two solutions r^x and xr^x. If the characteristic equation has a triple root r, then also $x^2 r^x$ is a solution. If r is a root of multiplicity k, then $P(x)r^x$ is a solution, where $P(x)$ is an arbitrary polynomial of degree $\leq k - 1$.

Example 1. The Fibonacci numbers are defined by $y(0) = 0$; $y(1) = 1$; $y(x + 2) = y(x + 1) + y(x)$ (hence 0, 1, 1, 2, 3, 5, 8, 13, 21, 34, ...). The characteristic equation is $\lambda^2 - \lambda - 1 = 0$ with the roots $\frac{1}{2}(1 \pm \sqrt{5})$. The "general" solution is

$$y(x) = a\left(\frac{1 + \sqrt{5}}{2}\right)^x + b\left(\frac{1 - \sqrt{5}}{2}\right)^x.$$

The initial conditions give

$$\begin{cases} a + b = 0 \\ (a - b)\sqrt{5} = 2 \end{cases} \quad \text{and} \quad a = -b = \frac{1}{\sqrt{5}}.$$

Hence

$$y(x) = \frac{1}{2^{x-1}}\left[\binom{x}{1} + \binom{x}{3}5 + \binom{x}{5}5^2 + \binom{x}{7}5^3 + \cdots + \binom{x}{2m + 1}5^m + \cdots\right].$$

Example 2. $y(x) = py(x + 1) + qy(x - 1)$ with $p + q = 1$, and $y(0) = 1$, $y_N = 0$. The characteristic equation is $\lambda = p\lambda^2 + q$, with the roots 1 and q/p. If $p \neq q$, we find the solution

$$y(x) = \frac{p^{N-x}q^x - q^N}{p^N - q^N};$$

if $p = q = \frac{1}{2}$, we get $y = 1 - x/N$.

We shall also indicate briefly how inhomogeneous linear difference equations can be handled. The technique is best demonstrated in a few examples.

Example 3. $\Delta y = (4x - 2) \cdot 3^x$.
Since $\Delta y = y(x + 1) - y(x)$, we have the homogeneous solution $y = w(x) \cdot 1^x$. The particular solution is obtained by putting $y = (ax + b) \cdot 3^x$. We get $\Delta y = (2ax + 3a + 2b) \cdot 3^x$, and identifying, we find the solution $y = (2x - 4) \cdot 3^x + w(x)$.

Example 4. $y(x + 2) - 3y(x + 1) + 2y(x) = 6 \cdot 2^x$.

The solution of the homogeneous equation is $y = w_1(x) \cdot 2^x + w_2(x)$. Putting $y = a \cdot x \cdot 2^x$, we easily find $a = 3$. Hence the general solution is

$$y = (w_1(x) + 3x) \cdot 2^x + w_2(x).$$

---

§6.3.   DIFFERENCES OF ELEMENTARY FUNCTIONS

The equation $\Delta f(x) = g(x)$, where $g(x)$ is a given function, was discussed in Section 5.4 in the special case when $g(x)$ is a factorial. A formal solution can be obtained in the following way. Setting $D^{-1}g(x) = \int_a^x g(t)\, dt = G(x)$, we find (with $h = 1$):

$$f(x) = \frac{g(x)}{e^D - 1} = \frac{D}{e^D - 1}\, G(x),$$

and hence, adding a periodic function $w(x)$, we get (see Section 3.5)

$$f(x) = w(x) + G(x) - \frac{1}{2}g(x) + \frac{1}{6}\frac{g'(x)}{2!} - \frac{1}{30}\frac{g'''(x)}{4!}$$

$$+ \frac{1}{42}\frac{g^{(v)}(x)}{6!} - \frac{1}{30}\frac{g^{(vii)}(x)}{8!} + \cdots.$$

The coefficient of $g^{(2p-1)}(x)$ for large p behaves like $(-1)^{p-1} \cdot 2/(2\pi)^{2p}$ (cf. (3.5.14)), and this fact gives some information concerning the convergence of the series. However, we shall now look at the problem from a point of view that corresponds to the well-known technique for performing integration by recognizing differentiation formulas. First, we indicate a few general rules for calculating differences, and then we shall give explicit formulas for the differences of some elementary functions.

$$(6.3.1) \quad \begin{cases} \Delta(u_i \pm v_i) = \Delta u_i \pm \Delta v_i, \\ \quad \Delta(u_i v_i) = u_{i+1}\,\Delta v_i + v_i\,\Delta u_i = u_i\,\Delta v_i + v_{i+1}\,\Delta u_i, \\ \Delta\left(\dfrac{u_i}{v_i}\right) = \dfrac{v_i\,\Delta u_i - u_i\,\Delta v_i}{v_i v_{i+1}}. \end{cases}$$

$$(6.3.2) \quad \begin{cases} \Delta x^{(n)} = n x^{(n-1)}, \\ \Delta a^x = (a - 1)a^x, \\ \Delta \ln x = \ln(1 + 1/x), \\ \Delta \sin px = 2\sin(p/2)\cos(px + p/2), \\ \Delta \cos px = -2\sin(p/2)\sin(px + p/2), \\ \Delta(u_m + u_{m+1} + \cdots + u_n) = u_{n+1} - u_m, \\ \Delta(u_m u_{m+1} \cdots u_n) = u_{m+1} u_{m+2} \cdots u_n(u_{n+1} - u_m). \end{cases}$$

Returning to the equation $\Delta f(x) = g(x)$, we will now show how a solution can be obtained when $g(x)$ is a rational function. We restrict ourselves to the case when the linear factors of the denominator are different; then $g(x)$ can be written as a sum of terms of the form $a/(x - \alpha)$. Hence the problem has been reduced to solving the equation

$$\Delta f(x) = \frac{a}{x - \alpha}.$$

Now there exists a function $\Gamma(x)$ which is continuous for $x > 0$, and takes the value $(x - 1)!$ for integer, positive x (see further 7.2.10–11). The following fundamental functional relation is fulfilled:

$$\Gamma(x + 1) = x\Gamma(x).$$

Differentiating, we get $\Gamma'(x + 1) = x\Gamma'(x) + \Gamma(x)$ and hence

$$\frac{\Gamma'(x + 1)}{\Gamma(x + 1)} = \frac{\Gamma'(x)}{\Gamma(x)} + \frac{1}{x}.$$

The function $\Gamma'(x)/\Gamma(x)$ is called the digamma function and is denoted by $\psi(x)$. Thus we have the following difference formulas:

(6.3.3)
$$\begin{cases} \Delta\Gamma(x) = (x - 1)\Gamma(x), \\ \Delta\psi(x) = \dfrac{1}{x}. \end{cases}$$

Differentiating the second formula, we get

(6.3.4)
$$\begin{cases} \Delta\psi'(x) = -\dfrac{1}{x^2}, \\[2mm] \Delta\psi''(x) = \dfrac{2}{x^3}, \\[2mm] \Delta\psi'''(x) = -\dfrac{6}{x^4}, \\[1mm] \quad\vdots \end{cases}$$

Example 1. Compute $S = \sum_{k=0}^{\infty} 1/((k + \alpha)(k + \beta))$, where $0 \le \alpha \le \beta$. Without difficulty, we find:

$$S = \frac{1}{\beta - \alpha} \lim_{N \to \infty} \sum_{k=0}^{N-1} \left(\frac{1}{k + \alpha} - \frac{1}{k + \beta} \right) = \frac{1}{\beta - \alpha} \lim_{N \to \infty} \sum_{k=0}^{N-1} [\Delta\psi(k + \alpha) - \Delta\psi(k + \beta)]$$

$$= \frac{1}{\beta - \alpha} \lim_{N \to \infty} \{\psi(N + \alpha) - \psi(N + \beta) - \psi(\alpha) + \psi(\beta)\} = \frac{\psi(\beta) - \psi(\alpha)}{\beta - \alpha}.$$

As a matter of fact, for large x we have $\psi(x) \sim \ln x - 1/2x - 1/12x^2 + \cdots$, and from this we get $\lim_{N \to \infty} \{\psi(N + \alpha) - \psi(N + \beta)\} = 0$. If $\alpha = \beta$, we find simply that $\sum_{k=1}^{\infty} 1/(k + \alpha)^2 = \psi'(\alpha)$ and, in particular, $\sum_{k=1}^{\infty} 1/k^2 = \psi'(1) = \pi^2/6$. The Γ function, as well as the functions ψ, ψ', ψ'', ..., has been tabulated extensively.

In some cases a solution $f(x)$ of $\Delta f(x) = g(x)$ can be obtained by trying a series expansion in factorials, possibly multiplied by some other suitable functions. Although power-series expansions are most convenient for differential equations, we find expansions in factorials more suitable for difference equations.

---

§6.4. PARTIAL DIFFERENCE EQUATIONS

A partial difference equation contains several independent variables, and is usually written as a recurrence equation. We shall here restrict ourselves to two independent variables, which will be denoted by x and y; the dependent variable is denoted by u.

We shall discuss only linear equations with constant coefficients, and first we shall demonstrate a direct technique. Consider the equation

$$u(x + 1, y) - u(x, y + 1) = 0.$$

Introducing shift operators in both directions, we get

$$\begin{cases} E_x u(x, y) = u(x + 1, y), \\ E_y u(x, y) = u(x, y + 1), \end{cases}$$

and hence we can write

$$E_x u = E_y u.$$

Previously we have found that a formal solution of $E_x u = au$ is $u = ca^x$; here c should be a function of y only. Hence we get $u = (E_y)^x f(y) = f(x + y)$ [a was replaced by E_y and c by $f(y)$]. The general solution is $u = w(x, y)f(x + y)$, where $w(x, y)$ is periodic in x and y with period 1.

Another method is due to Laplace. Here we try to obtain a generating function whose coefficients should be the values of the desired function at the grid points on a straight line parallel to one of the axes. We demonstrate the technique by an example.

We take the equation $u(x, y) = pu(x - 1, y) + qu(x, y - 1)$ with the boundary conditions $u(x, 0) = 0$ for $x > 0$; $u(0, y) = q^y$ [hence $u(0, 0) = 1$]. Introducing the generating function $\varphi_y(\xi) = u(0, y) + u(1, y)\xi + u(2, y)\xi^2 + \cdots$, we find

$$q\varphi_{y-1}(\xi) = qu(0, y - 1) + \sum_{x=1}^{\infty} qu(x, y - 1)\xi^x.$$

Further,

$$p\xi\varphi_y(\xi) = \sum_{x=1}^{\infty} pu(x - 1, y)\xi^x.$$

Adding together, we get

$$q\varphi_{y-1}(\xi) + p\xi\varphi_y(\xi) = qu(0, y - 1) + \sum_{x=1}^{\infty} u(x, y)\xi^x$$

$$= qu(0, y - 1) - u(0, y) + \varphi_y(\xi).$$

Here we have used the initial equation; if we make use of the boundary values as well, we get

$$\varphi_y(\xi) = \left(\frac{q}{1 - p\xi}\right)^y.$$

This completes the computation of the generating function. The desired function $u(x, y)$ is now obtained as the coefficient of the ξ^x-term in the power-series expansion of $\varphi_y(\xi)$. We obtain directly:

$$\varphi_y(\xi) = q^y(1 - p\xi)^{-y} = q^y\left(1 + y \cdot p\xi + \frac{y(y + 1)}{1 \cdot 2}p^2\xi^2 + \cdots\right),$$

and hence

$$u(x, y) = \frac{y(y + 1) \cdots (y + x - 1)}{1 \cdot 2 \cdots x}p^x q^y = \binom{x + y - 1}{x}p^x q^y.$$

Still another elegant method is due to Lagrange. Here we shall only sketch the method very briefly, and for this purpose, we use the same example. We try to find a solution in the form $u = \alpha^x \beta^y$ and get the condition $\alpha\beta = p\beta + q\alpha$. Hence $u(x, y) = \alpha^x q^y(1 - p\alpha^{-1})^{-y}$ with arbitrary α is a solution. Obviously, we can multiply with an arbitrary function $\varphi(\alpha)$ and integrate between arbitrary limits, and again we obtain new solutions of the equation. Now we have $\varphi(\alpha)$, as well as the integration limits, at our disposal for satisfying the boundary condition. The computation is performed with complex integration in a way similar to that used for solving ordinary differential equations by means of Laplace transformations.

Ordinary difference equations are of great importance for the numerical solution of ordinary differential equations, and analogously, partial difference equations play an important role in connection with the numerical treatment of partial differential equations. These matters will be discussed more thoroughly in Chapters 19 and 20.

EXERCISES

1. The difference equation $x_{n+2} = 4x_{n+1} - x_n$ is given with $x_0 = 2$ and $x_1 = 4$. Show that $x_{2n} = x_n^2 - 2$.

2. Suppose that

$$x_n = \frac{1}{6}[(3 + \sqrt{3})(1 + \sqrt{3})^n + (3 - \sqrt{3})(1 - \sqrt{3})^n].$$

Form a difference equation for x_n and compute $x_0, x_1, \ldots, x_6$.

3. Solve the difference equation $y_{n+1} = py_n + qn + r$ with $y_0 = a$. The case when $p = 1$ should also be considered.

4. Solve the difference equation $y_{n+2} - 2y_{n+1} + y_n = n^2$ with $y_0 = 0$, $y_1 = 1$.

5. The integral $\int_0^1 x^n e^{-x}\, dx$, where n is an integer ≥ 0, can be written in the form $a_n - b_n \cdot e^{-1}$, where a_n and b_n are integers. Prove that for $n > 1$, a_n and b_n satisfy the difference equations $a_n = na_{n-1}$; $b_n = nb_{n-1} + 1$, with $a_0 = b_0 = 1$, and find $\lim_{n \to \infty} (b_n/a_n)$.

6. Let a_0, a_1, and a_2 be given real numbers. We form $a_n = \frac{1}{3}(a_{n-1} + a_{n-2} + a_{n-3})$ for $n = 3, 4, 5, \ldots$ and write $a_n = A_n a_0 + B_n a_1 + C_n a_2$. Further, we put $A = \lim_{n \to \infty} A_n$; $B = \lim_{n \to \infty} B_n$; $C = \lim_{n \to \infty} C_n$. Find A, B, and C.

7. Solve the difference equation $u_{n+1} - 2 \cos x \cdot u_n + u_{n-1} = 0$, when

$$\text{(a)} \quad \begin{cases} u_0 = 1, \\ u_1 = \cos x, \end{cases} \qquad \text{(b)} \quad \begin{cases} u_0 = 0, \\ u_1 = 1. \end{cases}$$

8. Let α, β, μ_0, and μ_1 be given numbers. For $k = 0, 1, 2, \ldots$, we have $u_{k+2} = \alpha u_{k+1} + \beta u_k$. Putting $U_k = u_k u_{k+2} - u_{k+1}^2$, show that $U_k U_{k+2} - U_{k+1}^2 = 0$.

9. Find all values of k for which the difference equation $y_n - 2y_{n+1} + y_{n+2} + (k^2/N^2)y_n = 0$ has nontrivial solutions such that $y_0 = y_N = 0$. Also give the form of the solutions.

10. A_n and B_n are defined through

$$\begin{cases} A_n = A_{n-1} + xB_{n-1}, \\ B_n = B_{n-1} - xA_{n-1}, \end{cases} \quad n = 1, 2, 3, \ldots, \quad A_0 = 0, \quad B_0 = 1.$$

Determine A_n/B_n as a function of n and x.

11. Solve the system of difference equations

$$\begin{cases} x_{n+1} = 7x_n + 10y_n, \\ y_{n+1} = x_n + 4y_n, \end{cases}$$

with $x_0 = 3$, $y_0 = 2$.

12. A Fibonacci series a_n, $n = 0, 1, 2, \ldots$, is given. Show that

$$a_{r+2n} + (-1)^n a_r = k_n a_{r+n},$$

where k_n are integers forming a new Fibonacci series, and state the first terms in this series.

CHAPTER 7

Special Functions

I am so glad I am a Beta,
*   the Alphas work so hard.*
And we are much better than the Gammas
*   and Deltas.*

Aldous Huxley

§7.1. ANALYTIC FUNCTIONS AND COMPLEX INTEGRATION

The theory of analytic functions was mainly developed during the 19th century, and it has turned out to be a most powerful tool for solving many types of numerical problems. Methods of this kind are particularly useful in evaluating certain integrals. A complete theory would take much space, so we limit ourselves to those concepts which are of importance for our treatment.

We start with a simple example. Consider the infinite series

$$f_1(z) = 1 + (z - 1) + (z - 1)^2 + (z - 1)^3 + \cdots$$

Obviously it converges for all z such that $|z - 1| < 1$, and it has the sum $1/(2 - z)$. This function in turn is defined for all z except $z = 2$. Now suppose that we want to restructure the series so that we get fast convergence close to $z = -3$. Writing $z = -3 + w$, we get

$$\frac{1}{2 - z} = \frac{1}{5 - w} = \frac{1}{5(1 - w/5)} = \frac{1}{5}\left(1 + \frac{w}{5} + \left(\frac{w}{5}\right)^2 + \cdots\right)$$
$$= \frac{1}{5}\left(1 + \frac{z + 3}{5} + \left(\frac{z + 3}{5}\right)^2 + \cdots\right).$$

Calling this new series $f_2(z)$, we find that it converges when $|z + 3| < 5$. In a similar way we get, for an arbitrary point $z = a$:

$$f_3(z) = \frac{1}{2 - a}\left(1 + \frac{z - a}{2 - a} + \left(\frac{z - a}{2 - a}\right)^2 + \cdots\right); \qquad |z - a| < |2 - a|.$$

We see that these three series inside their convergence domains represent the same function, namely $f(z) = 1/(2 - z)$. Note that all convergence circles pass through the point $z = 2$. Together all these power series define an *analytic function,* which in this case can be written in closed form as shown above. The function is unique; however, this is by no means always the case, as is demonstrated by the function $w = z^{1/2}$. In the following we will consider only the simple case when the functions treated are unique.

We now turn to integration of analytic functions. A series

$$c_0 + c_1(z - a) + c_2(z - a)^2 + \cdots$$

can be integrated termwise to give

$$c + c_0(z - a) + c_1(z - a)^2/2 + c_2(z - a)^3/3 + \cdots$$

where c is an arbitrary constant; obviously the new series has the same radius of convergence as the first series.

Now let $f(z)$ be an analytic function and suppose that we want to compute the integral

$$\int_a^b f(z)\, dz$$

with the integration performed along a given curve C that lies completely inside the regularity domain. Further assume that there is a function $g(z)$ such that $g'(z) = f(z)$, and choose a series of points $z_0 = a, z_1, z_2, \ldots, z_n = b$ on C with two consecutive points close together. Then

$$\frac{g(z_1) - g(z_0)}{z_1 - z_0} = f(\zeta_1) + \delta_1$$

$$\frac{g(z_2) - g(z_1)}{z_2 - z_1} = f(\zeta_2) + \delta_2$$

$$\vdots$$

where all δ_k are absolutely smaller than a given ε, and ζ_k lies between z_{k-1} and z_k. Hence with easily understood notations we get

$$g(b) - g(a) = \sum f(z)\, \Delta z + R$$

where

$$|R| = |\delta_1(z_1 - z_0) + \delta_2(z_2 - z_1) + \cdots + \delta_n(z_n - z_{n-1})|$$
$$< \varepsilon\{|z_1 - z_0| + \cdots + |z_n - z_{n-1}|\} < \varepsilon L$$

if L is the length of the curve. Thus, in the limit of finer and finer partitions we get

$$\lim \sum f(z)\, \Delta z = g(b) - g(a)$$

which can also be written

$$\int_a^b f(z)\, dz = g(b) - g(a)$$

independently of the integration path. Our only condition is that the function $f(z)$ must be regular in the domain under consideration. If, in particular, $a = b$ (i.e., the integration is performed along a closed curve C with only regular points inside and on the boundary), then

$$\int_C f(z)\, dz = 0.$$

We now turn to the case in which there are one or several singularities of the form $c/(z - a)$ inside the integration curve. Let us start with a simple example, namely $I = \int_C dz/z$ where C is a circle with radius r and the origin as center. Putting $z = re^{it}$, we find $dz = re^{it}i\, dt$ and $dz/z = i\, dt$. Hence

$$I = \int_0^{2\pi} i\, dt = 2\pi i.$$

It is easy to see that only first order terms in the denominator give rise to such contributions. In contrast, for instance, we have

$$\int_C dz/z^2 = ir^{-1} \int_0^{2\pi} e^{-it}\, dt = 0.$$

In general we get

$$\int_C dz/z^n = 0 \qquad \text{for } n = 2, 3, 4, \dots.$$

The value of $(1/2\pi i) \int_C f(z)\, dz$, where C is a closed contour around the point $z = a$, is called the *residue* of $f(z)$ in the point $z = a$. If there exists an expansion

$$f(z) = \sum_{n = -\infty}^{\infty} c_n(z - a)^n,$$

then the residue is equal to c_{-1}. If the highest order term with negative exponent is $c_{-N}(z - a)^{-N}$, the point $z = a$ is said to be a *pole* of order N.

More generally, if we consider the integral

$$I = (1/2\pi i) \int_C f(z)\, dz$$

where C contains simple poles (of order 1) in the points $a_1, a_2, \dots, a_m$ and the corresponding residues are $r_1, r_2, \dots, r_m$, we obtain

$$I = r_1 + r_2 + \cdots + r_m.$$

This formula was first obtained by Cauchy.

Example 1. Find the value of $I = \int_{-\infty}^{\infty} dx/(x^2 + 1)^n$.

We consider instead a complex integral of the same form, taken along a closed curve C consisting of the straight line between $-R$ and $+R$ and the half-circle with center at the origin and of radius R. It is easily seen that the contribution from the half-circle will disappear in the limit when $R \to \infty$. The only singular point appears when $z^2 + 1 = 0$, i.e., $z = i$ which is inside C. Putting $z - i = u$ we get:

$$(z^2 + 1)^{-n} = ((u + i)^2 + 1)^{-n} = u^{-n}(2i + u)^{-n} = (2ui)^{-n}(1 - ui/2)^{-n}$$

$$= (2ui)^{-n}\left[1 + \frac{niu}{2} + \frac{n(n + 1)}{2!}(iu/2)^2 + \frac{n(n + 1)(n + 2)}{3!}(iu/2)^3 + \cdots \right].$$

We are now looking for the coefficient of u^{-1} in this expansion, and we find directly:

$$(2i)^{-n}\frac{n(n + 1)(n + 2)\cdots(2n - 2)}{(n - 1)!}(i/2)^{n-1} = i^{-1}2^{-(2n-1)}\frac{(2n - 2)!}{((n - 1)!)^2}$$

Multiplying with $2\pi i$ we find:

$$I = \frac{\pi}{2^{2n-2}}\frac{(2n - 2)!}{(n - 1)!(n - 1)!}.$$

Example 2.   Determine the value of

$$I = \int_{-\infty}^{\infty}\frac{\cos px}{x^2 + a^2}\,dx.$$

Obviously we can write

$$I = \int_{-\infty}^{\infty}\frac{\exp(ipz)}{z^2 + a^2}\,dz$$

where we can assume $p > 0$. We add a large semicircle $z = R\exp(it)$, $0 \leq t \leq \pi$, noting that $\exp(ipz) = \exp(ipR\cos t)\exp(-pR\sin t)$ where the last factor approaches zero when $R \to \infty$. Close to the pole $z = ai$, which lies inside the contour, the integrand is approximately $\exp(-ap)/[2ai(z - ai)]$; hence the residue is $2\pi i\exp(-ap)/2ai$, giving

$$I = \frac{\pi e^{-ap}}{a}.$$

In the following we shall treat a few of the most common special functions. Primarily we shall choose functions that are of great importance from a theoretical point of view, or in physical and technical sciences. Among these functions there are two main groups, one associated with simple linear homogeneous differential equations, usually of second order, and one with more complicated functions. In the latter case a rather thorough knowledge of the theory of analytic functions is indispensable for a comprehensive and logical treatment. However, it does not seem to be reasonable to assume such a knowledge, and for this reason it has been necessary to keep the discussion at a more elementary level, omitting some proofs and other rather vital parts. This is especially the case with the gamma function, but also for other functions with certain formulas depending on, for example, complex integration and residue calculus. Since the gamma and beta functions are of such a great importance in many applications, a rather careful description of their properties is desirable even if a complete foundation cannot be given for the reasons mentioned.

§7.2. THE GAMMA FUNCTION

For a long time a function called the factorial function has been known. It is defined for positive integer values of n through $n! = 1 \cdot 2 \cdot 3 \cdots n$. With this starting point, we try to find a more general function which coincides with the factorial function for integer values of the argument. Now we have trivially

$$n! = n \cdot (n - 1)!, \qquad n = 2, 3, 4, \ldots,$$

and if we prescribe the validity of this relation also for $n = 1$ we find $0! = 1$. For different (mostly historical) reasons, one prefers to modify the functional equation slightly, and instead we shall consider functions $g(z)$ satisfying

$$g(z + 1) = zg(z)$$

for arbitrary complex z, and the condition $g(n + 1) = n!$ for $n = 0, 1, 2, \ldots$ Taking logarithms and differentiating twice, we get

$$\frac{d^2}{dz^2} \ln g(z + 1) = -\frac{1}{z^2} + \frac{d^2}{dz^2} \ln g(z).$$

From $g(z + n + 1) = (z + n)(z + n - 1) \cdots zg(z)$, we find in a similar way for $n = 0, 1, 2, \ldots,$

$$\frac{d^2}{dz^2} \ln g(z) = \sum_{k=0}^{n} \frac{1}{(z + k)^2} + \frac{d^2}{dz^2} \ln g(z + n + 1).$$

Now we put

$$f(z) = \sum_{k=0}^{\infty} \frac{1}{(z + k)^2}$$

and find immediately

$$f(z) = \sum_{k=0}^{n} \frac{1}{(z + k)^2} + f(z + n + 1).$$

It now seems natural indeed to identify $f(z)$ with $(d^2/dz^2) \ln g(z)$, and in this way define a function $\Gamma(z)$ through

(7.2.1) $$\frac{d^2}{dz^2} \ln \Gamma(z) = \sum_{n=0}^{\infty} \frac{1}{(z + n)^2}$$

with

$$\Gamma(z + 1) = z\Gamma(z); \qquad \Gamma(1) = 1.$$

Integrating (7.2.1) from 1 to $z + 1$, we find

$$\frac{d}{dz} \ln \Gamma(z + 1) = \Gamma'(1) + \sum_{n=1}^{\infty} \left(\frac{1}{n} - \frac{1}{z + n} \right).$$

One more integration from 0 to z gives

$$\ln \Gamma(z + 1) = z \cdot \Gamma'(1) + \sum_{n=1}^{\infty} \left\{ \frac{z}{n} - \ln\left(1 + \frac{z}{n}\right) \right\}.$$

Putting $z = 1$ we get, because of $\Gamma(2) = 1 \cdot \Gamma(1) = 1$:

$$-\Gamma'(1) = \sum_{n=1}^{\infty} \left\{ \frac{1}{n} - \ln\left(1 + \frac{1}{n}\right) \right\}$$

$$= \lim_{n \to \infty} \left(1 + \frac{1}{2} + \frac{1}{3} + \cdots + \frac{1}{n} - \ln n\right) = \gamma.$$

Here, as usual, $\gamma \sim 0.5772156649$ is Euler's constant. Hence we obtain the representation

(7.2.2)
$$\Gamma(z) = \frac{1}{z} e^{-\gamma z} \prod_{n=1}^{\infty} e^{z/n} \Big/ \left(1 + \frac{z}{n}\right).$$

If we use $\sum_{n=1}^{\infty} \{1/n - \ln(1 + 1/n)\}$ instead of Euler's constant we get the equivalent formula

(7.2.3)
$$\Gamma(z) = \frac{1}{z} \prod_{n=1}^{\infty} \frac{(1 + 1/n)^z}{1 + z/n}.$$

From this formula we see that the Γ-function has poles (simple infinities) for $z = 0, -1, -2, \ldots$ Another representation which is frequently used is the following:

(7.2.4)
$$\Gamma(z) = \int_0^{\infty} e^{-t} t^{z-1} \, dt, \qquad \text{Re } z > 0.$$

By partial integration it is easily verified that the condition $\Gamma(z + 1) = z\Gamma(z)$ is satisfied.
    We are now going to derive some important formulas for the gamma function. First we consider $\Gamma(1 - z)$:

$$\Gamma(1 - z) = -z\Gamma(-z) = e^{\gamma z} \prod_{n=1}^{\infty} \left(1 - \frac{z}{n}\right)^{-1} e^{-z/n}.$$

Multiplying with $\Gamma(z)$ we get

$$\Gamma(z)\Gamma(1 - z) = z^{-1} \prod_{n=1}^{\infty} \left(1 - \frac{z^2}{n^2}\right)^{-1}.$$

In the chapter on Laplace transformation we will prove that this is equal to $\pi/\sin \pi z$ (see (8.2.5)). Hence, we have the important result:

(7.2.5)
$$\Gamma(z)\Gamma(1 - z) = \frac{\pi}{\sin \pi z}.$$

In particular we find for $z = \frac{1}{2}$ that

$$[\Gamma(\tfrac{1}{2})]^2 = \pi \qquad \text{and} \qquad \Gamma(\tfrac{1}{2}) = \sqrt{\pi}$$

since $\Gamma(x) > 0$ when $x > 0$. If the formula is used for a purely imaginary argument $z = iy$, we get

$$\Gamma(iy)\Gamma(1 - iy) = -iy\Gamma(iy)\Gamma(-iy) = \frac{\pi}{\sin \pi iy} = \frac{\pi}{i \sinh \pi y},$$

and hence

(7.2.6) $$|\Gamma(iy)|^2 = \frac{\pi}{y \sinh \pi y},$$

since $|\Gamma(iy)| = |\Gamma(-iy)|$. Using (7.2.2) we can also compute the argument from the well-known relations $\arg(z_1 z_2) = \arg z_1 + \arg z_2$; $\arg(x + iy) = \arctan(y/x)$; and $\arg e^{i\alpha} = \alpha$. In this way the following formula is obtained:

(7.2.7) $$\arg \Gamma(iy) = \sum_{n=1}^{\infty} \left\{ \frac{y}{n} - \arctan\left(\frac{y}{n}\right) \right\} - \gamma y - \left(\frac{\pi}{2}\right) \text{sign}(y).$$

We shall now state an important asymptotic formula for the gamma function, namely

(7.2.8) $$\ln \Gamma(z) \sim \frac{1}{2} \ln (2\pi) + \left(z - \frac{1}{2}\right) \ln z - z + \frac{1}{12z} - \frac{1}{360z^3} + \frac{1}{1260z^5} - \cdots$$

This is the well-known *Stirling formula*. Suitably truncated, it can be used for large values of $|z|$. A complete proof together with the remainder term is given in Section 18.3. For the gamma function itself we have:

(7.2.9) $$\Gamma(z) \sim (2\pi)^{1/2} z^{z-1/2} e^{-z} \left[1 + \frac{1}{12z} + \frac{1}{288z^2} - \frac{139}{51840z^3} - \cdots \right].$$

The formulas can be used only if $|\arg z| < \pi$, but it is advisable to keep to the right half-plane and apply the functional relation when necessary.

The Function $\psi(z)$

The logarithmic derivative of the Γ-function

(7.2.10) $$\psi(z) = \frac{d}{dz} \ln \Gamma(z) = \Gamma'(z)/\Gamma(z)$$

is sometimes called the digamma function; it has several interesting properties. From the functional relation for the Γ-function we have directly $\psi(z + 1) - \psi(z) = 1/z$. Further, from (7.2.2) we obtain:

(7.2.11) $$\psi(z) = -\gamma + \sum_{n=0}^{\infty} \left(\frac{1}{n+1} - \frac{1}{z+n} \right).$$

As is the case for the Γ-function, the ψ-function also has poles for $z = -n$, $n = 0, 1, 2, 3, \ldots$. The function values for positive integer arguments can easily be computed:

$$\psi(1) = -\gamma; \qquad \psi(2) = 1 - \gamma; \qquad \psi(3) = 1 + \tfrac{1}{2} - \gamma;$$
$$\psi(4) = 1 + \tfrac{1}{2} + \tfrac{1}{3} - \gamma, \qquad \text{and so on.}$$

It is immediately recognized that for large values of n we have $\psi(n) \sim \ln n$. An asymptotic expansion can be obtained from (7.2.8):

$$\psi(z) \sim \ln z - \frac{1}{2z} - \frac{1}{12z^2} + \frac{1}{120z^4} - \frac{1}{252z^6} + \cdots.$$

Due to the relation $\Delta\psi(z) = 1/z$, the ψ function is of some importance in the theory of difference equations (cf. Section 6.3).

The derivative of the Γ function can also be computed from (7.2.4), which allows differentiation under the integral sign:

$$(7.2.12) \qquad\qquad \Gamma'(z) = \int_0^\infty e^{-t} t^{z-1} \ln t \, dt.$$

For $z = 1$ and $z = 2$ we find in particular

$$(7.2.13) \qquad \begin{aligned} \Gamma'(1) &= -\gamma = \int_0^\infty e^{-t} \ln t \, dt, \\ \Gamma'(2) &= 1 - \gamma = \int_0^\infty t e^{-t} \ln t \, dt. \end{aligned}$$

As an example we shall now show how the properties of the Γ-function can be used for computation of infinite products of the form $\prod_{n=1}^\infty u_n$ where u_n is a rational function of n. First we can split u_n in factors in such a way that the product can be written

$$P = \prod_{n=1}^\infty \frac{A(n - a_1)(n - a_2) \cdots (n - a_r)}{(n - b_1)(n - b_2) \cdots (n - b_s)}.$$

A necessary condition for convergence is $\lim_{n \to \infty} u_n = 1$, which implies $A = 1$ and $r = s$. The general factor then has the form

$$P_n = \left(1 - \frac{a_1}{n}\right) \cdots \left(1 - \frac{a_r}{n}\right)\left(1 - \frac{b_1}{n}\right)^{-1} \cdots \left(1 - \frac{b_r}{n}\right)^{-1}$$

$$= 1 - \frac{\sum a_i - \sum b_i}{n} + O(n^{-2}).$$

But a product of the form $\prod_{n=1}^\infty (1 - a/n)$ is convergent if and only if $a = 0$ (if $a > 0$ the product tends to 0, if $a < 0$ to ∞). Hence we must have $\sum a_i = \sum b_i$ and we may add the factor $\exp\{n^{-1}(\sum a_i - \sum b_i)\}$ without changing the value of P. In this way we get

$$P = \prod_{n=1}^\infty \left(1 - \frac{a_1}{n}\right) e^{a_1/n} \cdots \left(1 - \frac{a_r}{n}\right) e^{a_r/n} \left(1 - \frac{b_1}{n}\right)^{-1} e^{-b_1/n} \cdots \left(1 - \frac{b_r}{n}\right)^{-1} e^{-b_r/n}.$$

But according to (7.2.2) we have

$$\prod_{n=1}^{\infty}\left(1 - \frac{z}{n}\right)e^{z/n} = \frac{e^{\gamma z}}{[-z\Gamma(-z)]} = \frac{e^{\gamma z}}{\Gamma(1 - z)}$$

and from this we get the final result:

$$(7.2.14) \qquad P = \prod_{i=1}^{r} \frac{\Gamma(1 - b_i)}{\Gamma(1 - a_i)}.$$

Finally, we give a few numerical values of the gamma function.

x	$\Gamma(x)$	x	$\Gamma(x)$
0.1	9.51350 76987	1/2	1.77245 38509
0.2	4.59084 37120	1/3	2.67893 85347
0.3	2.99156 89877	1/4	3.62560 99082
0.4	2.21815 95438	1/6	5.56631 60018
0.5	1.77245 38509	1/7	6.54806 29402
0.6	1.48919 22488	1/8	7.53394 15988
0.7	1.29805 53326	1/9	8.52268 81392
0.8	1.16422 97137	2/3	1.35411 79394
0.9	1.06862 87021	3/4	1.22541 67025

In addition, we also give the value of a definite integral:

$$\int_{0}^{\infty} dx/\Gamma(x) = 2.80777 \, 02420.$$

§7.3.   THE BETA FUNCTION

Many integrals can be brought to the form

$$(7.3.1) \qquad B(x, y) = \int_{0}^{1} t^{x-1}(1 - t)^{y-1}\, dt.$$

(Unfortunately, capital Beta cannot be distinguished from an ordinary capital B.) This relation defines the beta function $B(x, y)$ which, obviously, is a function of two variables x and y. For convergence of the integral we must claim Re $x > 0$, Re $y > 0$. First it will be shown that the beta function can be expressed in terms of gamma functions. Starting from (7.3.1) we perform the transformation $t = v/(1 + v)$, obtaining

$$(7.3.2) \qquad B(x, y) = \int_{0}^{\infty} v^{x-1}(1 + v)^{-x-y}\, dv.$$

Multiplying by $\Gamma(x + y)$ and observing that

$$\int_0^\infty e^{-(1+v)\xi}\xi^{x+y-1}\,d\xi = (1 + v)^{-x-y}\Gamma(x + y),$$

which follows through the transformation $(1 + v)\xi = z$, we obtain

$$\Gamma(x + y)B(x, y) = \int_0^\infty dv \cdot v^{x-1}\int_0^\infty e^{-(1+v)\xi}\xi^{x+y-1}\,d\xi$$

$$= \int_0^\infty d\xi e^{-\xi}\xi^{x+y-1}\int_0^\infty v^{x-1}e^{-v\xi}\,dv.$$

The inner integral has the value $\xi^{-x}\Gamma(x)$, and this gives

$$\Gamma(x + y)B(x, y) = \Gamma(x)\int_0^\infty e^{-\xi}\xi^{y-1}\,d\xi = \Gamma(x)\Gamma(y).$$

Hence we get the desired formula

(7.3.3) $$B(x, y) = \frac{\Gamma(x)\Gamma(y)}{\Gamma(x + y)}.$$

We can now make use of this relation in deriving another important formula. If the integration interval in (7.3.2) is divided into two parts we obtain

$$B(x, y) = \int_0^1 + \int_1^\infty \frac{v^{x-1}}{(1 + v)^{x+y}}\,dv.$$

The second integral is transformed through $v = 1/t$ and hence

$$B(x, y) = \int_0^1 \frac{v^{x-1} + v^{y-1}}{(1 + v)^{x+y}}\,dv.$$

Putting $y = x$ we get

$$B(x, x) = \frac{\Gamma(x)^2}{\Gamma(2x)} = 2\int_0^1 \frac{v^{x-1}}{(1 + v)^{2x}}\,dv.$$

If we put $v = (1 - t)/(1 + t)$, the integral is transformed to $2^{2-2x}\int_0^1 (1 - t^2)^{x-1}\,dt$, and then $t^2 = z$ gives

$$\frac{\Gamma(x)^2}{\Gamma(2x)} = \frac{1}{2^{2x-1}}\int_0^1 z^{-1/2}(1 - z)^{x-1}\,dz = \frac{B(x, \frac{1}{2})}{2^{2x-1}}.$$

From this we finally get

(7.3.4) $$\Gamma(2x) = \pi^{-1/2}2^{2x-1}\Gamma(x)\Gamma(x + \tfrac{1}{2}).$$

This duplication formula is initially due to Legendre.

Example 1

(7.3.5) $$\int_0^{\pi/2} \sin^{2x} t \cos^{2y} t\,dt = \frac{\Gamma(x + \frac{1}{2})\Gamma(y + \frac{1}{2})}{2\Gamma(x + y + 1)}, \qquad x > -\tfrac{1}{2}, \quad y > -\tfrac{1}{2}.$$

This important formula is obtained if we put $\sin^2 t = z$. In particular, we have for $k = 0, 1, 2, \ldots$:

(7.3.6)
$$\int_0^{\pi/2} \sin^{2k} t \, dt = \int_0^{\pi/2} \cos^{2k} t \, dt = \frac{\pi}{2} \frac{(2k)!}{(2^k k!)^2}$$
$$\int_0^{\pi/2} \sin^{2k+1} t \, dt = \int_0^{\pi/2} \cos^{2k+1} t \, dt = \frac{(2^k k!)^2}{(2k+1)!}.$$

§7.4. SOME FUNCTIONS DEFINED BY DEFINITE INTEGRALS

In this section we shall very briefly touch upon a few functions other than the gamma and beta functions (which could also be covered here since they can be defined through definite integrals). The *error integral* is used extensively in statistics and probability theory and is defined through

(7.4.1)
$$\text{erf}(x) = \frac{2}{\sqrt{\pi}} \int_0^x e^{-t^2} \, dt,$$

where Erf is an abbreviation for Error function. For large values of x the function converges quickly toward 1 and then an asymptotic expansion can be used conveniently:

(7.4.2)
$$\sqrt{\pi}(1 - \text{erf}(x)) \sim \frac{e^{-x^2}}{x}\left(1 - \frac{1}{2x^2} + \frac{1 \cdot 3}{(2x^2)^2} - \frac{1 \cdot 3 \cdot 5}{(2x^2)^3} + \cdots\right).$$

The *exponential integrals* are defined through

(7.4.3)
$$\text{Ei}(x) = \int_{-\infty}^x \frac{e^t}{t} \, dt, \qquad x > 0,$$

(7.4.4)
$$E_1(x) = \int_x^\infty \frac{e^{-t}}{t} \, dt, \qquad x > 0.$$

In the first integral the integrand is singular in the origin, and the integral should be interpreted as the Cauchy principal value

$$\lim_{\varepsilon \to 0} \int_{-\infty}^{-\varepsilon} + \int_{\varepsilon}^x \frac{e^t}{t} \, dt.$$

Closely related are the *sine* and *cosine integrals*

(7.4.5)
$$\text{Si}(x) = \int_0^x \frac{\sin t}{t} \, dt,$$

(7.4.6)
$$\text{Ci}(x) = -\int_x^\infty \frac{\cos t}{t} \, dt = \gamma + \ln x + \int_0^x \frac{\cos t - 1}{t} \, dt.$$

The last relation is proved by considering the integral

$$\int_C e^{it}t^{-1}\,dt$$

where C is a path from ε to R, from R to iR along a circle of radius R, from iR to $i\varepsilon$, and from $i\varepsilon$ to ε along a circle with radius ε. Since there are no poles inside this contour, we find after easy reductions (when $R \to \infty$):

$$\int_\varepsilon^\infty \frac{\cos t}{t}\,dt + i\int_\varepsilon^\infty \frac{\sin t}{t}\,dt - \int_\varepsilon^\infty \frac{e^{-t}}{t}\,dt - i\frac{\pi}{2} + O(\varepsilon) = 0.$$

Hence, when $\varepsilon \to 0$, we get

$$\int_0^\infty \frac{\cos t - e^{-t}}{t}\,dt = 0$$

and the well-known result

$$\int_0^\infty \frac{\sin t}{t}\,dt = \frac{\pi}{2}.$$

As can be proved differentiating both sides of (7.4.6), we have

$$\mathrm{Ci}(x) = -\int_x^\infty \frac{\cos t}{t}\,dt = a + \ln x + \int_0^x \frac{\cos t - 1}{t}\,dt$$

where a is a constant to be determined. We find:

$$a = -\ln x - \lim_{\varepsilon \to 0}\left\{\int_\varepsilon^\infty \frac{\cos t}{t}\,dt - \int_\varepsilon^x \frac{dt}{t}\right\}$$

$$= -\ln x - \lim_{\varepsilon \to 0}\left\{\int_\varepsilon^\infty \frac{e^{-t}}{t}\,dt - \ln x + \ln \varepsilon\right\}$$

$$= \lim_{\varepsilon \to 0}\left\{-\ln \varepsilon - [e^{-t}\ln t]_\varepsilon^\infty - \int_\varepsilon^\infty e^{-t}\ln t\,dt\right\}$$

$$= -\int_0^\infty e^{-t}\ln t\,dt = \gamma \qquad \text{(see 7.2.13)}.$$

The *complete elliptic integrals* are defined through

(7.4.7) $\quad K(k) = \int_0^{\pi/2} dt/\sqrt{(1 - k^2\sin^2 t)} = \int_0^1 dx/\sqrt{[(1 - x^2)(1 - k^2x^2)]},$

(7.4.8) $\quad E(k) = \int_0^{\pi/2} \sqrt{(1 - k^2\sin^2 t)}\,dt = \int_0^1 \sqrt{[(1 - k^2x^2)/(1 - x^2)]}\,dx.$

A large variety of integrals can be reduced to this form.

Example 1

$$I = \int_0^\infty dt/\sqrt{(1 + t^4)} = \int_0^1 + \int_1^\infty dt/\sqrt{(1 + t^4)} = 2\int_0^1 dt/\sqrt{(1 + t^4)}$$

(put $t = u^{-1}$ in the second integral). Then transform by $t = \tan(x/2)$:

$$I = \int_0^{\pi/2} dx/\sqrt{[\sin^4(x/2) + \cos^4(x/2)]} = \int_0^{\pi/2} dx/\sqrt{(1 - \tfrac{1}{2}\sin^2 x)}$$
$$= K(1/\sqrt{2}) = 1.85407\,46773.$$

§7.5. THE RIEMANN ZETA FUNCTION

The Riemann zeta function $\zeta(s)$ can be defined in two equivalent ways:

(7.5.1)
$$\zeta(s) = \sum_{n=1}^{\infty} n^{-s}$$

and

(7.5.2)
$$\zeta(s) = \prod_p (1 - p^{-s})^{-1}$$

where the product is taken over all primes p. Let s be a complex variable $s = \sigma + it$; then the series (which is, incidentally, called a Dirichlet series) converges when $\sigma > 1$ and defines an analytic function in this region. If we expand the product we get

$$\prod_p (1 + p^{-s} + p^{-2s} + \cdots)$$

and hence, since each natural number $n > 1$ can be expressed as the product of primes in exactly one way, we are back at the first expression. Already this connection with primes indicates the importance of the zeta function in prime number theory.

From the definition as a product of nonzero factors, it is at once clear that the zeta function has no zeros in the region $\sigma > 1$.

We now introduce the function $\pi(x)$, defined as the number of primes less than or equal to x (e.g., $\pi(2) = 1$, $\pi(3) = 2$, $\pi(8.3) = 4$). From the product representation we have:

$$\ln \zeta(s) = -\sum_p \ln(1 - p^{-s}) = -\sum_{n=2}^{\infty} [\pi(n) - \pi(n-1)] \ln(1 - n^{-s})$$
$$= -\sum_{n=2}^{\infty} \pi(n)[\ln(1 - n^{-s}) - \ln(1 - (n+1)^{-s})]$$
$$= \sum_{n=2}^{\infty} \pi(n) \int_n^{n+1} \frac{s}{x(x^s - 1)}\, dx = s \int_2^{\infty} \frac{\pi(x)}{x(x^s - 1)}\, dx$$

and we have a direct link between the zeta function and the number-of-primes function $\pi(x)$. It is in fact possible to estimate $\pi(x)$ from the relation above; the result is

$$\pi(x) \sim x/\ln x,$$

which is the famous prime number theorem.

From the product representation we also have

$$[\zeta(s)]^{-1} = \prod_p (1 - p^{-s}),$$

and carrying out the multiplications we see that the denominators are of the form n^s where the numbers n are products of primes with each prime appearing at most once. These numbers are 1, 2, 3, 5, 6, 7, 10, 11, 13, 14, 15, 17, ... We can write this:

(7.5.3)
$$[\zeta(s)]^{-1} = \sum_{n=1}^{\infty} \mu(n)/n^s$$

with

$$\mu(n) = \begin{cases} 1 & \text{if } n = 1; \\ 0 & \text{if } n \text{ contains a square factor;} \\ (-1)^k & \text{if } n \text{ is the product of } k \text{ different primes.} \end{cases}$$

The function $\mu(n)$ is known as the *Möbius function*.

As defined above, the ζ function can be computed only when $\sigma > 1$. However, this can be improved slightly by the following trick.

$$\zeta(s) = 1 + \frac{1}{2^s} + \frac{1}{3^s} + \frac{1}{4^s} + \frac{1}{5^s} + \frac{1}{6^s} + \cdots$$

$$-2^{1-s}\zeta(s) = \qquad -\frac{2}{2^s} \qquad -\frac{2}{4^s} \qquad -\frac{2}{6^s} - \cdots$$

Hence

$$(1 - 2^{1-s})\zeta(s) = 1 - 2^{-s} + 3^{-s} - 4^{-s} + 5^{-s} - 6^{-s} + \cdots,$$

a series which converges for $\sigma > 0$, i.e., also in the strip $0 < \sigma \le 1$. By use of more advanced methods the following functional relation can be obtained:

(7.5.4)
$$\zeta(s) = 2^s \pi^{s-1} \sin(s\pi/2)\, \Gamma(1 - s)\zeta(1 - s);$$

through this relation, the function can be computed when $\sigma < 0$. Another useful representation is

(7.5.5)
$$\zeta(s) = \frac{1}{\Gamma(s)} \int_0^{\infty} \frac{x^{s-1}}{e^x - 1}\, dx \qquad (\sigma > 1).$$

To prove this we have only to expand the integrand to the form

$$\sum_{n=1}^{\infty} e^{-nx} x^{s-1}$$

and the result follows directly. The formula can be transformed to

(7.5.6)
$$\zeta(s) = \frac{e^{-i\pi s}\Gamma(1 - s)}{2\pi i} \int_C \frac{z^{s-1}}{e^z - 1}\, dz$$

where C is the real axis from ∞ to ε, the circle $|z| = \varepsilon$, and again the real axis from ε to ∞. Using the expansion

$$\frac{z}{e^z - 1} = 1 - \frac{z}{2} + B_1 \frac{z^2}{2!} - B_2 \frac{z^4}{4!} + B_3 \frac{z^6}{6!} - \cdots$$

we find:

$$\zeta(0) = \frac{1}{2\pi i} \int_C \frac{dz}{z^2} \left[1 - \frac{z}{2} + \cdots \right] = -\frac{1}{2}$$

and in a similar way

$$\zeta(-2m) = 0; \qquad \zeta(1 - 2m) = (-1)^m B_m / 2m$$

where $m = 1, 2, 3, \ldots$ Further, multiplying the functional relation with $\Gamma(s)$, using $\Gamma(s)\Gamma(1 - s) = \pi/\sin \pi s$, and putting $s = 2m$, we get:

(7.5.7)
$$\zeta(2m) = \frac{1}{2} \cdot \frac{B_m (2\pi)^{2m}}{(2m)!}$$

(cf. (3.5.13)).

The zeros $-2, -4, -6, \ldots$ are called the *trivial zeros* of the zeta function. There is a famous conjecture by Riemann that all *nontrivial* zeros lie on the so-called critical line $\sigma = \frac{1}{2}$. The conjecture has been verified numerically for at least the first 200 million zeros, and it seems potentially possible to reach the first 1000 million zeros in the near future. A mathematical proof was announced in 1984 by Japanese mathematician H. Matsumoto in Paris. The proof is now being checked by specialists.

Finally, we give some numerical values of the zeta function.

s	$\zeta(s)$	s	$\zeta(s)$
5/2	1.34148 72573	1/3	$-0.97336\ 02485$
3/2	2.61237 53487	0	-0.5
4/3	3.60093 77506	$-1/2$	$-0.20788\ 62250$
1/2	$-1.46035\ 45088$	-1	$-0.08333\ 33333$

Additional values of $z_s = \zeta(s) - 1$ for $s = 2(1)33$ can be found in Section 18.5. The first nontrivial zeros $\frac{1}{2} \pm it_n$ are:

$$t_1 = 14.134725, \qquad t_2 = 21.022040, \qquad t_3 = 25.010858,$$
$$t_4 = 30.424876, \qquad t_5 = 32.935062.$$

The number of zeros N below a certain limit T is asymptotically

$$N \sim (T/2\pi)[\ln(T/2\pi) - 1].$$

§7.6.   BESSEL FUNCTIONS

In the following we will to a large extent use power series expansions for solving different linear differential equations. For this reason we start by giving a short account of the technique to be used. In general, we make the attempt

$$y = \sum_{n=r}^{\infty} a_n x^n,$$

where the numbers a_n so far represent unknown coefficients; also the starting level r remains to be established. By insertion into the differential equation we get relations between adjacent coefficients through the condition that the total coefficient for each power x^n must vanish. In the beginning of the series we obtain a special condition for determining r (the *index equation*).

Example 1

$$y'' + y = 0.$$

We suppose that the series starts with the term x^r. Introducing this single term and collecting terms of lowest possible degree, we are left with the expression $r(r - 1)x^{r-2}$. The index equation is $r(r - 1) = 0$, and we get two possible values of $r: r = 0$ and $r = 1$.

(a) $r = 0$

$$y = \sum_{n=0}^{\infty} a_n x^n.$$

$$y'' = \sum_{n=2}^{\infty} n(n - 1)a_n x^{n-2} = \sum_{n=0}^{\infty} (n + 1)(n + 2)a_{n+2} x^n.$$

Hence we obtain the relation $(n + 1)(n + 2)a_{n+2} + a_n = 0$, and choosing $a_0 = 1$ we find successively

$$a_2 = -\frac{1}{1 \cdot 2}; \qquad a_4 = \frac{1}{1 \cdot 2 \cdot 3 \cdot 4}; \qquad \cdots$$

and the solution

$$y = 1 - \frac{x^2}{2!} + \frac{x^4}{4!} - \frac{x^6}{6!} + \cdots = \cos x.$$

(b) $r = 1$. Here we have the same relation for the coefficients, and choosing $a_1 = 1$ we find

$$y = x - \frac{x^3}{3!} + \frac{x^5}{5!} - \frac{x^7}{7!} + \cdots = \sin x.$$

The general solution is $y = A \cos x + B \sin x$.

Similar series expansions cannot generally be established for all linear differential equations. If the equation has the form $y'' + P(x)y' + Q(x)y = 0$, then $P(x)$ and $Q(x)$ as a rule must not contain powers of degree lower than -1 and -2, respectively.

We shall now treat the differential equation

(7.6.1)
$$\frac{d^2u}{dx^2} + \frac{1}{x}\frac{du}{dx} + \left(1 - \frac{v^2}{x^2}\right)u = 0.$$

Putting $u = x^r$ we find the index equation $r(r-1) + r - v^2 = 0$ with the two roots $r = \pm v$. Trying the expansion $u = \sum_{k=v}^{\infty} a_k x^k$, we find

$$a_{k+2} = -\frac{a_k}{(k+2+v)(k+2-v)}.$$

We now start with $k = v$ and $a_v = 1$, obtaining

$$\begin{cases} a_{v+2} = -\dfrac{1}{2(2v+2)}, \\[2mm] a_{v+4} = \dfrac{1}{2\cdot 4(2v+2)(2v+4)}, \\[2mm] \quad\vdots \\[2mm] a_{v+2k} = \dfrac{(-1)^k}{2^{2k}\cdot k!(v+1)(v+2)\cdots(v+k)} = \dfrac{(-1)^k\Gamma(v+1)\cdot 2^v}{2^{v+2k}\cdot k!\Gamma(v+k+1)}. \end{cases}$$

The factor $2^v\Gamma(v+1)$ is constant and can be removed, and so we define a function $J_v(x)$ through the formula

(7.6.2)
$$J_v(x) = \sum_{k=0}^{\infty} \frac{(-1)^k(x/2)^{v+2k}}{k!\Gamma(v+k+1)}.$$

The function $J_v(x)$ is called a *Bessel function of the first kind*. Replacing v by $-v$ we get still another solution, which we denote by $J_{-v}(x)$, and a linear combination of these two will give the general solution. However, if v is an integer n, the terms corresponding to $k = 0, 1, \ldots, n-1$ will disappear since the gamma function in the denominator becomes infinite. Then we are left with

$$J_{-n}(x) = \sum_{k=n}^{\infty} \frac{(-1)^k(x/2)^{-n+2k}}{k!\Gamma(-n+k+1)} = \sum_{r=0}^{\infty} \frac{(-1)^n(-1)^r(x/2)^{-n+2n+2r}}{(n+r)!\Gamma(r+1)} = (-1)^n J_n(x),$$

where we have put $k = n + r$. Note that $(n+r)! = \Gamma(n+r+1)$ and $\Gamma(r+1) = r!$. Whenever v is an integer we only get one solution by direct series expansion, and we must try to find a second solution in some other way. A very general technique is demonstrated in the following example. Consider the differential equation $y'' + \alpha^2 y = 0$ with the solutions $y_1 = e^{\alpha x}$, $y_2 = e^{-\alpha x}$. If $\alpha = 0$ we get $y_1 = y_2 = 1$ and one solution has been lost. It is now clear that also $y_0 = (y_2 - y_1)/2\alpha$ is a solution of $y'' + \alpha^2 y = 0$, and this holds true even if $\alpha \to 0$. Since $\lim_{\alpha\to 0} y_0 = x$, the general solution of $y'' = 0$ becomes $y = A + Bx$.

Along the same lines, we now form

$$Y_n(x) = \frac{1}{\pi}\lim_{v\to n}\frac{J_v(x) - (-1)^n J_{-v}(x)}{v - n}$$

or

$$Y_n(x) = \frac{1}{\pi} \lim_{v \to n} \left\{ \frac{J_v(x) - J_n(x)}{v - n} - (-1)^n \frac{J_{-v}(x) - J_{-n}(x)}{v - n} \right\}$$

$$= \frac{1}{\pi} \left\{ \frac{\partial J_v(x)}{\partial v} - (-1)^n \frac{\partial J_{-v}(x)}{\partial v} \right\}_{v = n}.$$

We shall now prove that $Y_n(x)$ satisfies the Bessel equation

$$x^2 \frac{d^2 u}{dx^2} + x \frac{du}{dx} + (x^2 - v^2)u = 0.$$

Differentiating with respect to v, and with u replaced by $J_v(x)$ and $J_{-v}(x)$, we get

$$\begin{cases} x^2 \frac{d^2}{dx^2} \frac{\partial J_v(x)}{\partial v} + x \frac{d}{dx} \frac{\partial J_v(x)}{\partial v} + (x^2 - v^2) \frac{\partial J_v(x)}{\partial v} - 2v J_v(x) = 0 \\ x^2 \frac{d^2}{dx^2} \frac{\partial J_{-v}(x)}{\partial v} + x \frac{d}{dx} \frac{\partial J_{-v}(x)}{\partial v} + (x^2 - v^2) \frac{\partial J_{-v}(x)}{\partial v} - 2v J_{-v}(x) = 0. \end{cases}$$

We multiply the second equation by $(-1)^n$ and subtract:

$$\left(x^2 \frac{d^2}{dx^2} + x \frac{d}{dx} + x^2 - v^2 \right) \left(\frac{\partial J_v(x)}{\partial v} - (-1)^n \frac{\partial J_{-v}(x)}{\partial v} \right) = 2v(J_v(x) - (-1)^n J_{-v}(x)).$$

Now letting $v \to n$, the right-hand side will tend to zero, and we find after multiplication with $1/\pi$:

$$x^2 \frac{d^2 Y_n(x)}{dx^2} + x \frac{d Y_n(x)}{dx} + (x^2 - n^2) Y_n(x) = 0;$$

that is, $Y_n(x)$ satisfies the Bessel differential equation.

It is also possible to define a function $Y_v(x)$ even if v is not an integer:

$$Y_v(x) = \frac{J_v(x) \cos v\pi - J_{-v}(x)}{\sin v\pi},$$

and by use of l'Hospital's rule, for example, one can easily establish that

$$\lim_{v \to n} Y_v(x) = Y_n(x).$$

The general solution can now be written in a more consistent form:

$$u = A J_v(x) + B Y_v(x).$$

The Y-functions are usually called *Bessel functions of the second kind* or *Neumann functions*. In particular we find for $n = 0$:

$$Y_0(x) = \frac{2}{\pi} \left(\frac{\partial J_v(x)}{\partial v} \right)_{v = 0} = \frac{2}{\pi} \left\{ \frac{\partial}{\partial v} \sum_{k=0}^{\infty} \frac{(-1)^k (x/2)^{v + 2k}}{k! \Gamma(v + k + 1)} \right\}_{v = 0}$$

$$= \frac{2}{\pi} \left\{ \sum_{k=0}^{\infty} \frac{(-1)^k (x/2)^{v + 2k}}{k! \Gamma(v + k + 1)} \left(\ln \frac{x}{2} - \frac{\partial}{\partial v} \ln \Gamma(v + k + 1) \right) \right\}_{v = 0}.$$

Hence

(7.6.3)
$$Y_0(x) = \frac{2}{\pi} \sum_{k=0}^{\infty} \frac{(-1)^k(x/2)^{2k}}{(k!)^2} \left(\ln \frac{x}{2} - \psi(k+1) \right).$$

If $n > 0$ the calculations get slightly more complicated, and we only give the final result:

(7.6.4)
$$Y_n(x) = \frac{2}{\pi} \ln \frac{x}{2} J_n(x) - \frac{1}{\pi} \left(\frac{x}{2} \right)^{-n} \sum_{k=0}^{n-1} \frac{(n-k-1)!}{k!} \left(\frac{x}{2} \right)^{2k}$$
$$- \frac{1}{\pi} \sum_{k=0}^{\infty} \frac{(-1)^k(x/2)^{n+2k}}{k!(n+k)!} (\psi(k+1) + \psi(n+k+1)).$$

We are now going to prove a general relation for linear, homogeneous second-order equations. Consider the equation

$$y'' + P(x)y' + Q(x)y = 0,$$

and suppose that two linearly independent solutions y_1 and y_2 are known:

$$\begin{cases} y_1'' + P(x)y_1' + Q(x)y_1 = 0, \\ y_2'' + P(x)y_2' + Q(x)y_2 = 0. \end{cases}$$

The first equation is multiplied by y_2, the second by $-y_1$, and then the equations are added:

$$\frac{d}{dx}(y_1'y_2 - y_1y_2') + P(x)(y_1'y_2 - y_1y_2') = 0,$$

and hence we have $y_1'y_2 - y_1y_2' = \exp\{-\int P(x)\,dx\}$. For the Bessel equation we have $P(x) = 1/x$, and using this we get

$$J_\nu'(x)J_{-\nu}(x) - J_\nu(x)J_{-\nu}'(x) = \frac{C}{x}.$$

The constant C can be determined from the first terms in the series expansions:

$$J_\nu(x) = \frac{(x/2)^\nu}{\Gamma(\nu+1)}(1 + O(x^2)); \qquad J_\nu'(x) = \frac{(x/2)^{\nu-1}}{2\Gamma(\nu)}(1 + O(x^2)).$$

In this way we obtain

$$J_\nu'(x)J_{-\nu}(x) - J_\nu(x)J_{-\nu}'(x) = \frac{1}{x}\left\{ -\frac{1}{\Gamma(\nu+1)\Gamma(-\nu)} + \frac{1}{\Gamma(\nu)\Gamma(-\nu+1)} \right\} + O(x)$$

and $C = 2\sin\nu\pi/\pi$. If $\nu = n$ is an integer we get $C = 0$; that is, $J_n(x)$ and $J_{-n}(x)$ are proportional since $J_n'/J_n = J_{-n}'/J_{-n}$.

Relations of this kind are called Wronski relations and are of great importance. An alternative expression can be constructed from J_ν and Y_ν and we find

$$J_\nu'(x)Y_\nu(x) - J_\nu(x)Y_\nu'(x) = -\frac{2}{\pi x}.$$

In connection with numerical computation of functions such relations offer extremely good checking possibilities.

Recursion Formulas

Between Bessel functions of adjacent orders there are certain simple linear relationships from which new functions can be computed by the aid of old ones. First we shall prove that

$$(7.6.5) \qquad J_{\nu-1}(x) + J_{\nu+1}(x) = \frac{2\nu}{x} J_\nu(x).$$

Choosing the term (k) from the first expansion, $(k-1)$ from the second, and (k) from the third, we obtain the same power $(x/2)^{\nu+2k-1}$. The coefficients become

$$\frac{(-1)^k}{k!\Gamma(\nu+k)}, \qquad \frac{(-1)^{k-1}}{(k-1)!\Gamma(\nu+k+1)}, \qquad \text{and} \qquad \frac{(-1)^k \nu}{k!\Gamma(\nu+k+1)},$$

and it is easily seen that the sum of the first two is equal to the third. However, we must also investigate the case $k = 0$ separately by comparing powers $(x/2)^{\nu-1}$, since there is no contribution from the second term. The coefficients are $1/\Gamma(\nu)$ and $\nu/\Gamma(\nu+1)$ and so the whole formula is proved. In a similar way we can prove

$$(7.6.6) \qquad J_{\nu-1}(x) - J_{\nu+1}(x) = 2J_\nu'(x).$$

Eliminating either $J_{\nu-1}(x)$ or $J_{\nu+1}(x)$, one can construct other relations:

$$\begin{cases} J_\nu'(x) + \dfrac{\nu}{x} J_\nu(x) = J_{\nu-1}(x), \\[2mm] J_\nu'(x) - \dfrac{\nu}{x} J_\nu(x) = -J_{\nu+1}(x). \end{cases}$$

A corresponding set of formulas can be obtained also for the functions $Y_\nu(x)$. In particular we find for $\nu = 0$:

$$J_0'(x) = -J_1(x); \qquad Y_0'(x) = -Y_1(x).$$

Integral Representations

Consider the Bessel equation in the special case $\nu = 0$:

$$xu'' + u' + xu = 0.$$

If the coefficients had been constant, we could have tried $u = e^{\lambda x}$, leading to a number of discrete particular solutions from which the general solution would have been obtained as a linear combination. When the coefficients are functions of x we cannot hope for such a simple solution, but we still have a chance to try a solution that takes account of the properties of the exponential functions. Instead of discrete "frequencies" we must expect a whole band of frequencies ranging over a suitable interval, and the summation must be replaced by integration. In this way we are led to the attempt

$$u = \int_a^b \phi(\alpha)e^{\alpha x}\, d\alpha.$$

We now have large funds at our disposal, consisting of the function $\phi(\alpha)$ and the limits a and b. The differential equation is satisfied if

$$\int_a^b \{(\alpha^2 + 1)\phi(\alpha)xe^{\alpha x} + \alpha\phi(\alpha)e^{\alpha x}\}\, d\alpha = 0.$$

Observing that $xe^{\alpha x} = (\partial/\partial\alpha)e^{\alpha x}$, we can perform a partial integration:

$$[(\alpha^2 + 1)\phi(\alpha)e^{\alpha x}]_a^b - \int_a^b \{(\alpha^2 + 1)\phi'(\alpha) + \alpha\phi(\alpha)\}e^{\alpha x}\, d\alpha = 0.$$

We can now waste part of our funds by claiming that the integrated part is zero; this is the case if we choose $a = -i$, $b = i$ (also $-\infty$ will work if $x > 0$). Then the integral is also equal to zero, and we spend still another part of our funds by claiming that the integrand is zero:

$$\frac{d\phi}{\phi} + \frac{\alpha\, d\alpha}{\alpha^2 + 1} = 0$$

or $\phi(\alpha) = 1/\sqrt{(\alpha^2 + 1)}$, since a constant factor can be neglected here. In this way we have obtained

$$u = \int_{-i}^{+i} \frac{e^{\alpha x}}{\sqrt{(\alpha^2 + 1)}}\, d\alpha.$$

Putting $\alpha = i\beta$ and neglecting a trivial factor i, we get

$$u = \int_{-1}^{+1} \frac{e^{i\beta x}}{\sqrt{(1 - \beta^2)}}\, d\beta.$$

Writing $e^{i\beta x} = \cos\beta x + i\sin\beta x$, we can discard the second term since the integral of an odd function over the interval $(-1, 1)$ will disappear. Putting $\beta = \cos\theta$, we finally get $u = \int_0^\pi \cos(x\cos\theta)\, d\theta$.

We now state that

(7.6.7)
$$J_0(x) = \frac{1}{\pi}\int_0^\pi \cos(x\cos\theta)\, d\theta.$$

This follows because

$$\frac{1}{\pi}\int_0^\pi \cos(x\cos\theta)\, d\theta = \frac{2}{\pi}\int_0^{\pi/2} \sum_{k=0}^\infty \frac{(-1)^k x^{2k}\cos^{2k}\theta}{(2k)!}\, d\theta$$

$$= \frac{2}{\pi}\sum_{k=0}^\infty \frac{(-1)^k x^{2k}}{(2k)!}\frac{\Gamma(\tfrac{1}{2})\Gamma(k + \tfrac{1}{2})}{2\Gamma(k + 1)}$$

$$= \frac{1}{\pi}\sum_{k=0}^\infty \frac{(-1)^k x^{2k}\pi\,\Gamma(k + \tfrac{1}{2})}{2k \cdot 2^{2k-1}\Gamma(k)\Gamma(k + \tfrac{1}{2}) \cdot \Gamma(k + 1)}$$

$$= \sum_{k=0}^\infty \frac{(-1)^k (x/2)^{2k}}{(k!)^2} = J_0(x),$$

where we have made use of $(2k)! = 2k\Gamma(2k)$ and the duplication formula (7.3.4) for $\Gamma(2k)$.

The technique we have worked with here is a tool that can be used on a large variety of problems in numerical analysis and applied mathematics; it is called *Laplace transformation*, and it is treated in Chapter 8.

We shall now by other means derive a similar integral representation for the Bessel function $J_\nu(x)$. The general term in the series expansion can be written:

$$\frac{(-1)^k(x/2)^{\nu+2k}}{k!\Gamma(\nu+k+1)} = \frac{(-1)^k(x/2)^\nu}{\Gamma(\nu+\frac{1}{2})\Gamma(\frac{1}{2})} \frac{x^{2k}}{(2k)!} \frac{\Gamma(\nu+\frac{1}{2})\Gamma(k+\frac{1}{2})}{\Gamma(\nu+k+1)}$$

$$= \frac{(-1)^k(x/2)^\nu}{\Gamma(\nu+\frac{1}{2})\Gamma(\frac{1}{2})} \frac{x^{2k}}{(2k)!} \int_0^1 t^{\nu-1/2}(1-t)^{k-1/2}\,dt,$$

where again we have made use of formula (7.3.4). Summing the whole series, we now get

$$J_\nu(x) = \frac{(x/2)^\nu}{\Gamma(\nu+\frac{1}{2})\Gamma(\frac{1}{2})} \int_0^1 t^{\nu-1/2} \left\{ \sum_{k=0}^\infty \frac{(-1)^k x^{2k}(1-t)^{k-1/2}}{(2k)!} \right\} dt.$$

After the transformation $t = \sin^2\theta$, we obtain

$$J_\nu(x) = \frac{(x/2)^\nu}{\Gamma(\nu+\frac{1}{2})\Gamma(\frac{1}{2})} \int_0^{\pi/2} \sin^{2\nu-1}\theta \left\{ \sum_{k=0}^\infty \frac{(-1)^k x^{2k}\cos^{2k-1}\theta}{(2k)!} \right\} 2\sin\theta\cos\theta\,d\theta$$

or

$$(7.6.8) \qquad J_\nu(x) = \frac{(x/2)^\nu}{\Gamma(\nu+\frac{1}{2})\Gamma(\frac{1}{2})} \int_0^\pi \sin^{2\nu}\theta\cos(x\cos\theta)\,d\theta.$$

The factor 2 has been used for extending the integration interval from $(0, \pi/2)$ to $(0, \pi)$. Putting $\nu = 0$ we get back our preceding formula (7.6.7).

The formula that has just been derived can be used, for example, in the case when ν is half-integer, that is, $\nu = n + \frac{1}{2}$ where n is integer. It is then suitable to perform the transformation $t = \cos\theta$;

$$J_{n+1/2}(x) = \frac{(x/2)^{n+1/2}}{n!\sqrt{\pi}} \int_{-1}^{+1} (1-t^2)^n \cos xt\,dt.$$

This integral can be evaluated in closed form, and one finds, for example,

$$J_{1/2}(x) = \sqrt{(2/\pi x)}\sin x; \qquad\qquad J_{-1/2}(x) = \sqrt{(2/\pi x)}\cos x,$$

$$J_{3/2}(x) = \sqrt{(2/\pi x)}\left(\frac{\sin x}{x} - \cos x\right); \qquad J_{-3/2}(x) = -\sqrt{(2/\pi x)}\left(\frac{\cos x}{x} + \sin x\right).$$

Finally we quote a few useful summation formulas:

$$(7.6.9) \qquad \begin{cases} J_0(x) + 2J_2(x) + 2J_4(x) + \cdots = 1, \\ J_0(x) - 2J_2(x) + 2J_4(x) - \cdots = \cos x, \\ 2J_1(x) - 2J_3(x) + 2J_5(x) - \cdots = \sin x. \end{cases}$$

For proof of the relations, see Exercise 21 at the end of this chapter.

Numerical values of Bessel functions can be found in several excellent tables. In particular we mention here a monumental work from Harvard Computation Laboratory. In 12 huge volumes $J_n(x)$ is tabulated for $0 \le x \le 100$ and for $n \le 135$; for higher values of n we have $|J_n(x)| < \frac{1}{2} \cdot 10^{-10}$ in the given interval for x. There is also a rich literature on the theory of Bessel functions, but we restrict ourselves to mentioning the standard work by Watson.

Modified Bessel Functions

We start from the differential equation

(7.6.10)
$$x^2 \frac{d^2u}{dx^2} + x \frac{du}{dx} - (x^2 + v^2)u = 0.$$

If we put $x = -it$ the equation goes over into the usual Bessel equation, and for example, we have the solutions $J_v(ix)$ and $J_{-v}(ix)$. But

$$J_v(ix) = \sum_{k=0}^{\infty} \frac{(-1)^k (x/2)^{v+2k} i^{2k+v}}{k! \Gamma(v+k+1)} = e^{v\pi i/2} \sum_{k=0}^{\infty} \frac{(x/2)^{v+2k}}{k! \Gamma(v+k+1)}.$$

Thus $e^{-v\pi i/2} J_v(ix)$ is real for real values of x, and we now define

(7.6.11)
$$I_v(x) = \sum_{k=0}^{\infty} \frac{(x/2)^{v+2k}}{k! \Gamma(v+k+1)},$$

where the function $I_v(x)$ is called a *modified Bessel function*. In a similar way we define $I_{-v}(x)$. When v is an integer we encounter the same difficulties as before, and a second solution is defined through

$$K_v(x) = \frac{\pi}{2} \frac{I_{-v}(x) - I_v(x)}{\sin \pi v}.$$

For integer values of $v = n$ the limit value is taken instead, and we obtain

(7.6.12)
$$K_n(x) = \frac{1}{2} \sum_{k=0}^{n-1} \frac{(-1)^k (n-k-1)! (x/2)^{-n+2k}}{k!} + (-1)^{n+1} \sum_{k=0}^{\infty} \frac{(x/2)^{n+2k}}{k!(n+k)!}$$
$$\times \left[\ln \frac{x}{2} - \frac{1}{2} \psi(k+1) - \frac{1}{2} \psi(n+k+1) \right].$$

In exactly the same way as for $J_v(x)$, we can derive an integral representation for $I_v(x)$:

(7.6.13)
$$I_v(x) = \frac{(x/2)^v}{\Gamma(\frac{1}{2})\Gamma(v+\frac{1}{2})} \int_0^\pi \sin^{2v} \theta \cosh(x \cos \theta)\, d\theta.$$

Also for $K_v(x)$ there exists a simple integral representation:

(7.6.14)
$$K_v(x) = \int_0^\infty \exp(-x \cosh t) \cosh vt\, dt.$$

In particular, if $v = \frac{1}{2}$ the integral can be evaluated directly and we get

$$K_{1/2}(x) = \sqrt{(\pi/2x)}\, e^{-x}.$$

The modified Bessel functions also satisfy simple summation formulas:

$$(7.6.15) \qquad \begin{cases} I_0(x) - 2I_2(x) + 2I_4(x) - \cdots = 1, \\ I_0(x) + 2I_1(x) + 2I_2(x) + \cdots = e^x, \\ I_0(x) - 2I_1(x) + 2I_2(x) - \cdots = e^{-x}. \end{cases}$$

There are excellent tables also for the functions $I_n(x)$ and $K_n(x)$.

§7.7.   SPHERICAL HARMONICS

Many physical situations are governed by the *Laplace* equation

$$\Delta\psi = \frac{\partial^2\psi}{\partial x^2} + \frac{\partial^2\psi}{\partial y^2} + \frac{\partial^2\psi}{\partial z^2} = 0$$

or by the *Poisson* equation $\Delta\psi + \kappa\psi = 0$ where, as a rule, a unique and finite solution is wanted. Very often it is suitable to go over to polar coordinates:

$$\begin{cases} x = r \sin\theta \cos\varphi, \\ y = r \sin\theta \sin\varphi, \\ z = r \cos\theta. \end{cases}$$

If now $\kappa = \kappa(r)$ is a function of the radius only, the variables can be separated if we assume that ψ can be written $\psi = f(r)Y(\theta, \varphi)$. The following equation is then obtained for Y:

$$\frac{1}{\sin\theta}\frac{\partial}{\partial\theta}\left(\sin\theta\frac{\partial Y}{\partial\theta}\right) + \frac{1}{\sin^2\theta}\frac{\partial^2 Y}{\partial\varphi^2} + \lambda Y = 0,$$

where λ is the so-called separation constant. One more separation $Y(\theta, \varphi) = u(\theta)v(\varphi)$ gives

$$\frac{d^2v}{d\varphi^2} + \beta v = 0.$$

The uniqueness requirement implies $v(\varphi + 2\pi) = v(\varphi)$ and hence $\beta = m^2$, $m = 0, \pm 1, \pm 2, \ldots$ The remaining equation then becomes

$$\frac{1}{\sin\theta}\frac{d}{d\theta}\left(\sin\theta\frac{du}{d\theta}\right) - \frac{m^2 u}{\sin^2\theta} + \lambda u = 0.$$

Here we shall take only the case $m = 0$ into account. After the transformation $\cos\theta = x$, we obtain

$$(7.7.1) \qquad \frac{d}{dx}\left[(1 - x^2)\frac{du}{dx}\right] + \lambda u = 0.$$

The index equation is $r(r-1) = 0$, and further from $u = \sum a_r x^r$ we get:

$$a_{r+2} = \frac{r(r+1) - \lambda}{(r+1)(r+2)} a_r.$$

For large values of r the constant λ can be neglected and we find approximately $(r+2)a_{r+2} \simeq ra_r$, that is, $a_r \sim C/r$. But the series

$$1 + \frac{x^2}{2} + \frac{x^4}{4} + \cdots \quad \text{and} \quad x + \frac{x^3}{3} + \frac{x^5}{5} + \cdots$$

are both divergent when $x = 1$, while we are looking for a finite solution. The only way out of this difficulty is that λ has such a value that the series expansion breaks off, that is, $\lambda = n(n+1)$ where n is an integer. In this case the solutions are simply the well-known Legendre polynomials, treated in Section 4.3. Here we just remind of the fact that they form a family of orthogonal polynomials, satisfying the differential equation

$$(1 - x^2)P_n'' - 2xP_n' + n(n+1)P_n = 0$$

where n is an integer, $n \geq 0$. The explicit formula for P_n is:

$$P_n(x) = \frac{1}{2^n n!} \frac{d^n}{dx^n} (x^2 - 1)^n.$$

We are now going to look for another solution to (7.7.1), and we should then keep in mind that the factor $(1 - x^2)$ enters in an essential way. Hence it becomes natural to try with the following expression

$$u = \frac{1}{2} \ln \frac{1+x}{1-x} P_n - W_n.$$

since the derivative of

$$\frac{1}{2} \ln \frac{1+x}{1-x}$$

is exactly $1/(1 - x^2)$. One further finds

$$u' = \frac{1}{2} \ln \frac{1+x}{1-x} P_n' + \frac{1}{1-x^2} P_n - W_n',$$

$$u'' = \frac{1}{2} \ln \frac{1+x}{1-x} P_n'' + \frac{2}{1-x^2} P_n' + \frac{2x}{(1-x^2)^2} P_n - W_n'',$$

and after insertion:

$$(1 - x^2)W_n'' - 2xW_n' + n(n+1)W_n = 2P_n'.$$

It is easy to see that the equation is satisfied if W_n is a polynomial of degree $n - 1$, and by the aid of indeterminate coefficients the solution W_n can be computed explicitly.

For $n = 3$, for example, we have

$$P_3 = \frac{1}{2}(5x^3 - 3x); \qquad 2P'_3 = 15x^2 - 3;$$

and hence

$$(1 - x^2)W''_3 - 2xW'_3 + 12W_3 = 15x^2 - 3.$$

Putting $W_3 = \alpha x^2 + \beta x + \gamma$ we get $\alpha = \frac{5}{2}$, $\beta = 0$, and $\gamma = -\frac{2}{3}$. In this way we obtain $W_3 = \frac{5}{2}x^2 - \frac{2}{3}$ and the complete solution

$$Q_3(x) = \frac{1}{4}(5x^3 - 3x) \ln \frac{1 + x}{1 - x} - \frac{5}{2}x^2 + \frac{2}{3}.$$

In an analogous way we find

$$Q_0(x) = \frac{1}{2} \ln \frac{1 + x}{1 - x},$$

$$Q_1(x) = \frac{x}{2} \ln \frac{1 + x}{1 - x} - 1,$$

$$Q_2(x) = \frac{1}{4}(3x^2 - 1) \ln \frac{1 + x}{1 - x} - \frac{3}{2}x.$$

Finally we also give two integral representations:

$$P_n(x) = \frac{1}{\pi} \int_0^\pi (x + \sqrt{(x^2 - 1)} \cos t)^n \, dt,$$

$$Q_n(x) = \int_0^\infty (x + \sqrt{(x^2 - 1)} \cosh t)^{-n-1} \, dt.$$

EXERCISES

1. Compute by complex integration
$$\int_{-\infty}^{+\infty} dx/(x^4 + 4).$$

2. Compute by complex integration
$$\int_{-\infty}^{+\infty} (x^2 + 1)^{-1}(x^2 + 4)^{-1}(x^2 + 9)^{-1} \, dx.$$

3. Using Stirling's formula, compute $\Gamma(1/4)$ to 6 decimal places.

4. Compute the integral
$$\int_1^\infty \ln x (x^2 + x^{-2})^{-1} \, dx.$$

5. Compute

$$\int_0^\infty (x/\sinh x)\, dx$$

exactly.

6. Show that

$$\int_0^1 x^x\, dx = \sum_{n=1}^\infty (-1)^{n+1} n^{-n}.$$

7. Compute

$$\int_0^1 x(1 - x^3)^{1/3}\, dx.$$

8. Show that when $n > 1$,

$$\int_0^\infty dx/(1 + x^n) = \int_1^0 (1 - x^n)^{-1/n}\, dx.$$

Also find the exact value.

9. Show that

$$\int_0^\infty (1 + x^2/s)^{-(s+1)/2}\, dx = \tfrac{1}{2}(\pi s)^{1/2}\Gamma(s/2)/\Gamma((s+1)/2)$$

where $s > 0$. Also find the numerical value when $s = 4$.

10. Compute

$$\int_0^{\pi/2} (\tan x)^n\, dx.$$

For which values of n do we have convergence?

11. Compute

$$\int_0^{\pi/2} \cos(\cos x)\, dx.$$

12. Compute

$$\int_0^\pi \exp(2 \cos x)\, dx.$$

13. Compute

$$\int_0^1 \int_0^1 x^{13/3} (1 - x^2)^{1/2} y^{1/6} (1 - y)^{1/2}\, dx\, dy.$$

14. Compute

$$\prod_{n=1}^\infty \frac{n(n+1)}{(n + \frac{1}{2})^2}.$$

15. Let n be an integer > 0. Show that

$$\sum_{k=1}^n \frac{(-1)^{k-1}}{k} \binom{n}{k} = \sum_{k=1}^n \frac{1}{k}$$

by considering the integral

$$I = \int_0^\infty [1 - (1 - e^{-x})^n]\, dx.$$

16. Compute

$$\prod_{n=1}^\infty \frac{(10n - 1)(10n - 9)}{(10n - 5)^2}.$$

17. Compute

$$\prod_{n=1}^{\infty} \frac{n(n + 1)(n + 2)(n + 3)}{(n + \frac{3}{2})^4}.$$

18. Compute

$$\prod_{n=1}^{\infty} n^3/(n^3 + 1)$$

exactly.

19. Find the area T_p bounded by the curve $x^p + y^p = 1, p > 0$, and the positive x- and y-axes. Then compute $\lim_{p \to \infty} p^2(1 - T_p)$.

20. Using the formula

$$\sin \pi z = \pi z \prod_{n=1}^{\infty} (1 - z^2/n^2),$$

find a product representation for the function $e^z - 1$.

21. Prove the formulas $J_0(x) + 2J_2(x) + 2J_4(x) + \cdots = 1$ and $J_0(x) - 2J_2(x) + 2J_4(x) - \cdots = \cos x$ by comparing corresponding terms in the power series expansions.

22. Prove that $(d/dx)(x^n J_n(x)) = x^n J_{n-1}(x)$ (n positive integer).

23. Develop $\cos \alpha x$ and $\cosh \alpha x$ in series of Chebyshev polynomials.

24. Prove the formula

$$(-1)^n x^{n/2} (d^n/dx^n)(J_0(2\sqrt{x})) = J_n(2\sqrt{x}).$$

25. Develop $\cos(x \sin \theta)$ and $\sin(x \sin \theta)$ in Fourier series. Which formulas are obtained for $\theta = 0$ and $\theta = \pi/2$?

26. Find

$$\int_{-1}^{1} x^n P_n(x) \, dx$$

($P_n(x)$ are the Legendre polynomials).

27. Using a suitable recursion formula, show that $P_{2k+1}(0) = 0$ and that

$$P_{2k}(0) = (-1)^k \frac{(2k)!}{(2^k k!)^2}.$$

28. Using Rodrigues' formula, show that

$$(2n + 1)P_n(x) = P'_{n+1}(x) - P'_{n-1}(x).$$

Then compute $\int_0^1 P_{2k+1}(x) \, dx$.

29. Using the formula $P'_n - P'_{n-2} = (2n - 1)P_{n-1}$, express P'_n as a sum of Legendre polynomials. Then compute $\int_{-1}^{1} P'_n(x)^2 \, dx$.

where L_n are the Laguerre polynomials.

31. Compute

   (a) $\int_{-1}^{1} x^r P_n(x)\, dx,$ $\qquad r \geq n, \quad r - n$ even,

   (b) $\int_{0}^{\infty} x^r e^{-x} L_n(x)\, dx,$ $\qquad r \geq n,$

   (c) $\int_{-\infty}^{\infty} \exp(-x^2)\, x^r H_n(x)\, dx,$ $\qquad r \geq n, \quad r - n$ even.

CHAPTER 8

Laplace Transformation

"Do you really believe in this superstition?"
"I have been told that it is effective even if you do
not believe in it."

> Niels Bohr
> (commenting on a horseshoe by the
> entrance to his laboratory)

§8.1. BASIC THEORY

Assume that $f(x)$ and $K(t, x)$ are given functions and define

$$I(t) = \int_a^b f(x)K(t, x)\, dx.$$

Then $I(t)$ is said to be the *integral transform* of the function $f(x)$ with the kernel $K(t, x)$. If we choose $K(t, x) = e^{-tx}$ we obtain the *Laplace transform*, while $K(t, x) = x^{t-1}$ gives the *Mellin transform*. The theory of integral transforms originated in Fourier's theorem for continuous, integrable functions (see (3.4.6)):

(8.1.1)
$$\begin{cases} F(t) = (2\pi)^{-1/2} \int_{-\infty}^{\infty} f(x)e^{itx}\, dx \\ f(x) = (2\pi)^{-1/2} \int_{-\infty}^{\infty} F(t)e^{-itx}\, dt. \end{cases}$$

If we consider the first equation as an integral equation with $f(x)$ the unknown function, then the solution is given by the second equation, and e^{-itx} is called the *solving kernel*. In particular, if the solving kernel is equal to the original kernel, it is called a *Fourier kernel*. Under very mild conditions, as mentioned in (3.4.7), we have:

(8.1.2)
$$F(t) = (2/\pi)^{1/2} \int_0^{\infty} f(x) \cos tx\, dx$$
$$f(x) = (2/\pi)^{1/2} \int_0^{\infty} F(t) \cos tx\, dt$$

where $\cos tx$ could also be replaced by $\sin tx$; both are Fourier kernels. We now return to the Laplace transform defined by $K(t, x) = e^{-tx}$ with the interval in general $(0, \infty)$ (one-sided integrals), or $(-\infty, \infty)$ (two-sided integrals). In the following we shall prefer two-sided integrals, which are slightly more general.

The main idea is now as follows. If we want to compute some kind of expression (e.g., an integral) or to solve some kind of equation (containing differences, derivatives,

128

integrals, and so forth), we transform it and get an image expression or an image equation that we hope may be simpler to treat or to solve. After simplification we then transform back again. In particular we shall show that there is a correspondence between differentiation (integration) in the original plane, and multiplication (division) with a simple factor in the image plane. These properties are quite attractive in many applications.

We start from the following pair of equations:

(8.1.3)
$$\begin{cases} F(p) = (2\pi)^{-1/2} \int_{-\infty}^{\infty} e^{-ipx} H(x)\, dx \\ H(x) = (2\pi)^{-1/2} \int_{-\infty}^{\infty} e^{ipx} F(p)\, dp. \end{cases}$$

Now put $s = q + ip$ and $e^{qx}H(x) = h(x)$. Hence

$$F(p) = F(iq - is) = (2\pi)^{-1/2} \int_{-\infty}^{\infty} e^{-sx} h(x)\, dx.$$

The inverse formula can then be written:

$$H(x) = h(x)e^{-qx} = -i(2\pi)^{-1/2} \int_{q-i\infty}^{q+i\infty} e^{sx} e^{-qx} F(iq - is)\, ds.$$

Finally putting $F(iq - is) = (2\pi)^{-1/2} f(s)/s$ and replacing s by p we find:

(8.1.4)
$$\begin{cases} \dfrac{f(p)}{p} = \int_{-\infty}^{\infty} e^{-px} h(x)\, dx \qquad \mathrm{Re}(p) = q, \quad x \text{ real} \\ h(x) = \dfrac{1}{2\pi i} \int_{q-i\infty}^{q+i\infty} e^{px} \dfrac{f(p)}{p}\, dp. \end{cases}$$

In the following we will use the notation

$$h(x) \Rightarrow f(p)$$

to indicate that the image function $f(p)$ will result from the original function $h(x)$ as defined above.

The situation now is to some extent analogous to what we have in integral calculus: we can construct a table of formulas by performing the simple operation of differentiation, and in this way solve a number of integrals by identification. However, it seems reasonable to start with some simple general rules.

First we introduce a device for facilitating the passage between one- and two-sided transforms, namely the *unit function* $U(x)$ defined through

$$U(x) = \begin{cases} 0, & x < 0 \\ \frac{1}{2}, & x = 0 \\ 1, & x > 0. \end{cases}$$

Taking $U(x - a)$ as the original function we get

$$\frac{f(p)}{p} = \int_{a}^{\infty} e^{-px}\, dx = e^{-ap}/p,$$

i.e.

$$U(x - a) \Rightarrow e^{-ap},$$

and hence

$$U(x) \Rightarrow 1.$$

The last formula explains the name unit function. The derivative of the unit function is an improper function of great importance, particularly in theoretical physics. It is usually called Dirac's delta function and is defined through

$$\delta(x) = 0, \qquad x \neq 0$$

$$\int_{-\infty}^{\infty} \delta(x) \, dx = 1.$$

It is possible to treat this function in a strictly consistent way from a mathematical point of view, but we refrain from doing so. We only mention that derivatives of different orders can also be defined: $\delta'(x)$, $\delta''(x)$, and so forth. We now compute the Laplace transform:

$$f(p) = p \int_{-\infty}^{\infty} e^{-px} \delta(x) \, dx = p.$$

Similar computations can be done for the higher derivatives, and we find the general result:

$$\delta^{(n)}(x) \Rightarrow p^{n+1}.$$

We now proceed by computing the image function of $h(cx)$ when $h(x) \Rightarrow f(p)$. The result is immediate from the definition:

$$\begin{cases} h(cx) \Rightarrow f(p/c), & c \text{ real}, > 0, \\ h(cx) \Rightarrow -f(p/c), & c \text{ real}, < 0. \end{cases}$$

Next we prove the so-called shift rule, which states:

$$\text{If} \quad h(x) \Rightarrow f(p), \quad \text{then} \quad h(x + c) \Rightarrow e^{cp} f(p).$$

This formula follows because

$$p \int_{-\infty}^{\infty} e^{-px} h(x + c) \, dx = pe^{cp} \int_{-\infty}^{\infty} e^{-pz} h(z) \, dz = e^{cp} f(p).$$

Example 1. The rectangle function is defined through

$$h(x) = \begin{cases} 0, & x < -a \text{ and } x > a \\ b, & -a \leq x \leq a. \end{cases}$$

Apart from the isolated points $\pm a$ we have:

$$h(x) = b[U(x + a) - U(x - a)]$$

and hence

$$h(x) \Rightarrow b(e^{ap} - e^{-ap}) = 2b \sinh ap.$$

Example 2. Find the function $h(x)$ from the relation

$$H(x) = \sum_{n=1}^{\infty} h(x/n), \qquad x > 0.$$

Putting $x = e^t$ we get: $H(e^t) = \sum_{n=1}^{\infty} h(e^{t-\ln n})$. Further suppose

$$H(e^t) \Rightarrow F(p),$$
$$h(e^t) \Rightarrow f(p).$$

Hence

$$h(e^{t-\ln n}) \Rightarrow e^{-p \ln n} f(p) = n^{-p} f(p)$$

and

$$F(p) = \sum_{n=1}^{\infty} n^{-p} f(p) = \zeta(p) f(p) \qquad \text{(see (7.5.1))}$$

or $f(p) = F(p)/\zeta(p)$. But $[\zeta(p)]^{-1} = \sum_{n=1}^{\infty} \mu(n) n^{-p}$ where $\mu(n)$, defined in (7.5.3), is the Möbius function. Thus we get:

$$f(p) = \sum_{n=1}^{\infty} \mu(n) F(p) e^{-p \ln n}.$$

Using the shift rule we obtain:

$$H(e^t/n) = H(e^{t-\ln n}) \Rightarrow e^{-p \ln n} F(p)$$

and returning to the original plane:

$$h(x) = \sum_{n=1}^{\infty} \mu(n) H(x/n).$$

There is also another shift rule, which can be obtained as follows. Suppose that $h(x) \Rightarrow f(p)$. Hence

$$e^{-cx} h(x) \Rightarrow p \int_{-\infty}^{\infty} e^{-(p+c)x} h(x)\, dx = \frac{p}{p+c}(p+c) \int_{-\infty}^{\infty} e^{-(p+c)x} h(x)\, dx$$

$$= \frac{p}{p+c} f(p+c).$$

Example 1. $U(x) \Rightarrow 1; \ e^{-cx} U(x) \Rightarrow \dfrac{p}{p+c}.$

So far we have not discussed the legitimacy of our operations. In general it can be said that either the two-sided Laplace integral is divergent, or it converges in a strip $a < \operatorname{Re}(p) < b$. On further operations the convergence region may decrease, increase, or remain constant. However, we must refrain from treating these questions in more detail.

We shall now consider Laplace transformations of derivatives and integrals. Starting from $h(x) \Rightarrow f(p)$ and using the shift rule, we get:

$$\frac{h(x + \varepsilon) - h(x)}{\varepsilon} \Rightarrow \frac{e^{\varepsilon p} - 1}{\varepsilon} f(p).$$

Supposing that it is legitimate to pass to the limit $\varepsilon \to 0$, we find:

(8.1.5) $$h'(x) \Rightarrow p f(p).$$

Of course, this formula can also be proved directly from the definition, provided $h(x)$ disappears sufficiently strongly in $\pm \infty$. Generalization to higher order derivatives is immediate:

$$h''(x) \Rightarrow p^2 f(p)$$
$$\vdots$$
$$h^{(n)}(x) \Rightarrow p^n f(p).$$

We now seek the Laplace transform of the integral $\int_{-\infty}^{x} h(s)\, ds$ and find:

$$p \int_{-\infty}^{\infty} e^{-px}\, dx \int_{-\infty}^{x} h(s)\, ds = p \int_{-\infty}^{\infty} h(s)\, ds \int_{s}^{\infty} e^{-px}\, dx$$
$$= p \frac{1}{p} \int_{-\infty}^{\infty} e^{-ps} h(s)\, ds = f(p)/p,$$

provided $\text{Re}(p) > 0$. Similarly, if $\text{Re}(p) < 0$:

$$\int_{\infty}^{x} h(s)\, ds \Rightarrow f(p)/p.$$

Starting from

$$f(p)/p = \int_{-\infty}^{\infty} e^{-px} h(x)\, dx$$

we get, differentiating n times with respect to p:

$$p(-d/dp)^n \, (f(p)/p) = p \int_{-\infty}^{\infty} e^{-px} x^n h(x)\, dx,$$

which can also be written

$$x^n h(x) \Rightarrow p(-d/dp)^n \, (f(p)/p).$$

Replacing $h(x)$ by $h'(x)$ and f by pf, and putting $n = 1$, we get:

$$xh'(x) \Rightarrow \left(-p \frac{d}{dp}\right) f(p)$$

which can be generalized directly to give

$$\left(x \frac{d}{dx}\right)^n h(x) \Rightarrow \left(-p \frac{d}{dp}\right)^n f(p).$$

If we replace $h(x)$ by $h^{(s)}(x)$ in the formula starting with $x^n h(x)$ above, we obtain:

(8.1.6) $$x^r h^{(s)}(x) \Rightarrow p(-d/dp)^r(p^{s-1}f(p)).$$

Convolution

Let the original function be the so-called convolution integral, i.e.,

$$h(x) = \int_{-\infty}^{\infty} h_1(t)h_2(x-t)\,dt.$$

Then we obtain

$$f(p) = p\int_{-\infty}^{\infty} e^{-px}\,dx \int_{-\infty}^{\infty} h_1(t)h_2(x-t)\,dt$$

$$= \int_{-\infty}^{\infty} h_1(t)\,dt\left[p\int_{-\infty}^{\infty} e^{-px}h_2(x-t)\,dx \right].$$

The expression within brackets is (letting $x - t = s$):

$$pe^{-pt}\int_{-\infty}^{\infty} e^{-ps}h_2(s)\,ds = f_2(p)e^{-pt}.$$

Hence

$$f(p) = f_2(p)\int_{-\infty}^{\infty} e^{-pt}h_1(t)\,dt = \frac{1}{p}f_1(p)f_2(p)$$

and we have the final formula:

(8.1.7) $$\int_{-\infty}^{\infty} h_1(t)h_2(x-t)\,dt \Rightarrow \frac{1}{p}f_1(p)f_2(p)$$

and similarly

$$U(x)\int_0^x h_1(t)h_2(x-t)\,dt \Rightarrow \frac{1}{p}f_1(p)f_2(p).$$

A straightforward but somewhat lengthy calculation gives the following result. If $U(x)h(x) \Rightarrow f(p)$, then

$$U(x)h(x^2) \Rightarrow p\pi^{-1/2}\int_0^{\infty} \exp(-p^2 t^2/4)f(1/t^2)\,dt.$$

---

§8.2. LAPLACE TRANSFORMS OF SOME SPECIAL FUNCTIONS

We start with $h(x) = x^n U(x)$ and find:

(8.2.1) $$x^n U(x) \Rightarrow p\int_0^{\infty} e^{-px}x^n\,dx = \frac{\Gamma(n+1)}{p^n}.$$

Similarly:

(8.2.2)
$$e^{-ax}U(x) \Rightarrow p\int_{-\infty}^{\infty} e^{-(p+a)x}\,dx = \frac{p}{p+a},$$

a result which also follows directly from the second shift rule. Further:

$$\cosh(ax)\,U(x) \Rightarrow \frac{1}{2}\left[\frac{p}{p-a} + \frac{p}{p+a}\right] = \frac{p^2}{p^2 - a^2}$$

$$\sinh(ax)\,U(x) \Rightarrow \frac{1}{2}\left[\frac{p}{p-a} - \frac{p}{p+a}\right] = \frac{ap}{p^2 - a^2}$$

$$\cos(ax)\,U(x) \Rightarrow \frac{1}{2}\left[\frac{p}{p-ia} + \frac{p}{p+ia}\right] = \frac{p^2}{p^2 + a^2}$$

(8.2.3)
$$\sin(ax)\,U(x) \Rightarrow \frac{1}{2i}\left[\frac{p}{p-ia} - \frac{p}{p+ia}\right] = \frac{ap}{p^2 + a^2}$$

$$\exp(-x^2) \Rightarrow p\int_{-\infty}^{\infty} \exp(-x^2 - px)\,dx$$

$$= p\exp(p^2/4)\int_{-\infty}^{\infty} \exp(-(x+p/2)^2)\,dx$$

$$= p\exp(p^2/4)\int_{-\infty}^{\infty} \exp(-z^2)\,dz$$

$$= \pi^{1/2}p\exp(p^2/4).$$

Using the second shift rule we get:

$$U(x)x^n e^{ax} \Rightarrow \Gamma(n+1)/(p-a)^n$$

and further:

$$(\ln x)U(x) \Rightarrow p\int_0^{\infty} e^{-px}\ln x\,dx = \int_0^{\infty} e^{-t}(\ln t - \ln p)\,dt$$

$$= -\gamma - \ln p$$

where γ is Euler's constant, $0.5772156649\ldots$

The Bessel functions of order n are given by

$$J_n(x) = \sum_{k=0}^{\infty} \frac{(-1)^k(x/2)^{n+2k}}{k!\,\Gamma(n+k+1)}.$$

Hence

$$x^{n/2}J_n(2\sqrt{x}) = \sum_{k=0}^{\infty} \frac{(-1)^k x^{n+k}}{k!\,\Gamma(n+k+1)}$$

and since $x^{n+k}U(x) \Rightarrow \Gamma(n+k+1)/p^{n+k}$, we get:

$$x^{n/2}J_n(2\sqrt{x})U(x) \Rightarrow \frac{1}{p^n}\sum_{k=0}^{\infty} \frac{(-1)^k(1/p)^k}{k!} = \exp(-1/p)/p^n.$$

Using the formula giving the image function of $h(x^2)$, we can deduce:

$$(8.2.4) \qquad J_n(x)U(x) \Rightarrow \frac{p}{\sqrt{(p^2+1)}} (\sqrt{(p^2+1)} - p)^n.$$

In particular, $J_0(x)U(x) \Rightarrow p/\sqrt{(p^2+1)}$.

We shall now give some typical examples from different areas to show the power of Laplace transformation.

Example 1. $y'' - 5y' + 6y = 2e^x$ with $y(0) = 0$, $y'(0) = 1$.
Put $h(x) = y(x)U(x)$. Then

$$h'(x) = y'(x)U(x) + y(x)\,\delta(x) = y'(x)U(x)$$

(since $y(0) = 0$). Further:

$$h''(x) = y''(x)U(x) + y'(x)\,\delta(x) = y''(x)U(x) + \delta(x)$$

(since $y'(0) = 1$). Hence:

$$h'' - 5h' + 6h = (y'' - 5y' + 6y)U(x) + \delta(x) = 2e^xU(x) + \delta(x).$$

Transforming, we find:

$$(p^2 - 5p + 6)f = \frac{2p}{p-1} + p = \frac{p(p+1)}{p-1}$$

$$f = \frac{p(p+1)}{(p-1)(p-2)(p-3)} = \frac{p}{p-1} - \frac{3p}{p-2} + \frac{2p}{p-3}.$$

Transforming back, we get:

$$h(x) = e^x - 3e^{2x} + 2e^{3x}.$$

Example 2.   Solve the integral equation

$$\int_0^x y(t)y(x-t)\,dt = \sin 4x.$$

By used of the convolution integral formula, the equation transforms to:

$$\frac{1}{p}f(p)^2 = \frac{4p}{p^2+16} \qquad \text{giving} \qquad f(p) = 2\,\frac{p/4}{\sqrt{\{1 + (p/4)^2\}}}.$$

Transforming back, we get $y = 2J_0(4x)$.

Example 3.   The formula

$$\pi/4 = \sin x + \sin 3x/3 + \sin 5x/5 + \cdots$$

is valid for $0 < x < \pi$. For $\pi < x < 2\pi$ the sum is equal to $-\pi/4$ instead, and so forth. The general formula, valid except when $x = n\pi$, has the form:

$$\frac{\pi}{4}\left[U(x) - 2U(x-\pi) + 2U(x-2\pi) - \cdots\right]$$

$$= U(x)\left(\sin x + \frac{1}{3}\sin 3x + \frac{1}{5}\sin 5x + \cdots\right)$$

Transforming the left hand side, we get:

$$\frac{\pi}{4}(1 - 2e^{-\pi p} + 2e^{-2\pi p} - \cdots) = \frac{\pi}{4}\tanh\frac{\pi p}{2}.$$

Similarly for the right hand side:

$$\frac{p}{p^2 + 1} + \frac{1}{3}\frac{3p}{p^2 + 9} + \frac{1}{5}\frac{5p}{p^2 + 25} + \cdots$$

Hence:

$$\frac{\pi}{4p}\tanh\frac{\pi p}{2} = \frac{1}{p^2 + 1} + \frac{1}{p^2 + 9} + \frac{1}{p^2 + 25} + \cdots$$

Putting $p = iq$ we get:

$$\frac{\pi}{2}\tan\frac{\pi q}{2} = \frac{1}{1 - q} - \frac{1}{1 + q} + \frac{1}{3 - q} - \frac{1}{3 + q} + \frac{1}{5 - q} - \frac{1}{5 + q} + \cdots$$
$$(q \neq \pm 1, \pm 3, \pm 5, \ldots).$$

Example 4.   Find a solution to the equation

$$y''(x) + 2y'(x - 1) + y(x - 2) = 1 \qquad \text{with } y(x) = 0 \text{ for } x < 0.$$

Laplace transformation yields:

$$(p^2 + 2pe^{-p} + e^{-2p})f(p) = 1$$

and hence

$$f(p) = (p + e^{-p})^{-2} = p^{-2}(1 - 2e^{-p}/p + 3e^{-2p}/p^2 - 4e^{-3p}/p^3 + \cdots)$$
$$= \sum_{k=0}^{\infty}\frac{(-1)^k(k + 1)e^{-kp}}{p^{k+2}}.$$

The corresponding original function is:

$$\sum_{k=0}^{\lfloor x\rfloor}\frac{(-1)^k(k + 1)}{(k + 2)!}(x - k)^{k+2}U(x - k).$$

Splitting up into different intervals, we obtain:

$$
\begin{array}{ll}
0 < x < 1 & y = x^2/2 \\
1 < x < 2 & y = x^2/2 - (x - 1)^3/3 \\
2 < x < 3 & y = x^2/2 - (x - 1)^3/3 + (x - 2)^4/8
\end{array}
$$

and so forth.

Example 5.   We develop the function $(\pi - x)/2$ in Fourier series for $0 < x < 2\pi$ and find the result:

$$\frac{\pi - x}{2} = \sin x + \frac{1}{2}\sin 2x + \frac{1}{3}\sin 3x + \cdots$$

To obtain a generally valid formula for $x > 0$ we must add π in the interval $(2\pi, 4\pi)$, 2π in the interval $(4\pi, 6\pi)$, and so forth. In this way we get the general result:

$$\frac{\pi - x}{2} U(x) + \pi[U(x - 2\pi) + U(x - 4\pi) + \cdots]$$

$$= \left[\sin x + \frac{1}{2} \sin 2x + \frac{1}{3} \sin 3x + \cdots \right] U(x).$$

Then Laplace transformation gives:

$$\frac{\pi}{2} [1 + 2e^{-\pi p} + 2e^{-4\pi p} + \cdots] - \frac{1}{2p} = \frac{p}{p^2 + 1} + \frac{p}{p^2 + 4} + \frac{p}{p^2 + 9} + \cdots$$

Since the expression within brackets equals coth $p\pi$, we put $p = iz$ and obtain:

$$\frac{\pi}{2i} \cot \pi z - \frac{1}{2iz} = -iz \left[\frac{1}{z^2 - 1} + \frac{1}{z^2 - 4} + \frac{1}{z^2 - 9} + \cdots \right]$$

and hence

$$\pi \cot \pi z = \frac{1}{z} + \sum_{n=1}^{\infty} \left(\frac{1}{z - n} + \frac{1}{z + n} \right).$$

Integrating this relation we get:

$$\ln \sin \pi z = A + \ln z + \sum_{n=1}^{\infty} (\ln(1 - z/n) + \ln(1 + z/n)).$$

Letting $z \to 0$, we find $A = \ln \pi$ and obtain finally:

(8.2.5)
$$\frac{\sin \pi z}{\pi z} = \prod_{n=1}^{\infty} (1 - z^2/n^2).$$

This result is useful in the theory of the Γ-function, since it proves the relation $\Gamma(z)\Gamma(1 - z) = \pi/\sin \pi z$ (cf. (7.2.5)).

EXERCISES

1. An original function is defined to be 0 when $x \leq 0$; for positive values of x it is defined through a train of straight line segments joining the points $(0, 0)$, $(1, 1)$, $(2, 0)$, $(3, 1)$, $(4, 0)$, ... Compute the image function.

2. Find the image function of

$$h(x) = U(x) \int_x^{\infty} (e^{-t}/t) \, dt.$$

3. Compute the image function of

$$h(x) = U(x) \int_x^{\infty} (\cos t/t) \, dt.$$

4. The error function is defined through

$$\mathrm{erf}(x) = (2/\sqrt{\pi}) \int_0^x \exp(-t^2)\,dt.$$

Compute the image function of $\mathrm{erf}(\sqrt{x})$.

5. The original function is defined through:

$$h(x) = \begin{cases} 0, & x < 0 \\ 1, & 2k\pi < x < (2k+1)\pi, \quad k = 0, 1, 2, \ldots \\ -1, & (2k+1)\pi < x < (2k+2)\pi \end{cases}$$

Find the image function.

6. Find the image function of

$$h(x) = U(x) \int_0^x (\sin t/t)\,dt.$$

7. Find the image function of $h(x)$ when

$$h(x) = \begin{cases} 0, & x < 0 \\ (-1)^n/(n+1), & n < x < n+1, \quad n = 0, 1, 2, \ldots \end{cases}$$

8. Find the original functions of the image functions

$$[(p-1)(p-2)\cdots(p-n)]^{-1} \quad \text{and} \quad [(p+1)(p+2)\cdots(p+n)]^{-1}.$$

9. Find the original function of the image function

$$f(p) = [(p^2+1)(p^2+4)\cdots(p^2+n^2)]^{-1}.$$

10. Find the original function of the image function $f(p) = (p+1)^{-n}$.

11. Which formula is obtained on Laplace transformation of the identity

$$(2/\pi) \sum_{k=0}^{\infty} \cos(2k+1)x/(2k+1)^2 = \begin{cases} \pi/4 - x/2, & 0 < x < \pi \\ -3\pi/4 + x/2, & \pi \le x < 2\pi \end{cases}$$

suitably generalized to include all values $x \ge 0$?

12. An original function is defined through

$$h(x) = \begin{cases} 0, & x < 0 \\ x, & 0 < x < \pi \\ x - 2n\pi, & (2n-1)\pi < x < (2n+1)\pi. \end{cases}$$

This function has the Fourier expansion

$$2U(x)\left(\sin x - \frac{1}{2}\sin 2x + \frac{1}{3}\sin 3x - \cdots\right).$$

Which relation is obtained if these two identical representations are Laplace-transformed?

13. We know that

$$|\sin x| = (2/\pi)\left[1 - 2\sum_{n=1}^{\infty}(\cos 2nx/(4n^2-1))\right].$$

Which relation is obtained on Laplace-transformation of this formula (multiplied by $U(x)$) and replacing p by ip?

14. Compute $\int_0^x \cos(x - t)J_0(t)\,dt$ and $\int_0^x \sin(x - t)J_0(t)\,dt$ and thereafter

$$\int_0^x \cos(t)J_0(t)\,dt \quad \text{and} \quad \int_0^x \sin(t)J_0(t)\,dt.$$

15. Show that

$$(1/x)J_n(x)U(x) \Rightarrow (p/n)(\sqrt{(p^2 + 1)} - p)^n$$

and then compute

$$\int_0^x (J_m(t)/t)(J_n(x - t)/(x - t))\,dt$$

when m and n are integers > 0.

16. Find the image function of $U(x)\sin(2\sqrt{(kx)})/\sqrt{(\pi k)}$ and use the result to solve

$$\int_0^x y(t)y(x - t)\,dt = xJ_2(2\sqrt{x}).$$

17. Solve the integral equation

$$\int_0^x y(t)y(x - t)\,dt = 2(\sin x - x \cos x).$$

18. Using Laplace-transformation, find a solution of the differential equation
$xy'' + (2 - x)y' + 2y = 0$.

19. Using Laplace-transformation, solve the differential equation

$$y^{(iv)} - 4y''' + 6y'' - 4y' + y = 24e^x;$$
$$y(0) = y'(0) = y''(0) = 0; \qquad y'''(0) = 30.$$

20. Using Laplace-transformation, solve the system

$$\begin{cases} y' = 2y - 3z \\ z' = -2y + z \end{cases} \quad \text{with} \quad y(0) = 8,\, z(0) = 3.$$

21. Using Laplace-transformation, solve the system

$$\begin{cases} u' + v' \quad\;\;\; = w \\ \quad\;\; v' - w' = u \\ 7u' \quad\;\; + 4w' = 4v \end{cases} \quad \text{with} \quad \begin{cases} u(0) = 2 \\ v(0) = 5 \\ w(0) = 4. \end{cases}$$

22. Using Laplace-transformation, solve the integral equation

$$g(x) = 1 + x + \int_0^x (x - t)g(t)\,dt.$$

23. Solve the integral equation

$$2y = x^3 + \int_0^x (x - t)^2 y(t)\,dt.$$

24. The Laguerre polynomials $L_n(x)$ have the image function
$$U(x)L_n(x) \Rightarrow n!(1 - p^{-1})^n.$$

Using this, prove the relation

$$L_n'(x) + nL_{n-1} + n(n - 1)L_{n-2} + \cdots + n!L_0 = 0.$$

25. Solve the difference equation

$$y(x) - 2y(x - 1) + 2y(x - 2) - 2y(x - 3) + \cdots = x$$

with $y(x) = 0$ when $x < 0$.

26. Using Laplace-transformation, solve the equation

$$\int_0^x y(t)\, dt + y'(x) = 1 \qquad \text{when } y(0) = 3.$$

27. Solve the integro-differential equation

$$\int_0^x t^2 y(x - t)\, dt - 2y'(x) + 4 = 0 \qquad \text{with } y(0) = 0.$$

28. The spherical harmonics have the following image functions:

$$P_{2n}(\cos x) \Rightarrow \prod_{k=1}^{n} \frac{p^2 + (2k-1)^2}{p^2 + (2k)^2}; \qquad P_{2n-1}(\cos x) \Rightarrow \prod_{k=1}^{n-1} \frac{p^2 + (2k)^2}{p^2 + (2k+1)^2}.$$

Compute the value of

$$\int_0^x P_{2n-1}(\cos t) P_{2n}(\cos(x - t))\, dt.$$

29. Using Laplace-transformation, find a solution in integral form of the differential equation $xy'' - y = 0$.

30. Find the general solution of the equation in Exercise 29 by performing the transformations $x = t^2/4$ and $y = tz$. Then use the result to compute the exact value of the integral obtained in Exercise 29.

CHAPTER 9

Calculus of Variations

Slalom: to find the silliest path between two given points.

The main problem of the calculus of variations is to find functions that minimize or maximize certain integrals, possibly under given extra conditions. If a special function is entered into the integral a certain value will result, and in a way we can say that the independent variable is a function while the dependent variable is a numerical value. We call such a relationship a *functional*.

Here we can treat only some simple classical cases, and we have to refrain from discussing whether a found solution really attains a minimum or a maximum. In its simplest form the main problem can be formulated in the following way.

Find a function $y = y(x)$ such that

$$J = \int_a^b F(x, y, y') \, dx$$

becomes as small as possible. The classical method to solve this problem goes like this. We introduce a comparison function $y + \varepsilon f(x)$ where we claim $f(a) = f(b) = 0$. With $f(x)$ a given function and y the desired solution, J becomes an ordinary function of ε, which must obviously have an extremum for $\varepsilon = 0$.

$$J(\varepsilon) = \int_a^b F(x, y + \varepsilon f(x), y' + \varepsilon f'(x)) \, dx,$$

$$\left(\frac{dJ}{d\varepsilon}\right)_{\varepsilon=0} = \int_a^b \left[\frac{\partial F}{\partial y} f(x) + \frac{\partial F}{\partial y'} f'(x) \right] dx$$

$$= \left[\frac{\partial F}{\partial y'} f(x) \right]_a^b + \int_a^b \left[\frac{\partial F}{\partial y} - \frac{d}{dx} \frac{\partial F}{\partial y'} \right] f(x) \, dx$$

$$= \int_a^b M(x) f(x) \, dx.$$

Hence, since the first term disappears due to the boundary values, we get

$$\int_a^b M(x) f(x) \, dx = 0$$

which can hold only if $M(x) = 0$. In this way we obtain the so-called *Euler equation*:

$$(9.1) \qquad \frac{\partial F}{\partial y} - \frac{d}{dx}\frac{\partial F}{\partial y'} = 0.$$

This condition is necessary but not sufficient; the solutions are called *extremals*.

There is a special case which is quite common, namely that the function F does not contain the variable x explicitly. If we perform the differentiation in Euler's equation we get:

$$\frac{\partial F}{\partial y} - \frac{\partial^2 F}{\partial y \partial y'} y' - \frac{\partial^2 F}{\partial y'^2} y'' = 0.$$

On the other hand we have:

$$\frac{d}{dx}\left(F - y'\frac{\partial F}{\partial y'}\right) = \frac{\partial F}{\partial y} y' + \frac{\partial F}{\partial y'} y'' - y''\frac{\partial F}{\partial y'} - y'\frac{\partial^2 F}{\partial y \, \partial y'} y' - y'\frac{\partial^2 F}{\partial y'^2} y''$$

$$= y'\left[\frac{\partial F}{\partial y} - \frac{\partial^2 F}{\partial y \, \partial y'} y' - \frac{\partial^2 F}{\partial y'^2} y''\right] = 0.$$

Hence we can integrate one step to obtain:

$$(9.2) \qquad F - y'\frac{\partial F}{\partial y'} = \text{const.}$$

The history of the development of the calculus of variations is interesting. Its beginning can be dated to 1662, when Fermat used his elegant principle of least time to solve the problem of the refraction of light between two media of different densities. J. Bernoulli then adapted this method to solve the brachistochrone problem (see below).*

We now quote some classical examples, which may also serve as illustrations.

1. Find a curve $y = y(x)$, passing through two given points above the x-axis, which on rotation around this axis generates a surface with minimum area.

According to Guldin's rule an arc element ds will produce the area $2\pi y \, ds$. Hence the problem can be formulated:

Find $y = y(x)$ such that

$$J = \int_{x=a}^{x=b} y \, ds = \int_a^b y(1 + y'^2)^{1/2} \, dx$$

is minimized. This means that $F = y(1 + y'^2)^{1/2}$ is independent of x, and from $\partial F/\partial y' = yy'(1 + y'^2)^{-1/2}$ we get:

$$y(1 + y'^2)^{1/2} - yy'^2(1 + y'^2)^{-1/2} = c$$

or

$$y = c(1 + y'^2)^{1/2}; \qquad y' = [(y/c)^2 - 1]^{1/2}.$$

---

* Those interested in the historical aspects are referred to Herman H. Goldstine's book: A History of the Calculus of Variations from the 17th through the 19th Century (Springer–Verlag, 1980).

Putting $y = c \cosh z$ and $y' = c \sinh z \, (dz/dx)$, we get $c \, dz = dx$ and $z = (x - \alpha)/c$. The final result is:

$$y = c \cosh \frac{x - \alpha}{c}$$

which is a *catenary*. There is a simple model of the solution to this problem by soap bubbles, connected with two circular rings which are parallel and perpendicular to the line between the centers. If the rings are too far from each other the surface will break up into two disjoint plane surfaces, due to the fact that no catenary through the given points exists. Minimizing the height of the center of gravity for a homogeneous chain leads to the same equations.

2. *The Brachistochrone.* This word is Greek for "shortest time," and it describes the problem of finding a curve between two points A and B in a vertical plane such that a particle will travel between the two points, influenced only by gravity (no friction) and with zero initial velocity, in the shortest possible time.

We introduce a coordinate system with the x-axis horizontal and the y-axis vertical and directed downward. Denoting the velocity by v, we get from the energy principle:

$$\tfrac{1}{2}mv^2 = mgy + \text{constant}$$

where m is the mass of the particle and g the gravitational constant. With $v = ds/dt$ (where as usual $ds^2 = dx^2 + dy^2$), we obtain:

$$(ds/dt)^2 = 2g(y - \alpha).$$

Let T be the total time:

$$T = (2g)^{-1/2} \int_0^s ds/(y - \alpha)^{1/2} = (2g)^{-1/2} \int_a^b \left[\frac{1 + y'^2}{y - \alpha} \right]^{1/2} dx.$$

Hence $F = (1 + y'^2)^{1/2}/(y - \alpha)^{1/2}$, and

$$F - y' \frac{\partial F}{\partial y'} = (1 + y'^2)^{-1/2}(y - \alpha)^{-1/2} = \text{const.} = (2b)^{-1/2}.$$

This gives the differential equation:

$$(1 + y'^2)(y - \alpha) = 2b.$$

(Note that y' is infinite, i.e., the direction is vertical when $y = \alpha$ and the motion starts.) Now we put $y - \alpha = 2b \sin^2 z = b(1 - \cos 2z)$; then $dy = 2b \sin 2z \, dz$, and the differential equation yields:

$$y'^2 = \frac{2b}{y - \alpha} - 1 = 1/\sin^2 z - 1 = \cot^2 z,$$

i.e., $y' = dy/dx = \cot z$ or

$$dx = dy/\cot z = 4b \sin^2 z \, dz = 2b(1 - \cos 2z) \, dz.$$

Integrating, we get $x - a = 2b(z - \frac{1}{2}\sin 2z)$, or putting $u = 2z$:

$$\begin{cases} x - a = b(u - \sin u) \\ y - \alpha = b(1 - \cos u) \end{cases}$$

which is the usual parametric equation of a *cycloid*. It can be shown that there is always exactly one cycloid passing through two given points (note that the constant α is given from the beginning). The total time is

$$T = (b/g)^{1/2}(u_2 - u_1)$$

where $u_1 = 0$. If the points are $(0, 0)$ and $(2, 2)$ (meters), then $a = \alpha = 0$, $b = 1.1458$. $u_2 = 2.412$, and $T = 0.824$ (seconds) (where $g = 9.81$). For comparison, if the particle glides along a straight line the time will be $(8/g)^{1/2} = 0.903$ seconds.

We will now generalize this discussion in different directions. First we consider the case when the integral also contains, e.g., the second derivative. Proceeding along the same lines as before, we obtain the Euler equation:

(9.3)
$$\frac{d^2}{dx^2}\frac{\partial F}{\partial y''} - \frac{d}{dx}\frac{\partial F}{\partial y'} + \frac{\partial F}{\partial y} = 0.$$

However, another generalization is perhaps of greater interest, namely to the case of several independent variables. We exemplify this by treating a function u of two variables x and y. We introduce the notations $u_x(x, y) = \partial u/\partial x$ and $u_y(x, y) = \partial u/\partial y$. We then consider the integral

$$I(u) = \iint_R F(x, y, u, u_x, u_y)\, dx\, dy$$

where u is supposed to take on prescribed values on the boundary of R, often denoted ∂R. Using similar methods as in the one-dimensional case, we obtain the necessary condition:

(9.4)
$$\frac{\partial}{\partial x}\frac{\partial F}{\partial u_x} + \frac{\partial}{\partial y}\frac{\partial F}{\partial u_y} - \frac{\partial F}{\partial u} = 0.$$

Examples. Let

$$I(u) = \iint_R \left[\left(\frac{\partial u}{\partial x}\right)^2 + \left(\frac{\partial u}{\partial y}\right)^2\right] dx\, dy.$$

Then the Euler equation takes the form:

$$\frac{\partial^2 u}{\partial x^2} + \frac{\partial^2 u}{\partial y^2} = 0;$$

this partial differential equation is known as Laplace's equation and will be discussed extensively in a subsequent chapter. If we add the term $2uf(x, y)$ to the integrand above, we get instead

$$\frac{\partial^2 u}{\partial x^2} + \frac{\partial^2 u}{\partial y^2} = f(x, y)$$

i.e., Poisson's equation. For equations originating in physics, the minimization of certain integrals often has something to do with basic physical principles, such as the energy principle. This is probably why methods of this kind were first introduced by people working on problems in mathematical physics.

It is now interesting to observe that there is another minimization method, which is similar to the classical variational approach. We consider the class V of functions v such that $v(x)$ is continuous in the interval $(0, 1)$ with $v(0) = v(1) = 0$ and $v'(x)$ is bounded and piecewise continuous. Then we pose the problem to find a function $u(x)$ such that

$$\int_0^1 u'(x)v'(x)\,dx = \int_0^1 f(x)v(x)\,dx$$

valid for all functions $v \in V$. For simplicity we use the notation

$$(v, w) = \int_0^1 v(x)w(x)\,dx.$$

Hence we have $(u', v') - (f, v) = 0$ for all $v \in V$, and by partial integration we obtain:

$$[u'v]_0^1 - (u'', v) - (f, v) = 0,$$

where the first term vanishes. From this we conclude that

$$(u'' + f, v) = 0,$$

and since v can be chosen arbitrarily we must have $u'' + f = 0$. It is easy to show that if u is a solution of this equation, then $(u', v') - (f, v) = 0$ for all $v \in V$.

The minimization principle discussed above is particularly useful in the finite element method, which will be treated to some extent in the chapters on ordinary and partial differential equations.

Applications

We have seen that problems within the calculus of variations give rise to differential equations with certain boundary conditions. In many applications it is, in fact, possible to work the other way round. This means that with a certain differential equation given, we try to find an integral whose Euler equation is exactly the given equation. If this is possible we have transformed the initial problem into a minimization problem, which can conveniently be solved by expanding the desired function into a series of orthogonal polynomials. The integral can then be expressed in the coefficients, and it is an easy matter to find the minimum value. This value is often remarkably close to the theoretical value, even if we use just a few coefficients. This general method is called the *Ritz principle*, and in the chapter on differential equations we will give some examples as illustrations.

EXERCISES

1. Find all catenaries passing through the points (1, 1) and (2, 3):

$$y = c \cosh \frac{x - a}{c}.$$

2. A cycloid arc passes through the origin where the tangent is vertical, and through the point (5, 1) (the y-axis is directed downward). Find the lowest point on the arc.

3. Find the extremals corresponding to the integral

$$\int_a^b (1 + y'^2)^{1/2} \, dx.$$

4. Find the extremals of the integral

$$\int_a^b [y - c(1 + y'^2)^{1/2}] \, dx.$$

Part Two

EQUATIONS

CHAPTER 10

Systems of Linear Equations

"Mine is a long and sad tale" said the Mouse,
turning to Alice and sighing.
"It is a long tail, certainly" said Alice, looking
down with wonder at the Mouse's tail, "but why do
you call it sad?"

Lewis Carroll

In this chapter we treat the problem of solving systems of linear equations. We assume that we have m equations in n unknowns. If $m > n$, there is in general no solution. In a subsequent chapter we will discuss how to find an approximate "best" solution by minimizing a suitably defined error. If instead $m < n$, there are in general infinitely many solutions. Here we will treat only the case $m = n$. The system can then be written $Ax = b$ where A is the coefficient matrix and b the known right-hand side vector. Suppose that $D = \det A$ with $D \neq 0$, and further that D_r is the determinant obtained when the rth column of A is replaced by b. Then according to Cramer's rule we have $x_r = D_r/D$. If $D = 0$ and $b \neq 0$, there are in general no solutions; however, it is still possible to find "best solutions" even if they are not unique (cf. the section on pseudo-inverses). If $b = 0$ and $D \neq 0$, there is only the trivial solution $x = 0$. A necessary condition for nontrivial solutions when $b = 0$ is then $D = 0$.

Cramer's rule, which of course gives the solution $x = A^{-1}b$, should never be used in practice due to the excessive amount of computation needed for the determination of $n + 1$ determinants; in fact each determinant requires $O(n^3)$ operations, giving a total of $O(n^4)$. It is of course completely prohibitive to use the basic definition of determinants, which would take about $(n - 1)n!$ multiplications and $n!$ additions.

In the following we will distinguish between *direct* and *iterative* methods. Among the direct methods the classical Gauss elimination has a well-deserved reputation and is widely used, sometimes in a slightly modified form (e.g., as suggested by Crout). We will also briefly mention Jordan's method, although it requires considerably more computing effort.

§10.1. GAUSSIAN ELIMINATION

We will now demonstrate Gaussian elimination on a simple explicit example. Consider the system

$$
\begin{aligned}
x - 2y + 3z - 4u &= 0 \\
3x - 2y + 3z - 7u &= 5 \\
5x - 18y + 29z - 23u &= 1 \\
4x - 4y \qquad\;\; - 29u &= -25
\end{aligned}
$$

or in matrix form $Aw = b$, where we write $A \,|\, b$ together:

$$
\left[
\begin{array}{rrrr|r}
① & -2 & 3 & -4 & 0 \\
3 & -2 & 3 & -7 & 5 \\
5 & -18 & 29 & -23 & 1 \\
4 & -4 & 0 & -29 & -25
\end{array}
\right]
$$

We now eliminate x from the second, third, and fourth rows by subtracting suitable multiples of the first row. In this special case we use the multipliers 3, 5, and 4, which we then place as a column of its own to the left:

$$
\begin{array}{r}
\\
3 \\
5 \\
4
\end{array}
\left[
\begin{array}{rrrr|r}
1 & -2 & 3 & -4 & 0 \\
0 & ④ & -6 & 5 & 5 \\
0 & -8 & 14 & -3 & 1 \\
0 & 4 & -12 & -13 & -25
\end{array}
\right]
$$

Obviously the problem has now been reduced to three dimensions, and the same procedure is repeated while the first row and the first column are left aside. Hence, the second row multiplied with -2 and 1 is subtracted from the third and fourth rows, the new multipliers written down in an analogous manner:

$$
\begin{array}{rr}
\\
3 & \\
5 & -2 \\
4 & 1
\end{array}
\left[
\begin{array}{rrrr|r}
1 & -2 & 3 & -4 & 0 \\
0 & 4 & -6 & 5 & 5 \\
0 & 0 & ② & 7 & 11 \\
0 & 0 & -6 & -18 & -30
\end{array}
\right]
$$

One more step using the multiplier -3 gives the result:

$$
\begin{array}{rrr}
\\
3 & & \\
5 & -2 & \\
4 & 1 & -3
\end{array}
\left[
\begin{array}{rrrr|r}
1 & -2 & 3 & -4 & 0 \\
0 & 4 & -6 & 5 & 5 \\
0 & 0 & 2 & 7 & 11 \\
0 & 0 & 0 & 3 & 3
\end{array}
\right]
$$

We have in fact obtained a triangular system, which can be solved directly by back-substitution:

$$\left.\begin{array}{r} x - 2y + 3z - 4u = 0 \\ 4y - 6z + 5u = 5 \\ 2z + 7u = 11 \\ 3u = 3 \end{array}\right\}$$ giving in turn $u = 1$, $z = 2$, $y = 3$, and $x = 4$.

It should be easy to construct a general algorithm from this scheme. However, some caution must be exercised. The circled elements play a decisive role when the elimination is performed: the elements below in the same column divided by this *pivot element* determine the multipliers used. From this we understand that the pivot elements must never be zero. If a zero appears we simply let two equations change place so that we get a nonzero pivot element. This is always possible provided that A is nonsingular. But even the case when the pivot elements get small must be considered carefully because the relative errors have a tendency to become large. The remedy is the same: among all possible candidates, select the largest element as pivot. This means that we have to perform row-wise permutations, i.e., change the order of the equations. It is obvious that these changes need not be recorded. The process just described is known as *row-wise* or *partial* pivoting.

On the other hand, we can also search for the largest element (in absolute value) in the whole remaining matrix and perform permutations both row-wise and column-wise. Since the latter operation means a change of order between the unknowns, such permutations must be recorded so that the final solutions can be assigned to the right variables. This total pivoting is costly and time-consuming, and is in general not considered to be worthwhile.

We now turn to the multipliers, which we arranged to form a triangular matrix. Placing ones in the main diagonal and zeros elsewhere we obtain in this special case:

$$L = \begin{pmatrix} 1 & 0 & 0 & 0 \\ 3 & 1 & 0 & 0 \\ 5 & -2 & 1 & 0 \\ 4 & 1 & -3 & 1 \end{pmatrix}$$

Denoting our final upper triangular matrix by R we have:

$$R = \begin{pmatrix} 1 & -2 & 3 & -4 \\ 0 & 4 & -6 & 5 \\ 0 & 0 & 2 & 7 \\ 0 & 0 & 0 & 3 \end{pmatrix}$$

If we execute the multiplication LR we see that we actually perform the elimination steps backwards, which means that $LR = A$. Hence we can interpret the elimination process as a step-by-step multiplication from the left by the inverse L^{-1} (which, of course, is not computed explicitly!) to give the much simpler system $Rx = c$, where $c = L^{-1}b$ is computed simultaneously (here we get $c^T = (0, 5, 11, 3)$). In this discussion we have ignored possible permutations, but at least in principle they could be performed before we start the elimination process. It is obvious that the algorithm works also in the case when we have several right-hand sides. In particular, the inverse of a matrix can be determined by using the identity matrix as right-hand sides. However, in practical work the inverse is computed explicitly only in exceptional cases.

The Gauss elimination produces a decomposition of the matrix A in the form $A = LR$ where L is lower triangular with ones in the main diagonal, while R is upper triangular. It is self-evident that by a trivial division with the diagonal elements in R we obtain the decomposition $A = LDR$, where the new R also has ones in the main diagonal while D is diagonal. In our example we get:

$$
A = \begin{pmatrix} 1 & 0 & 0 & 0 \\ 3 & 1 & 0 & 0 \\ 5 & -2 & 1 & 0 \\ 4 & 1 & -3 & 1 \end{pmatrix} \begin{pmatrix} 1 & 0 & 0 & 0 \\ 0 & 4 & 0 & 0 \\ 0 & 0 & 2 & 0 \\ 0 & 0 & 0 & 3 \end{pmatrix} \begin{pmatrix} 1 & -2 & 3 & -4 \\ 0 & 1 & -3/2 & 5/4 \\ 0 & 0 & 1 & 7/2 \\ 0 & 0 & 0 & 1 \end{pmatrix}
$$

From the construction it follows immediately that the decomposition is unique.

If, in particular, A is symmetric and positive definite, then except in some very special cases there exists a decomposition $A = LL^T$, where L may also contain purely imaginary elements (Cholesky decomposition).

Jordan's Method

In the Gaussian method the elimination was performed only downward, and in this way a new coefficient matrix, upper triangular in form, was produced. Following Jordan we now eliminate both downward and upward, first dividing the elements in the pivot row by the pivot element. In this way the coefficient matrix finally becomes just the unit matrix. We demonstrate the method on a matrix inversion; i.e., we solve the system $AX = I$ where the n right-hand sides are the column vectors of I. We choose $n = 4$ and A as the so-called Wilson matrix; the active part is the 4 by 4 matrix within the frame.

10	7	8	7	1	0	0	0
7	5	6	5	0	1	0	0
8	6	10	9	0	0	1	0
7	5	9	10	0	0	0	1

1	0.7	0.8	0.7	0.1	0	0	0
0	0.1	0.4	0.1	-0.7	1	0	0
0	0.4	3.6	3.4	-0.8	0	1	0
0	0.1	3.4	5.1	-0.7	0	0	1

1	0	-2	0	5	-7	0	0
0	1	4	1	-7	10	0	0
0	0	2	3	2	-4	1	0
0	0	3	5	0	-1	0	1

$$
\begin{array}{ccc|cccc|c}
1 & 0 & 0 & 3 & 7 & -11 & 1 & 0 \\
0 & 1 & 0 & -5 & -11 & 18 & -2 & 0 \\
0 & 0 & 1 & 1.5 & 1 & -2 & 0.5 & 0 \\
0 & 0 & 0 & 0.5 & -3 & 5 & -1.5 & 1
\end{array}
$$

$$
\begin{array}{cccc|cccc}
1 & 0 & 0 & 0 & 25 & -41 & 10 & -6 \\
0 & 1 & 0 & 0 & -41 & 68 & -17 & 10 \\
0 & 0 & 1 & 0 & 10 & -17 & 5 & -3 \\
0 & 0 & 0 & 1 & -6 & 10 & -3 & 2
\end{array}
$$

When the elements of both A and A^{-1} are integers as in this example, it follows that det $A = +1$ or -1 (here it is $+1$).

Operation Counts

We consider the Gaussian elimination at a stage when k rows remain to be handled. If we have p right-hand sides, we must perform $(k-1)$ divisions and $(k-1)^2$ multiplications and additions in the coefficient matrix, and $p(k-1)$ multiplications and additions on the right-hand sides. Hence we have in total

$$\sum_{k=1}^{n-1} k = n(n-1)/2 \quad \text{divisions}$$

$$\sum_{k=1}^{n-1} (k^2 + pk) = n(n-1)(2n-1)/6 + pn(n-1)/2 \quad \text{multiplications and additions.}$$

Further, for the back-substitution we need $pn(n-1)/2$ multiplications and additions, and pn divisions. If n is large we need in total

$$\sim n^3/3 + pn^2 \quad \text{additions and multiplications}$$
$$\sim n^2/2 + pn \quad \text{divisions.}$$

A similar analysis for Jordan's method gives about $\frac{1}{2}n^3 + pn^2$ additions and multiplications, and $\frac{1}{2}n^2 + pn$ divisions, and hence this method is less efficient.

---

§10.2. ERROR ANALYSIS

We will now turn to the question of how errors of different kinds influence the final solution obtained. Three sources should be taken into account:

1. Errors in the coefficient matrix, i.e., because we work with a matrix $A + \delta A$ instead of A.

2. Errors in the right-hand sides, i.e., because we are working with the vector $b + \delta b$ instead of b.

3. Rounding errors because we are performing the computations with values rounded to s significant digits instead of the exact values.

We shall start by analyzing the first two error types. Suppose that the system of equations is written:

$$(A + \delta A)(x + \delta x) = b + \delta b$$

where x is the solution of the "exact" system, i.e., $Ax = b$. We will now try to find an estimate of the error δx. Subtracting $Ax = b$ we get:

$$(A + \delta A)\,\delta x = \delta b - \delta A \cdot x \quad \text{or} \quad \delta x = (A + \delta A)^{-1}(\delta b - \delta A \cdot x).$$

Introducing the notation $A^{-1}\,\delta A = E$, we have $A + \delta A = A(I + E)$. Further if we assume $\|\delta A\| < 1/\|A^{-1}\|$, we obtain

$$\|E\| = \|A^{-1}\,\delta A\| \le \|A^{-1}\| \cdot \|\delta A\| < 1.$$

Hence we can use the theorem below, which was proved in (2.4.1):

$$1/(1 + \|E\|) \le \|(I \pm E)^{-1}\| \le 1/(1 - \|E\|)$$

and so we obtain

$$\|(I + E)^{-1}\| \le 1/(1 - \|E\|) \le 1/(1 - \|A^{-1}\| \cdot \|\delta A\|).$$

Observing that $\delta x = (A(I + E))^{-1}(\delta b - \delta A \cdot x)$, we find for the norm of δx:

$$\|\delta x\| \le \frac{\|A^{-1}\|}{1 - \|A^{-1}\| \cdot \|\delta A\|}(\|\delta b\| + \|\delta A\| \cdot \|x\|).$$

From $Ax = b$ we have $\|b\| = \|Ax\| \le \|A\| \cdot \|x\|$ or $1/\|x\| \le \|A\|/\|b\|$, and

$$\frac{\|\delta x\|}{\|x\|} \le \frac{\|A^{-1}\|}{1 - \|A^{-1}\| \cdot \|\delta A\|}\left(\frac{\|\delta b\|}{\|x\|} + \|\delta A\|\right)$$

$$\le \frac{\|A^{-1}\|}{1 - \|A^{-1}\| \cdot \|\delta A\|}\left(\frac{\|A\| \cdot \|\delta b\|}{\|b\|} + \|\delta A\|\right).$$

We take out the factor $\|A\|$ and introduce the *condition number* of A:

$$\text{cond}(A) = \mu = \|A\| \cdot \|A^{-1}\|$$

which gives

$$\frac{\|\delta x\|}{\|x\|} \le \frac{\mu}{1 - \mu\|\delta A\|/\|A\|}\left(\frac{\|\delta A\|}{\|A\|} + \frac{\|\delta b\|}{\|b\|}\right).$$

Denoting the "relative" errors

$$\varepsilon_1 = \|\delta A\|/\|A\|, \qquad \varepsilon_2 = \|\delta b\|/\|b\|, \qquad \varepsilon = \|\delta x\|/\|x\|$$

we find

(10.2.1)
$$\varepsilon \le \frac{\mu}{1 - \varepsilon_1\mu}(\varepsilon_1 + \varepsilon_2).$$

Note in particular that μ cannot be treated as a small quantity, since

$$1 = \|I\| = \|AA^{-1}\| \le \|A\| \cdot \|A^{-1}\| = \mu, \qquad \text{i.e.,} \qquad \mu \ge 1.$$

For the errors introduced through rounding, the situation is completely different. It has turned out that no sensible results could be obtained in the traditional way. Instead the following question should be asked: Which original system would give the obtained answer if all computations were done *exactly*? We know that the system $Ax = b$ gives the solution ξ when all computations are performed in the usual approximate way. On the other hand, a certain disturbed system $(A + \delta A)x = b + \delta b$ will give the same answer $x = \xi$ after exact computations. What estimates can be given for δA and δb? This question was answered in the 1950's by Wilkinson through his famous *backward analysis*. Here we will only quote the result, which is as follows if the machine precision is $\eta = \frac{1}{2}N^{-s}$ and all elements of A and b are less than 1 in absolute value:

$$(10.2.2) \quad (\delta A \,|\, \delta b) \le \eta
\begin{pmatrix}
0 & 0 & 0 & 0 & \cdots & 0 & 0 & 0 & 1 \\
1 & 1 & 1 & 1 & \cdots & 1 & 1 & 1 & 3 \\
1 & 2 & 2 & 2 & \cdots & 2 & 2 & 2 & 5 \\
1 & 2 & 3 & 3 & \cdots & 3 & 3 & 3 & 7 \\
\vdots & & & & & & & & \\
1 & 2 & 3 & 4 & \cdots & n-2 & n-2 & n-2 & 2n-3 \\
1 & 2 & 3 & 4 & \cdots & n-2 & n-1 & n-1 & 2n-1
\end{pmatrix}$$

Taking norms, we find after some straight-forward computation:

$$\|\delta A\|_E < \eta n^2/\sqrt{6}; \qquad \|\delta b\| < 2\eta n^{3/2}/\sqrt{3}.$$

If the condition number is large, this does not necessarily indicate that the relative error in x becomes large, only that there exist right-hand sides such that $\|\delta x\|/\|x\|$ becomes large. If so, the problem is *ill-conditioned*, which is often a hint that it has been formulated in an awkward manner.

§10.3. OVERDETERMINED SYSTEMS OF EQUATIONS

An overdetermined system has more equations than unknowns. We will start with a simple example and choose the system

$$3x + 2y = 23$$
$$-x + 4y = 11$$
$$5x + y = 43$$

or

$$At = b \qquad \text{where} \qquad A = \begin{pmatrix} 3 & 2 \\ -1 & 4 \\ 5 & 1 \end{pmatrix}, \quad t = \begin{pmatrix} x \\ y \end{pmatrix}, \quad b = \begin{pmatrix} 23 \\ 11 \\ 43 \end{pmatrix}.$$

For given values of x and y we obtain a residual vector $r = At - b$ or

$$r = \begin{pmatrix} 3x + 2y - 23 \\ -x + 4y - 11 \\ 5x + y - 43 \end{pmatrix}.$$

We now try to minimize

$$\|r\|_2^2 = (3x + 2y - 23)^2 + (-x + 4y - 11)^2 + (5x + y - 43)^2,$$

and after simple calculations we get the system

$$\begin{cases} 5x + y = 39 \\ x + 3y = 19 \end{cases}$$

with the solution $t = \begin{pmatrix} x \\ y \end{pmatrix} = \begin{pmatrix} 7 \\ 4 \end{pmatrix}$. Any solution (x, y) can be interpreted as a linear combination of the two column vectors of A:

$$u = \begin{pmatrix} 3 \\ -1 \\ 5 \end{pmatrix}, \qquad v = \begin{pmatrix} 2 \\ 4 \\ 1 \end{pmatrix}.$$

These two vectors cannot span a three-dimensional space. To attain this we add a third vector w, linearly independent of u and v and orthogonal to both of them. After an easy calculation we find $w = \begin{pmatrix} 3 \\ -1 \\ -2 \end{pmatrix}$. The residual vector for the minimum solution is easily computed, and we get $r = \begin{pmatrix} 6 \\ -2 \\ -4 \end{pmatrix} = 2w$. Hence r is orthogonal to u and v. An exact solution of our system cannot be obtained in terms of u and v, since we also need the vector w, orthogonal to u and v. We see that the minimum solution can be viewed as the projection in the (u, v)-plane of the complete solution $7u + 4v - 2w$.

It is now easy to treat the general case. Let the system be $Ax = y$ where A is $m \times n$, with $m > n$, and x and y are vectors of length n and m, respectively. We form the residual vector $r = Ax - y$ and try to find a vector x that minimizes $F = \|r\|_2^2$. We form

$$F = (Ax - y)^T(Ax - y) = x^T A^T A x - 2x^T A^T y + y^T y$$

(note that $x^T A^T y = y^T Ax$ is a pure number). Writing $B = A^T A$ we have

$$\frac{1}{2} F = \frac{1}{2} \sum_{i,k} b_{ik} x_i x_k - \sum_{i,k} x_i a_{ki} y_k + \frac{1}{2} \sum_k y_k^2.$$

Differentiating with respect to x_i, we get:

$$\frac{1}{2} \partial F / \partial x_i = \sum_k b_{ik} x_k - \sum_k a_{ki} y_k = 0$$

or in matrix form $Bx = A^T Ax = A^T y$ with the solution

$$x = (A^T A)^{-1} A^T y = A^+ y \qquad \text{(see Section 2.8).}$$

Now let S be the m-dimensional subspace spanned by the column vectors of A, and let R be the orthogonal complement. Then the vector y can be uniquely represented as $y = s + r$ where $s \in S$ and $r \in R$. We have actually tried to represent y as a vector $s \in S$, but since this is impossible we have instead determined a vector s such that the norm of the residual vector r is minimized. It is immediately clear that $A^T r = 0$ since $x = (A^T A)^{-1} A^T y$ and $r = Ax - y = A(A^T A)^{-1} A^T y - y$. Hence $A^T r = (A^T A)(A^T A)^{-1} A^T y - A^T y = 0$. This means that r is orthogonal to the column vectors of A, as shown previously in the example.

We will treat a greater variety of problems on data fitting using similar methods in Chapter 15.

§10.4. ITERATIVE METHODS

We are now going to discuss how to solve the linear system $Ax = y$ where we assume that A is of type $n \times n$ and not singular. Further it is supposed that n is so large that an iterative method is more attractive than a direct one. However, it turns out that, in general, iterative methods can be applied only when A is *diagonally dominant*. More precisely, we demand that

$$|a_{ii}| \geq \sum_{j \neq i} |a_{ij}|$$

with strict inequality for at least one value i. Any iterative method will produce a series of vectors approximating the exact solution $x = A^{-1} y$. We denote a certain approximation vector by ξ and define the *residual vector* r and the *error vector* e as follows:

$$(10.4.1) \qquad \begin{cases} r = A\xi - y \\ e = \xi - A^{-1} y. \end{cases}$$

Obviously, $e = A^{-1} r$, and we get the estimate $\|e\| \leq \|A^{-1}\| \cdot \|r\|$. Unfortunately, this is not very useful, since even if r is easily computed we can say nothing about $\|A^{-1}\|$. In fact, it is quite possible that $\|r\|$ is small and $\|e\|$ is large.

Suppose now that we start by choosing an initial approximation vector x_0. Hence we have $r_0 = Ax_0 - y$. We will then construct an iterative method depending on an approximate inverse $B \sim A^{-1}$. Assume that we have already obtained x_{k-1}. Then the corresponding residual vector will be $r_{k-1} = Ax_{k-1} - y$. Putting $y = Ax$, we find the exact solution

$$x = x_{k-1} - A^{-1} r_{k-1}.$$

Replacing A^{-1} by B we instead obtain a new and (we hope) better approximation x_k: $x_k = x_{k-1} - Br_{k-1}$. Premultiplying by A and subtracting y we get:

$$Ax_k - y = Ax_{k-1} - y - ABr_{k-1} \quad \text{or} \quad r_k = (I - AB)r_{k-1}.$$

Taking $x_0 = By$ we find $r_0 = -(I - AB)y$, and denoting $I - AB$ by E we have $r_k = -E^{k+1}y$.

Summing the relations

$$\begin{cases} x_0 = By \\ x_1 = x_0 - Br_0 \\ x_2 = x_1 - Br_1 \\ \quad\vdots \\ x_k = x_{k-1} - Br_{k-1} \end{cases}$$

we find:

$$\begin{aligned} x_k &= B(y - r_0 - r_1 - \cdots - r_{k-1}) \\ &= B(I + E + E^2 + \cdots + E^k)y. \end{aligned}$$

If $\|E\| < 1$, the infinite series $I + E + E^2 + \cdots$ converges toward $(I - E)^{-1} = (AB)^{-1} = B^{-1}A^{-1}$, and

$$\lim_{k \to \infty} x_k = B(I - E)^{-1}y = A^{-1}y$$

as it should be.

In the following we shall discuss three particular methods, namely the Jacobi, Gauss–Seidel, and Successive Over-Relaxation (SOR) methods. In the Jacobi method we split the matrix A into $D + C$ where D consists of the diagonal of A with zeros elsewhere. Starting from $x_0 = 0$, for example, we form successively $x_1, x_2, \ldots$ from

$$x_{k+1} = -D^{-1}Cx_k + D^{-1}y = Bx_k + c.$$

Thus we get

$$x_1 = c; \quad x_2 = (I + B)c; \quad x_3 = (I + B + B^2)c; \ldots$$

Assuming $\|B\| < 1$, we obtain

$$\begin{aligned} \lim_{k \to \infty} x_k &= (I - B)^{-1}c = (I - B)^{-1}D^{-1}y = \{D(I - B)\}^{-1}y \\ &= (D + C)^{-1}y = A^{-1}y. \end{aligned}$$

In general, without giving precise conditions, we conclude that convergence is obtained if the matrix A has a pronounced diagonal dominance.

We shall now discuss the method of Gauss–Seidel. Starting from the system $Ax = y$, we construct a series of approximate solutions in the following way. The first vector $x^{(0)}$ is chosen to be $\mathbf{0}$. We put for example $x_2 = x_3 = \cdots = x_n = 0$ in the first equation, and solve for $x_1 = x_1^{(1)}$. In the second equation, we put $x_3 = x_4 = \cdots = x_n = 0$ and further $x_1 = x_1^{(1)}$, and then we solve for $x_2 = x_2^{(1)}$, and so on.

In this way we obtain a vector

$$x^{(1)} = \begin{pmatrix} x_1^{(1)} \\ \vdots \\ x_n^{(1)} \end{pmatrix},$$

which ought to be a better approximation than $x^{(0)}$, and the whole procedure can be repeated. Introducing the matrices

$$A_1 = \begin{pmatrix} a_{11} & 0 & 0 & \cdots & 0 \\ a_{21} & a_{22} & 0 & \cdots & 0 \\ \vdots & & & & \\ a_{n1} & a_{n2} & a_{n3} & \cdots & a_{nn} \end{pmatrix}; \quad A_2 = \begin{pmatrix} 0 & a_{12} & a_{13} & \cdots & a_{1n} \\ 0 & 0 & a_{23} & \cdots & a_{2n} \\ \vdots & & & & \\ 0 & 0 & 0 & \cdots & 0 \end{pmatrix};$$

$$A_1 + A_2 = A,$$

we can formulate the Gauss–Seidel iteration method: $A_1 x^{(p+1)} = y - A_2 x^{(p)}$.

Choosing $x^{(0)} = 0$, for $p = 0$ we get $x^{(1)} = A_1^{-1} y$. Further, for $p = 1$, we have $x^{(2)} = A_1^{-1}(y - A_2 A_1^{-1} y) = (I - A_1^{-1} A_2) A_1^{-1} y$ and analogously for $p = 2$, we have $x^{(3)} = (I - A_1^{-1} A_2 + A_1^{-1} A_2 A_1^{-1} A_2) A_1^{-1} y$. Putting $A_1^{-1} A_2 = E$, we find by induction that $x^{(p)} = (I - E + E^2 - \cdots + (-1)^{p-1} E^{p-1}) A_1^{-1} y$. When $p \to \infty$ the series within parentheses converges if $\|E\| < 1$, and the final value is

$$x = (I + E)^{-1} A_1^{-1} y = \{A_1(I + E)\}^{-1} y = (A_1 + A_2)^{-1} y = A^{-1} y.$$

The condition for convergence of Gauss–Seidel's method is that all eigenvalues of $A_1^{-1} A_2$ must lie inside the unit circle. For most practical purposes the following criterion can be used: We have convergence if for $i = 1, 2, \ldots, n$, $|a_{ii}| > S_i$, where $S_i = \sum_{k \neq i} |a_{ik}|$. We also form $\rho_i = S_i/(|a_{ii}| - S_i)$ and put $\rho = \max_i \rho_i$. Then the method converges if all $\rho_i > 0$, and it converges more rapidly the smaller ρ is. An empirical rule is that ρ should be < 2 in order to produce sufficiently fast convergence. Essentially, it is important that the matrix has a clear diagonal dominance.

On the whole we can say that Gauss–Seidel's method converges twice as fast as Jacobi's; a proof for this is given at the end of this section.

Example 1

$$\begin{cases} 8x - 3y + 2z = 20, \\ 4x + 11y - z = 33, \\ 6x + 3y + 12z = 36. \end{cases}$$

i	x_i	y_i	z_i
1	2.5	2.1	1.2
2	2.988	2.023	1.000
3	3.0086	1.9969	0.9965
4	2.99971	1.99979	1.00020
5	2.999871	2.000065	1.000048

Systems of equations with a large number of unknowns appear essentially when one wants to solve ordinary or partial differential equations by difference technique. The coefficient matrices are often sparse, and moreover they usually possess a special characteristic which we call property A. This property is defined in the following way. Let n be the order of the matrix and W the set $\{1, 2, 3, \ldots, n\}$. Then there must exist two disjoint subsets S and T such that $S \cup T = W$ and further $a_{ik} \neq 0$ implies $i = k$ or $i \in S$, $k \in T$, or $i \in T$, $k \in S$. This means that after suitable row-permutations, each followed by the corresponding column permutation, the matrix can be written in the form

$$\begin{pmatrix} D_1 & E \\ F & D_2 \end{pmatrix},$$

where D_1 and D_2 are diagonal matrices. For example, the matrix

$$\begin{pmatrix} 2 & -1 & 0 & 0 \\ -1 & 2 & -1 & 0 \\ 0 & -1 & 2 & -1 \\ 0 & 0 & -1 & 2 \end{pmatrix}$$

has property A, and we can choose $S = \{1, 3\}$, $T = \{2, 4\}$. Permutation of columns 1 and 4 followed by a similar row-permutation will produce a matrix of the mentioned form.

For systems of equations with coefficient matrices of this kind, D. M. Young has developed a special technique, *Successive Over-Relaxation, SOR*. We shall further assume A symmetric and $a_{ii} > 0$; $a_{ii} > \sum_{k \neq i} |a_{ik}|$ for all values of i. As before we split A into two parts $A = D + C$, D containing the diagonal elements. The equation $(D + C)x = y$ is now rewritten in the form $(I + D^{-1}C)x = D^{-1}y$ or $x = Bx + c$, where $B = -D^{-1}C$ and $c = D^{-1}y$. Thus all diagonal elements of B are equal to zero, and we can split B into one upper and one lower triangular matrix: $B = R + L$. Then the SOR-method is defined through

$$x^{(k+1)} = (1 - \omega)x^{(k)} + \omega\{Lx^{(k+1)} + Rx^{(k)} + c\},$$

where ω is the so-called relaxation factor. When $0 < \omega < 1$ we have *under-relaxation*; when $\omega > 1$ we have *over-relaxation*. If $\omega = 1$, we are back with Gauss–Seidel's method. Solving for $x^{(k+1)}$, we get

$$x^{(k+1)} = (I - \omega L)^{-1}\{(1 - \omega)I + \omega R\}x^{(k)} + (I - \omega L)^{-1}\omega c.$$

It is then clear that Jacobi's, Gauss–Seidel's, and Young's methods can be represented in the form $x^{(k+1)} = Mx^{(k)} + d$ according to the following:

Jacobi	$M = -D^{-1}C = B$	$d = D^{-1}y = c$
(10.4.2) Gauss–Seidel	$M = (I - L)^{-1}R$	$d = (I - L)^{-1}c$
SOR	$M = (I - \omega L)^{-1}\{(1 - \omega)I + \omega R\}$	$d = (I - \omega L)^{-1}\omega c.$

The convergence speed will essentially depend upon the properties of the matrix M, convergence being obtained only if $\rho(M) < 1$, where $\rho(M)$ is the spectral radius of

M. For the SOR-method we encounter the important question of how ω should be chosen to make $\rho(M)$ as small as possible.

It is easy to see that the eigenvalues of A do not change during the permutations that are performed for creating B. We notice, for example, that $\mu I - B$ is of the following type (in the special case $n = 8$):

$$
\begin{vmatrix}
\mu & 0 & 0 & 0 & 0 & * & * & * \\
0 & \mu & 0 & 0 & 0 & * & * & * \\
0 & 0 & \mu & 0 & 0 & * & * & * \\
0 & 0 & 0 & \mu & 0 & * & * & * \\
0 & 0 & 0 & 0 & \mu & * & * & * \\
* & * & * & * & * & \mu & 0 & 0 \\
* & * & * & * & * & 0 & \mu & 0 \\
* & * & * & * & * & 0 & 0 & \mu
\end{vmatrix}
=
\left(
\begin{array}{c|c}
\mu I_1 & E \\ \hline
F & \mu I_2
\end{array}
\right).
$$

From well-known determinantal rules it is clear that the same number of elements from E and from F must enter an arbitrary term in the determinant. Two conclusions can be drawn from this. First: the characteristic equation in the example above must be $\mu^8 + a_1\mu^6 + a_2\mu^4 + a_3\mu^2 = 0$ and in general $\mu^r P_s(\mu^2) = 0$, where $s = (n - r)/2$. Now B need not be symmetric, but

$$
B' = D^{1/2}BD^{-1/2} = -D^{1/2}(D^{-1}C)D^{-1/2} = -D^{-1/2}CD^{-1/2}
$$

is symmetric. In fact, putting for a moment $S = D^{-1/2}$, we have

$$
(SCS)^T = S^T C^T S^T = SCS,
$$

since S is diagonal and C symmetric (because A was supposed to be symmetric). Hence, B and B' have the same eigenvalues, which must be real: $0, 0, \ldots, \pm\mu_1, \pm\mu_2, \ldots$ The second conclusion which can be made is the following. If all elements of E are multiplied by a factor $k \neq 0$ and all elements of F are divided by the same factor, the value of the determinant is unchanged.

We now derive an important relationship between an eigenvalue λ of the matrix $M = (I - \omega L)^{-1}\{(1 - \omega)I + \omega R\}$ and an eigenvalue μ of B. The equation $\det(M - \lambda I) = 0$ can be written

$$
\det\{(I - \omega L)^{-1}[\omega R - (\omega - 1)I - (I - \omega L)\lambda]\} = 0,
$$

that is,

$$
\det\left([R + \lambda L] - \frac{\lambda + \omega - 1}{\omega} I \right) = 0.
$$

As has just been discussed, we move a factor $\lambda^{1/2}$ from L to R which does not change the value of the determinant, and then we divide all elements by $\lambda^{1/2}$:

$$
\det\left([R + L] - \frac{\lambda + \omega - 1}{\omega\lambda^{1/2}} I \right) = 0.
$$

But $R + L = B$ and if μ is an eigenvalue of B, we must have

$$\mu = \frac{\lambda + \omega - 1}{\omega \lambda^{1/2}}.$$

Since $b_{ii} = 0$ and $\sum_{k \neq i} |b_{ik}| < 1$ (note that $a_{ii} > \sum_{k \neq i} |a_{ik}|$), we have according to Gershgorin $\mu^2 < 1$. We can now express λ in terms of μ; conveniently we put $z = \lambda^{1/2}$ and obtain

$$z = \tfrac{1}{2}\omega\mu \pm \sqrt{(\tfrac{1}{4}\omega^2\mu^2 - \omega + 1)}.$$

For certain values of ω we get real solutions; for other values, complex solutions. The limit between these two possibilities is determined by the equation $\tfrac{1}{4}\mu^2\omega^2 - \omega + 1 = 0$ with the two roots:

$$\omega_1 = 2(1 - \sqrt{(1 - \mu^2)})/\mu^2; \qquad \omega_2 = 2(1 + \sqrt{(1 - \mu^2)})/\mu^2.$$

Real solutions z are obtained if $\omega \leq \omega_1$ or $\omega \geq \omega_2$, and complex solutions z if $\omega_1 < \omega < \omega_2$. In the real case only the *greater* solution

$$z_1 = \tfrac{1}{2}\omega\mu + \sqrt{(\tfrac{1}{4}\omega^2\mu^2 - \omega + 1)}$$

is of interest. A simple calculation shows that $dz_1/d\omega < 0$ for $\omega < \omega_1$, while $dz_1/d\omega > 0$ for $\omega > \omega_2$. When $\omega \to \omega_1$ from the left, the derivative goes toward $-\infty$, and when $\omega \to \omega_2$ from the right, the derivative will approach $+\infty$. When $\omega_1 < \omega < \omega_2$, we have

$$z = \tfrac{1}{2}\omega\mu \pm i\sqrt{(\omega - 1 - \tfrac{1}{4}\omega^2\mu^2)},$$

that is, $|z|^2 = \omega - 1$. If $|\lambda| = |z|^2$ is represented as a function of ω, we get the result shown in Figure 10.4.1. Using this figure, we can draw all the conclusions we need. First it is obvious that the optimal value of the relaxation-factor is $\omega_b = \omega_1$, that is,

(10.4.3)
$$\omega_b = \frac{2(1 - \sqrt{(1 - \mu^2)})}{\mu^2},$$

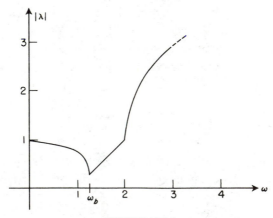

FIGURE 10.4.1

where μ is the spectral radius of B. It is also easy to understand that a value ω which is a little too big is far less disastrous than a value which is a little too small, since the left derivative is infinite. For convergence we must have $|\lambda| < 1$, and hence $0 < \omega < 2$, but we have SOR only when $1 < \omega < 2$.

In more complicated cases when μ is not known, one must resort to guesses or experiments. A correct choice of ω may imply considerable savings in computation effort, and there are realistic examples which indicate that the work may be reduced by a factor of 10–100 compared to Gauss–Seidel's method.

We shall now illustrate the method on a simple example.

$$
\begin{aligned}
4x_1 \qquad\quad + \ x_3 + \ x_4 &= 1 \\
4x_2 \qquad\quad + \ x_4 &= 2 \\
x_1 \qquad\quad + 4x_3 \qquad\quad &= 3 \\
x_1 + \ x_2 \qquad\quad + 4x_4 &= 4
\end{aligned}
$$

Exact solution:
$$
\begin{aligned}
\xi_1 &= -41/209 = -0.196172 \\
\xi_2 &= 53/209 = 0.253589 \\
\xi_3 &= 167/209 = 0.799043 \\
\xi_4 &= 206/209 = 0.985646.
\end{aligned}
$$

The coefficient matrix obviously fulfills all conditions posed before. Now μ is determined from the equation

$$
\begin{vmatrix}
\mu & 0 & -\frac{1}{4} & -\frac{1}{4} \\
0 & \mu & 0 & -\frac{1}{4} \\
-\frac{1}{4} & 0 & \mu & 0 \\
-\frac{1}{4} & -\frac{1}{4} & 0 & \mu
\end{vmatrix} = 0,
$$

that is, $16\mu^4 - 3\mu^2 + \frac{1}{16} = 0$. The largest positive root is $\mu = (\sqrt{5} + 1)/8$, giving $\omega = 2(1 - \sqrt{(1 - \mu^2)})/\mu^2 = 1.04464$ and $\lambda = \omega - 1 = 0.04464$. In the table below we compare the fifth approximation in different methods. It turns out that lower approximations become best for larger values of ω, and the error, defined as $e = \sum |x_i - \xi_i|$ in the fifth approximation, is actually smallest for $\omega = 1.050$, when we get an accuracy of exactly 6 decimals.

	x_1	x_2	x_3	x_4	e
Jacobi	−0.184570	0.260742	0.798828	0.985352	0.019265
Gauss–Seidel	−0.195862	0.253780	0.798965	0.985520	0.000705
SOR ($\omega = 1.04464$)	−0.196163	0.253594	0.799042	0.985644	0.000018
Exact	−0.196172	0.253589	0.799043	0.985646	—

The convergence speed for the iteration method $x^{(n+1)} = Mx^{(n)} + c$ is defined through $R = -\ln \rho(M)$, where $\rho(M)$ is the spectral radius for M. In the example above we find:

$$
R_1 = -\ln \frac{\sqrt{5} + 1}{8} = 0.9 \quad \text{for Jacobi's method,}
$$

$$
R_2 = -2 \ln \frac{\sqrt{5} + 1}{8} = 1.8 \quad \text{for Gauss–Seidel's method,}
$$

$$
R_3 = -\ln 0.04464 = 3.1 \quad \text{for SOR.}
$$

The relationship $\mu = (\lambda + \omega - 1)/(\lambda^{1/2}\omega)$ used for $\omega = 1$ gives $\lambda = \mu^2$, where λ is the spectral radius for Gauss–Seidel's matrix, and μ the spectral radius for B, that is, the matrix of Jacobi's method. Hence we have the general result that the convergence for Gauss–Seidel's method is twice as fast as for Jacobi's method.

---

§10.5. GRADIENT METHODS

The iterative methods we have used so far are *stationary*, since we apply the same iteration matrix throughout. We will now consider a nonstationary method called the *conjugate gradient method* (CG). It was suggested around 1950 by Stiefel, Lanczos, and Hestenes. When constructing this method the designers used a geometric model, an idea that has proved quite successful in many cases. In fact, we will reshape the whole problem to a minimization problem.

We assume that the matrix A is *symmetric, positive definite* and introduce the function

$$F(x_1, x_2, \cdots x_n) = \tfrac{1}{2}x^T A x - y^T x$$

which can also be written

$$F = \tfrac{1}{2}\sum a_{ik}x_i x_k - \sum y_i x_i; \qquad a_{ki} = a_{ik}.$$

Trying to minimize F, we obtain through differentiation:

$$\partial F/\partial x_i = \sum_k a_{ik}x_k - y_i = 0, \qquad i = 1(1)n$$

which can also be written $Ax = y$. The technique of considering a certain equation as a condition for minimization of some suitable function is often successful. It is used extensively in physics, and we will meet it again in connection with the treatment of differential equations.

The vector with components $\partial F/\partial x_1, \partial F/\partial x_2, \ldots, \partial F/\partial x_n$ is usually called the *gradient vector* and denoted grad F. Sometimes, particularly in theoretical physics, it is practical to use a symbolic differentiation vector called ∇ (nabla):

(10.5.1) $$\nabla = (\partial/\partial x_1, \partial/\partial x_2, \ldots, \partial/\partial x_n).$$

(This should not be confused with the backward difference operator.) Then $\nabla F = $ grad F; further, we can also form the scalar product of ∇ and a vector X, called the divergence of X:

(10.5.2) $$\text{div } X = \nabla X = \sum \partial X_k/\partial x_k$$

which is obviously a scalar quantity.

The gradient vector points in the direction of the greatest variation of F (the temperature gradient is a well-known example).

We start by choosing a point z_0 and compute the residual vector $r_0 = Az_0 - y$. Since $dF = \sum (\partial F/\partial x_k)\, dx_k = (\text{grad } F, dt) = 0$ where dt is the tangent vector, grad F is orthogonal to the surface $F(z) = C$, which is a generalized ellipsoid in n dimensions. We now move in the negative direction along the gradient vector as long as the length of the residual vector decreases; i.e., we search for an ellipsoid in the family as close to the center as possible. Hence we take next point $z_1 = z_0 + \lambda_0 p_0$ where p_0 is still at our disposal; we could choose $p_0 = -r_0$, or more generally $p = -r$ (this is called the "method of steepest descent"). However, first we determine the value λ_0 from the condition that p_0 must be a tangent of the smallest ellipsoid concerned, i.e., $(r_1, p_0) = 0$ where

$$r_1 = Az_1 - y = A(z_0 + \lambda_0 p_0) - y = r_0 + \lambda_0 A p_0.$$

This gives

$$\lambda_0 = -\frac{r_0^T p_0}{p_0^T A p_0}.$$

As just mentioned, we can choose p in the direction of the residual vector, but there is a better choice: since we want to reach an ellipsoid as far in as possible, we take in general the new direction *conjugate* to the old one, i.e.,

$$p_k^T A p_{k-1} = 0.$$

Hence we put

$$p_k = -r_k + \varepsilon_{k-1} p_{k-1}, \qquad k = 1, 2, 3, \ldots$$

and obtain the condition

$$-r_k^T A p_{k-1} + \varepsilon_{k-1} p_{k-1}^T A p_{k-1} = 0.$$

This gives ε_{k-1}:

$$\varepsilon_{k-1} = \frac{r_k^T A p_{k-1}}{p_{k-1}^T A p_{k-1}}.$$

Further $z_{k+1} = z_k + \lambda_k p_k$ gives $r_{k+1} = r_k + \lambda_k A p_k$, where $r_{k+1}^T p_k = 0$ implies $\lambda_k = -r_k^T p_k / p_k^T A p_k$. With the convention to choose z_0 arbitrarily and $p_0 = -r_0$ we compute the different quantities in the following order: $\lambda_0, z_1, r_1, \varepsilon_0, p_1, \lambda_1, z_2, r_2, \varepsilon_1, p_2, \ldots$

Example 1

$$\begin{cases} 4x - y + 2z = 12 \\ -x + 5y + 3z = 10 \\ 2x + 3y + 6z = 18 \end{cases}$$

$$A = \begin{pmatrix} 4 & -1 & 2 \\ -1 & 5 & 3 \\ 2 & 3 & 6 \end{pmatrix}; \quad \text{taking } z_0 = \begin{pmatrix} 1 \\ 0 \\ 0 \end{pmatrix} \quad \text{we get} \quad r_0 = \begin{pmatrix} -8 \\ -11 \\ -16 \end{pmatrix} = -p_0$$

and $\lambda_0 = 0.11639$. After this we get in turn:

$$z_1 = \begin{pmatrix} 1.93112 \\ 1.28029 \\ 1.86223 \end{pmatrix} \quad r_1 = \begin{pmatrix} -1.83135 \\ 0.05701 \\ 0.87648 \end{pmatrix} \quad \varepsilon_0 = 0.009354$$

$$p_1 = \begin{pmatrix} 1.90619 \\ 0.04589 \\ -0.72681 \end{pmatrix} \quad \lambda_1 = 0.34968; \quad z_2 = \begin{pmatrix} 2.59767 \\ 1.29634 \\ 1.60808 \end{pmatrix} \quad r_2 = \begin{pmatrix} 0.31052 \\ -1.29176 \\ 0.73283 \end{pmatrix}$$

$$\varepsilon_1 = 0.55804; \quad p_2 = \begin{pmatrix} 0.75322 \\ 1.31737 \\ -1.13842 \end{pmatrix} \quad \lambda_2 = 0.53414; \quad z_3 = \begin{pmatrix} 2.999997 \\ 2.000004 \\ 1.000001 \end{pmatrix}.$$

The final solution $x = 3$, $y = 2$, $z = 1$ is reached after exactly $n = 3$ sweeps. However, to get rid of rounding errors, one extra iteration might be recommended.

In certain cases, particularly when A contains many zero elements, the CG method will be faster than Gaussian elimination.

EXERCISES

1. Solve by Gaussian elimination $Ax = b$ where

$$A = \begin{pmatrix} 3 & 7 & -6 & 2 \\ 9 & 23 & -13 & 10 \\ 15 & 31 & -36 & 10 \\ -12 & -24 & 38 & 9 \end{pmatrix} \quad b = \begin{pmatrix} 4 \\ 9 \\ 30 \\ -17 \end{pmatrix}.$$

As a by-product, factorize $(A, b) = L(R, c)$.

2. Factorize $Ax = b$ as $LD(L^Tx - c)$ where

$$A = \begin{pmatrix} 4 & 8 & -12 & -4 \\ 8 & 19 & -24 & 4 \\ -12 & -24 & 38 & 8 \\ -4 & 4 & 8 & 61 \end{pmatrix} \quad b = \begin{pmatrix} 16 \\ 26 \\ -54 \\ -28 \end{pmatrix}$$

and find the solution vector. L is supposed to be lower triangular with unit diagonal, and D is a diagonal matrix.

3. Solve by Gaussian elimination

$$\begin{cases} x + 2y + 3z = 13 \\ 2x + 4y + 2z = 14 \\ 3x + 5y - z = 11. \end{cases}$$

4. Perform the factorization $A = LDL^T$ where

$$A = \begin{pmatrix} 2 & 6 & -4 & 2 \\ 6 & 17 & -17 & 9 \\ -4 & -17 & -20 & -1 \\ 2 & 9 & -1 & -51 \end{pmatrix} \qquad b = \begin{pmatrix} 8 \\ 20 \\ -24 \\ 64 \end{pmatrix}.$$

Then solve the system $Ax = b$.

5. Factorize

$$A = \begin{pmatrix} 10 & 7 & 8 & 7 \\ 7 & 5 & 6 & 5 \\ 8 & 6 & 10 & 9 \\ 7 & 5 & 9 & 10 \end{pmatrix}$$

in the form LL^T where L is a lower triangular matrix (Cholesky decomposition).

6. Pascal's matrix of order n is defined by

$$(P_n)_{ik} = \binom{i + k - 2}{i - 1} = \binom{i + k - 2}{k - 1}, \qquad i, k = 1(1)n.$$

Factorize P_5 in the form LL^T where L is lower triangular.

7. Find the inverse of Pascal's matrix P_5.

8. A is an $n \times n$ matrix with all elements equal to 1. Find a number p such that $I + pA$ becomes the inverse of $I - A$.

9. A is a band matrix with the elements in the main diagonal equal to 2 and the elements in the diagonals next to it equal to -1. Show that the (i, k)-element in A^{-1} is equal to $\min(i, k)(n + 1 - \max(i, k))/(n + 1)$.

10. (a) The 5×5 matrix A has all elements in the main diagonal equal to 6 and the elements in the two closest diagonals on both sides equal to -1. Using Gauss–Seidel's method, solve the system $Ax = b$ where $b^T = (0, 1, 2, 3, 4)$.

(b) Forming successive sub-matrices starting in the upper left corner and denoting the corresponding determinants by D_k, we have $D_1 = 6$, $D_2 = 35$, and so on, and in general $D_{n+2} = 6D_{n+1} - D_n$. We put conveniently $D_0 = 1$ and $D_{-1} = 0$. Compute $D_5 = \det A$.

(c) The solutions of the system of equations are obviously of the form $x_k = m_k/D_5$ where the m_k are integers. Find these using the x-values obtained in (a).

11. Show that a band matrix with diagonal elements $a_1, a_2, \ldots, a_n$ and all elements in the two diagonals on each side of it equal to 1 can be written $A = LDL^T$, where L has the diagonal elements equal to 1, the elements just below equal to $c_2, c_3, \ldots, c_n$, and the remaining elements 0. With diag $D = d_1, d_2, \ldots, d_n$ (all $d_k \neq 0$), verify that $d_1 = a_1$, $c_{k+1} = 1/d_k$, and $d_{k+1} = a_{k+1} - c_{k+1}$. Perform the computations explicitly in the case $a_k = k$.

12. Using Gauss–Seidel's method, solve the system

$$\begin{cases} 8x + y - z = 12 \\ x + 9y + z = 13 \\ x + y + 10z = 14. \end{cases}$$

13. Find a solution (to 5 decimals) of the system

$$\begin{cases} 10x + y^2 + z^3 = 5 \\ x^4 + 12y^2 + (z^2 + 10)^{-1} = 6 \\ e^{-12x} + \sin^2 y + 20z^3 = 7. \end{cases}$$

14. Find the optimal relaxation factor for the system

$$\begin{cases} 2x - y = 7 \\ -x + 2y - z = 1 \\ - y + 2z = 1. \end{cases}$$

Perform 4 iterations with the starting vector $(3, 2, 1)^T$ and the relaxation parameter ω rounded to one decimal.

15. Using the conjugate gradient method, solve the system $Ax = b$ where

$$A = \begin{pmatrix} 4 & -1 & 2 \\ -1 & 5 & 3 \\ 2 & 3 & 6 \end{pmatrix} \qquad b = \begin{pmatrix} 12 \\ 10 \\ 18 \end{pmatrix}$$

Start with the vector $z_0^T = (1, 0, 0)$.

CHAPTER 11

Nonlinear Equations and Systems of Equations

When it is known that x is the same as 6 (which by
the way is understood from the pronunciation) all
algebraic equations with 1 or 19 unknowns are easily
solved by inserting x, substituting 6, elimination of 6
by x, and so on.

Falstaff, fakir*

§11.1. EQUATIONS IN ONE VARIABLE

It is rather natural that historically algebraic equations were first on the scene. Equations of first degree were solved in ancient times, while quadratic equations were mastered in the 10th century. Cubic equations were treated successfully by Tartaglia (1535), and quartic equations by Ferrari (1545). Great efforts were spent also to solve equations of fifth and higher degree, but in vain. Finally the Norwegian mathematician N. H. Abel in 1823 proved that the solution of such equations could only occasionally be given as algebraic expressions in the coefficients.

With these facts in mind, it is obvious that only in some rare cases will it be possible to find the solutions of an algebraic equation by direct mathematical methods. For all other algebraic equations we must resort to numerical methods. In fact, such methods are usually preferred even for cubic and quartic equations.

Passing to transcendental equations, we understand that in general only numerical methods can be used. The task of constructing such methods has attracted much attention from many of the greatest mathematicians, and here we mention only Newton, Leibniz, Euler, Gauss, Chebyshev, Laguerre, and Laplace. Modern computers have made several new methods feasible during the last few decades.

§11.2. HORNER'S SCHEME

For a while we return to algebraic equations, with the intention to give some theoretical background and some important computational means. The fundamental

---

* Famous Swedish humorist-author (1865–1896), influenced, for one, by Mark Twain. His real name was Axel Wallengren, and the quotation above comes from the book, *Everyone His Own Professor*.

theorem of algebra states that each polynomial

$$P(x) = x^n + a_1 x^{n-1} + \cdots + a_n$$

has at least one root. Then it is practically trivial to show that there are exactly n roots. To show this, let r be a root; then $(x - r)$ divides $P(x)$. This is proved by performing the division $P(x)/(x - r)$, which gives the quotient $Q(x)$ and the constant remainder term R:

$$P(x) = Q(x)(x - r) + R.$$

Putting $x = r$ we get $R = 0$, since $P(r) = 0$. After that the fundamental theorem can be used on $Q(x)$ instead, and so forth. Denoting the roots by $x_1, x_2, \ldots, x_n$ we get:

$$(x - x_1)(x - x_2) \cdots (x - x_n) \equiv x^n + a_1 x^{n-1} + \cdots + a_n$$

and identifying:

$$\begin{cases} a_1 = -\sum x_i \\ a_2 = \sum x_i x_k \\ \vdots \\ a_n = (-1)^n x_1 x_2 \cdots x_n. \end{cases}$$

We shall now consider how the division of $P(x)$ by the linear factor $(x - r)$ when r is a root should be carried out in practice. Putting

$$Q(x) = x^{n-1} + b_1 x^{n-2} + \cdots + b_{n-1}$$

we have the identity:

$$x^n + a_1 x^{n-1} + \cdots + a_n \equiv (x - r)(x^{n-1} + b_1 x^{n-2} + \cdots + b_{n-1}).$$

Comparing coefficients for $x^{n-1}, x^{n-2}, \ldots, x^0$ we get:

$$b_1 = a_1 + r$$
$$b_2 = a_2 + rb_1$$
$$\vdots$$
$$b_{n-1} = a_{n-1} + rb_{n-2}$$
$$0 = a_n + rb_{n-1}.$$

Each value b_k obtained in line k is used as input in the next line, which means that the computation is performed successively. If all b_k are eliminated, we obtain the condition $P(r) = 0$. We now construct a simple scheme for this computation:

1	a_1	a_2	$\cdots$	a_n	$\underline{\quad r}$
	r	rb_1	$\cdots$	rb_{n-1}	
1	$a_1 + r$	$a_2 + rb_1$	$\cdots$	$a_n + rb_{n-1}$	
	$(= b_1)$	$(= b_2)$	$\cdots$	$(= 0)$	

The second row is obtained by successive multiplication of the elements in the third row by r and shifting one step to the right, while the third row is obtained by adding the first and second rows (these operations alternate). Note that if r is not a root of the equation, then the last value in the third row will become $P(r)$. This nested multiplication method takes fewer operations than most other methods.

The algorithm above did in fact produce the result $P(r)$. We shall now show that repeated use of the same algorithm will in turn produce the results $P'(r)$, $P''(r)/2!$, $P'''(r)/3!$, and so forth, ending with $P^{(n)}(r)/n! = a_0$ where we suppose $P(x) = a_0x^n + a_1x^{n-1} + \cdots + a_n$. For simplicity we shall treat the case $n = 4$ explicitly; generalization to arbitrary n is immediate.

Below we display the complete Horner scheme for $n = 4$.

	a_0	a_1	a_2	a_3	a_4	r
(P_4)	a_0	a_1	a_2	a_3	a_4	
		a_0r	$a_0r^2+a_1r$	$a_0r^3+a_1r^2+a_2r$	$a_0r^4+a_1r^3+a_2r^2+a_3r$	
(P_3)	a_0	a_0r+a_1	$a_0r^2+a_1r+a_2$	$a_0r^3+a_1r^2+a_2r+a_3$	$a_0r^4+a_1r^3+a_2r^2+a_3r+a_4$ $= P(r)$	
		a_0r	$2a_0r^2+a_1r$	$3a_0r^3+2a_1r^2+a_2r$		
(P_2)	a_0	$2a_0r+a_1$	$3a_0r^2+2a_1r+a_2$	$4a_0r^3+3a_1r^2+2a_2r+a_3$		
		a_0r	$3a_0r^2+a_1r$	$= P'(r)$		
(P_1)	a_0	$3a_0r+a_1$	$6a_0r^2+3a_1r+a_2$			
		a_0r	$= P''(r)/2!$			
(P_0)	a_0	$4a_0r+a_1$	$= P'''(r)/3!$			
	a_0	$= P^{(iv)}(r)/4!$				

Let us now form polynomials with the coefficients taken from the different levels to the left of the "stair" and denote them $P_4(x) (= P(x))$, $P_3(x)$, $P_2(x)$, $P_1(x)$, and $P_0(x)$. Hence:

$$P_4(x) = a_0x^4 + a_1x^3 + a_2x^2 + a_3x + a_4$$
$$P_3(x) = a_0x^3 + (a_0r + a_1)x^2 + (a_0r^2 + a_1r + a_2)x + (a_0r^3 + a_1r^2 + a_2r + a_3)$$
$$P_2(x) = a_0x^2 + (2a_0r + a_1)x + (3a_0r^2 + 2a_1r + a_2)$$
$$P_1(x) = a_0x + (3a_0r + a_1)$$
$$P_0(x) = a_0.$$

Putting $x = r$ we obtain:

$$P_4(r) = a_0r^4 + a_1r^3 + a_2r^2 + a_3r + a_4 = P(r)$$
$$P_3(r) = 4a_0r^3 + 3a_1r^2 + 2a_2r + a_3 \quad = P'(r)$$
$$P_2(r) = 6a_0r^2 + 3a_1r + a_2 \quad = P''(r)/2!$$
$$P_1(r) = 4a_0r + a_1 \quad = P'''(r)/3!$$
$$P_0(r) = a_0 \quad = P^{(iv)}(r)/4!$$

Since we have used the Horner technique for the successive construction of the polynomials $P_m(x)$, we have:

$$P_m(x) = (x - r)P_{m-1}(x) + P_m(r).$$

Differentiation of this formula k times gives:

$$P_m^{(k)}(x) = (x - r)P_{m-1}^{(k)}(x) + kP_{m-1}^{(k-1)}(x).$$

Hence

$$P_m^{(k)}(r) = kP_{m-1}^{(k-1)}(r) = k(k - 1)P_{m-2}^{(k-2)}(r) = \cdots = k!P_{m-k}(r).$$

Finally, $m = n$ gives

$$P_{n-k}(r) = P_n^{(k)}(r)/k!.$$

In particular, $n = 4$ gives for $k = 1, 2, 3, 4$:

$$P_3(r) = P'(r), \qquad P_2(r) = P''(r)/2!, \qquad P_1(r) = P'''(r)/3!, \qquad P_0(r) = P^{(iv)}(r)/4!$$

in accordance with our previous results.

Example 1.   Move the origin to $x = 2$ for the function

$$f(x) = 2x^4 - 5x^3 + 6x^2 + 8x - 10,$$

i.e., find coefficients s_k such that

$$f(x) = s_0(x-2)^4 + s_1(x-2)^3 + s_2(x-2)^2 + s_3(x-2) + s_4.$$

Obviously, $s_4 = f(2)$ and similarly $s_3 = f'(2)$, $s_2 = f''(2)/2!$, $s_1 = f'''(2)/3!$ and $s_0 = f^{(iv)}(2)/4!$. Horner's scheme becomes (with $r = 2$):

```
2    -5     6     8    -10 | 2
      4    -2     8     32
2    -1     4    16     22
      4     6    20
2     3    10    36
      4    14
2     7    24
      4
2    11
|2
```

Hence we obtain:

$$f(x) = 2(x-2)^4 + 11(x-2)^3 + 24(x-2)^2 + 36(x-2) + 22.$$

§11.3.   INTERPOLATION METHODS FOR REAL ROOTS

We start by describing a few classical methods, and after that we will develop a more theoretical treatment of the subject. Suppose that the real function $f(x)$ is given together with two values a and b such that $f(a)$ and $f(b)$ have opposite signs. Then putting $c = \frac{1}{2}(a + b)$, we replace a by c if $f(a)f(c) > 0$; otherwise we replace b by c. Proceeding in this way we see that the root will be contained in an interval of length $|a - b| \cdot 2^{-m}$ after m steps. Denoting the exact root by ξ and the nth approximation by x_n, with $\varepsilon_n = |x_n - \xi|$ we have:

$$\varepsilon_{n+1} \simeq \tfrac{1}{2}\varepsilon_n$$

(on the average). We say that the convergence rate of this *bisection algorithm* is *linear* or *geometrical*.

It is obvious that better methods can be constructed; we have in fact not used all of the information available, since we employed only the sign of $f(c)$. If we also take the numerical values into account, we can define a class of better methods using a simple geometrical model. The main idea is to replace the function curve by a suitable straight line (secant or tangent). If we know one point (x_0, y_0) or two points $(x_0, y_0), (x_1, y_1)$ on the curve $y = f(x)$ in the vicinity of the root, we conveniently choose the slope of the line as $k = f'(x_0)$ or $k = (y_1 - y_0)/(x_1 - x_0)$. It is easy to show that the convergence then will become linear, but better than for the bisection method. On the other hand, the slope k does not have to be recomputed. If we construct a sequence of points by drawing secants between two points defining an interval containing the root, we also get linear convergence. This classical method is known as *Regula falsi* (method of false position). An obvious improvement is obtained by choosing

$$k = (y_n - y_{n-1})/(x_n - x_{n-1}),$$

which is a linear interpolation method. The corresponding iteration formula has the form:

(11.3.1) $$x_{n+1} = x_n - y_n \frac{x_n - x_{n-1}}{y_n - y_{n-1}} = \frac{x_{n-1}y_n - x_ny_{n-1}}{y_n - y_{n-1}}.$$

Here the first expression should be preferred, although the quotient

$$k^{-1} = (x_n - x_{n-1})/(y_n - y_{n-1})$$

will give rise to large errors: both the numerator and the denominator will produce cancellation. But k^{-1} is multiplied with a small quantity y_n to produce a correction term, and the accuracy lost will not be critical. The last expression, on the other hand, will be strongly influenced by cancellation in the denominator and should not be used.

We will now examine the convergence rate of the method. Let ξ be the exact root and put $x_n = \xi + \varepsilon_n$. Supposing both the first and the second derivative $\neq 0$, we find by Taylor expansion:

$$\varepsilon_{n+1} = \frac{\varepsilon_{n-1}f(\xi + \varepsilon_n) - \varepsilon_nf(\xi + \varepsilon_{n-1})}{f(\xi + \varepsilon_n) - f(\xi + \varepsilon_{n-1})} = \varepsilon_{n-1}\,\varepsilon_n\,\frac{f''(\xi)}{2f'(\xi)} + \cdots$$

If higher order terms are neglected, the error formula has the form

$$\varepsilon_{n+1} \simeq A\varepsilon_n\varepsilon_{n-1}.$$

We try to satisfy this by putting $\varepsilon_n = K\varepsilon_{n-1}^m$ and hence $\varepsilon_{n+1} = K\varepsilon_n^m$. This leads to $\varepsilon_{n+1} = A\varepsilon_n(\varepsilon_n/K)^{1/m} = K\varepsilon_n^m$ or $1 + 1/m = m$. Thus we get $m = (1 \pm \sqrt{5})/2$, where the plus sign should be chosen. We call this convergence rate *superlinear*.

We now try to determine an optimal value of the slope k by examining the error in general. From $y - y_n = k(x - x_n)$, putting $y = 0$, we get

$$x = x_{n+1} = x_n - y_n/k$$

and hence

$$\xi + \varepsilon_{n+1} = \xi + \varepsilon_n - \frac{1}{k}[f(\xi) + \varepsilon_nf'(\xi) + \tfrac{1}{2}\varepsilon_n^2f''(\xi) + \cdots].$$

Observing that $f(\xi) = 0$, we find

$$\varepsilon_{n+1} = \varepsilon_n[1 - f'(\xi)/k] - \varepsilon_n^2 f''(\xi)/2k - \cdots$$

We see that an optimal choice would be $k = f'(\xi)$; however, since ξ is not known we choose $k = f'(x_n)$ instead. This will result in the classical *Newton–Raphson formula*:

(11.3.2) $$x_{n+1} = x_n - f(x_n)/f'(x_n).$$

The convergence speed can be found from

$$\xi + \varepsilon_{n+1} = \xi + \varepsilon_n - f(\xi + \varepsilon_n)/f'(\xi + \varepsilon_n).$$

An easy computation gives

$$\varepsilon_{n+1} \simeq (f''(\xi)/2f'(\xi))\varepsilon_n^2$$

i.e., quadratic convergence, provided $f'(\xi) \neq 0$. If $f'(\xi) = 0$, then the root ξ is multiple, and convergence is only linear. In this case we should instead try to find the zero ξ of the function $f(x)/f'(x)$, since this zero will now be simple. In this way we find the formula

(11.3.3) $$x_{n+1} = x_n - f_n f'_n/(f'^2_n - f_n f''_n)$$

where quadratic convergence has been restored.

For the secant method, it has been suggested that the function $g(x) = f(x)^2/[f(x + f(x)) - f(x)]$ be used instead; it has only simple roots (Steffensen's method).

If $f(x)$ is twice continuously differentiable and $f(a)f(b) < 0$, $a < b$, it is easy to show that Newton–Raphson's method converges from an arbitrary starting value in the interval $[a, b]$ provided that $f'(x) \neq 0$ and $f''(x)$ has the same sign everywhere in the interval, and further

$$\max\{|f(a)/f'(a)|, |f(b)/f'(b)|\} \leq b - a$$

(global convergence). The geometrical interpretation of the conditions are left to the reader.

Finally we give an error estimate which can be used for any method. We suppose that $x_n = \xi + \varepsilon_n$ is an approximation to the root ξ of the equation $f(x) = 0$, where $f(x)$ is analytic. We also suppose that x_n is so close to ξ that $\varphi(x) = f(x)/f'(x)$ varies monotonically in the interval between ξ and x_n, and further that $f'(x) \neq 0$ in this same interval. Since $\varphi(\xi) = 0$, we get using the mean value theorem

$$\varphi(x_n) = \varphi(x_n) - \varphi(x_n - \varepsilon_n) = \varepsilon_n \varphi'(x_n - \theta\varepsilon_n),$$

where $0 < \theta < 1$. Hence $\varepsilon_n = \varphi(x_n)/\varphi'(x_n - \theta\varepsilon_n)$ and

$$\varepsilon_n = \frac{(f/f')_{x_n}}{1 - [(f/f') \cdot (f''/f')]_{x_n - \theta\varepsilon_n}}.$$

Putting $K = \sup|f''/f'|$ and $h = f(x_n)/f'(x_n)$, we find

(11.3.4) $$|\varepsilon_n| < \frac{|h|}{1 - K|h|};$$

where, of course, we have supposed that $|h| < 1/K$.

For the error estimate (11.3.4), we must know x_n, $f(x_n)$, and $f'(x_n)$. Hence, if we use Newton–Raphson's method, we could as well compute x_{n+1}, and we would then be interested in an estimate of the error ε_{n+1} expressed in known quantities. Let ξ be the exact root, as before. Expanding in Taylor series and using Lagrange's remainder term, we get

$$f(\xi) = f(x_n - \varepsilon_n) = f(x_n) - \varepsilon_n f'(x_n) + \tfrac{1}{2}\varepsilon_n^2 f''(x_n - \theta\varepsilon_n) = 0$$

with $0 < \theta < 1$. Putting

$$h = f(x_n)/f'(x_n) \qquad \text{and} \qquad Q = f''(x_n - \theta\varepsilon_n)/f'(x_n),$$

we find

$$2h - 2\varepsilon_n + Q\varepsilon_n^2 = 0 \qquad \text{or} \qquad \varepsilon_n = \frac{2h}{1 + \sqrt{(1 - 2Qh)}}.$$

(The other root is discarded since we are looking for a value of ε_n close to h.) We put $K = \sup|Q|$ and suppose that we are close enough to the root ξ to be sure that $K|h| < s < \tfrac{1}{2}$. But $\varepsilon_{n+1} = \varepsilon_n - h$ and hence

$$|\varepsilon_{n+1}| = |h| \cdot \left| \frac{2}{1 + \sqrt{(1 - 2Qh)}} - 1 \right| < |h|\left(\frac{1}{1 - \alpha K|h|} - 1\right) = \frac{\alpha K h^2}{1 - \alpha K|h|}$$

where $\alpha \geq (1 - \sqrt{(1 - 2s)})/2s$. Naturally, we must assume that $|h| < (2K)^{-1}$. For $s = \tfrac{1}{2}$ we get

(11.3.5)
$$|\varepsilon_{n+1}| < \frac{Kh^2}{1 - K|h|};$$

however, in normal cases s is much smaller, and when $s \to 0$ we have $\alpha \to \tfrac{1}{2}$. If we neglect the term $\alpha K|h|$ in the denominator, we are essentially back at our previous result.

The Newton–Raphson technique is widely used on computers for calculation of various simple functions. Older computers sometimes did not have built-in division, and this operation had to be programmed. The quantity a^{-1} can be interpreted as a root of the equation $1/x - a = 0$. From this we obtain the simple recursion formula

$$x_{n+1} = x_n + \frac{1/x_n - a}{1/x_n^2}$$

or

(11.3.6)
$$x_{n+1} = x_n(2 - ax_n).$$

This relation can also be written $1 - ax_{n+1} = (1 - ax_n)^2$, showing the quadratic convergence clearly. It is easy to construct formulas which converge still faster, for example, $1 - ax_{n+1} = (1 - ax_n)^3$, or $x_{n+1} = x_n(3 - ax_n + a^2x_n^2)$, but the improved convergence has to be bought at the price of a more complicated formula, and there is no real advantage.

If one wants to compute $\sqrt{a}$, one starts from the equation $x^2 - a = 0$ to obtain

$$(11.3.7) \qquad x_{n+1} = \frac{1}{2}\left(x_n + \frac{a}{x_n}\right).$$

This formula has been widely used for automatic computation of square roots. A corresponding formula can easily be deduced for Nth roots. From $y = x^N - a$, we find

$$(11.3.8) \qquad x_{n+1} = x_n - (x_n^N - a)/Nx_n^{N-1} = [(N-1)x_n^N + a]/Nx_n^{N-1}.$$

Especially for $a^{1/3}$ we have

$$(11.3.9) \qquad x_{n+1} = \frac{1}{3}\left(2x_n + \frac{a}{x_n^2}\right)$$

and for $1/\sqrt{a}$

$$(11.3.10) \qquad x_{n+1} = \tfrac{1}{2}x_n(3 - ax_n^2).$$

Newton–Raphson's formula is the first approximation of a more general expression, which will now be derived. Let ξ be the exact root and x_0 the starting value with $x_0 = \xi + h$. Further we suppose that $f'(x_0) \neq 0$ and put $f/f' = \alpha$, $f''/f' = a_2$, $f'''/f' = a_3$, $f^{(iv)}/f' = a_4, \ldots$ Hence

$$f(\xi) = f(x_0 - h) = f - hf' + \frac{h^2}{2}f'' - \cdots = 0$$

or after division by $f'(x_0)$,

$$\alpha - h + a_2\frac{h^2}{2} - a_3\frac{h^3}{6} + a_4\frac{h^4}{24} - \cdots = 0.$$

Now we write h as a power-series expansion in α,

$$h = \alpha + c_2\alpha^2 + c_3\alpha^3 + c_4\alpha^4 + \cdots,$$

and inserting this value in the preceding equation, we have

$$c_2 = a_2/2; \qquad c_3 = \tfrac{1}{6}(3a_2^2 - a_3); \qquad c_4 = \tfrac{1}{24}(15a_2^3 - 10a_2a_3 + a_4); \qquad \cdots$$

Thus we find

$$\xi = x_0 - [\alpha + \tfrac{1}{2}a_2\alpha^2 + \tfrac{1}{6}(3a_2^2 - a_3)\alpha^3 + \tfrac{1}{24}(15a_2^3 - 10a_2a_3 + a_4)\alpha^4 + \cdots].$$

This formula is particularly useful if one is looking for zeros of a function defined by a differential equation (usually of second order).

Example 1.   Find the least positive zero of the Bessel function $J_0(x)$ satisfying the differential equation $xy'' + y' + xy = 0$ (see 7.6.1–2). The zero is close to $x = 2.4$, where we have

$$y(2.4) = 0.00250\,76833, \qquad y'(2.4) = -0.52018\,52682.$$

From the differential equation, we easily get

$$y''(2.4) = 0.21423\ 61785; \qquad y'''(2.4) = 0.34061\ 02514;$$
$$y^{(iv)}(2.4) = -0.20651\ 12692$$

and further

$$a_2 = -0.41184\ 59, \qquad a_3 = -0.65478\ 64, \qquad a_4 = 0.39699\ 56;$$
$$\alpha = -0.00482\ 07503, \qquad \alpha^2 = 0.00002\ 32396, \qquad \alpha^3 = -0.00000\ 01120,$$
$$\alpha^4 = 0.00000\ 00005.$$

Hence

$$
\begin{array}{r}
h = -0.00482\ 07503 \\
- \qquad 47856 \\
- \qquad 217 \\
- \qquad 1 \\
\hline
-0.00482\ 55577
\end{array}
$$

and $\xi = x_0 - h = 2.40482\ 55577$.

Example 2. $x^4 - x = 10$.

By Newton–Raphson's formula, setting $f(x) = x^4 - x - 10$, we get

$$x_{n+1} = x_n - \frac{x_n^4 - x_n - 10}{4x_n^3 - 1} = \frac{3x_n^4 + 10}{4x_n^3 - 1},$$

$$x_0 = 2,$$
$$x_1 = 1.871,$$
$$x_2 = 1.85578,$$
$$x_3 = 1.855585,$$
$$x_4 = 1.85558452522,$$
$$\vdots$$

Example 3.

$$f(z) = z^5 + (7 - 2i)z^4 + (20 - 12i)z^3 + (20 - 28i)z^2 + (19 - 12i)z + (13 - 26i)$$
$$= 0.$$

Choosing $z_0 = 3i$, we find

$$z_1 = -0.293133 + 2.505945i,$$
$$z_2 = -0.548506 + 2.131282i,$$
$$z_3 = -0.819358 + 1.902395i,$$
$$z_4 = -1.038654 + 1.965626i,$$
$$z_5 = -0.997344 + 1.999548i,$$
$$z_6 = -1.000002 + 1.999993i,$$
$$z_7 = -1.000000 + 2.000000i.$$

Example 4. $e^{-x} = \sin x$.

$$x_{n+1} = x_n + \frac{e^{-x_n} - \sin x_n}{e^{-x_n} + \cos x_n}.$$

$$x_0 = 0.6,$$
$$x_1 = 0.5885,$$
$$x_2 = 0.58853274,$$
$$\vdots$$

This converges toward the smallest root; the equation has an infinite number of roots lying close to $\pi, 2\pi, 3\pi, \ldots$, which can be obtained if other starting values are used.

§11.4. FIXED-POINT METHODS

It is often convenient to replace an equation $f(x) = 0$ by another equation of the form $x = g(x)$. If this relation is considered as a mapping, then under rather general conditions there exists a fixed point, i.e., a point ξ that is mapped onto itself (Brouwer's fixed point theorem). An iteration procedure $x_{n+1} = g(x_n)$ can then be constructed, and if it converges we have $\xi = \lim_{n \to \infty} x_n$.

The mapping function $g(x)$ can in fact be constructed in an infinite number of ways. In order to show that the choice is not insignificant, we consider the equation $x^3 - 2 = 0$. It is rewritten on the one hand (a) as $x = x^3 + x - 2$, and on the other (b) as $x = (2 + 5x - x^3)/5$. Starting with $x_0 = 1.2$, we get the results:

	(a)	(b)
$x_1 =$	0.928	1.2544
$x_2 =$	-0.273	1.2596
$x_3 =$	-2.293	1.2599
$x_4 =$	-16.349	1.25992

The correct value $2^{1/3} \simeq 1.259921$ was completely lost in (a) but rapidly attained in (b). Figure 11.1 will help to explain the performance of the method in different cases: when (a) $0 < k < 1$, (b) $k > 1$, (c) $-1 < k < 0$, and (d) $k < -1$ where $k = f'(\xi)$. Obviously, we have convergence in cases (a) and (c), i.e., when $|k| < 1$.

We will now examine the convergence analytically. Starting with $x_1 = g(x_0)$, we find that $x - x_1 = g(x) - g(x_0) = (x - x_0)g'(\xi_0)$, with $x_0 < \xi_0 < x$. Analogously $x - x_2 = (x - x_1)g'(\xi_1); \ldots; x - x_n = (x - x_{n-1})g'(\xi_{n-1})$. Here x is the wanted root of the equation $x = g(x)$. Multiplying, we obtain: $x - x_n = (x - x_0)g'(\xi_0)g'(\xi_1) \cdots g'(\xi_{n-1})$.

Now suppose that $|g'(\xi_k)| \le m$. Then we get $|x - x_n| \le m^n \cdot |x - x_0|$, and hence we have convergence if $m < 1$, that is, if $|g'(x)| < 1$ in the interval of interest. We attain the same result by simple geometric considerations (see Figure 11.4.1).

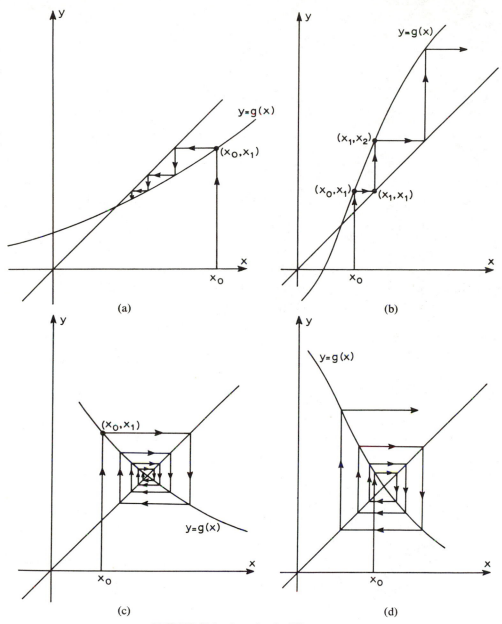

FIGURE 11.1. Iteration in different cases.

It is also easy to give an error estimate in a more useful form. From $x - x_{n-1} = x - x_n + x_n - x_{n-1}$, we get, using $x - x_n = g(x) - g(x_{n-1})$,

$$\left| x - x_{n-1} \right| \leq \left| x - x_n \right| + \left| x_n - x_{n-1} \right| \leq \left| x - x_{n-1} \right| \cdot \left| g'(\xi_{n-1}) \right|$$
$$+ \left| x_n - x_{n-1} \right| \leq m \cdot \left| x - x_{n-1} \right| + \left| x_n - x_{n-1} \right|.$$

Hence

$$|x - x_{n-1}| \leq \frac{|x_n - x_{n-1}|}{1 - m},$$

and

$$|x - x_n| \leq \frac{m}{1 - m} |x_n - x_{n-1}|.$$

If in practical computation it is difficult to estimate m from the derivative, we can use

$$m \simeq \frac{|x_{n+1} - x_n|}{|x_n - x_{n-1}|}.$$

In particular, choosing $g(x) = x - f(x)/f'(x)$ (Newton–Raphson's method) we find:

$$x_{n+1} = x_n - \frac{f(x_n)}{f'(x_n)} = x_n - h \quad \text{and} \quad g'(x) = \frac{f(x)}{f'(x)} \cdot \frac{f''(x)}{f'(x)}.$$

Using the same arguments as before we get $|x - x_n| \leq |h|/(1 - M)$ where $M = |g'(\xi_n)|$ and ξ_n lies between x and x_n. But $|x - x_{n+1}| = M|x - x_n|$ and hence

$$|x - x_{n+1}| < \frac{M|h|}{1 - M}.$$

If we now assume that $f(x)/f'(x)$ is monotonic we have

$$\left| \frac{f(\xi_n)}{f'(\xi_n)} \right| < \left| \frac{f(x_n)}{f'(x_n)} \right| = |h|.$$

Further, $|f''(\xi_n)/f'(\xi_n)| \leq K$, where $K = \sup|f''(x)/f'(x)|$ taken over the interval in question. Hence we have $M/(1 - M) \leq K|h|/(1 - K|h|)$ and we finally obtain

$$|x - x_{n+1}| \leq \frac{Kh^2}{1 - K|h|}$$

in accordance with previous results.

The procedure above can be speeded up by Aitken acceleration (cf. Section 14.4):

$$y_n = f(x_n); \quad z_n = f(y_n); \quad x_{n+1} = x_n - \frac{(y_n - x_n)^2}{z_n - 2y_n + x_n}$$

which corresponds to another iteration function g:

$$x_{n+1} = g(x_n) \quad \text{where} \quad g(x) = \frac{xf(f(x)) - f(x)^2}{f(f(x)) - 2f(x) + x}.$$

Writing $x = f_0$, $f(x) = f_1$, and $f(f(x)) = f_2$, we obtain the more suggestive expression:

$$g = \frac{f_0 f_2 - f_1^2}{f_2 - 2f_1 + f_0}.$$

This method is due to Steffensen.

Example 1. $x = \frac{1}{2} + \sin x$.
The iteration $x_{n+1} = \frac{1}{2} + \sin x_n$, with $x_0 = 1$, gives the following values:

$$x_1 = 1.34, \qquad x_2 = 1.47, \qquad x_3 = 1.495,$$
$$x_4 = 1.4971, \qquad x_5 = 1.49729.$$

For m we can choose the value $m = \cos 1.4971 \simeq 0.074$, and hence $|x - x_5| \leq (0.074/0.926) \cdot 0.00019 = 0.000015$. In this case it is obvious that $x > x_5$, and this leads to the estimate

$$1.497290 < x < 1.497305.$$

The correct value to six decimal places is $x = 1.497300$.
The convergence of a certain method is said to be of *order p* if the limit value

$$C = \lim_{n \to \infty} \frac{|x_{n+1} - \xi|}{|x_n - \xi|^p}$$

exists with $p \geq 1$ and $C > 0$. Then C is called the *asymptotic error constant*. The efficiency of a certain method depends not only on p but also on the number of function evaluations in each step. Further, it should be observed that we have discussed only *local* convergence, in the sense that we have assumed a sufficiently good initial approximation. In the following section we will also give some attention to other properties.

---

§11.5. LAGUERRE'S METHOD

While most methods described so far have good local convergence properties, little can be said about their global convergence. In addition, as a rule we do not get convergence to non-real roots from real starting values. However, for polynomial equations there is a method with global convergence for real roots, local cubic convergence to a simple root, and local linear convergence to a multiple root. Further, non-real roots can be obtained with real initial values. This method is known as *Laguerre's method*.

Let $f(z)$ be a complex polynomial with zeros $r_1, r_2, \ldots, r_n$, and let z be close to the root r_j for some fixed j. Define:

$$P(z) = \frac{f'(z)}{f(z)} = \sum_{i=1}^{n} (z - r_i)^{-1},$$

$$Q(z) = -\frac{dP}{dz} = \frac{f'^2 - ff''}{f^2} = \sum_{i=1}^{n} (z - r_i)^{-2}.$$

Put $\alpha(z) = (z - r_j)^{-1}$ and $\beta(z) + \gamma_i(z) = (z - r_i)^{-1}$ for $i = 1(1)n$, $i \neq j$. Here $\beta(z)$ is the average of $(z - r_i)^{-1}$, $i \neq j$, and $\sum_{i \neq j} \gamma_i(z) = 0$. Defining $\delta^2 = \sum_{i \neq j} \gamma_i^2$ we can rewrite the equations for P and Q:

$$P = \alpha + (n-1)\beta,$$
$$Q = \alpha^2 + (n-1)\beta^2 + \delta^2.$$

Eliminating β and solving for α we get:

$$\alpha = \{p \pm \sqrt{[(n-1)(nQ - P^2 - n\delta^2)]}\}/n.$$

Since $\alpha = (z - r_j)^{-1}$ we find

$$r_j = z - n/\{P \pm \sqrt{[(n-1)(nQ - P^2 - n\delta^2)]}\}$$

where naturally P, Q, and δ^2 are evaluated at the point z. So far, the expression is exact, but only the values P and Q can be obtained without knowledge of the roots. It seems reasonable to approximate by putting $\delta^2 = 0$, and doing so we obtain Laguerre's iteration formula:

$$(11.5.1) \qquad z_j^{(k+1)} = z_j^{(k)} - n/\{P \pm \sqrt{[(n-1)(nQ - P^2)]}\}$$

where the sign should be chosen so as to make the denominator as large as possible.

Example 1

$f(z) = z^3 - 4z^2 + 3z;$ $\quad z^{(0)} = 4;$ $\quad f = 12,$ $\quad f' = 19,$ $\quad f'' = 16;$ $\quad P = 19/12;$
$Q = (13/12)^2;$ $\quad z^{(1)} = 4 - 36/(19 + \sqrt{292}) = 3.0024;$ $\quad z^{(2)} = 3.00000\,00001.$

Example 2

$f(z) = z^3 + z^2 + z + 1;$ $\quad z^{(0)} = 1;$ $\quad f = 4,$ $\quad f' = 6,$ $\quad f'' = 8;$
$P = 3/2;$ $\quad Q = 1/4;$ $\quad z^{(1)} = (1 + 4i\sqrt{3})/7;$ $\quad P = (30 - i\sqrt{3})/4;$
$Q = (769 + 60i\sqrt{3})/16;$ $\quad z^{(2)} = 0.00016\,63264 + i0.99999\,99862.$

Previously we have discussed two main groups of methods: interpolation methods and fixed-point methods. It is obvious that there is no sharp dividing line between them. Newton's method, for example, can be considered as an interpolation method where the curve is replaced by the tangent, but also as a fixed-point method where we use the iteration function $g(x) = x - f(x)/f'(x)$. What is more important, however, is the fact that in practical computation there will always be problems that behave in peculiar and unpredictable ways, and all methods require a large amount of skill and good judgment for isolation and separation of the roots. With this in mind, we will mention briefly a technique, suggested by Lehmer, which is highly insensitive to phenomena appearing as a result of clustering of roots. It depends on the well-known fact that the integral

$$\frac{1}{2\pi i} \int_C \frac{P'(z)}{P(z)} \, dz,$$

where $P(z)$ is a polynomial and C a closed contour, gives the number of roots inside C (see Chapter 7). This is a special case of a general theorem stating that if $P(z)$ is replaced by an ordinary analytic function $f(z)$, then the integral above gives the value $N - P$ where N is the number of zeros and P the number of poles inside C. In any case, since the value of the integral must be an integer, it can be computed numerically by rather rough methods.

Summing up, there are many good algorithms for computing zeros of general functions with good local convergence, but the situation becomes much more difficult when we are concerned with global convergence. We shall see later that this problem is still more serious when we turn to systems of nonlinear equations.

§11.6. SYSTEMS OF NONLINEAR EQUATIONS

We consider the system

$$\begin{cases} f_1(x_1, x_2, \ldots, x_n) = 0 \\ \quad \vdots \\ f_n(x_1, x_2, \ldots, x_n) = 0. \end{cases}$$

We denote by x the vector with components $(x_1, x_2, \ldots, x_n)$ and by f the vector with components $(f_1, f_2, \ldots, f_n)$. Then the system can be written in a simpler way as one vector equation,

$$f(x) = 0.$$

If we take the gradients of the different components, we obtain a function matrix called the *Jacobian*, which is defined as follows:

(11.6.1) $$J(f, x) = \left(\frac{\partial f_i}{\partial x_k} \right) = \begin{pmatrix} \dfrac{\partial f_1}{\partial x_1} & \dfrac{\partial f_1}{\partial x_2} & \cdots & \dfrac{\partial f_1}{\partial x_n} \\ \vdots & & & \\ \dfrac{\partial f_n}{\partial x_1} & \dfrac{\partial f_n}{\partial x_2} & \cdots & \dfrac{\partial f_n}{\partial x_n} \end{pmatrix}.$$

Our problem is now to solve the equation $f(x) = 0$. We will first discuss two families of methods with close analogies to the one-dimensional case.

Fixed-Point Methods

We rewrite the system in the following way:

$$\begin{cases} x_1 = g_1(x_1, x_2, \ldots, x_n) \\ x_2 = g_2(x_1, x_2, \ldots, x_n) \\ \quad \vdots \\ x_n = g_n(x_1, x_2, \ldots, x_n) \end{cases}$$

or in shorter form as $x = g(x)$. If x and g were scalar, we could try the well known iteration $x_{n+1} = g(x_n)$. Then we know that if $\lim_{n \to \infty} x_n$ exists and is equal to ξ, then

$\xi = g(\xi)$ and ξ is called a fixed point of g. This idea can now be translated to our case, and immediately gives rise to two different methods.

A. Gauss–Jacobi:

$$x_i^{(m+1)} = g_i(x_1^{(m)}, x_2^{(m)}, \ldots, x_n^{(m)})$$

with the following convergence conditions: $\|A\|_\infty < 1$ where $A = (a_{ik})$ and $a_{ik} = \sup_x |\partial g_i/\partial x_k|$, the initial approximation supposed to be sufficiently good.

B. Gauss–Seidel:

$$x_i^{(m+1)} = g_i(x_1^{(m+1)}, \ldots, x_{i-1}^{(m+1)}, x_i^{(m)}, \ldots, x_n^{(m)}).$$

The convergence conditions are more complicated in this case, and we refrain from details. In real applications one would hardly encounter any difficulties in deciding whether convergence occurs or not.

Newton Methods

Consider again the vector equation $f(x) = 0$. We suppose that the vector ξ is the exact solution, and further that our present approximation x can be written $x = \xi + h$. We now compute $f_i(x_1, \ldots, x_n)$ and call these known values g_i. Hence

$$f_i(\xi_1 + h_1, \xi_2 + h_2, \ldots, \xi_n + h_n) = g_i$$

or in vector notation $f(\xi + h) = g$. Developing this in a Taylor series, we find

$$f_i(\xi_1, \xi_2, \ldots, \xi_n) + \sum h_k(\partial f_i/\partial \xi_k) + (\text{higher order terms}) = g_i.$$

The first term is zero by definition, and neglecting higher order terms we get, since $J_{ik} = (\partial f_i/\partial x_k)_{x=\xi}$:

$$Jh = g \qquad \text{and} \qquad h = J^{-1}g$$

where we suppose J to be nonsingular. The vector h found in this way in general does not give us an exact solution, and instead we obtain the iteration formula:

$$x^{(m+1)} = x^{(m)} - J^{-1}f(x^{(m)}).$$

In one dimension this formula becomes the usual Newton–Raphson formula. It is obviously essential that the matrix J is nonsingular. Note, however, that J changes from step to step, which suggests a simplification to reduce computational efforts: replace $J(x^{(m)})$ by $J(x^{(0)})$, for instance.

§11.7. MINIMIZATION METHODS

We first discuss this problem in general, with the intention later to specialize in the direction of solution of systems of equations. Let ·

$$y = F(x_1, x_2, \ldots, x_n) = F(x)$$

be a given function. The problem is then to find a point $\boldsymbol{\xi}$ which gives a local minimum of F. Consider the following Taylor series expansion:

$$F(\boldsymbol{x} + \boldsymbol{h}) = F(\boldsymbol{x}) + \sum h_i \frac{\partial F}{\partial x_i} + \frac{1}{2} \sum h_i h_k \frac{\partial^2 F}{\partial x_i \, \partial x_k} + \cdots$$

The coefficients of the second order terms can be organized as a matrix $\boldsymbol{H}$ with elements $H_{ik} = \partial^2 F / \partial x_i \, \partial x_k$; $\boldsymbol{H}$ is usually called the *Hessian*. The series expansion can then be written

$$F(\boldsymbol{x} + \boldsymbol{h}) = F(\boldsymbol{x}) + \boldsymbol{h}^T \text{ grad } F(\boldsymbol{x}) + \frac{1}{2} \boldsymbol{h}^T \boldsymbol{H} \boldsymbol{h} + \cdots$$

Neglecting higher order terms, we try to choose $\boldsymbol{h} = (h_1, h_2, \ldots, h_n)$ in such a way that F is minimized. Then we get

$$\frac{\partial F}{\partial h_i} \simeq \frac{\partial F}{\partial x_i} + \sum_k \frac{\partial^2 F}{\partial x_i \, \partial x_k} h_k = 0$$

or in vector form

$$\frac{\partial F}{\partial \boldsymbol{h}} \simeq \text{grad } F + \boldsymbol{H}\boldsymbol{h} = 0.$$

In this way we come out with the iteration method

$$\boldsymbol{x}^{(n+1)} = \boldsymbol{x}^{(n)} - \boldsymbol{H}(\boldsymbol{x}^{(n)})^{-1} \text{ grad } F(\boldsymbol{x}^{(n)}).$$

However, the inverse is not computed explicitly in practical work. This is called the Newton method for solution of the minimum problem; the corresponding maximum problem is treated by minimizing $-F$.

In this connection *gradient methods* are also of interest. If we start from a certain point $\boldsymbol{x}^{(n)}$, we should move in the negative gradient direction as long as the function values decrease:

$$\boldsymbol{x}^{(n+1)} = \boldsymbol{x}^{(n)} - t^{(n)} \text{ grad } F(\boldsymbol{x}^{(n)}).$$

Here t indicates how far we move in the chosen direction.

Gradient methods can also be applied for solving systems of equations

$$f_i(x_1, x_2, \ldots, x_n) = 0, \qquad i = 1(1)n.$$

We form $F(\boldsymbol{x}) = \sum_{i=1}^{n} f_i(\boldsymbol{x})^2$ and see at once that $F(\boldsymbol{x}) \geq 0$, and hence we are looking for minima $\boldsymbol{x}$ such that $F(\boldsymbol{x}) = 0$. If again we move in the negative gradient direction we are certain to reach a local minimum. However, it could happen that this minimum value is >0, and if so we have to look for solutions of the system in some other region.

Here we also mention that reformulation to a minimum problem is a powerful method that can be applied in much more general cases. For example, differential equations can often be considered as conditions for minimizing certain integrals (Ritz' principle). An elegant method of finding global maxima and minima makes use of the *range* of a function when the variables are kept within given limits. Generalization of concepts such as function, interpolation, Taylor series expansion, and convergence to interval arithmetics has provided a powerful tool for solving these problems. However, we must refrain from going into more detail.

Example 1

$$\begin{cases} x = x^2 + y^3 + z^5 \\ y = x^3 + y^5 + z^7 \\ z = x^5 + y^7 + z^{11} \end{cases} \qquad \begin{cases} x_0 = 0.8 \\ y_0 = 0.5 \\ z_0 = 0.3 \end{cases}$$

We put

$$\begin{cases} F = x^2 - x + y^3 + z^5 \\ G = x^3 + y^5 - y + z^7 \\ H = x^5 + y^7 + z^{11} - z \end{cases} \qquad J = \begin{pmatrix} 2x - 1 & 3y^2 & 5z^4 \\ 3x^2 & 5y^4 - 1 & 7z^6 \\ 5x^4 & 7y^6 & 11z^{10} - 1 \end{pmatrix}$$

$$J_0 = \begin{pmatrix} 0.6 & 0.75 & 0.0405 \\ 1.92 & -0.6875 & 0.0051 \\ 2.048 & 0.1094 & -1 \end{pmatrix} \qquad \begin{cases} F_0 = -0.03257 \\ G_0 = 0.0434687 \\ H_0 = 0.0354943 \end{cases}$$

Further with

$$\begin{cases} x_1 = x_0 - a \\ y_1 = y_0 - b \\ z_1 = z_0 - c \end{cases} \qquad \begin{aligned} g &= (F, G, H)^T \\ h &= (a, b, c)^T \end{aligned}$$

we obtain by Newton–Raphson's method: $Jh = -g$ and $h = -J^{-1}g$ with $x^{(m+1)} = x^{(m)} + h^{(m)}$.

m	0	1	2	3		∞
x_m	0.8	0.7940	0.7917	0.7916681	$\cdots$	0.7916675703
y_m	0.5	0.5467	0.5444	0.5443466	$\cdots$	0.5443461301
z_m	0.3	0.3283	0.3252	0.3251335	$\cdots$	0.3251333166

It turns out that Gauss–Jacobi's method does not converge in this example.

EXERCISES

1. The equation $x^6 = x^4 + x^3 + 1$ has one root between 1 and 2. Find this root to six decimals.

2. The equation $x^4 - 5x^3 - 12x^2 + 76x - 79 = 0$ has two roots close to $x = 2$. Compute these roots to four decimals.

3. Find the smallest positive root of the equation $\tan x + \tanh x = 0$ to five correct decimals.

4. What positive values of x make the function $y = \tan x/x^2$ a minimum?

5. Find the smallest value of a (to five decimals) such that $a\sqrt{x} \geq \sin x$ for all positive values of x.

6. Compute to five decimals the smallest constant C such that $\cosh x > Cx$ for all $x \geq 0$.

7. Find the smallest positive root of the equation

$$1 - x + \frac{x^2}{(2!)^2} - \frac{x^3}{(3!)^2} + \frac{x^4}{(4!)^2} - \cdots = 0.$$

8. The k-fold zero $x = a$ of the equation $f(x) = 0$ is to be determined. If $k = 1$, Newton–Raphson's method gives quadratic convergence, while $k > 1$ makes the convergence rate geometrical. Show that the modified formula

$$x_{n+1} = x_n - kf(x_n)/f'(x_n)$$

restores quadratic convergence in the neighborhood of a. Use the formula for finding a double root close to zero of the equation

$$x^5 - 7x^4 + 10x^3 + 10x^2 - 7x + 1 = 0.$$

9. One wants to compute the positive root of the equation $x = a - bx^2$ (a and b positive) by using the iterative method $x_{k+1} = a - bx_k^2$. What is the condition for convergence?

10. If one attempts to solve the equation $x = 1.4 \cos x$ by using the formula $x_{n+1} = 1.4 \cos x_n$, then in the limit x_n will oscillate between two well-defined values a and b. Find these and the correct solution to four decimal places.

11. Find the smallest value of p (to six decimals) such that $e^{-px} \leq (1 + x^2)^{-1}$ for all positive values of x.

12. The equation $1 - x + x^2/2! - x^3/3! + \cdots - x^n/n! = 0$ (n odd integer) has one real solution z which is positive and, of course, depends upon n. Evaluate $\lim_{n \to \infty} (z/n)$ by using Stirling's asymptotic formula $(n - 1)! \simeq (2\pi)^{1/2} e^{-n} n^{n-1/2}$.

13. The expression $(45 + 29\sqrt{2})^{1/3} + (45 - 29\sqrt{2})^{1/3}$ is rational. Prove this and compute the value.

14. Using Horner's scheme, move the origin to $x = 1$ for the polynomial

$$f(x) = 2x^4 - 7x^3 + 12x^2 - 15x + 10.$$

15. The equation $z = e^z$ where $z = x + iy$ has one solution such that $0 < x < 1$ and $1 < y < 2$. Find this solution to 7 decimals.

16. The equation $z^2 + e^z = 0$ has an infinite number of solutions. Find the four solutions closest to the origin ($y > 0$).

17. Determine maximum and minimum of $f(x, y) = x^2 + 4y^2$ in the domain

$$x + y - 3 \geq 0, \quad 5x + 7y - 35 \leq 0, \quad x \geq 0, \quad y \geq 0.$$

18. Find the minimum of the function $f = x^2 + 4y^2 + (x + 1)^{-1}(y + 1)^{-1}$ when x and y are both positive.

19. The function $u = 100/(x^2 + y^2 + 3) + 2x^2 - y^2 + 8y$ should be studied inside and on the boundary of a square with corners $(0, 0)$, $(3, 0)$, $(3, 3)$, and $(0, 3)$. Find all maxima, minima, and saddle points.

20. The equation $x^{32} = e^x$ has one negative and two positive solutions. One tries to solve the equation using the iteration $x_{n+1} = g(x_n)$, $x_0 = 2$, and (a) $g(x) = e^{x/32}$, (b) $g(x) = 32 \ln x$, (c) $g(x) = -e^{x/32}$. Which values are obtained? Explain the result!

21. One has
$$\begin{cases} x + y = 2 \\ x^{16} + y^{16} = 3. \end{cases}$$
Compute $x^{64} + y^{64}$ (to six decimals).

22. The factorial function $n!$ is primarily defined only for positive integers and zero. However, for reasonably large values of n we have with good accuracy $n! \simeq f(n)$ where
$$f(x) = (2\pi)^{1/2} x^{x+1/2} e^{-x} (1 + 1/12x + 1/288x^2).$$
Compute x to four decimals so that $f(x) = 1000$.

CHAPTER 12

Algebraic Eigenvalue Problems

Das, also, war des Pudels Kern!

Goethe

In this section we will treat the problem of how actually to compute eigenvalues and eigenvectors for a given square matrix A. The eigenvalues are roots of the characteristic equation

$$\det(A - \lambda I) = 0.$$

However, it has been observed that explicit evaluation of this equation and solving for the eigenvalues is a rather unstable procedure, even if the matrix A is well-behaved. Instead we have to rely on other methods. In principle we distinguish between iteration methods and transformation methods.

§12.1. ITERATIVE METHODS

The Power Method

We assume that the eigenvalues of A are $\lambda_1, \lambda_2, \ldots, \lambda_n$, where $|\lambda_1| > |\lambda_2| \geq \cdots \geq |\lambda_n|$. Now we let A operate repeatedly on a vector v, which we express as a linear combination of the eigenvectors

(12.1.1) $$v = c_1 v_1 + c_2 v_2 + \cdots + c_n v_n.$$

Then we have

$$Av = c_1 Av_1 + c_2 Av_2 + \cdots + c_n Av_n = \lambda_1 \left(c_1 v_1 + c_2 \frac{\lambda_2}{\lambda_1} v_2 + \cdots + c_n \frac{\lambda_n}{\lambda_1} v_n \right)$$

and through iteration we obtain

(12.1.2) $$A^p v = \lambda_1^p \left\{ c_1 v_1 + c_2 \left(\frac{\lambda_2}{\lambda_1} \right)^p v_2 + \cdots + c_n \left(\frac{\lambda_n}{\lambda_1} \right)^p v_n \right\}.$$

For large values of p, the vector

$$c_1 v_1 + c_2 \left(\frac{\lambda_2}{\lambda_1} \right)^p v_2 + \cdots + c_n \left(\frac{\lambda_n}{\lambda_1} \right)^p v_n$$

will converge toward $c_1 v_1$, that is, the eigenvector of λ_1. The eigenvalue is obtained as

(12.1.3) $$\lambda_1 = \lim_{p \to \infty} \frac{(A^{p+1} v)_r}{(A^p v)_r}, \qquad r = 1, 2, \ldots, n,$$

where the index r signifies the rth component in the corresponding vector. The rate of convergence is determined by the quotient λ_2/λ_1; convergence is faster the smaller $|\lambda_2/\lambda_1|$ is. For numerical purposes the algorithm just described can be formulated in the following way. Given a vector y_k, we form two other vectors, y_{k+1} and z_{k+1}:

(12.1.4) $$\begin{aligned} z_{k+1} &= A y_k, \\ \alpha_{k+1} &= \max_r |(z_{k+1})_r|, \\ y_{k+1} &= z_{k+1}/\alpha_{k+1}. \end{aligned}$$

The initial vector y_0 should be chosen in a convenient way; often one tries a vector with all components equal to 1.

Example 1

$$A = \begin{pmatrix} 1 & -3 & 2 \\ 4 & 4 & -1 \\ 6 & 3 & 5 \end{pmatrix}.$$

Starting from

$$y_0 = \begin{pmatrix} 1 \\ 1 \\ 1 \end{pmatrix},$$

we find that

$$y_5 = \begin{pmatrix} 0.3276 \\ 0.0597 \\ 1 \end{pmatrix}, \qquad y_{10} = \begin{pmatrix} 0.3007 \\ 0.0661 \\ 1 \end{pmatrix},$$

$$y_{15} = \begin{pmatrix} 0.3000 \\ 0.0667 \\ 1 \end{pmatrix}, \qquad \text{and} \qquad z_{16} = \begin{pmatrix} 2.0999 \\ 0.4668 \\ 7.0001 \end{pmatrix}.$$

After round-off, we get

$$\lambda_1 = 7 \quad \text{and} \quad v_1 = \begin{pmatrix} 9 \\ 2 \\ 30 \end{pmatrix}.$$

If the matrix A is Hermitian and all eigenvalues are different, the eigenvectors, as shown before, are orthogonal. Let x be the vector obtained after p iterations:

$$x = c_1 v_1 + c_2 \left(\frac{\lambda_2}{\lambda_1}\right)^p v_2 + \cdots + c_n \left(\frac{\lambda_n}{\lambda_1}\right)^p v_n = c_1 v_1 + \varepsilon_2 v_2 + \cdots + \varepsilon_n v_n.$$

We suppose that all v_i are normalized:

$$v_i^H v_k = \delta_{ik}.$$

Then we have

$$A x = c_1 \lambda_1 v_1 + \varepsilon_2 \lambda_2 v_2 + \cdots + \varepsilon_n \lambda_n v_n$$

and

$$x^H A x = \lambda_1 |c_1|^2 + \lambda_2 |\varepsilon_2|^2 + \cdots + \lambda_n |\varepsilon_n|^2.$$

Further, $x^H x = |c_1|^2 + |\varepsilon_2|^2 + \cdots + |\varepsilon_n|^2$. When p increases, all ε_i tend to zero, and with $x = A^p (\sum c_i v_i)$, we get Rayleigh's quotient

$$(12.1.5) \qquad \lambda_1 = \lim_{p \to \infty} \frac{x^H A x}{x^H x}.$$

Example 2. With Wilson's matrix

$$A = \begin{pmatrix} 10 & 7 & 8 & 7 \\ 7 & 5 & 6 & 5 \\ 8 & 6 & 10 & 9 \\ 7 & 5 & 9 & 10 \end{pmatrix} \quad \text{and} \quad x_0 = \begin{pmatrix} 1 \\ 1 \\ 1 \\ 1 \end{pmatrix},$$

we obtain for $p = 1, 2,$ and 3, $\lambda_1 = 29.75$, 30.287, and 30.288662, respectively, compared with the correct value 30.28868. The corresponding eigenvector is

$$\begin{pmatrix} 0.95761 \\ 0.68892 \\ 1 \\ 0.94379 \end{pmatrix}.$$

The quotients of the individual vector components give much slower convergence; for example, $(x_3)_1/(x_2)_1 = 30.25987$.

The power method can easily be modified in such a way that certain other eigenvalues can also be computed. If, for example, A has an eigenvalue λ, then $A - qI$ has an eigenvalue $\lambda - q$. Using this principle, we can produce the two outermost eigenvalues. Further, we know that λ^{-1} is an eigenvalue of A^{-1} and analogously that $(\lambda - q)^{-1}$ is an eigenvalue of $(A - qI)^{-1}$. If we know that an eigenvalue is close to q, we can concentrate on that, since $(\lambda - q)^{-1}$ becomes large as soon as λ is close to q. We will return to this problem in some more detail later (see 12.1.8).

We will now discuss how the next largest eigenvalue (in absolute value) can be calculated if we know the largest eigenvalue λ_1 and the corresponding eigenvector x_1. Let a^T be the first row vector of A and form

(12.1.6)
$$A_1 = A - x_1 a^T.$$

Here x_1 is supposed to be normalized in such a way that the first component is 1. Hence the first row of A_1 is zero. Now let λ_2 and x_2 be an eigenvalue and the corresponding eigenvector with the first component of x_2 equal to 1. Then we have

$$A_1(x_1 - x_2) = A(x_1 - x_2) - x_1 a^T(x_1 - x_2) = \lambda_1 x_1 - \lambda_2 x_2 - (\lambda_1 - \lambda_2)x_1$$
$$= \lambda_2(x_1 - x_2),$$

since $a^T x_1 = \lambda_1$ and $a^T x_2 = \lambda_2$ (note that the first component of x_1 as well as of x_2 is 1).

Thus λ_2 is an eigenvalue and $x_1 - x_2$ is an eigenvector of A_1. Since $x_1 - x_2$ has the first component equal to 0, the first column of A_1 is irrelevant, and in fact we need consider only the $(n-1, n-1)$-matrix that is obtained when the first row and first column of A are removed. We determine an eigenvector of this matix, and by adding a zero as first component, we get a vector z. Then we obtain x_2 from the relation

$$x_2 = x_1 + cz.$$

Multiplying with a^T we find $a^T x_2 = a^T x_1 + ca^T z$, and hence $c = (\lambda_2 - \lambda_1)/a^T z$. When λ_2 and x_2 have been determined, the process, which is called *deflation*, can be repeated.

Example 3. The matrix

$$A = \begin{pmatrix} -306 & -198 & 426 \\ 104 & 67 & -147 \\ -176 & -114 & 244 \end{pmatrix}$$

has an eigenvalue $\lambda_1 = 6$ and the corresponding eigenvector

$$\begin{pmatrix} 2 \\ -1 \\ 1 \end{pmatrix},$$

or normalized,

$$x_1 = \begin{pmatrix} 1 \\ -\frac{1}{2} \\ \frac{1}{2} \end{pmatrix}.$$

Without difficulty we find

$$A_1 = \begin{pmatrix} 0 & 0 & 0 \\ -49 & -32 & 66 \\ -23 & -15 & 31 \end{pmatrix}.$$

Now we need consider only

$$B_1 = \begin{pmatrix} -32 & 66 \\ -15 & 31 \end{pmatrix},$$

and we find the eigenvalues $\lambda_2 = -2$ and $\lambda_3 = 1$, which are also eigenvalues of the original matrix A. The two-dimensional eigenvector belonging to $\lambda_2 = -2$ is

$$\begin{pmatrix} 11 \\ 5 \end{pmatrix}.$$

and hence

$$x_2 = x_1 + cz = \begin{pmatrix} 1 \\ -\frac{1}{2} \\ \frac{1}{2} \end{pmatrix} + c \begin{pmatrix} 0 \\ 11 \\ 5 \end{pmatrix}.$$

Since $a^T z = -48$, we get $c = \frac{1}{6}$ and

$$x_2 = \begin{pmatrix} 1 \\ \frac{4}{3} \\ \frac{4}{3} \end{pmatrix} \quad \text{or} \quad \begin{pmatrix} 3 \\ 4 \\ 4 \end{pmatrix}.$$

With $\lambda_3 = 1$, we find

$$x_3 = \begin{pmatrix} 1 \\ -\frac{1}{2} \\ \frac{1}{2} \end{pmatrix} + c \begin{pmatrix} 0 \\ 2 \\ 1 \end{pmatrix}$$

and $a^T z = 30$. Hence $c = -\frac{1}{6}$ and

$$x_3 = \begin{pmatrix} 1 \\ -\frac{5}{6} \\ \frac{1}{3} \end{pmatrix} \quad \text{or} \quad \begin{pmatrix} 6 \\ -5 \\ 2 \end{pmatrix},$$

and all eigenvalues and eigenvectors are known.

If A is Hermitian, we have $x_1^H x_2 = 0$ when $\lambda_1 \neq \lambda_2$. Now suppose that $x_1^H x_1 = 1$, and form

(12.1.7) $$A_1 = A - \lambda_1 x_1 x_1^H.$$

It is easily understood that the matrix A_1 has the same eigenvalues and eigenvectors as A except λ_1, which has been replaced by zero. In fact, we have $A_1 x_1 = A x_1 - \lambda_1 x_1 x_1^H x_1 = \lambda_1 x_1 - \lambda_1 x_1 = 0$ and $A_1 x_2 = A x_2 - \lambda_1 x_1 x_1^H x_2 = \lambda_2 x_2$, and so on. Then we can again use the power method on the matrix A_1.

Example 4

$$A = \begin{pmatrix} 10 & 7 & 8 & 7 \\ 7 & 5 & 6 & 5 \\ 8 & 6 & 10 & 9 \\ 7 & 5 & 9 & 10 \end{pmatrix}; \quad \lambda_1 = 30.288686; \quad x_1 = \begin{pmatrix} 0.528561 \\ 0.380255 \\ 0.551959 \\ 0.520933 \end{pmatrix};$$

$$A_1 = \begin{pmatrix} 1.53804 & 0.91234 & -0.83654 & -1.33984 \\ 0.91234 & 0.62044 & -0.35714 & -0.99979 \\ -0.83654 & -0.35714 & 0.77228 & 0.29097 \\ -1.33984 & -0.99979 & 0.29097 & 1.78053 \end{pmatrix}.$$

With the starting vector

$$y_0 = \begin{pmatrix} 1 \\ 1 \\ -1 \\ -1 \end{pmatrix},$$

we find the following values for Rayleigh's quotient: $\lambda_2 = 3.546$, 3.8516, and 3.85774 compared with the correct value 3.858057.

If the numerically largest eigenvalue of a real matrix A is complex, $\lambda e^{i\varphi}$, then $\lambda e^{-i\varphi}$ must also be an eigenvalue. It is also clear that if x_1 is the eigenvector belonging to $\lambda e^{i\varphi}$, then x_1^* is the eigenvector belonging to $\lambda e^{-i\varphi}$. Denoting a certain component of $A^m x$ by p_m, we get after easy computations:

$$\lim_{m \to \infty} \frac{p_m p_{m+2} - p_{m+1}^2}{p_{m-1}p_{m+1} - p_m^2} = \lambda^2.$$

$$\lim_{m \to \infty} \frac{\lambda^2 p_m + p_{m+2}}{2\lambda p_{m+1}} = \cos \varphi.$$

Inverse Iteration

As mentioned above, if we know an eigenvalue approximately, we can improve this by using a simple device. Suppose that λ is close to an eigenvalue of A. Then $A - \lambda I$ is almost singular, and so we can construct an iteration that is essentially the power method applied to $(A - \lambda I)^{-1}$. Making a reasonable guess for an eigenvector x_0, we compute a better approximation x_1, and in general we compute x_{n+1} from x_n:

(12.1.8) $(A - \lambda I)x_{n+1} = x_n.$

Here x_{n+1} is obtained by Gaussian elimination, and then we find an approximate value of the absolutely largest eigenvalue μ of $(A - \lambda I)^{-1}$. Keeping λ constant, only the right-hand side changes, and the Gaussian elimination has to be performed just once. Hence a better approximation for the desired eigenvalue of A is $\lambda + \mu^{-1}$. In particular, if A is symmetric we form the Rayleigh quotient

$$\mu \simeq \frac{x_{n+1}^T x_n}{x_n^T x_n}$$

which gives a much better approximation.

Example 5. Again we consider the symmetric Wilson matrix, which has an eigenvalue close to 4. We form $B = A - 4I$:

$$B = \begin{pmatrix} 6 & 7 & 8 & 7 \\ 7 & 1 & 6 & 5 \\ 8 & 6 & 6 & 9 \\ 7 & 5 & 9 & 6 \end{pmatrix}$$

and start by choosing $x_0 = (-1, -1, 1, 1)^T$. This gives the system

$$B x_1 = x_0$$

and after Gaussian elimination we get the solution

$$x_1^T = \frac{1}{47}(388, 254, -177, -391).$$

The Rayleigh quotient is

$$\mu_1 = -605/94 = -6.43617 \quad \text{and} \quad \lambda_1 = 4 - 1/6.43617 = 3.844628.$$

One more iteration yields

$$\mu_2 = -7.0440903 \quad \text{and} \quad \lambda_2 = 3.858037$$

as compared with the correct value 3.858057456.

Special Matrices

We will briefly comment upon a simple, rather direct method that can be used on band matrices in which the elements in the diagonals are constant. We will demonstrate the technique on a tridiagonal matrix $C = aI + bA$ where $(A)_{ik} = 1$ if $i - k = \pm 1$ and 0 otherwise. Suppose that x is an eigenvector of A with components $x_1, x_2, \ldots, x_n$ and λ the corresponding eigenvalue. Then we obtain the system:

$$\begin{cases} x_0 + x_2 = \lambda x_1 \\ x_1 + x_3 = \lambda x_2 \\ \vdots \\ x_{n-2} + x_n = \lambda x_{n-1} \\ x_{n-1} + x_{n+1} = \lambda x_n \end{cases}$$

where we have added the dummy variables x_0 and x_{n+1}, which should be made zero. We rewrite the system in the more compact form

$$x_{r-1} + x_{r+1} = \lambda x_r, \quad r = 1(1)n$$

with the extra conditions $x_0 = x_{n+1} = 0$. The characteristic equation is

$$\mu^2 - \lambda\mu + 1 = 0.$$

From Gershgorin's theorem we have $|\lambda| \leq 2$, and putting $\lambda = 2\cos\varphi$ we get

$$\mu = \cos\varphi \pm i\sin\varphi = \exp(\pm i\varphi)$$

and hence

$$x_r = A\exp(ir\varphi) + B\exp(-ir\varphi).$$

Now $r = 0$ gives $x_0 = A + B = 0$, and apart from a trivial factor we find $x_r = \sin r\varphi$. Then φ can be determined from the condition $x_{n+1} = 0$ or $\sin(n+1)\varphi = 0$, giving $\varphi = \pi r/(n+1)$, $r = 1(1)n$. Hence the eigenvalues are

$$(12.1.9) \qquad\qquad \lambda_r = 2\cos\frac{\pi r}{n+1}$$

and the corresponding eigenvectors are

$$(\sin\varphi, \sin 2\varphi, \ldots, \sin n\varphi)^T.$$

Taking $n = 5$, for example, we get the eigenvalues $\pm\sqrt{3}$, ± 1, 0 and the eigenvectors

$$(1, \pm\sqrt{3}, 2, \pm\sqrt{3}, 1)^T, \qquad (1, \pm 1, 0, \mp 1, -1)^T, \qquad (1, 0, -1, 0, 1)^T.$$

The eigenvalues of C are, of course, $a + b\lambda$.

---

§12.2. TRANSFORMATION METHODS

All transformation methods depend on the fact that the characteristic equation and hence also the eigenvalues are invariant under similarity transformations. The general idea is now to construct a series of very simple transformations that will gradually transform the original matrix to diagonal, tridiagonal, or triangular form.

First of all we must establish whether the initial matrix has any simple properties. We know that Hermitian matrices can be diagonalized by unitary transformation matrices. However, we will restrict ourselves to the somewhat simpler case of real symmetric matrices that can be diagonalized by orthogonal transformations; the eigenvalues will then appear in the main diagonal. We will also demonstrate that a real symmetric matrix can be brought to tridiagonal form by so-called Householder transformations. The eigenvalues can then be computed by use of a simple recursive procedure, and the eigenvectors belonging to the tridiagonal matrix are found by a trivial computation.

For matrices without symmetry properties it is recommended first to transform them to *Hessenberg* form. This is either upper Hessenberg, characterized by $a_{ik} = 0$

when $i > k + 1$, or lower Hessenberg, characterized by $a_{ik} = 0$ when $k > i + 1$. From there on the matrix is transformed to triangular form by unitary transformations. This so-called QR-method was introduced by Francis; it has proved very efficient and robust and has practically outperformed all other methods.

Jacobi's Method

In many applications we meet the problem of diagonalizing real, symmetric matrices. This problem is particularly important in quantum mechanics.

In Section 2.3 we proved that for a real symmetric matrix A, all eigenvalues are real, and that there exists a real orthogonal matrix Q such that $Q^{-1}AQ$ is diagonal. We shall now try to produce the desired orthogonal matrix as a product of very special orthogonal matrices. Among the off-diagonal elements we choose the numerically largest element: $|a_{ik}| = \max$. The elements a_{ii}, a_{ik}, $a_{ki}(=a_{ik})$, and a_{kk} form a $(2, 2)$-submatrix, which can easily be transformed to diagonal form. We put

$$Q = \begin{pmatrix} \cos \varphi & -\sin \varphi \\ \sin \varphi & \cos \varphi \end{pmatrix},$$

and get

(12.2.1) $\quad D = Q^{-1}AQ = \begin{pmatrix} \cos \varphi & \sin \varphi \\ -\sin \varphi & \cos \varphi \end{pmatrix} \begin{pmatrix} a_{ii} & a_{ik} \\ a_{ik} & a_{kk} \end{pmatrix} \begin{pmatrix} \cos \varphi & -\sin \varphi \\ \sin \varphi & \cos \varphi \end{pmatrix}.$

$$d_{ii} = a_{ii} \cos^2 \varphi + 2a_{ik} \sin \varphi \cos \varphi + a_{kk} \sin^2 \varphi,$$
$$d_{ik} = d_{ki} = -(a_{ii} - a_{kk}) \sin \varphi \cos \varphi + a_{ik}(\cos^2 \varphi - \sin^2 \varphi),$$
$$d_{kk} = a_{ii} \sin^2 \varphi - 2a_{ik} \sin \varphi \cos \varphi + a_{kk} \cos^2 \varphi.$$

Now choose the angle φ such that $d_{ik} = d_{ki} = 0$, that is, $\tan 2\varphi = 2a_{ik}/(a_{ii} - a_{kk})$. This equation gives four different values of φ, and in order to get as small rotations as possible we claim $-\pi/4 \le \varphi \le \pi/4$. Putting

$$R = ((a_{ii} - a_{kk})^2 + 4a_{ik}^2)^{1/2} \quad \text{and} \quad \sigma = \begin{cases} 1 & \text{if } a_{ii} \ge a_{kk}, \\ -1 & \text{if } a_{ii} < a_{kk}, \end{cases}$$

we obtain:

$$\begin{cases} \sin 2\varphi = 2\sigma a_{ik}/R, \\ \cos 2\varphi = \sigma(a_{ii} - a_{kk})/R, \end{cases}$$

since the angle 2φ must belong to the first quadrant if $\tan 2\varphi > 0$ and to the fourth quadrant if $\tan 2\varphi < 0$. Hence we have for the angle φ:

$$\varphi = \tfrac{1}{2} \arctan(2a_{ik}/(a_{ii} - a_{kk})) \quad \text{if } a_{ii} \ne a_{kk},$$

$$\varphi = \begin{cases} \pi/4 & \text{when } a_{ik} > 0 \\ -\pi/4 & \text{when } a_{ik} < 0 \end{cases} \quad \text{if } a_{ii} = a_{kk},$$

where the value of the arctan-function is chosen between $-\pi/2$ and $\pi/2$. After a few simple calculations we get finally:

(12.2.2)
$$\begin{cases} d_{ii} = \frac{1}{2}(a_{ii} + a_{kk} + \sigma R), \\ d_{kk} = \frac{1}{2}(a_{ii} + a_{kk} - \sigma R), \\ d_{ik} = d_{ki} = 0. \end{cases}$$

(Note that $d_{ii} + d_{kk} = a_{ii} + a_{kk}$ and $d_{ii}d_{kk} = a_{ii}a_{kk} - a_{ik}^2$.)

We perform a series of such two-dimensional rotations; the transformation matrices have the form given above in the elements (i, i), (i, k), (k, i), and (k, k) and are identical with the unit matrix elsewhere. Each time we choose such values i and k that $|a_{ik}| = $ max. We shall show that with the notation $P_r = Q_1 Q_2 \cdots Q_r$ the matrix $B_r = P_r^{-1} A P_r$ for increasing r will approach a diagonal matrix D with the eigenvalues of A along the main diagonal. Then it is obvious that we get the eigenvectors as the corresponding columns of $Q = \lim_{r \to \infty} P_r$ since we have $Q^{-1}AQ = D$, that is, $AQ = QD$. Let x_k be the kth column vector of Q and λ_k the kth diagonal element of D. Then we have

$$A x_k = \lambda_k x_k.$$

If $\sum_{k \neq i} |a_{ik}|$ is denoted by ε_i, we know from Gershgorin's theorem that $|a_{ii} - \lambda| < \varepsilon_i$ for some value of i, and if the process has been brought sufficiently far, *every* circle defined in this way contains exactly one eigenvalue, provided all eigenvalues are different. Thus it is easy to see when sufficient accuracy has been attained and the procedure can be discontinued.

The convergence of the method has been examined by von Neumann and Goldstine in the following way. We put $\tau^2(A) = \sum_i \sum_{k \neq i} a_{ik}^2 = N^2(A) - \sum_i a_{ii}^2$ and, as before, $B = Q^{-1}AQ$. The orthogonal transformation affects only the ith row and column and the kth row and column. Taking only off-diagonal elements into account, we find for $r \neq i$ and $r \neq k$ relations of the form

$$\begin{cases} a'_{ir} = a_{ir} \cos \varphi + a_{kr} \sin \varphi, \\ a'_{kr} = -a_{ir} \sin \varphi + a_{kr} \cos \varphi, \end{cases}$$

and hence $a'^2_{ir} + a'^2_{kr} = a^2_{ir} + a^2_{kr}$. Thus $\tau^2(A)$ will be changed only through the cancellation of the elements a_{ik} and a_{ki}, that is,

$$\tau^2(A') = \tau^2(A) - 2a_{ik}^2.$$

Since a_{ik} was the absolutely largest of all $n(n - 1)$ off-diagonal elements, we have

$$a_{ik}^2 \geq \frac{\tau^2(A)}{n(n - 1)},$$

and

$$\tau^2(A') \leq \tau^2(A) \cdot \left(1 - \frac{2}{n(n - 1)}\right) < \tau^2(A) \cdot \exp\left(-\frac{2}{n(n - 1)}\right).$$

Hence we get the final estimate,

$$(12.2.3) \qquad \tau(A') < \tau(A) \cdot \exp\left(-\frac{1}{n(n-1)}\right).$$

After N iterations, $\tau(A)$ has decreased with at least the factor $\exp(-N/n(n-1))$, and for a sufficiently large N we come arbitrarily close to the diagonal matrix containing the eigenvalues.

In a slightly different modification, we go through the matrix row by row, performing a rotation as soon as $|a_{ik}| > \varepsilon$. Here ε is a prescribed tolerance which, of course, has to be changed for each sweep. This modification seems to be more powerful than the preceding one.

The method was first suggested by Jacobi in connection with computations on planetary movements. It has proved very efficient for diagonalization of real symmetric matrices of moderate size.

Example 1

$$A = \begin{pmatrix} 10 & 7 & 8 & 7 \\ 7 & 5 & 6 & 5 \\ 8 & 6 & 10 & 9 \\ 7 & 5 & 9 & 10 \end{pmatrix}.$$

Choosing $i = 3$, $k = 4$, we obtain $\tan 2\varphi = 18/(10 - 10) = \infty$ and $\varphi = 45°$. After the first rotation, we have

$$A_1 = \begin{pmatrix} 10 & 7 & 15/\sqrt{2} & -1/\sqrt{2} \\ 7 & 5 & 11/\sqrt{2} & -1/\sqrt{2} \\ 15/\sqrt{2} & 11/\sqrt{2} & 19 & 0 \\ -1/\sqrt{2} & -1/\sqrt{2} & 0 & 1 \end{pmatrix}.$$

Here we take $i = 1$, $k = 3$, and obtain $\tan 2\varphi = 15\sqrt{2}/(10 - 19)$ and $\varphi = -33°.5051$. After the second rotation we have

$$A_2 = \begin{pmatrix} 2.978281 & 1.543214 & 0 & -0.589611 \\ 1.543214 & 5 & 10.349806 & -0.707107 \\ 0. & 10.349806 & 26.021719 & -0.390331 \\ -0.589611 & -0.707107 & -0.390331 & 1 \end{pmatrix}$$

and after 10 rotations we have

$$A_{10} = \begin{pmatrix} 3.858056 & 0 & -0.000656 & -0.001723 \\ 0 & 0.010150 & 0.000396 & 0.000026 \\ -0.000656 & 0.000396 & 30.288685 & 0.001570 \\ -0.001723 & 0.000026 & 0.001570 & 0.843108 \end{pmatrix}.$$

After 17 rotations the diagonal elements are 3.85805745, 0.01015005, 30.28868533, and 0.84310715, while the remaining elements are equal to 0 to 8 decimals accuracy.

The sum of the diagonal elements is 34.99999999 and the product 1.00000015 in good agreement with the exact characteristic equation:

$$\lambda^4 - 35\lambda^3 + 146\lambda^2 - 100\lambda + 1 = 0.$$

Householder's Method

This method, also, has been designed for real, symmetric matrices. We shall essentially follow the presentation given by Wilkinson. The first step consists of reducing the given matrix A to a band matrix. This is done by orthogonal transformations representing *reflections*. The orthogonal matrices will be denoted by P_r with the general structure

(12.2.4)
$$P = I - 2ww^T.$$

Here w is a column vector such that

(12.2.5)
$$w^T w = 1.$$

It is evident that P is symmetric. Further, we have

$$P^T P = (I - 2ww^T)(I - 2ww^T) = I - 4ww^T + 4ww^T ww^T = I;$$

that is, P is also orthogonal.

The matrix P acting as an operator can be given a simple geometric interpretation. Let P operate on a vector x from the left:

$$Px = (I - 2ww^T)x = x - 2(w^T x)w.$$

In Figure 12.1 the line L is perpendicular to the unit vector w in a plane defined by w and x. The distance from the endpoint of x to L is $|x| \cos(x, w) = w^T x$, and the mapping P means a reflection in a plane perpendicular to w.

Those vectors w which will be used are constructed with the first $(r - 1)$ components zero, or

$$w_r^T = (0, 0, \ldots, 0, x_r, x_{r+1}, \ldots, x_n).$$

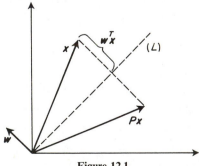

Figure 12.1

With this choice we form $P_r = I - 2w_r w_r^T$. Further, by (12.2.5) we have

$$x_r^2 + x_{r+1}^2 + \cdots + x_n^2 = 1.$$

Now put $A = A_1$ and form successively

(12.2.6) $$A_r = P_r A_{r-1} P_r,$$

$r = 2, 3, \ldots, n - 1$. At the first transformation, we get zeros in the positions $(1, 3)$, $(1, 4), \ldots, (1, n)$ and in the corresponding places in the first column. The final result will then become a band matrix. The matrix A_{r-1} contains $n - r$ elements in the row $(r - 1)$, which must be reduced to zero by transformation with P_r; this gives $n - r$ equations for the $n - r + 1$ elements $x_r, x_{r+1}, \ldots, x_n$, and further we have the condition that the sum of the squares must be 1.

We carry through one step in the computation in an example:

$$A = \begin{pmatrix} a_1 & b_1 & c_1 & d_1 \\ b_1 & b_2 & c_2 & d_2 \\ c_1 & c_2 & c_3 & d_3 \\ d_1 & d_2 & d_3 & d_4 \end{pmatrix},$$

$$w_2^T = (0, x_2, x_3, x_4); \qquad x_2^2 + x_3^2 + x_4^2 = 1.$$

The transformation $P_2 A P_2$ must now produce zeros instead of c_1 and d_1. Obviously, the matrix P_2 has the following form:

$$P_2 = \begin{pmatrix} 1 & 0 & 0 & 0 \\ 0 & 1 - 2x_2^2 & -2x_2 x_3 & -2x_2 x_4 \\ 0 & -2x_2 x_3 & 1 - 2x_3^2 & -2x_3 x_4 \\ 0 & -2x_2 x_4 & -2x_3 x_4 & 1 - 2x_4^2 \end{pmatrix}.$$

Since in the first row of P_2 only the first element is not zero, for example, the $(1, 3)$-element of $P_2 A P_2$ can become zero only if the corresponding element is zero already in $A P_2$. Putting $p_1 = b_1 x_2 + c_1 x_3 + d_1 x_4$, we find that the first row of $A P_2$ has the following elements:

$$a_1, \qquad b_1 - 2p_1 x_2, \qquad c_1 - 2p_1 x_3, \qquad d_1 - 2p_1 x_4.$$

Now we claim that

(12.2.7) $$\begin{cases} c_1 - 2p_1 x_3 = 0, \\ d_1 - 2p_1 x_4 = 0. \end{cases}$$

Since we are performing an orthogonal transformation, the sum of the squares of the elements in a row is invariant, and hence

$$a_1^2 + (b_1 - 2p_1 x_2)^2 = a_1^2 + b_1^2 + c_1^2 + d_1^2.$$

Putting $S = (b_1^2 + c_1^2 + d_1^2)^{1/2}$, we obtain

(12.2.8) $$b_1 - 2p_1 x_2 = \pm S.$$

Multiplying (12.2.8) by x_2 and (12.2.7) by x_3 and x_4, we get

$$b_1 x_2 + c_1 x_3 + d_1 x_4 - 2p_1(x_2^2 + x_3^2 + x_4^2) = \pm S x_2.$$

The sum of the first three terms is p_1 and further $x_2^2 + x_3^2 + x_4^2 = 1$. Hence

(12.2.9) $$p_1 = \mp S x_2.$$

Inserting this into (12.2.8), we find that $x_2^2 = \frac{1}{2}(1 \mp b_1/S)$, and from (12.2.7), $x_3 = \mp c_1/2S x_2$ and $x_4 = \mp d_1/2S x_2$.

In the general case, two square roots have to be evaluated, one for S and one for x_2. Since we have x_2 in the denominator, we obtain the best accuracy if x_2 is large. This is accomplished by choosing a suitable sign for the square-root extraction for S. Thus the quantities ought to be defined as follows:

(12.2.10) $$x_2^2 = \frac{1}{2}\left(1 + \frac{b_1 \cdot \text{sign } b_1}{S}\right).$$

The sign for this square root is irrelevant and we choose plus. Hence we obtain for x_3 and x_4:

(12.2.11) $$x_3 = \frac{c_1 \text{ sign } b_1}{2S x_2}; \qquad x_4 = \frac{d_1 \text{ sign } b_1}{2S x_2}.$$

The end result is a band matrix B with the form:

$$B = \begin{pmatrix} \alpha_1 & \beta_1 & 0 & \cdots & & 0 \\ \beta_1 & \alpha_2 & \beta_2 & \cdots & & 0 \\ 0 & \beta_2 & \alpha_3 & \cdots & & 0 \\ \vdots & & & & & \beta_{n-1} \\ 0 & 0 & \cdots & & \beta_{n-1} & \alpha_n \end{pmatrix}.$$

We now form a sequence of polynomials which are the successive determinants of the matrix $\lambda I - B$:

$$f_i(\lambda) = (\lambda - \alpha_i)f_{i-1}(\lambda) - \beta_{i-1}^2 f_{i-2}(\lambda) \qquad \text{with } f_{-1} = 0 \text{ and } f_0 = 1.$$

By induction it is easy to prove that $f_n(\lambda)$ is the characteristic polynomial. Similarly, it can be proved that $f_i(\lambda)$ has i real roots, separated by the roots of $f_{i-1}(\lambda)$. The sequence $f_i(z)$ is called a *Sturm sequence* and has some very interesting properties. Denote by $V(z)$ the number of sign changes in the sequence $f_0(z), f_1(z), \ldots, f_n(z)$; evidently $V(-\infty) = n$, $V(\infty) = 0$. Suppose that the roots of $f_n(z) = 0$ are $z_1, z_2, \ldots, z_n$; then a careful analysis shows that $V(z)$ decreases by one unit when z passes from left to right through a zero z_k, $k = 1(1)n$. Hence, if $V(u) - V(v) = 1$ then we know that there is a root of $f_n(z) = 0$ between u and v, i.e., an eigenvalue in this interval. Taking $w = \frac{1}{2}(u + v)$ and computing $V(w)$, we can easily decide in which half the eigenvalue is situated. In this way we have defined a *bisection method*, which is very fast since the sequence $f_k(z)$ and hence $V(z)$ is computed quite quickly. Computation of the eigenvectors of B is trivial in principle, but occasionally some numerical difficulties may appear. In order to get an eigenvector v of A we should multiply an eigenvector x of the band matrix B by the matrices $P_2, P_3, \ldots, P_{n-1}$, and conveniently this should be done successively:

$$\begin{cases} x_{n-1} = P_{n-1}x \\ x_{n-2} = P_{n-2}x_{n-1} \\ \vdots \\ v \equiv x_2 = P_2 x_3. \end{cases}$$

The *QR* Method of Francis

If the original matrix is not symmetric, it is recommended first to transform it to Hessenberg form. (In the symmetric case we would instead obtain a tridiagonal matrix.) We will first show how the reduction to upper Hessenberg form is carried out. Following Householder, we describe one step in the reduction leading from A_{r-1} to A_r, where $A_0 = A$ and

$$A_{r-1} = \left(\begin{array}{c|c|c} H_{r-1} & \multicolumn{1}{c}{} & C_{r-1} \\ \hline 0 & b_{r-1} & B_{r-1} \end{array} \right) \begin{array}{l} \} r \\ \} n-r \end{array}$$
$$\underbrace{\phantom{H_{r-1}}}_{r-1} \underbrace{}_{1} \underbrace{\phantom{B_{r-1}}}_{n-r}$$

and where

$$H_{r-1} = \begin{pmatrix} * & * & * & * & \cdots & * & * \\ * & * & * & * & \cdots & * & * \\ & * & * & * & \cdots & * & * \\ & & & * & * & \cdots & * & * \\ & & & & \mathbf{0} & & & \\ & & & & & & * & * \end{pmatrix}$$

is of upper Hessenberg form. The matrix A_r will then be produced through $A_r = P_r A_{r-1} P_r$, leaving H_{r-1} as well as the null-matrix of dimension $(n-r) \times (r-1)$ unchanged while the vector b_{r-1} must be annihilated except for the first component. In this way a new Hessenberg matrix H_r, of dimension $(r+1) \times (r+1)$ is formed by moving one new row and one new column to H_{r-1}. Now we choose

$$P_r = \left(\begin{array}{c|c} I & 0 \\ \hline 0 & Q_r \end{array} \right) \begin{array}{l} \\ \} n-r \end{array}$$
$$\underbrace{}_{n-r}$$

with $Q_r = I - 2w_r w_r^H$ and $w_r^H w_r = 1$, w_r being a column vector with $(n-r)$ elements. A simple computation gives

$$P_r A_{r-1} P_r = \left(\begin{array}{c|c} H_{r-1} & C_{r-1}Q_r \\ \hline 0 \quad a_{r-1} & Q_r B_{r-1} Q_r \end{array} \right) \quad \text{with} \quad a_{r-1} = \begin{pmatrix} \alpha_r \\ 0 \\ 0 \\ \vdots \\ 0 \end{pmatrix}.$$

Hence $Q_r b_{r-1} = a_{r-1}$ and $a_{r-1}^H a_{r-1} = b_{r-1}^H Q_r^H Q_r b_{r-1} = b_{r-1}^H b_{r-1}$, that is, $|\alpha_r| = \|b_{r-1}\|$ (we suppose Euclidean vector norm). Further $b_{r-1} = Q_r^H a_{r-1}$ and $e_1^T b_{r-1} = e_1^T Q_r^H a_{r-1} = (1 - \|w_1\|^2)\alpha_r$, and since $1 - \|w_1\|^2$ is real, $\arg \alpha_r = \arg(e_1^T b_{r-1})$. Here e_1 is a vector with the first component $= 1$ and all other components $= 0$. Thus the argument of α_r is equal to the argument of the top element of the vector b_{r-1}. Finally, since

$$Q_r b_{r-1} = (I - 2w_r w_r^H)b_{r-1} = b_{r-1} - (2w_r^H b_{r-1})w_r = a_{r-1},$$

we get

$$w_r = \frac{b_{r-1} - a_{r-1}}{\|b_{r-1} - a_{r-1}\|},$$

and A_r is completely determined. If this procedure is repeated we finally reach the matrix A_{n-2}, which is of upper Hessenberg form.

We now turn to the QR method itself. The following theorem is essential in this connection: Every regular square matrix A can be written as $A = QR$ where Q is unitary and R upper triangular, and both are uniquely determined. The existence of such a partition can be proved by a method resembling Gram–Schmidt's orthogonalization. Suppose that A is nonsingular and put

$$QR = A.$$

The column vectors of Q and A are denoted by q_i and a_i. If we now multiply all the rows in Q by the first column of R we get (since only the first element in the R-column is not equal to 0) $r_{11} q_1 = a_1$. But since Q is unitary, we also have $q_i^H q_k = \delta_{ik}$; hence $r_{11} = |a_1|$ (apart from a trivial factor $e^{i\theta_1}$); then also q_1 is uniquely determined. Now we assume that the vectors $q_1, q_2, \ldots, q_{k-1}$ are known. Multiplying all rows of Q by the kth column of R, we find

$$\sum_{i=1}^{k} r_{ik} q_i = a_k.$$

Next, for $j = 1, 2, \ldots, k - 1$, we multiply from the left with q_j^H and obtain $r_{jk} = q_j^H a_k$. Further we also have

$$r_{kk} q_k = a_k - \sum_{i=1}^{k-1} r_{ik} q_i.$$

The right-hand side is certainly not equal to 0, because otherwise we would have linear dependence between the vectors a_i. Hence we find

$$r_{kk} = \left\| a_k - \sum_{i=1}^{k-1} r_{ik} q_i \right\|,$$

and the vector q_k is also determined. Obviously, uniqueness is secured apart from factors $e^{i\theta_k}$ in the diagonal elements of R, and hence the condition $r_{kk} > 0$ will determine Q and R completely.

Now we start from $A = A_1$ and form sequences of matrices A_s, Q_s, and R_s by use of the following algorithm:

$$A_s = Q_s R_s; \qquad A_{s+1} = Q_s^H A_s Q_s = Q_s^H Q_s R_s Q_s = R_s Q_s.$$

This means that first A_s is partitioned in a product of a unitary and an upper triangular matrix, and then A_{s+1} is computed as $R_s Q_s$. It also means that A_{s+1} is formed from A_s through a similarity transformation with a unitary matrix. Next, we put

$$P_s = Q_1 Q_2 \cdots Q_s \qquad \text{and} \qquad R_s R_{s-1} \cdots R_1 = U_s,$$

and so we obtain

$$A_{s+1} = Q_s^{-1} A_s Q_s = Q_s^{-1}(Q_{s-1}^{-1} A_{s-1} Q_{s-1})Q_s = \cdots = P_s^{-1} A_1 P_s.$$

Then we form

$$P_s U_s = Q_1 Q_2 \cdots Q_s R_s R_{s-1} \cdots R_1 = Q_1 Q_2 \cdots Q_{s-1} A_s R_{s-1} \cdots R_1$$
$$= P_{s-1} A_s U_{s-1}.$$

But $A_s = P_{s-1}^{-1} A_1 P_{s-1}$ and consequently $P_{s-1} A_s = A_1 P_{s-1}$ which gives

$$P_s U_s = A_1 P_{s-1} U_{s-1} = \cdots = A_1^{s-1} P_1 U_1 = A_1^{s-1} Q_1 R_1 = A_1^s.$$

Here P_s is unitary and U_s upper triangular, and in principle they could be computed from A_1^s by partition in a product of a unitary and an upper triangular matrix. In this way we would also obtain A_{s+1} through a similarity transformation:

$$A_{s+1} = P_s^{-1} A_1 P_s.$$

We now assert that for increasing s the matrix A_{s+1} more and more will approach an upper triangular matrix. We do not give a complete proof but restrict ourselves to the main points. The following steps are needed for the proof:

1. $A = A_1$ is written $A_1 = XDX^{-1} = XDY$, which is always possible if all eigenvalues are different (this restriction can be removed afterward). Further we assume $d_{ii} = \lambda_i$ with $|\lambda_1| > |\lambda_2| > \cdots$

2. Then we also have $P_s U_s = A_1^s = XD^s Y$.

3. X is partitioned as $X = Q_X R_X$ and Y as $Y = L_Y R_Y$ where Q_X is unitary, R_X and R_Y upper triangular, and L_Y lower triangular with ones in the main diagonal (both partitions are unique). For the latter partition a permutation might be necessary.

4. Then we get $P_s U_s = Q_X R_X D^s L_Y D^{-s} D^s R_Y$. The decisive point is that $D^s L_Y D^{-s}$, as is easily shown, in subdiagonal elements will contain quotients $(\lambda_i/\lambda_1)^s$ so that this matrix actually will approach the identity matrix. If so, we are left with $P_s U_s \simeq Q_X R_X D^s R_Y$.

5. But P_s and Q_X are unitary while U_s and $R_X D^s R_Y$ are upper triangular. Since the partition is unique, we can draw the conclusion that

$$\lim_{s \to \infty} P_s = Q_X \qquad \text{and} \qquad U_s \simeq R_X D^s R_Y.$$

6. $A_{s+1} = P_s^H A_1 P_s = P_s^H X D X^{-1} P_s \to Q_X^H X D X^{-1} Q_X \to R_X D R_X^{-1}$ (since $X = Q_X R_X$ and $R_X = Q_X^H X$). The matrix $R_X D R_X^{-1}$ is an upper triangular matrix with the same eigenvalues as A.

The QR method can be used on arbitrary matrices but will become too laborious, and instead it is used on special matrices, preferably Hessenberg or symmetric band-matrices. The method has good stability properties and seems to be one of the most promising at present. Several algorithms treating this method have been published.

EXERCISES

1. Find the eigenvalues and eigenvectors of the matrix

$$A = \begin{pmatrix} 3 & 4 & 0 \\ 4 & 3 & 0 \\ 0 & 0 & 2 \end{pmatrix}.$$

2. Using the power method, find the largest eigenvalue of the matrix

$$A = \begin{pmatrix} -37 & 25 & 75 & -13 \\ 25 & -5 & -5 & 7 \\ 75 & -5 & 150 & -25 \\ -13 & 7 & -25 & 3 \end{pmatrix}.$$

3. Find the largest eigenvalue of the matrix

$$A = \begin{pmatrix} 25 & -41 & 10 & -6 \\ -41 & 68 & -17 & 10 \\ 10 & -17 & 5 & -3 \\ -6 & 10 & -3 & 2 \end{pmatrix}.$$

4. The matrix

$$A = \begin{pmatrix} 2 & 1 & 0 & 0 & 0 \\ 1 & 1 & 1 & 0 & 0 \\ 0 & 1 & 0 & 1 & 0 \\ 0 & 0 & 1 & -1 & 1 \\ 0 & 0 & 0 & 1 & -2 \end{pmatrix}$$

is given. If the power method is used we obtain a series of Rayleigh quotients converging toward a value which is a little too low (we suppose that we start with a vector close to an eigenvector of λ with a small admixture of an eigenvector belonging to $-\lambda$). Start with the vector $(67, 50, 20, 6, 1)^T$. Explain the result and suggest suitable measures.

5. A real symmetric matrix has its largest eigenvalue equal to λ_1 and the next largest equal to λ_2, both > 0. All other eigenvalues are much smaller in absolute value. Using the power method we have determined three consecutive Rayleigh quotients, R_1, R_2, and R_3. Determine λ_1 as accurately as possible from these values.

6. The Hilbert matrix of order $n = 4$ and its inverse are:

$$H_4 = \begin{pmatrix} 1 & 1/2 & 1/3 & 1/4 \\ 1/2 & 1/3 & 1/4 & 1/5 \\ 1/3 & 1/4 & 1/5 & 1/6 \\ 1/4 & 1/5 & 1/6 & 1/7 \end{pmatrix}; \quad H_4^{-1} = \begin{pmatrix} 16 & -120 & 240 & -140 \\ -120 & 1200 & -2700 & 1680 \\ 240 & -2700 & 6480 & -4200 \\ -140 & 1680 & -4200 & 2800 \end{pmatrix}.$$

Determine the largest eigenvalue μ_1 of H_4 and the largest eigenvalue λ_1 of H_4^{-1}. Then compute all eigenvalues of H_4^{-1} when it is known that $\det H_4^{-1} = 6048000$.

7. Find the largest eigenvalue and the corresponding eigenvector of the Hermitian matrix

$$H = \begin{pmatrix} 8 & -5i & 3-2i \\ 5i & 3 & 0 \\ 3+2i & 0 & 2 \end{pmatrix}.$$

8. Using the fact that $\lambda - a$ is an eigenvalue of $A - aI$ if λ is an eigenvalue of A, find the highest and the lowest eigenvalue of the matrix

$$A = \begin{pmatrix} 9 & 10 & 8 \\ 10 & 5 & -1 \\ 8 & -1 & 3 \end{pmatrix}.$$

Choose $a = 12$ and the starting vectors

$$\begin{pmatrix} 1 \\ 1 \\ 1 \end{pmatrix} \quad \text{and} \quad \begin{pmatrix} -1 \\ 1 \\ 1 \end{pmatrix},$$

respectively. (Desired accuracy: two decimal places.)

9. The matrix

$$A = \begin{pmatrix} 14 & 7 & 6 & 9 \\ 7 & 9 & 4 & 6 \\ 6 & 4 & 9 & 7 \\ 9 & 6 & 7 & 15 \end{pmatrix}$$

has an eigenvalue close to 4. Compute this eigenvalue to six places, using the matrix $B = (A - 4I)^{-1}$.

10. Find the largest eigenvalue of the modified eigenvalue problem $Ax = \lambda Bx$ when

$$A = \begin{pmatrix} 1 & 6 & 6 & 4 \\ 6 & 37 & 43 & 16 \\ 6 & 43 & 86 & -27 \\ 4 & 16 & -27 & 106 \end{pmatrix} \quad \text{and} \quad B = \begin{pmatrix} 1 & 2 & -1 & 4 \\ 2 & 5 & 1 & 6 \\ -1 & 1 & 11 & -11 \\ 4 & 6 & -11 & 22 \end{pmatrix},$$

11. In the matrix A of type (n, n) all diagonal elements are a, while all the others are b. Find such numbers p and q that $A^2 - pA + qI = 0$. Use this relation for finding the eigenvalues [one is simple, and is $(n-1)$-fold] and the eigenvectors of A.

12. Perform decomposition of $A_{k-1} = L_k U_k$ where L_k is lower triangular with unit diagonal and U_k is upper triangular; then form $A_k = U_k L_k$. Show that A_k has the same eigenvalues as A_{k-1}. For certain matrices $A = A_0$, we get convergence toward a diagonal matrix. Compute A_1, A_2 and A_3 when

$$A_0 = \begin{pmatrix} 7 & 6 \\ 3 & 4 \end{pmatrix}.$$

13. A is a matrix with eigenvalues $\lambda_1, \lambda_2, \ldots, \lambda_n$ (all different). Put

$$B_k = \frac{(A - \lambda_1 I)(A - \lambda_2 I) \cdots (A - \lambda_{k-1} I)(A - \lambda_{k+1} I) \cdots (A - \lambda_n I)}{(\lambda_k - \lambda_1)(\lambda_k - \lambda_2) \cdots (\lambda_k - \lambda_{k-1})(\lambda_k - \lambda_{k+1}) \cdots (\lambda_k - \lambda_n)}.$$

Show that $B_k^2 = B_k$.

CHAPTER 13

Linear Programming

Le mieux est l'ennemi du bien.

Voltaire

One main problem characterizes linear programming: to seek the maximum or the minimum of a linear expression when the variables of the problem are subject to restrictions in the form of certain linear equalities or inequalities. Problems of this kind are encountered when we have to exploit limited resources in an optimal way. Production and transport problems, which play an important role in industry, are of special significance in this respect.

This problem category came into the limelight during the 1940's with the background that during World War II, mathematical methods were used for planning portions of military activities in an optimal manner, especially in England and the United States. The methods developed during this period were then taken over by civilian industry, and in this way the theory of linear programming developed.

§13.1. THE SIMPLEX METHOD

We shall start by discussing a simple example. Find the maximum value of $y = 7x_1 + 5x_2$ under the conditions

$$\begin{cases} x_1 + 2x_2 \le 6, & x_1 \ge 0, \\ 4x_1 + 3x_2 \le 12, & x_2 \ge 0. \end{cases}$$

The point (x_1, x_2) must lie inside or on the boundary of the domain marked with lines in Figure 13.1. For different values of y the equation

$$7x_1 + 5x_2 = y$$

represents parallel straight lines, and in order to get useful solutions we must have the lines pass through the domain or the boundary. In this way it is easy to see that we only have to look at the values in the corners $(0, 0)$, $(0, 3)$, $(1.2, 2.4)$, and

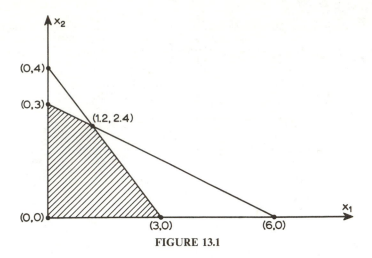

FIGURE 13.1

$(3, 0)$. We get $y = 0$, 15, 20.4, and 21 respectively, and hence $y_{max} = 21$ for $x_1 = 3$ and $x_2 = 0$.

We shall now formulate the main problem. We assume that we have n variables $x_1, x_2, \ldots, x_n$ subject to the conditions

$$\begin{cases} a_{11}x_1 + a_{12}x_2 + \cdots + a_{1n}x_n = b_1, \\ a_{21}x_1 + a_{22}x_2 + \cdots + a_{2n}x_n = b_2, \qquad m < n, \\ \vdots \\ a_{m1}x_1 + a_{m2}x_2 + \cdots + a_{mn}x_n = b_m, \end{cases}$$

and further that $x_i \geq 0$, $i = 1, 2, \ldots, n$. Find the minimum value of

$$y = c_1x_1 + c_2x_2 + \cdots + c_nx_n.$$

Using matrix notations, we get the alternative formulation: Minimize $y = c^T x$ under the conditions $x \geq 0$; $Ax = b$. Often the conditions $Ax = b$ are given as inequalities, but by adding so-called slack variables, the inequalities are transformed to equalities. Hence $a_{i1}x_1 + a_{i2}x_2 + \cdots + a_{in}x_n \leq b_i$ is replaced by

$$a_{i1}x_1 + a_{i2}x_2 + \cdots + a_{in}x_n + x_{n+i} = b_i, \qquad i = 1, 2, \ldots, m,$$

and then we can also replace $m + n$ by n.

By choosing the signs of the coefficients in a suitable way, we can always formulate the problem as has just been described with all $b_i > 0$. Clearly, the domain defined by the secondary conditions is a convex hyperpolyhedron, which can be either limited or unlimited. In some cases the conditions may be such that this domain vanishes; then the problem has no solution. For different values of y, the equation $y = c_1x_1 + c_2x_2 + \cdots + c_nx_n$ defines a family of hyperplanes; for suitable values of y they have points in common with the hyperpolyhedron. When y decreases, the intersection changes, and it is easily inferred that in general there exists a position where the intersection has contracted to a point (or possibly a straight line), and this represents the solution. As a rule, the solution is a corner of the domain, and it is now possible to construct a method which implies that we start at one corner and then proceed successively to other corners, simultaneously watching that y must decrease all the time. This method, which has been invented by Dantzig, is known as the *simplex method* (a simplex is a hyperpolyhedron with $n + 1$ corners in n dimensions, for example, a point, a limited straight line, a triangle, a tetrahedron, etc.).

Before discussing the solution technique we shall define a few important concepts. Each vector x satisfying the equation $Ax = b$ is called a *solution*. If no component is negative, the vector x is said to be a *feasible solution*. In general, none of these is the *optimal* solution which we are looking for. If m columns of the matrix A can be chosen in such a way that the determinant is not equal to 0, these vectors are said to form a *basis* and the corresponding variables are called *basic variables*. If all other variables are put equal to 0, the system can be solved, but naturally we cannot guarantee that the solution is feasible. If some of the basic variables vanish, the solution is said to be *degenerate*.

In the sequel we suppose that all slack variables needed have already been introduced so that we obtain a linear system of m equations in n unknowns ($m < n$). As already mentioned, the system represents m hyperplanes and normally the optimal solution is a corner in the corresponding hyperpolyhedron. In the typical case we have m equations with m slack variables and hence only $n - m$ proper variables. Thus the hyperplanes belong to an $(n - m)$-dimensional space. Normally, a corner is obtained as the intersection of $n - m$ hyperplanes, and for each hyperplane one slack variable will vanish, that is, we have in total $n - m$ variables equal to zero. Since the coordinate planes $x_i = 0$ may contribute in forming corners, proper variables may also vanish in a corner, but the total number of variables not equal to 0 will be the same. In degenerate cases *more* than $n - m$ hyperplanes may pass through the same corner, but the conclusion is still unchanged. Hence, in an optimal solution at most m variables are positive. The reasoning can easily be generalized to cover also the case when a number of equalities needing no slack variables are present already from the beginning.

For a description of the method, we introduce the column vectors

$$P_1 = \begin{pmatrix} a_{11} \\ a_{21} \\ \vdots \\ a_{m1} \end{pmatrix}; \quad P_2 = \begin{pmatrix} a_{12} \\ a_{22} \\ \vdots \\ a_{m2} \end{pmatrix}; \quad \dots; \quad P_n = \begin{pmatrix} a_{1n} \\ a_{2n} \\ \vdots \\ a_{mn} \end{pmatrix}; \quad P_0 = \begin{pmatrix} b_1 \\ b_2 \\ \vdots \\ b_m \end{pmatrix}, \quad P_0 > 0.$$

Then we have to minimize $c^T x$ under the secondary conditions $x \geq 0$ and

$$x_1 P_1 + x_2 P_2 + \dots + x_n P_n = P_0.$$

We assume that we know a feasible solution x with components $x_1, x_2, \dots, x_m$ ($x_{m+1} = x_{m+2} = \dots = x_n = 0$); $x_i \geq 0$, $i = 1, 2, \dots, m$. Then we have

(13.1.1) $$x_1 P_1 + x_2 P_2 + \dots + x_m P_m = P_0,$$

and the corresponding y-value is

(13.1.2) $$c_1 x_1 + c_2 x_2 + \dots + c_m x_m = y_0.$$

The vectors $P_1, P_2, \ldots, P_m$ are supposed to be linearly independent (in practice, they are often chosen in such a way that they belong to the slack variables, and consequently they become unit vectors), and then $P_{m+1}, P_{m+2}, \ldots, P_n$ can be expressed as linear combinations of the base vectors:

(13.1.3) $\quad P_j = x_{1j}P_1 + x_{2j}P_2 + \cdots + x_{mj}P_m, \qquad j = m+1, m+2, \ldots, n.$

Further, we define y_j by

(13.1.4) $\qquad\qquad\qquad y_j = c_1 x_{1j} + c_2 x_{2j} + \cdots + c_m x_{mj}.$

If for some value j the condition $y_j - c_j > 0$ is satisfied, then we can find a better solution. Multiplying (13.1.3) by a number p and subtracting from (13.1.1), we obtain

(13.1.5) $\quad (x_1 - px_{1j})P_1 + (x_2 - px_{2j})P_2 + \cdots + (x_m - px_{mj})P_m + pP_j = P_0.$

Analogously, from (13.1.2) and (13.1.4),

(13.1.6) $\quad (x_1 - px_{1j})c_1 + (x_2 - px_{2j})c_2 + \cdots + (x_m - px_{mj})c_m + pc_j$
$$= y_0 - p(y_j - c_j).$$

If the coefficients for $P_1, P_2, \ldots, P_m, P_j$ are all ≥ 0, then we have a new feasible solution with the corresponding y-value $y = y_0 - p(y_j - c_j) < y_0$, since $p > 0$ and $y_j - c_j > 0$. If for a fixed value of j at least one $x_{ij} > 0$, then the largest p-value we can choose is

$$p = \min_i \frac{x_i}{x_{ij}} > 0,$$

and if the problem has not degenerated, this condition determines i and p. The coefficient of P_i now becomes zero, and further we again have P_0 represented as a linear combination of m base vectors. We have also reached a lower value of y. The same process is repeated and continued until *either* all $y_j - c_j < 0$, *or* for some $y_j - c_j > 0$, we have all $x_{ij} \leq 0$. In the latter case we can choose p as large as we want, and the minimum value is $-\infty$.

Before we can proceed with the next step, all vectors must be expressed in terms of the new base vectors. Suppose that the base vector P_i has to be replaced by the base vector P_k. Our original basis was $P_1, P_2, \ldots, P_i, \ldots, P_m$, and our new basis is to be $P_1, P_2, \ldots, P_{i-1}, P_{i+1}, \ldots, P_m, P_k$. Now we have

$$\begin{cases} P_0 = x_1 P_1 + x_2 P_2 + \cdots + x_i P_i + \cdots + x_m P_m, \\ P_k = x_{1k}P_1 + x_{2k}P_2 + \cdots + x_{ik}P_i + \cdots + x_{mk}P_m, \\ P_j = x_{1j}P_1 + x_{2j}P_2 + \cdots + x_{ij}P_i + \cdots + x_{mj}P_m. \end{cases}$$

From the middle relation we solve P_i, which is inserted into the two others:

$$\begin{cases} P_0 = x_1' P_1 + \cdots + x_{i-1}' P_{i-1} + x_k' P_k + x_{i+1}' P_{i+1} + \cdots + x_m' P_m, \\ P_j = x_{1j}' P_1 + \cdots + x_{i-1,j}' P_{i-1} + x_{kj}' P_k + x_{i+1,j}' P_{i+1} + \cdots + x_{mj}' P_m, \end{cases}$$

where $x_r' = x_r - (x_i/x_{ik})x_{rk}$ for $r = 1, 2, \ldots, i-1, i+1, \ldots, m$, and $x_k' = x_i/x_{ik}$. Analogously, $x_{rj}' = x_{rj} - (x_{ij}/x_{ik})x_{rk}$ for $r \neq i$ and $x_{kj}' = x_{ij}/x_{ik}$.

Now we have

$$y'_j - c_j = x'_{1j}c_1 + \cdots + x'_{kj}c_k + \cdots + x'_{mj}c_m - c_j$$

$$= \sum_r x'_{rj}c_r - c_j = \sum_{r \neq i}\left(x_{rj} - \frac{x_{ij}}{x_{ik}}x_{rk}\right)c_r + \frac{x_{ij}}{x_{ik}}c_k - c_j$$

$$= \sum_r x_{rj}c_r - x_{ij}c_i - \frac{x_{ij}}{x_{ik}}\left(\sum_r x_{rk}c_r - x_{ik}c_i - c_k\right) - c_j$$

$$= y_j - c_j - \frac{x_{ij}}{x_{ik}}(y_k - c_k),$$

and further

$$y'_0 = c_1 x'_1 + \cdots + c_k x'_k + \cdots + c_m x'_m$$

$$= \sum_{r \neq i}\left(x_r - \frac{x_i}{x_{ik}}x_{rk}\right)c_r + \frac{x_i}{x_{ik}}c_k$$

$$= \sum_r c_r x_r - c_i x_i - \frac{x_i}{x_{ik}}\left(\sum_r x_{rk}c_r - x_{ik}c_i - c_k\right)$$

$$= y_0 - \frac{x_i}{x_{ik}}(y_k - c_k).$$

Example 1. Seek the maximum of $y = -5x_1 + 8x_2 + 3x_3$ under the conditions

$$\begin{cases} 2x_1 + 5x_2 - x_3 \leq 1, \\ -3x_1 - 8x_2 + 2x_3 \leq 4, \\ -2x_1 - 12x_2 + 3x_3 \leq 9, \\ x_1 \geq 0, \quad x_2 \geq 0, \quad x_3 \geq 0. \end{cases}$$

This problem is fairly difficult to solve by use of the direct technique, and for this reason we turn to the simplex method. We rewrite the problem in standard form as follows.

Find the minimum value of

$$y = 5x_1 - 8x_2 - 3x_3$$

when

$$\begin{cases} 2x_1 + 5x_2 - x_3 + x_4 \qquad\qquad = 1, \\ -3x_1 - 8x_2 + 2x_3 \qquad + x_5 \qquad = 4, \qquad x_i \geq 0. \\ -2x_1 - 12x_2 + 3x_3 \qquad\qquad + x_6 = 9, \end{cases}$$

We use the following scheme:

Basis	c	P_0	5 P_1	-8 P_2	-3 P_3	0 P_4	0 P_5	0 P_6
P_4	0	1	2	5	-1	1	0	0
P_5	0	4	-3	-8	2	0	1	0
P_6	0	9	-2	-12	3	0	0	1
		0	-5	8	3	0	0	0

Over the vectors $P_1, \ldots, P_6$ one places the coefficients c_i; the last line contains y_0 and $y_j - c_j$; thus

$$0 \cdot 1 + 0 \cdot 4 + 0 \cdot 9 = 0;$$
$$0 \cdot 2 + 0 \cdot (-3) + 0 \cdot (-2) - 5 = -5,$$

and so on. Now we choose a positive number in the last line (8), and in the corresponding column a positive number (5) (both numbers are printed in boldface). In the basis we shall now exchange P_4 and P_2, that is, we have $i = 4$, $k = 2$. The first line (P_4) holds an exceptional position, and in the reduction all elements (except c) are divided by 5; the new value of c becomes -8, belonging to the vector P_2.

			5	-8	-3	0	0	0
Basis	c	P_0	P_1	P_2	P_3	P_4	P_5	P_6
P_2	-8	$\frac{1}{5}$	$\frac{2}{5}$	1	$-\frac{1}{5}$	$\frac{1}{5}$	0	0
P_5	0	$\frac{28}{5}$	$\frac{1}{5}$	0	2/5	$\frac{8}{5}$	1	0
P_6	0	$\frac{57}{5}$	$\frac{14}{5}$	0	$\frac{3}{5}$	$\frac{12}{5}$	0	1
		$-\frac{8}{5}$	$-\frac{41}{5}$	0	23/5	$-\frac{8}{5}$	0	0

The other elements are obtained in the following way:

$$9 - \tfrac{1}{5}(-12) = \tfrac{57}{5};$$

the numbers 9, 1, 5, and -12 stand in the corners of a rectangle. Analogously, $-2 - \frac{2}{5}(-12) = \frac{14}{5}$, and so on. Note that also the elements in the last line can be obtained in this way, which makes checking simple. In the next step we must choose $k = 3$, since $\frac{23}{5} > 0$, but we have two possibilities for the index i, namely, $i = 5$ and $i = 6$, corresponding to the elements $\frac{2}{5}$ and $\frac{3}{5}$, respectively. But $\frac{28}{5}/\frac{2}{5} = 14$ is less than $\frac{57}{5}/\frac{3}{5} = 19$, and hence we must choose $i = 5$. This reduction gives the result

			5	-8	-3	0	0	0
Basis	c	P_0	P_1	P_2	P_3	P_4	P_5	P_6
P_2	-8	3	$\frac{1}{2}$	1	0	1	$\frac{1}{2}$	0
P_3	-3	14	$\frac{1}{2}$	0	1	4	$\frac{5}{2}$	0
P_6	0	3	$\frac{5}{2}$	0	0	0	$-\frac{3}{2}$	1
		-66	$-\frac{21}{2}$	0	0	-20	$-\frac{23}{2}$	0

Since all numbers in the last line are negative or zero, the problem is solved and we have $y_{\min} = -66$ for

$$\begin{cases} x_2 = 3, \\ x_3 = 14, \\ x_6 = 3, \end{cases} \quad \text{and} \quad \begin{cases} x_1 = 0, \\ x_4 = 0, \\ x_5 = 0. \end{cases}$$

since P_1, P_4 and P_5 do not enter the base.

We have now exclusively treated the case when all unit vectors entered the secondary conditions without special arrangements. If this is not the case, we can master the problem by introducing an artificial basis. Suppose that we have to minimize $c_1 x_1 + \cdots + c_n x_n$ under the conditions

$$
\begin{cases}
a_{11} x_1 + \cdots + a_{1n} x_n = b_1, \\
\vdots \\
a_{m1} x_1 + \cdots + a_{mn} x_n = b_m,
\end{cases}
$$

and $x_i \geq 0$ $(i = 1, 2, \ldots, n)$.

Instead we consider the problem of minimizing

$$
c_1 x_1 + \cdots + c_n x_n + w x_{n+1} + w x_{n+2} + \cdots + w x_{n+m}
$$

under the conditions

$$
\begin{cases}
a_{11} x_1 + \cdots + a_{1n} x_n + x_{n+1} & = b_1, \\
a_{21} x_1 + \cdots + a_{2n} x_n \quad\quad + x_{n+2} & = b_2, \\
\vdots \\
a_{m1} x_1 + \cdots + a_{mn} x_n \quad\quad\quad\quad + x_{n+m} = b_m,
\end{cases}
$$

and $x_i \geq 0$ $(i = 1, 2, \ldots, n + m)$.

Then we let w be a large positive number, which need not be specified. In this way the variables $x_{n+1}, \ldots, x_{n+m}$ are, in fact, eliminated from the secondary conditions, and we are back at our old problem. The expressions $y_j - c_j$ now become linear functions of w; the constant terms are written as before in a special line, while the coefficients of w are added in an extra line below. Of these, the largest positive coefficient determines a new base vector, and the old one, which should be replaced, is chosen along the same lines as before. An eliminated base vector can be disregarded in the following computations.

Example 2.   Minimize

$$
y = x_1 + x_2 + 2x_3
$$

under the conditions

$$
\begin{cases}
x_1 + x_2 + x_3 \leq 9, \\
2x_1 - 3x_2 + 3x_3 = 1, \qquad x_i \geq 0. \\
-3x_1 + 6x_2 - 4x_3 = 3,
\end{cases}
$$

First we solve the problem by conventional methods. In the inequality we add a slack variable x_4 and obtain

$$
x_1 + x_2 + x_3 + x_4 = 9.
$$

Regarding x_4 as a known quantity, we get

$$
\begin{cases}
x_1 = \frac{1}{2}(13 - 3x_4), \\
x_2 = \frac{1}{4}(13 - x_4), \\
x_3 = \frac{3}{4}(-1 + x_4).
\end{cases}
$$

Hence

$$y = x_1 + x_2 + 2x_3 = \tfrac{1}{4}(33 - x_4).$$

The conditions $x_1 \geq 0$, $x_2 \geq 0$, $x_3 \geq 0$ together give the limits $1 \leq x_4 \leq \tfrac{13}{3}$. Since y is going to be minimized, x_4 should be chosen as large as possible, that is, $x_4 = \tfrac{13}{3}$. Thus we get

$$x_1 = 0; \qquad x_2 = \tfrac{13}{6}; \qquad x_3 = \tfrac{5}{2}, \qquad \text{and} \qquad y_{\min} = \tfrac{43}{6}.$$

Now we pass to the simplex method.

| | | | 1 | 1 | 2 | 0 | w | w |
Basis	c	P_0	P_1	P_2	P_3	P_4	P_5	P_6
P_4	0	9	1	1	1	1	0	0
P_5	w	1	2	-3	3	0	1	0
P_6	w	3	-3	6	-4	0	0	1
		0	-1	-1	-2	0	0	0
		4	-1	3	-1	0	0	0
P_4	0	$\tfrac{17}{2}$	$\tfrac{3}{2}$	0	$\tfrac{5}{3}$	1	0	—
P_5	w	$\tfrac{5}{2}$	$\tfrac{1}{2}$	0	1	0	1	—
P_2	1	$\tfrac{1}{2}$	$-\tfrac{1}{2}$	1	$-\tfrac{2}{3}$	0	0	—
		$\tfrac{1}{2}$	$-\tfrac{3}{2}$	0	$-\tfrac{8}{3}$	0	0	—
		$\tfrac{5}{2}$	$\tfrac{1}{2}$	0	1	0	0	—
P_4	0	$\tfrac{13}{3}$	$\tfrac{2}{3}$	0	0	1	—	—
P_3	2	$\tfrac{5}{2}$	$\tfrac{1}{2}$	0	1	0	—	—
P_2	1	$\tfrac{13}{6}$	$-\tfrac{1}{6}$	1	0	0	—	—
		$\tfrac{43}{6}$	$-\tfrac{1}{6}$	0	0	0		

Hence we get $x_4 = \tfrac{13}{3}$; $x_3 = \tfrac{5}{2}$; $x_2 = \tfrac{13}{6}$, and $y_{\min} = \tfrac{43}{6}$, exactly as before.

Here we mention briefly the existence of the *dual* counterpart of a linear programming problem. Assume the following primary problem: Find a vector x such that $c^T x = \min$ under the conditions $x \geq 0$, $Ax = b$. Then the *dual unsymmetric* problem is the following: Find a vector y such that

$$b^T y = \max$$

under the condition $A^T y \leq c$. Here we do not require that y be ≥ 0. The following theorem has been proved by Dantzig and Orden: If one of the problems has a finite solution, then the same is true for the other problem, and further

$$\min c^T x = \max b^T y.$$

Alternatively for the primary problem: Find a vector x such that $c^T x = \min$ under the conditions $Ax \geq b$ and $x \geq 0$. We then have the following *dual symmetric* problem: Find a vector y such that $b^T y = \max$ under the conditions $A^T y \leq c$ and $y \geq 0$. The theorem just mentioned is valid also in this case.

Among possible complications we have already mentioned degeneration. This is not quite unusual but on the other hand not very difficult to master. An obvious measure is perturbation as suggested by Charnes. The vector b is replaced by another vector b':

$$b' = b + \sum_{k=1}^{N} \varepsilon^k P_k,$$

where N is the total number of vectors including possible artificial base vectors. Here ε can be understood as a small positive number which need not be specified closer, and when the solution has been obtained we put $\varepsilon = 0$.

§13.2. THE TRANSPORTATION PROBLEM

In many cases, we have linear programming problems of a special kind with a very simple structure, and among these the transportation problem occupies a dominant position. The problem can be formulated in the following way: An article is produced by m producers in the quantities $a_1, a_2, \ldots, a_m$, and it is consumed by n consumers in the quantities $b_1, b_2, \ldots, b_n$. To begin with, we assume that $\sum a_i = \sum b_j$. The transportation cost from producer i to consumer k is c_{ik} per unit, and we search for the quantities x_{ik} which should be delivered from i to k so that the total transportation cost will be as small as possible. The problem can be solved by the conventional simplex method, but usually one prefers a less involved iterative technique, introduced by Hitchcock.

In the usual simplex method, when we are dealing with m equations and n variables, the solution is, in general, a corner of a hyperpolyhedron. The solution contains at least $n - m$ variables which are zero. In our case the number of equations is $m + n - 1$ (namely, $\sum_j x_{ij} = a_i$ and $\sum_i x_{ij} = b_j$; however, we must take the identity $\sum a_i = \sum b_j$ into account), and the number of variables is mn. Thus a feasible solution must not contain more than $m + n - 1$ nonzero elements.

We now formulate the problem mathematically. Find numbers $x_{ij} \geq 0$ such that

(13.2.1)
$$f = \sum_{i=1}^{m} \sum_{j=1}^{n} c_{ij}x_{ij} = \min$$

under the conditions

(13.2.2)
$$\sum_{j=1}^{n} x_{ij} = a_i,$$

(13.2.3)
$$\sum_{i=1}^{m} x_{ij} = b_j,$$

(13.2.4)
$$\sum_{i=1}^{m} a_i = \sum_{j=1}^{n} b_j.$$

In order to make the discussion easier we shall consider a special case; the conclusions that can be drawn from this are then generalized without difficulty, Suppose $m = 3$ and $n = 4$ and write down the coefficient matrix A for (13.2.2) and (13.2.3) (the columns correspond in turn to $x_{11}, x_{12}, \ldots, x_{34}$):

$$A = \begin{pmatrix} 1 & 1 & 1 & 1 & 0 & 0 & 0 & 0 & 0 & 0 & 0 & 0 \\ 0 & 0 & 0 & 0 & 1 & 1 & 1 & 1 & 0 & 0 & 0 & 0 \\ 0 & 0 & 0 & 0 & 0 & 0 & 0 & 0 & 1 & 1 & 1 & 1 \\ 1 & 0 & 0 & 0 & 1 & 0 & 0 & 0 & 1 & 0 & 0 & 0 \\ 0 & 1 & 0 & 0 & 0 & 1 & 0 & 0 & 0 & 1 & 0 & 0 \\ 0 & 0 & 1 & 0 & 0 & 0 & 1 & 0 & 0 & 0 & 1 & 0 \\ 0 & 0 & 0 & 1 & 0 & 0 & 0 & 1 & 0 & 0 & 0 & 1 \end{pmatrix}.$$

We have $m + n$ rows and mn columns; the first m rows correspond to (13.2.2) and the last n to (13.2.3). Since the sum of the first m rows is equal to the sum of the last n rows, the rank is at most $m + n - 1$. We shall denote the columns of A by p_{ij}, which should correspond to the variables x_{ij} taken in the order $x_{11}, x_{12}, x_{13}, x_{14}, x_{21}, x_{22}, x_{23}, x_{24}, x_{31}, x_{32}, x_{33}, x_{34}$. Let us now compare two vectors p_{ij} and p_{rs}. We see at once that if $i = r$, then the first m components coincide; and if $j = s$, then the last n components coincide. This observation gives us simple means to examine linear dependence for vectors p_{ij}. If we form a cyclic sequence where two adjacent vectors alternately coincide in row-index and column-index, then the vectors must become linearly dependent. For example, we have $p_{12} - p_{32} + p_{34} - p_{24} + p_{23} - p_{13} = 0$. This fact is of great importance when a feasible initial solution is chosen; as has already been observed it must contain $m + n - 1$ elements not equal to 0. If the vectors p_{ij} are arranged in matrix form, we get the picture:

$$\begin{pmatrix} p_{11} & p_{12} & p_{13} & p_{14} \\ p_{21} & p_{22} & p_{23} & p_{24} \\ p_{31} & p_{32} & p_{33} & p_{34} \end{pmatrix}$$

or, using the example above,

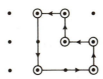

From this discussion it is clear that a feasible initial solution cannot be chosen in such a way that $x_{ij} \neq 0$ in points (i, j) which are corners in a closed polygon with only horizontal and vertical sides. For this would imply that the determinant corresponding to a number, possibly all, of the variables x_{ij} which are not equal to 0, would become zero because the vectors related to these x_{ij} are linearly dependent, and such an initial solution cannot exist.

Example 1

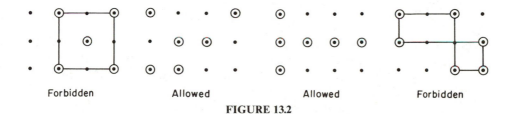

| Forbidden | Allowed | Allowed | Forbidden |

FIGURE 13.2

Thus we start by constructing a feasible solution satisfying the following conditions:

1. exactly $m + n - 1$ of the variables x_{ij} shall be positive, the others zero;

2. the boundary conditions (13.2.2) and (13.2.3) must be satisfied;

3. $x_{ij} \neq 0$ in such points that no closed polygon with only horizontal and vertical sides can appear.

How this construction should be best performed in practice will be discussed later.

First we determine α_i and β_j in such a way that $c_{ij} = \alpha_i + \beta_j$ for all such indices i and j that $x_{ij} > 0$; *one* value, for example, α_1, can be chosen arbitrarily. Then we *define* auxiliary quantities (also called fictitious transportation costs) $k_{ij} = \alpha_i + \beta_j$ in all remaining cases. One could, for example, imagine that the transport passes over some central storage which would account for the transport cost being split into two parts, one depending on the producer and one depending on the consumer. Thus we get

$$(13.2.5) \qquad f = \sum_{i,j} c_{ij}x_{ij} = \sum_{i,j} (c_{ij} - k_{ij})x_{ij} + \sum_i \alpha_i \sum_j x_{ij} + \sum_j \beta_j \sum_i x_{ij}$$
$$= \sum_{i,j} (c_{ij} - k_{ij})x_{ij} + \sum_i a_i\alpha_i + \sum_j b_j\beta_j.$$

But $c_{ij} = k_{ij}$ for all $x_{ij} \neq 0$, and hence we obtain

$$(13.2.6) \qquad f = \sum_i a_i\alpha_i + \sum_j b_j\beta_j.$$

If for some i, j we have $c_{ij} < k_{ij}$, it is possible to find a better solution. Suppose that we move a quantity ε to the place (i, j). Then the total cost will decrease with $\varepsilon(k_{ij} - c_{ij})$. This fact suggests the following procedure. We search for the minimum of $c_{ij} - k_{ij}$, and if this is <0, we choose ε as large as possible with regard to the conditions (13.2.2) and (13.2.3). Then we calculate new values α_i, β_j, and k_{ij} and repeat the whole procedure. When all $c_{ij} \geq k_{ij}$, we have attained an optimal solution; in exceptional cases several such solutions may exist.

The technique is best demonstrated on an example. We suppose that a certain commodity is produced in three factories in quantities of 8, 9, and 13 units, and it is used by four consumers in quantities of 6, 7, 7, and 10 units. The transportation costs are given in the following table:

$$(c_{ij}) = \begin{pmatrix} 3 & 8 & 9 & 16 \\ 6 & 11 & 14 & 9 \\ 5 & 13 & 10 & 12 \end{pmatrix}.$$

We start by constructing a feasible solution. To the left of the x_{ij}-matrix, which is so far unknown, we write down the column a_i, and above the matrix we write the row b_j. Then we fill in the elements, one by one, in such a way that the conditions (13.2.2) and (13.2.3) are not violated. First we take the element with the lowest cost, in this case $(1, 1)$, and in this place we put a number as large as possible (6).

	(6)0	7	7	10
(8)2	6			
9	0			
13	0			

Then we get zeros in the first column, while the row sum 8 is not fully exploited. From among the other elements in the first row, we choose the one with the lowest c_{ij}, in this case $(1, 2)$, and in this place we put a number as large as possible (2):

	0	(7)5	7	10
(2)0	6	2	0	0
9	0			
13	0			

In a similar way we obtain in the next step:

	0	5	7	(10)1
0	6	2	0	0
(9)0	0	0	0	9
13	0			

In the last step we have no choice, and hence we obtain

$$(x_{ij}) = \begin{pmatrix} 6 & 2 & 0 & 0 \\ 0 & 0 & 0 & 9 \\ 0 & 5 & 7 & 1 \end{pmatrix}.$$

We find $f = \sum c_{ij}x_{ij} = 262$. Then we compute α_i and β_j, choosing arbitrarily $\alpha_1 = 0$, and we get without difficulty $\beta_1 = 3$; $\beta_2 = 8$; $\alpha_3 = 5$; $\beta_3 = 5$; $\beta_4 = 7$; $\alpha_2 = 2$. These values are obtained from the elements $(1, 1)$, $(1, 2)$, $(2, 4)$, $(3, 2)$, $(3, 3)$, and $(3, 4)$ of (c_{ij}); they correspond to the x_{ij}-elements which are not zero. From $k_{ij} = \alpha_i + \beta_j$, we easily get

$$(k_{ij}) = \begin{pmatrix} 3 & 8 & 5 & 7 \\ 5 & 10 & 7 & 9 \\ 8 & 13 & 10 & 12 \end{pmatrix}; \qquad (c_{ij} - k_{ij}) = \begin{pmatrix} 0 & 0 & 4 & 9 \\ 1 & 1 & 7 & 0 \\ -3 & 0 & 0 & 0 \end{pmatrix}.$$

Thus we have $c_{31} - k_{31} < 0$, and it should be possible to reduce the transportation cost by moving as much as possible to this place. Hence, we modify (x_{ij}) in the following way:

$$(x_{ij}) = \begin{pmatrix} 6 - \varepsilon & 2 + \varepsilon & 0 & 0 \\ 0 & 0 & 0 & 9 \\ \varepsilon & 5 - \varepsilon & 7 & 1 \end{pmatrix}.$$

The elements which can be affected are obtained if we start at the chosen place (i, j), where $c_{ij} - k_{ij} < 0$, and draw a closed polygon with only horizontal and vertical sides, and with all corners in elements that are not zero.

Since all x_{ij} must be ≥ 0, and the number of nonzero elements should be $m + n - 1 = 6$, we find $\varepsilon = 5$ and

$$(x_{ij}) = \begin{pmatrix} 1 & 7 & 0 & 0 \\ 0 & 0 & 0 & 9 \\ 5 & 0 & 7 & 1 \end{pmatrix}.$$

As before, we determine α_i, β_j, and k_{ij} and obtain

$$(c_{ij} - k_{ij}) = \begin{pmatrix} 0 & 0 & 1 & 6 \\ 4 & 4 & 7 & 0 \\ 0 & 3 & 0 & 0 \end{pmatrix},$$

which shows that the solution is optimal. The total transportation cost becomes $f = 247$.

We see that the technique quarantees an integer solution if all initial values also are integers. Further we note that the matrix $(c_{ij} - k_{ij})$ normally contains $m + n - 1$ zeros at places corresponding to actual transports. If there is, for example, one more zero, we have a degenerate case that allows several optimal solutions. A transport can be moved to this place without the solution ceasing to be optimal. Combining two different optimal solutions, we can form an infinite number of optimal solutions. In this special case we can actually construct solutions with *more* than $m + n - 1$ elements not equal to 0, and, of course, there are also cases when we have less than $m + n - 1$ elements not equal to 0.

We shall now briefly indicate a few complications that may arise. So far we have assumed that the produced commodities are consumed, that is, there is no overproduction. The cases of over- and underproduction can easily be handled by introducing fictitious producers and consumers. For example, if we have a fictitious consumer, we make the transportation costs to him equal to zero. Then the calculation will show who is producing for the fictitious consumer; in other words, the overproduction is localized to specific producers. If we have a fictitious producer, we can make the transportation costs from him equal to zero, as in the previous case. We give an example to demonstrate the technique. Suppose that two producers manufacture 7 and 9 units, while four consumers ask for 2, 8, 5, and 5 units. Obviously, there is a deficit of 4 units, and hence we introduce a fictitious producer. We assume the following transportation cost table:

$$(c_{ij}) = \begin{pmatrix} 0 & 0 & 0 & 0 \\ 3 & 6 & 10 & 12 \\ 7 & 5 & 4 & 8 \end{pmatrix}.$$

We easily find

$$(x_{ij}) = \begin{pmatrix} 0 & 0 & 0 & 4 \\ 2 & 5 & 0 & 0 \\ 0 & 3 & 5 & 1 \end{pmatrix} \quad \text{and} \quad (k_{ij}) = \begin{pmatrix} -6 & -3 & -4 & 0 \\ 3 & 6 & 5 & 9 \\ 2 & 5 & 4 & 8 \end{pmatrix}.$$

Hence

$$(c_{ij} - k_{ij}) = \begin{pmatrix} 6 & 3 & 4 & 0 \\ 0 & 0 & 5 & 3 \\ 5 & 0 & 0 & 0 \end{pmatrix}.$$

Hence, the solution is optimal, and the fourth customer, who asked for 5 units, does not obtain more than 1 unit.

As a rule, $m + n - 1$ elements of our tranport table should differ from zero. However, there are two cases when this requirement cannot be met. It may sometimes happen that a remaining production capacity and a remaining consumer demand vanish simultaneously while the initial transport table is constructed. But also on transformation of the table as described above, extra zeros may appear. This difficulty can be overcome by the same perturbation technique as has been used before, and a quantity δ (some people use + instead) is placed in the table to obey the polygon rule, whereas δ is put equal to zero in all arithmetic calculations. This is motivated by the fact that the desired minimum value is a continuous function of the quantities a_i, b_j, and c_{ij}, and hence degeneration can be avoided by a slight change of some of them. The quantity (or quantities) δ are used also on computation of k_{ij}.

Example 2

	3	5	5
5	2	6	11
5	4	10	16
3	4	16	14

To begin, we get:

	3	5	5
5	3	2	
5		3	2
3			3

$$(k_{ij}) = \begin{pmatrix} 2 & 6 & 12 \\ (6) & 10 & 16 \\ 4 & 8 & 14 \end{pmatrix}, \qquad f = 122.$$

Obviously we ought to move as much as possible (3 units) to the (2, 1)-place, but then we get zeros in both (1, 1) and (2, 2). Let us therefore put, for example, the

$(1, 1)$-element $= 0$ and the $(2, 2)$-element $= \delta$:

$$\begin{array}{c|ccc} \hline 3 & 5 & 5 \\ \hline 5 & & 5 & \\ 5 & 3 & \delta & 2 \\ 3 & & & 3 \end{array} \qquad (k_{ij}) = \begin{pmatrix} 0 & 6 & (12) \\ 4 & 10 & 16 \\ 2 & 8 & 14 \end{pmatrix}, \qquad f = 116.$$

Now we move as much as possible (2 units) to the $(1, 3)$-place and get:

$$\begin{array}{c|ccc} \hline 3 & 5 & 5 \\ \hline 5 & & 3 & 2 \\ 5 & 3 & 2 & \\ 3 & & & 3 \end{array} \qquad (k_{ij}) = \begin{pmatrix} 0 & 6 & 11 \\ 4 & 10 & 15 \\ 3 & 9 & 14 \end{pmatrix}, \qquad f = 114.$$

This is the optimal solution.

---

§13.3.   QUADRATIC, INTEGER, AND DYNAMIC PROGRAMMING

   Linear programming is, in fact, only a special case (though a very important one), and generalizations in different directions are possible. Near at hand is the possibility of minimizing a quadratic expression instead of a linear one, adhering to the secondary linear conditions (equalities or inequalities). This problem is known as quadratic programming.

   If we have a process which occurs in several stages, where each subprocess is dependent on the strategy chosen, we have a dynamic programming problem. The theory of dynamic programming is essentially due to Bellman. As an illustration we give the following example, constructed by Vajda.

   We have n machines of a certain kind at our disposal, and these machines can perform two different kinds of work. If z machines are working in the first way, commodities worth $g(z)$ are produced, and if z machines are working in the second way, commodities worth $h(z)$ are produced. However, the machines are partly destroyed, and in the first case, $a(z)$ machines are left over and in the second, $b(z)$ machines. Here, a, b, g, and h are given functions. We assign x_1 machines for the first job, and $y_1 = n - x_1$ machines for the second job. After one stage we are left with $n_2 = a(x_1) + b(y_1)$ machines, of which we assign x_2 for the first job and y_2 for the second job. After N stages the total value of the produced goods amounts to

$$f = \sum_{i=1}^{N} [g(x_i) + h(y_i)],$$

with

$$\begin{aligned} x_i + y_i &= n_i, & n_1 &= n, \\ a(x_i) + b(y_i) &= n_{i+1}, & i &= 1, 2, \ldots, N - 1, \\ 0 \le x_i &\le n_i, & i &= 1, 2, \ldots, N. \end{aligned}$$

The problem is to maximize f. In particular, if the functions are linear, we have again a linear programming problem.

Let $f_N(n)$ be the maximum total value when we start with n machines and work in N stages using an optimal policy. Then we have

$$f_1(n) = \max_{0 \leq x \leq n} [g(x) + h(n - x)],$$

$$f_k(n) = \max_{0 \leq x \leq n} \{g(x) + h(n - x) + f_{k-1}[a(x) + b(n - x)]\}; \quad k > 1.$$

In this way the solution can be obtained by use of a recursive technique.

Last, we also mention that in certain programming problems, all quantities must be integers (integer programming). However, a closer account of this problem falls outside the scope of this book.

Finally we shall illustrate the solution technique in a numerical example simultaneously containing elements of dynamic and integer programming. Suppose that a ship is to be loaded with different goods and that every article is available only in units with definite weight and definite value. The problem is now to choose goods with regard to the weight restrictions (the total weight being given) so that the total value is maximized. Let the number of articles be N, the weight capacity z, and further the value, weight, and number of units of article i be v_i, w_i and x_i. Then we want to maximize

$$L_N(x) = \sum_{i=1}^{N} x_i v_i,$$

under the conditions $\sum_{i=1}^{N} x_i w_i \leq z$ with x_i integer and ≥ 0. Defining

$$f_N(z) = \max_{\{x_i\}} L_N(x),$$

we shall determine the maximum over combinations of x_i-values satisfying the conditions above. We can now derive a functional relation as follows. Let us first choose an arbitrary amount x_N leaving a remaining weight capacity $z - x_N w_N$. By definition, the best value we can get from this weight is $f_{N-1}(z - x_N w_N)$. Our choice of x_N gives the total value $x_N v_N + f_{N-1}(z - x_N w_N)$, and hence we must choose x_N so that this value is maximized. From this we get the fundamental and typical relationship

$$f_N(z) = \max_{x_N} \{x_N v_N + f_{N-1}(z - x_N w_N)\},$$

with $0 \leq x_N \leq \lfloor z/w_N \rfloor$. The initial function is trivially

$$f_1(z) = v_1 \lfloor z/w_1 \rfloor.$$

The solution in the case $N = 5$, $z = 20$ is derived below for the following values of v_i and w_i:

i	v_i	w_i	v_i/w_i
1	9	4	2.25
2	13	5	2.60
3	16	6	2.67
4	20	9	2.22
5	31	11	2.82

The computation can be performed through successive tabulation of $f_1, f_2, \ldots, f_5$ for $z = 1, 2, 3, \ldots, 20$. The results are presented in the following table.

z	f_1	f_2	f_3	f_4	f_5
1	0	0	0	0	0
2	0	0	0	0	0
3	0	0	0	0	0
4	9	9	9	9	9
5	9	13	13	13	13
6	9	13	16	16	16
7	9	13	16	16	16
8	18	18	18	18	18
9	18	22	22	22	22
10	18	26	26	26	26
11	18	26	29	29	31
12	27	27	32	32	32
13	27	31	32	32	32
14	27	35	35	35	35
15	27	39	39	39	40
16	36	39	42	42	44
17	36	40	45	45	47
18	36	44	48	48	48
19	36	48	48	48	49
20	45	52	52	52	53

The maximum value 53 is attained for $x_1 = 1$, $x_2 = 1$, $x_3 = 0$, $x_4 = 0$, and $x_5 = 1$.

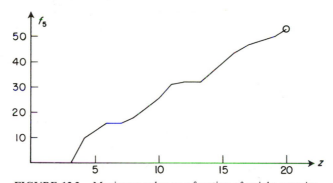

FIGURE 13.3. Maximum value as a function of weight capacity.

EXERCISES

1. Find the maximum of $y = x_1 - x_2 + 2x_3$ when

$$\begin{cases} x_1 + x_2 + 3x_3 + x_4 \le 5, \\ x_1 \qquad + x_3 - 4x_4 \le 2, \\ \qquad\qquad\qquad x_i \ge 0. \end{cases}$$

2. Find the minimum of $y = 5x_1 - 4x_2 + 3x_3$ when

$$\begin{cases} 2x_1 + x_2 - 6x_3 = 20, \\ 6x_1 + 5x_2 + 10x_3 \leq 76, \\ 8x_1 - 3x_2 + 6x_3 \leq 50, \\ \qquad\qquad x_i \geq 0. \end{cases}$$

3. Find the minimum of $f = \lambda x_1 - x_2$ as a function of λ $(-\infty < \lambda < \infty)$, when

$$\begin{cases} x_1 + x_2 \leq 6, \\ x_1 + 2x_2 \leq 10, \\ \quad\; x_i \geq 0. \end{cases}$$

4. Find the maximum of $f = -1 + x_2 - x_3$ when

$$\begin{cases} x_4 = x_1 - x_2 + x_3, \\ x_5 = 2 - x_1 - x_3, \\ x_i \geq 0. \end{cases}$$

5. Find the minimum of $f = x_1 + x_4$ when

$$\begin{cases} 2x_1 + 2x_2 + x_3 & \leq 7, \\ 2x_1 + x_2 + 2x_3 & \leq 4, \\ x_2 + \qquad x_4 \geq 1, \\ x_2 + x_3 + x_4 = 3, \\ \qquad\qquad x_i \geq 0. \end{cases}$$

6. Find the minimum of $f = 4x_1 + 2x_2 + 3x_3$ when

$$\begin{cases} 2x_1 + \qquad 4x_3 \geq 5, \\ 2x_1 + 3x_2 + x_3 \geq 4, \\ \qquad\qquad x_i \geq 0. \end{cases}$$

7. Maximize $f = 2x_1 + x_2$ when

$$\begin{cases} x_1 - x_2 \leq 2, \\ x_1 + x_2 \leq 6, \\ x_1 + 2x_2 \leq \alpha, \\ \qquad x_i \geq 0. \end{cases}$$

The maximum value of f should be given as a function of α when $0 \leq \alpha \leq 12$.

8. Minimize $f = -3x + y - 3z$ under the conditions

$$\begin{cases} -x + 2y + z \leq 0, \\ 2x - 2y - 3z = 9, \\ x - y - 2z \geq 6, \end{cases}$$

and $x \geq 0$, $y \geq 0$, $-\infty < z < \infty$.

9. The following linear programming problem is given: Maximize $c^T x$ under the conditions $Ax = b$, $x \geq d \geq 0$. Show how this problem can be transformed to the following type: Maximize $g^T y$ under the conditions $Fy = f$, $y \geq 0$, where the matrix F is of the same type (m, n) as the matrix A. Also solve the following

problem: Maximize $z = 3x_1 + 4x_2 + x_3 + 7x_4$ when

$$\begin{cases} 8x_1 + 3x_2 + 4x_3 + x_4 \le 42, \\ 6x_2 + x_3 + 2x_4 \le 20, \\ x_1 + 4x_2 + 5x_3 + 2x_4 \le 37, \end{cases}$$

and $x_1 \ge 2$; $x_2 \ge 1$; $x_3 \ge 3$; $x_4 \ge 4$.

10. A manufacturer uses three raw products, a, b, c, priced at 30, 50, 120 \$/lb, respectively. He can make three different products, A, B, and C, which can be sold at 90, 100, and 120 \$/lb, respectively. The raw products can be obtained only in limited quantities, namely, 20, 15, and 10 lb/day. Given: 2 lb of a plus 1 lb of b plus 1 lb of c will yield 4 lb of A; 3 lb of a plus 2 lb of b plus 2 lb of c will yield 7 lb of B; 2 lb of b plus 1 lb of c will yield 3 lb of C. Make a production plan, assuming that other costs are not influenced by the choice among the alternatives.

11. A mining company is taking a certain kind of ore from two mines, A and B. The ore is divided into three quality groups, a, b, and c. Every week the company has to deliver 240 tons of a, 160 tons of b, and 440 tons of c. The cost per day for running mine A is \$3000 and for running mine B \$2000. Each day A will produce 60 tons of a, 20 tons of b, and 40 tons of c. The corresponding figures for B are 20, 20, and 80. Construct the most economical production plan.

Solve the following transport problems:

12.

	3	3	4	5
4	13	11	15	20
6	17	14	12	13
9	18	18	15	12

13.

	75	75	75	75
100	19	15	19	20
100	20	23	17	31
100	14	25	20	18

14.

	10	5	7	6
18	2	11	6	3
9	12	10	15	5
11	4	9	13	10

(Give at least two different solutions.)

15.

	10	5	5
8	3	2	2
3	2	4	2
6	4	2	2

16.

	20	40	30	10	50	25
30	1	2	1	4	5	2
50	3	3	2	1	4	3
75	4	2	5	9	6	2
20	3	1	7	3	4	6

17. For three different services, *a*, *b*, and *c*, 100, 60, and 30 people, respectively, are needed. Three different categories, *A*, *B*, and *C*, are available with 90 of *A*, 70 of *B*, and 50 of *C*. The following table displays the "suitability numbers":

	a	*b*	*c*
A	6	4	4
B	6	10	8
C	9	8	8

Make a choice among the 210 people available for the 190 places so that the sum of the "suitability numbers" becomes as large as possible.

18. The following transport problem has a degenerate solution with only five transports. Find this solution and the minimum value!

	5	25	70
10	2	2	6
60	5	6	10
20	2	4	8
10	6	3	4

Part Three

APPROXIMATION

CHAPTER 14

Interpolation

Sir, In your otherwise beautiful poem (The Vision of
Sin) there is a verse which reads
   "Every moment dies a man,
   every moment one is born."
Obviously, this cannot be true and I suggest that in
the next edition you have it read
   "Every moment dies a man
   every moment $1\frac{1}{16}$ is born."
Even this value is slightly in error but should be
sufficiently accurate for poetry.

Charles Babbage
(in a letter to Lord Tennyson)

In a broad sense almost all numerical computations can be regarded as approximations. However, in this text we will limit ourselves to the classical concepts: data fitting and representation of given functions in terms of other functions. We will also consider interpolation and numerical differentiation, which in a way belong here. Further, we will treat some methods based on artificial sampling.

The main problem of interpolation in one dimension can be described as follows. Given a set of values $y_1, y_2, \ldots, y_n$ supposed to correspond to another set of values $x_1, x_2, \ldots, x_n$, determine in a reasonable way the value y corresponding to the given value x, preferably situated inside the interval defined by the coordinates x_k. We assume that y is some function of x, observing that the dependence can be quite complicated in nature. For example, x_k may represent different times while y_k may be the temperatures measured at these times.

There are many ways to perform an interpolation, and in order to make a proper choice of method one will have to use good judgment based on experience. When physical quantities are involved, knowledge of the physical laws behind the relationship will be most useful. In many cases polynomial approximation will be adequate, while in other cases techniques such as trigonometric or exponential interpolation will be more natural.

§14.1. POLYNOMIAL INTERPOLATION

We start by quoting a remarkable theorem by Weierstrass.

If $f(x)$ is a continuous function in the finite interval $[a, b]$, then for each $\varepsilon > 0$ there exists a polynomial $P(x)$ such that $|f(x) - P(x)| < \varepsilon$ for all x in the given interval.

We refrain from a complete proof and mention only that one well-known proof rests on the *Bernstein* polynomials defined by

$$(B_n f)(x) = \sum_{k=0}^{n} f\left(\frac{k}{n}\right)\binom{n}{k} x^k (1-x)^{n-k}.$$

If we use this approximation, the degree of the polynomial as a rule will be drastically overestimated.

We now turn to *Lagrange* interpolation, where we construct a polynomial of degree $n-1$ whose graph will pass exactly through n given points (x_k, y_k), $k = 1(1)n$. Here we suppose that all x_k are different. Putting $y = a_0 + a_1 x + \cdots + a_{n-1} x^{n-1}$ and inserting the given coordinates, we get a linear system of equations in the coefficients $a_0, a_1, \ldots, a_{n-1}$; it is easily proved that the determinant (called Vandermonde's determinant) has the value $\prod_{i>k}(x_i - x_k)$ and hence is $\neq 0$. This guarantees a unique solution; however, we prefer to write it in a different way:

$$y = P(x) = \sum_{k=1}^{n} L_k(x) y_k \quad \text{where}$$

(14.1.1) $$L_k(x) = \frac{(x-x_1)(x-x_2)\cdots(x-x_{k-1})(x-x_{k+1})\cdots(x-x_n)}{(x_k-x_1)(x_k-x_2)\cdots(x_k-x_{k-1})(x_k-x_{k+1})\cdots(x_k-x_n)}.$$

We immediately verify the property $L_k(x_i) = \delta_{ik}$ and hence we see that $P(x_i) = y_i$. Further it is clear that the degree of $L_k(x)$ is $n-1$, and hence $P(x)$ is the desired unique polynomial satisfied by the n coordinate pairs (x_k, y_k).

We shall now examine the difference between the given function $f(x)$ and the polynomial $P(x)$ for an arbitrary value x_0 of x. Then it is convenient to use the following functions:

(14.1.2) $$F(x) = \prod_{r=1}^{n} (x-x_r); \qquad F_k(x) = \prod_{r\neq k} (x-x_r).$$

Obviously, we have $F(x) = (x - x_k)F_k(x)$ and $F'(x) = (x - x_k)F_k'(x) + F_k(x)$, and hence $F'(x_k) = F_k(x_k)$. Thus we can also write

(14.1.3) $$P(x) = \sum_{k=1}^{n} \frac{F_k(x)}{F_k(x_k)} y_k = \sum_{k=1}^{n} \frac{F(x)}{(x - x_k)F'(x_k)} y_k.$$

We suppose that the point x_0 lies in the closed interval I bounded by the extreme points of $(x_1, x_2, \ldots, x_n)$ and further that $x_0 \neq x_k$, $k = 1, 2, \ldots, n$. We define the function $G(x) = f(x) - P(x) - RF(x)$, where R is a constant which is determined so that $G(x_0) = 0$. Obviously, we have $G(x) = 0$ for $x = x_0, x_1, x_2, \ldots, x_n$, and by using Rolle's theorem repeatedly, we conclude that $G^{(n)}(\xi) = 0$, for some value ξ where $\xi \in I$. But $G^{(n)}(\xi) = f^{(n)}(\xi) - R \cdot n!$, since $P(x)$ is of degree $n-1$, and $R = f^{(n)}(\xi)/n!$. Hence, replacing x_0 by x, we obtain (note that ξ is a function of x)

$$f(x) = P(x) + \frac{f^{(n)}(\xi)}{n!} F(x).$$

Example 1. We take $y = f(x) = 2\sin(\pi x/6)$ and choose the points $(0, 0)$, $(1, 1)$, $(2, \sqrt{3})$, and $(3, 2)$ as collocation points. The Lagrangian polynomial becomes (note that the term containing $y = 0$ vanishes!):

$$P(x) = \frac{x(x-2)(x-3)}{1(1-2)(1-3)} \cdot 1 + \frac{x(x-1)(x-3)}{2(2-1)(2-3)} \cdot \sqrt{3} + \frac{x(x-1)(x-2)}{3(3-1)(3-2)} \cdot 2$$

$$= 1.0686x - 0.0359x^2 - 0.0327x^3.$$

For the error we find $|P(x) - f(x)| < 0.0048$ when $0 \le x \le 3$, while the theoretical estimate is about 0.0063. If we put $x = 1.5$ we get $P(1.5) = 1.4118$, while $f(1.5) = \sqrt{2} = 1.4142$ with an error of 0.0024. However, outside the interval $(0, 3)$ the results quickly get worse: $P(-1) = -1.0718$ while $f(-1) = -1$; $P(4) = 1.6077$ while $f(4) = 1.73205$.

It should be emphasized that the Lagrangian technique must be used with discrimination. Suppose that we start from the points $(0, 0)$, $(1, 1)$, $(8, 2)$, $(27, 3)$ and $(64, 4)$ on the curve $y = x^{1/3}$ and try to compute $20^{1/3}$ by Lagrange interpolation. We then find $P(20) = -1.3139$ as compared with the correct value 2.7144. On the other hand, linear interpolation between $(8, 2)$ and $(27, 3)$ gives 2.63, which is in error by only 3%. This example shows clearly that a higher order formula does not necessarily give better results than a lower order formula.

A well-known example illustrating the difficulties just mentioned is the function

$$y = (1 + 25x^2)^{-1}, \qquad -1 \le x \le 1.$$

We construct a Lagrangian polynomial through the points (x, y), $x = -0.9(0.2)0.9$. Since y is an even function, all odd terms drop out, and we actually get a polynomial of degree 8. The curve behaves in a very irregular manner (see Figure 14.1),

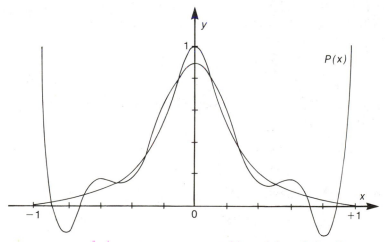

FIGURE 14.1. $y = (1 + 25x^2)^{-1}$ and the polynomial $P(x)$ of degree 8 through the points $x = -0.9(0.2)0.9$ and the corresponding y-values. $P(1) = 3.004$.

and the deviations for $x = 0(0.2)1$ are: $-0.1052, 0.06427, -0.04774, 0.07161, -0.2387,$ 2.965. This result of the Lagrangian interpolation is known as *Runge's phenomenon*.

Although Lagrangian interpolation is sometimes useful in theoretical investigations, it is rarely used in practical computations, possibly except for a special type of extrapolation. Suppose that we have an algorithm for computing a function $F(n)$, n integer, and that we have reason to believe that F approaches a finite value with an error, say $O(n^{-1/2})$, as $n \to \infty$. Then we compute, e.g., $a = F(9)$, $b = F(16)$, $c = F(25)$, and $d = F(36)$ and associate these values with $x = 1/3, 1/4, 1/5,$ and $1/6$. Since the computations are easier with integers, we multiply all x-values by a suitable constant, in this case 60, yielding 20, 15, 12, and 10, and finally perform an extrapolation to 0. This gives:

$$P(x) = \frac{(x-15)(x-12)(x-10)}{(20-15)(20-12)(20-10)} a + \frac{(x-20)(x-12)(x-10)}{(15-20)(15-12)(15-10)} b$$
$$+ \frac{(x-20)(x-15)(x-10)}{(12-20)(12-15)(12-10)} c + \frac{(x-20)(x-15)(x-12)}{(10-20)(10-15)(10-12)} d$$

and

$$F(\infty) \simeq P(0) = -\tfrac{9}{2}a + 32b - \tfrac{125}{2}c + 36d.$$

Note that $-9/2 + 32 - 125/2 + 36 = 1$, corresponding to interpolation of the function $f(x) = 1$ and providing a useful check. If the form of the error is not known, one has to try several hypotheses with, e.g., $O(n^{-r})$ until consistent values are obtained.

Hermite's Interpolation Formula

The Hermitian interpolation is rather similar to the Lagrangian. The difference is that we now seek a polynomial $P(x)$ of degree $2n - 1$ such that in the points $x_1, x_2, \ldots, x_n$, not only the functions $P(x)$ and $f(x)$, but also the derivatives $P'(x)$ and $f'(x)$, coincide. Thus we form

$$(14.1.4) \qquad P(x) = \sum_{k=1}^{n} U_k(x) f(x_k) + \sum_{k=1}^{n} V_k(x) f'(x_k).$$

Here $U_k(x)$ and $V_k(x)$ are supposed to be polynomials of degree $2n - 1$. Our requirements are fulfilled if we claim that

$$(14.1.5) \qquad \begin{cases} U_k(x_i) = \delta_{ik}, & V_k(x_i) = 0, \\ U'_k(x_i) = 0, & V'_k(x_i) = \delta_{ik}. \end{cases}$$

As is easily inferred, we may choose

$$(14.1.6) \qquad \begin{cases} U_k(x) = W_k(x) L_k(x)^2, \\ V_k(x) = Z_k(x) L_k(x)^2, \end{cases}$$

where $L_k(x)$ is defined in (14.1.1); clearly, we have $L_k(x_i) = \delta_{ik}$. Now $L_k(x)$ is of degree

$n - 1$, and hence W_k and Z_k must be linear functions. By using the conditions (14.1.5), we get

(14.1.7)
$$\begin{cases} W_k(x_k) = 1 & Z_k(x_k) = 0, \\ W'_k(x_k) = -2L'_k(x_k), & Z'_k(x_k) = 1. \end{cases}$$

Thus Hermite's interpolation formula takes the form

$$(14.1.8) \quad P(x) = \sum_{k=1}^{n} \{1 - 2L'_k(x_k)(x - x_k)\}L_k(x)^2 f(x_k) + \sum_{k=1}^{n} (x - x_k)L_k(x)^2 f'(x_k).$$

Interpolation of this kind is sometimes called osculating interpolation. The error term has the form

$$(14.1.9) \qquad\qquad R(x) = f^{(2n)}(\xi)F(x)^2/(2n)!$$

§14.2. DIVIDED DIFFERENCES

If one wants to interpolate by use of function values which are given for non-equidistant points, the Lagrangian scheme is impractical and requires much labor. In this situation the divided differences are more convenient.

Let $x_0, x_1, \ldots, x_n$ be $n + 1$ given points. Then we define the *first divided difference* of $f(x)$ between x_0 and x_1:

$$(14.2.1) \qquad\qquad f(x_0, x_1) = \frac{f(x_1) - f(x_0)}{x_1 - x_0} = f(x_1, x_0).$$

Analogously, the second divided difference is defined by

$$(14.2.2) \qquad\qquad f(x_0, x_1, x_2) = \frac{f(x_1, x_2) - f(x_0, x_1)}{x_2 - x_0},$$

and in a similar way the nth divided difference:

$$(14.2.3) \qquad f(x_0, x_1, \ldots, x_n) = \frac{f(x_1, x_2, \ldots, x_n) - f(x_0, x_1, \ldots, x_{n-1})}{x_n - x_0}.$$

By induction it is easy to prove that

$$(14.2.4) \quad f(x_0, x_1, \ldots, x_k)$$

$$= \sum_{p=0}^{k} \frac{f(x_p)}{(x_p - x_0)(x_p - x_1) \cdots (x_p - x_{p-1})(x_p - x_{p+1}) \cdots (x_p - x_k)}.$$

For equidistant arguments we have

$$f(x_0, x_1, \ldots, x_n) = \frac{1}{h^n \cdot n!} \Delta^n f_0,$$

where $h = x_{k+1} - x_k$. From this we see that $f(x_0, x_1, \ldots, x_k)$ is a symmetric function of the arguments $x_0, x_1, \ldots, x_k$. If two arguments are equal, we can still attribute a meaning to the difference; we have, for example,

$$f(x_0, x_0) = \lim_{x \to x_0} \frac{f(x) - f(x_0)}{x - x_0} = f'(x_0)$$

and analogously

$$\underbrace{f(x_0, x_0, \ldots, x_0)}_{r+1 \text{ arguments}} = \frac{f^{(r)}(x_0)}{r!}.$$

Finally, we also get

$$f(x, x, x_0, x_1, \ldots, x_n) = \frac{d}{dx} f(x, x_0, x_1, \ldots, x_n).$$

From the defining equations we have

$$\begin{cases} f(x) & = f(x_0) + (x - x_0) \cdot f(x, x_0), \\ f(x, x_0) & = f(x_0, x_1) + (x - x_1) \cdot f(x, x_0, x_1), \\ f(x, x_0, x_1) & = f(x_0, x_1, x_2) + (x - x_2) \cdot f(x, x_0, x_1, x_2), \\ \vdots \\ f(x, x_0, \ldots, x_{n-1}) & = f(x_0, x_1, \ldots, x_n) + (x - x_n) \cdot f(x, x_0, x_1, \ldots, x_n). \end{cases}$$

Multiplying the second equation by $(x - x_0)$, the third by $(x - x_0)(x - x_1)$, and so on, and finally the last equation by $(x - x_0)(x - x_1) \cdots (x - x_{n-1})$, and adding we find

$$(14.2.5) \quad f(x) = f(x_0) + (x - x_0) \cdot f(x_0, x_1) + (x - x_0)(x - x_1) \cdot f(x_0, x_1, x_2) + \cdots$$
$$+ (x - x_0)(x - x_1) \cdots (x - x_{n-1}) \cdot f(x_0, x_1, \ldots, x_n) + R,$$

where $R = f(x, x_0, x_1, \ldots, x_n) \prod_{i=0}^{n} (x - x_i)$. This is Newton's interpolation formula with divided differences.

For a moment we put $f(x) = P(x) + R$. Since $P(x)$ is a polynomial of degree n and, further, R vanishes for $x = x_0, x_1, \ldots, x_n$, we have $f(x_k) = P(x_k)$ for $k = 0, 1, 2, \ldots, n$, and clearly $P(x)$ must be identical with the Lagrangian interpolation polynomial. Hence

$$(14.2.6) \qquad R = \frac{f^{(n+1)}(\xi)}{(n+1)!} \prod_{i=0}^{n} (x - x_i).$$

We also find that

$$f(x, x_0, x_1, \ldots, x_n) = \frac{f^{(n+1)}(\xi)}{(n+1)!}.$$

Example 1.   Find a polynomial satisfied by $(-4, 1245)$, $(-1, 33)$, $(0, 5)$, $(2, 9)$, and $(5, 1335)$.

$$
\begin{array}{cc}
x & y \\
-4 & 1245 \\
& & -404 \\
-1 & 33 & & 94 \\
& & -28 & & -14 \\
0 & 5 & & 10 & & 3 \\
& & 2 & & 13 \\
2 & 9 & & 88 \\
& & 442 \\
5 & 1335
\end{array}
$$

(Note that this scheme contains divided differences.)

$$
\begin{aligned}
f(x) &= 1245 - 404(x + 4) + 94(x + 4)(x + 1) \\
&\quad - 14(x + 4)(x + 1)x + 3(x + 4)(x + 1)x(x - 2) \\
&= 3x^4 - 5x^3 + 6x^2 - 14x + 5.
\end{aligned}
$$

The practical computation is best done by a technique developed by Aitken. The different interpolation polynomials are denoted by $I(x)$, and first we form the linear expression

$$
I_{0,1}(x) = \frac{y_0(x_1 - x) - y_1(x_0 - x)}{x_1 - x_0} = \frac{1}{x_1 - x_0}\begin{vmatrix} y_0 & x_0 - x \\ y_1 & x_1 - x \end{vmatrix}.
$$

Obviously, $I_{0,1}(x_0) = y_0$ and $I_{0,1}(x_1) = y_1$. Next we form

$$
I_{0,1,2}(x) = \frac{I_{0,1}(x)(x_2 - x) - I_{0,2}(x)(x_1 - x)}{x_2 - x_1} = \frac{1}{x_2 - x_1}\begin{vmatrix} I_{0,1}(x) & x_1 - x \\ I_{0,2}(x) & x_2 - x \end{vmatrix}
$$

and observe that

$$
I_{0,1,2}(x_0) = \frac{y_0(x_2 - x_0) - y_0(x_1 - x_0)}{x_2 - x_1} = y_0,
$$

$I_{0,1,2}(x_1) = y_1$, and $I_{0,1,2}(x_2) = y_2$. In general, it is easy to prove that if

$$
I_{0,1,2,\ldots,n}(x) = \frac{I_{0,1,2,\ldots,n-2,n-1}(x_n - x) - I_{0,1,2,\ldots,n-2,n}(x_{n-1} - x)}{x_n - x_{n-1}},
$$

we have $I_{0,1,2,\ldots,n}(x_k) = y_k$; $k = 0, 1, 2, \ldots, n$. Hence nth degree interpolation can be performed by $n(n + 1)/2$ linear interpolations. Conveniently, this is done by aid of the scheme below.

$$
\begin{array}{lllll}
x_0 & y_0 & & & & x_0 - x \\
x_1 & y_1 & I_{0,1}(x) & & & x_1 - x \\
x_2 & y_2 & I_{0,2}(x) & I_{0,1,2}(x) & & x_2 - x \\
x_3 & y_3 & I_{0,3}(x) & I_{0,1,3}(x) & I_{0,1,2,3}(x) & x_3 - x
\end{array}
$$

Example 2

$$K(x) = \int_0^1 dt / \sqrt{[(1 - x^2t^2)(1 - t^2)]}$$

is to be computed for $x = 0.4142$. From a table the following values are obtained:

x	$y = K(x)$					
0.30	1.608049					-1142
0.35	1.622528	1.641119				-642
0.40	1.640000	1.644537	1.645508			-142
0.45	1.660886	1.648276	1.645714	1.645567		358
0.50	1.685750	1.652416	1.645954	1.645571	1.645563	858

This interpolation is identical with the Lagrangian, but it has two essential advantages. On the one hand, it is much simpler computationally; on the other, it gives a good idea of the accuracy obtained.

Difference Schemes

We now assume that all x_k are equidistant, i.e., $x_k = x_0 + kh$. A difference scheme is then constructed as shown in the following example.

y	Δ	Δ^2	Δ^3	Δ^4
0				
	1			
1		14		
	15		36	
16		50		24
	65		60	
81		110		24
	175		84	
256		194		24
	369		108	
625		302		
	671			
1296				

In general, we have the following picture showing the notations ($\Delta y_k = y_{k+1} - y_k$):

$$
\begin{array}{ccccccc}
y_0 \\
& \Delta y_0 \\
y_1 & & \Delta^2 y_0 \\
& \Delta y_1 & & \Delta^3 y_0 \\
y_2 & & \Delta^2 y_1 & & \Delta^4 y_0 \\
& \Delta y_2 & & \Delta^3 y_1 & & \Delta^5 y_0 \\
y_3 & & \Delta^2 y_2 & & \Delta^4 y_1 \\
& \Delta y_3 & & \Delta^3 y_2 \\
y_4 & & \Delta^2 y_3 \\
& \Delta y_4 \\
y_5
\end{array}
$$

We see directly that the quantities $\Delta^k y_0$ lie on a straight line sloping down to the right. On the other hand, since $\Delta = E\nabla$, we have, for example, $\Delta y_4 = \nabla y_5$; $\Delta^2 y_3 = \nabla^2 y_5$; $\Delta^3 y_2 = \nabla^3 y_5$; and so on; and we infer that the quantities $\nabla^k y_n$ lie on a straight line sloping upward to the right. Finally, we also have $\Delta = E^{1/2}\delta$ and hence, for example, $\Delta^2 y_1 = E\delta^2 y_1 = \delta^2 y_2$; $\Delta^4 y_0 = \delta^4 y_2$; and so on. In this way we find that the quantities $\delta^{2k} y_n$ lie on a *horizontal* line. Note that the difference scheme is exactly the same, and that it is only a question of notations what the differences are called. For example, we have

$$\Delta^3 y_1 = \nabla^3 y_4 = \delta^3 y_{5/2}.$$

When working with difference schemes, we observe a very characteristic kind of error propagation, which we shall now illustrate. Consider a function that is zero in all grid points except one, where it is ε. We obtain the following difference scheme.

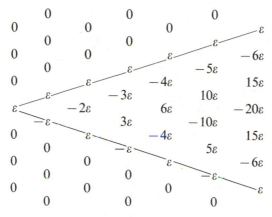

The error propagates in a triangular pattern and grows quickly; apart from the sign we recognize the binomial coefficients in the different columns. In higher-order differences the usual round-off errors appear as tangible irregular fluctuations. Gross errors are easily revealed by use of a difference scheme, and in such cases the picture above should be kept in mind.

The propagation of round-off errors is clearly demonstrated in the scheme below, where we have assumed as large variations as possible between two consecutive values.

ε		-4ε	16ε
	-2ε		8ε
$-\varepsilon$		4ε	-16ε
	2ε		-8ε
ε		-4ε	16ε
	-2ε		8ε
$-\varepsilon$		4ε	-16ε
	2ε		-8ε
ε		-4ε	16ε
	-2ε		8ε
$-\varepsilon$		4ε	-16ε

Hence, in the worst possible case, we can obtain a doubling of the error for every new difference introduced.

§14.3. INTERPOLATION BY USE OF DIFFERENCES

We shall now suppose that the points x_k are equidistant, i.e., $x_k = x_0 + kh$. We want to compute y for $x = x_0 + ph$, where in general $-1 < p < 1$; for convenience we also introduce the notation $q = 1 - p$. We intend to use operator techniques, so we remind you of the operators E and Δ defined through $Ey_n = y_{n+1}$ and $\Delta y_n = y_{n+1} - y_n$; hence we have $E = 1 + \Delta$. What we now really want to compute is $y_p = y(x_0 + ph)$:

$$y_p = E^p y_0 = (1 + \Delta)^p y_0 = y_0 + \binom{p}{1} \Delta y_0 + \binom{p}{2} \Delta^2 y_0 + \cdots$$

This formula, known as Newton's forward difference formula, has a finite number of terms if y is a polynomial; in other cases it must be truncated with a fair amount of good judgment. An obvious condition is that the successive terms must decrease reasonably rapidly.

Example 1

x	y	Δ	Δ^2	Δ^3	Δ^4	Δ^5
0	0					
		173648				
10	173648		-5276			
		168372		-5116		
20	342020		-10392		316	
		157980		-4800		144
30	500000		-15192		460	
		142788		-4340		137
40	642788		-19532		597	
		123256		-3743		
50	766044		-23275			
		99981				
60	866025					

The table above gives $10^6 \sin x$ with x expressed in degrees. Note that in hand computation it is rather convenient to transform all entries to integers. Now suppose that we want to compute y when $x = 14$. Obviously, $x_0 = 10$ and $p = 0.4$, and we get:

$$\binom{p}{1} = 0.4; \quad \binom{p}{2} = -0.12; \quad \binom{p}{3} = 0.064; \quad \binom{p}{4} = -0.0416; \quad \binom{p}{5} = 0.029952;$$

hence

$$y_p \simeq 173648 + 0.4 \cdot 168372 + 0.12 \cdot 10392 - 0.064 \cdot 4800 - 0.0416 \cdot 460 + 0.029952 \cdot 137$$
$$= 241922$$

where $\sin 14° = 0.2419218956$. We see that the computed interpolation value is correct to six digits. The differences used are underlined; they are all situated along a straight line sloping down to the right.

There are a large number of interpolation formulas, and rather than describing all of them we refer to Figure 14.2 and the table below.

Let φ and ψ (with or without indices) denote *even* and *odd* functions of δ, respectively. The best-known formulas have the following structure (note that δ and

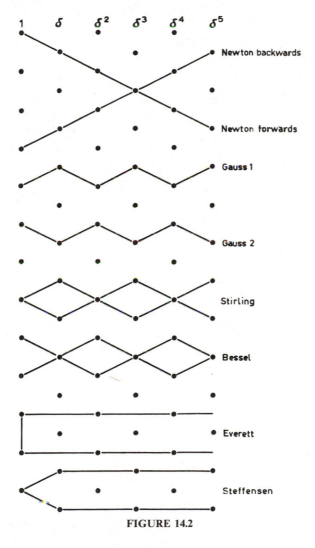

FIGURE 14.2

μ were defined in 5.2):

1. $y_p = \varphi(\delta)y_0 + \psi(\delta)y_{-1/2}$ (Gauss)
2. $y_p = \varphi(\delta)y_0 + \psi(\delta)y_{1/2}$ (Gauss)
3. $y_p = \varphi(\delta)y_0 + \mu\psi(\delta)y_0$ (Stirling)
4. $y_p = \mu\varphi(\delta)y_{1/2} + \psi(\delta)y_{1/2}$ (Bessel)
5. $y_p = \varphi_0(\delta)y_0 + \varphi_1(\delta)y_1$ (Everett)
6. $y_p = y_0 + \psi_0(\delta)y_{-1/2} + \psi_1(\delta)y_{1/2}$ (Steffensen)

Naturally, φ and ψ stand for different functions in different cases. We now select one of the formulas that is widely used, namely Everett's formula, and give a short derivation. We want to find an expression for y_p in terms of y_0 and y_1 together with the even central differences $\delta^{2k}y_0$ and $\delta^{2k}y_1$. Putting $y_p = \varphi_0(\delta)y_0 + \varphi_1(\delta)y_1$ where φ_0 and φ_1 are even functions (note that $y_1 = Ey_0 = e^U y_0$; $y_p = E^p y_0 = e^{pU} y_0$; $U = hD$) we get:

$$\begin{cases} e^{pU} = \varphi_0 + \varphi_1 e^U \\ e^{-pU} = \varphi_0 + \varphi_1 e^{-U}. \end{cases}$$

Here we have made use of the property that $\delta = 2\sinh(U/2)$ is an odd function of U. Solving for φ_0 and φ_1 we obtain (since $q = 1 - p$):

$$\varphi_0 = \frac{\sinh qU}{\sinh U}; \qquad \varphi_1 = \frac{\sinh pU}{\sinh U}$$

and using formula (5.3.3) we get the result:

$$y_p = qy_0 + \binom{q+1}{3}\delta^2 y_0 + \binom{q+2}{5}\delta^4 y_0 + \cdots$$
$$+ py_1 + \binom{p+1}{3}\delta^2 y_1 + \binom{p+2}{5}\delta^4 y_1 + \cdots$$

In many cases we want a simple formula that takes only the first and second order terms into account. Then we put $\delta^2 y_0 \simeq \delta^2 y_1 \simeq \delta^2 y$ and neglect the higher order terms to get:

$$y_p \simeq qy_0 + py_1 - \tfrac{1}{2}pq\,\delta^2 y.$$

§14.4. EXTRAPOLATION

In general it is obvious that extrapolation is a far more delicate process than interpolation. For example, one could point at the simple fact that a function might possess a very weak singularity which is hardly apparent in the given values but nevertheless might have a fatal effect in the extrapolation point. There are, however, a few cases when a fairly safe extrapolation can be performed, implying considerable gains in computing time and accuracy.

First we shall discuss *Aitken extrapolation*. Suppose that we have a sequence $y_0, y_1, y_2, \ldots$ converging toward a value y. We assume the convergence to be geometric, that is,

$$y - y_n = ah^n + \varepsilon_n h^n,$$

where $\varepsilon_n \to 0$ when $h \to 0$. This is often expressed through

$$y - y_n = ah^n + o(h^n).$$

From this we obtain

$$\frac{y - y_{n+1}}{y - y_n} = h + o(h)$$

and

$$\frac{y - y_n}{y - y_{n-1}} = h + o(h).$$

Subtracting we get

$$\frac{y - y_{n+1}}{y - y_n} - \frac{y - y_n}{y - y_{n-1}} = o(h)$$

and hence

$$y_{n-1} y_{n+1} - y_n^2 - y(y_{n-1} - 2y_n + y_{n+1}) = o(h^{2n}).$$

Since

$$y_{n-1} - 2y_n + y_{n+1} \simeq -ah^{n-1}(1 - h)^2 = O(h^{n-1}),$$

we get

$$y = \frac{y_{n-1} y_{n+1} - y_n^2}{y_{n-1} - 2y_n + y_{n+1}} + o(h^{n+1})$$

or

(14.4.1)
$$y = y_n^* = y_{n+1} - \frac{(\Delta y_n)^2}{\Delta^2 y_{n-1}} + o(h^{n+1}).$$

The procedure can then be repeated with the new series y_n^*.

Richardson Extrapolation

We will here briefly discuss the basic principles of Richardson extrapolation. Many numerical procedures depend on some kind of discretization where we use different interval lengths. As a rule these lengths can be written as ph, where p is a rational number. Often we use h, $h/2$, and $h/4$, or h, $h/2$, and $h/3$, but other combinations also appear. At least two different interval lengths are needed, but three, four, or five are not unusual. Below we give a few examples that should illustrate the technique.

Suppose that we want to compute a quantity $F(0) = F$, which is to be determined from the relation

$$F(h) = F + ah^2 + bh^4 + ch^6 + \cdots.$$

Further suppose that we have computed $F(h) = A$, $F(h/2) = B$, and $F(h/3) = C$. Then, neglecting terms of order 6 and higher, we get:

$$\begin{cases} F + ah^2 + bh^4 = A \\ F + \frac{1}{4}ah^2 + \frac{1}{16}bh^4 = B \\ F + \frac{1}{9}ah^2 + \frac{1}{81}bh^4 = C \end{cases}$$

and after simple computations

$$F = \frac{243C - 128B + 5A}{120}.$$

With only A and B known we have instead:

$$F = \frac{4B - A}{3}$$

and if $a = 0$, then $F = (16B - A)/15$. Similarly, with $a \neq 0$ and $D = F(h/4)$ we find

$$F = \frac{8192D - 6561C + 896B - 7A}{2520}.$$

In the chapter on integration we will describe in some detail how Richardson extrapolation is used in a very systematic way for numerical computation of integrals by use of Romberg's method. We will then use intervals $h/2^m$, $m = 0, 1, 2, \ldots$, and perform repeated Richardson extrapolation, eliminating in turn terms in h^2, h^4, $\ldots$ and so forth.

§14.5.   SPECIAL FEATURES

With easy access to calculators and computers, interpolation is nowadays practically never performed on elementary functions. However, there are more complex functions such as Bessel functions, the exponential and trigonometric integrals, and the gamma function for which interpolation is still of some interest (see, e.g., the standard table by Abramowitz and Stegun). It may then happen that the function cannot be interpolated directly, usually because of a singularity in the neighborhood. Often a simple transformation may remedy this, as shown in the following example. Consider the function (an exponential integral)

$$y = \int_x^\infty (e^{-t}/t)\, dt = E_1(x).$$

This function is represented in the table mentioned above, but for small values of x it varies very rapidly because of the logarithmic singularity in the origin. Instead we conveniently consider the function

$$z = y + \ln x = \int_1^\infty (e^{-t}/t)\, dt - \int_x^1 [(1 - e^{-t})/t]\, dt$$

where the first term is constant and the second term is regular even when $x = 0$. Hence the modified function can be written as a power series in x:

$$z = C + x - \frac{x^2}{2 \cdot 2!} + \frac{x^3}{3 \cdot 3!} - \cdots$$

and z can be interpolated without any problems. The numerical evidence is given below.

x	y	$z = y + \ln x$
0	∞	-0.5772
0.01	4.0379	-0.5673
0.02	3.3547	-0.5573
0.03	2.9591	-0.5475
0.04	2.6813	-0.5376
0.05	2.4679	-0.5278
0.06	2.2953	-0.5181

We can even compute the constant C exactly:

$$C = \lim_{\varepsilon \to 0} \left(\ln \varepsilon + \int_\varepsilon^\infty (e^{-t}/t)\, dt \right) = \lim \left(\ln \varepsilon + [e^{-t} \ln t]_\varepsilon^\infty + \int_\varepsilon^\infty e^{-t} \ln t\, dt \right)$$

(see 7.2.13). In this special case we might as well use the series expansion directly for small values of x.

In some cases when the analytical form is not known, it may still be possible to make a suitable transformation using a trial-and-error technique. The following numerical material suggests computation of $z = xy$ in order to get a smooth function that can be interpolated:

x	y	$z = xy$
0.01	98.4342	0.984342
0.02	48.4392	0.968784
0.03	31.7775	0.953325
0.04	23.4492	0.937968
0.05	18.4542	0.922710

Numerical Differentiation

This problem can actually be considered as a special kind of interpolation. Since the derivative is essentially approximated by a quotient of two small quantities, it is obvious that a considerable loss of accuracy is inherent in any numerical method for numerical differentiation. We restrict ourselves to equidistant grid-points and try to find an approximate value of $y' = Dy$ when $x = x_0$. Starting from $E = e^{hD} = 1 + \Delta$ we get formally:

$$hD = \ln(1 + \Delta) = \Delta - \Delta^2/2 + \Delta^3/3 - \cdots$$

or

(14.5.1)
$$y' = \frac{1}{h}\left(\Delta y - \frac{1}{2}\Delta^2 y + \frac{1}{3}\Delta^3 y - \cdots\right).$$

Using central differences, we can deduce slightly better formulas. In (5.3.4) we have obtained the power series

$$\frac{U}{\cosh(U/2)} = \frac{U}{\mu} = \delta - \frac{\delta^3}{6} + \frac{\delta^5}{30} - \frac{\delta^7}{140} + \cdots$$

where μ is the mean value operator defined by $\mu = \frac{1}{2}(E^{1/2} + E^{-1/2})$. Since $\mu\delta = \frac{1}{2}(E - E^{-1})$, we find:

(14.5.2)
$$y' = \frac{1}{2h}\left[y_1 - y_{-1} - \frac{1}{6}(\delta^2 y_1 - \delta^2 y_{-1}) + \frac{1}{30}(\delta^4 y_1 - \delta^4 y_{-1})\right.$$
$$\left. - \frac{1}{140}(\delta^6 y_1 - \delta^6 y_{-1}) + \cdots\right].$$

Similarly, using the formula for $z^2 = U^2$ we get directly:

(14.5.3)
$$y'' = \frac{1}{h^2}\left[\delta^2 y - \frac{1}{12}\delta^4 y + \frac{1}{90}\delta^6 y - \frac{1}{560}\delta^8 y + \cdots\right].$$

EXERCISES

1. The table below contains an error which should be located and corrected.

x	$f(x)$	x	$f(x)$
3.60	0.112046	3.65	0.152702
3.61	0.120204	3.66	0.160788
3.62	0.128350	3.67	0.168857
3.63	0.136462	3.68	0.176908
3.64	0.144600		

2. The function $y = f(x)$ is given in the points (7, 3), (8, 1), (9, 1), and (10, 9). Find the value of y for $x = 9.5$, using Lagrange's interpolation formula.

3. From the beginning of a table, the following values are reproduced:

x	$f(x)$
0.001	0.54483 33726
0.002	0.55438 29800
0.003	0.56394 21418
0.004	0.57351 08675
0.005	0.58308 91667

Find the function value for $x = 0.00180$ as accurately as possible.

4. A function $f(x)$ is known in three points, x_1, x_2, and x_3, in the vicinity of an extreme point x_0. Show that

$$x_0 \simeq \frac{x_1 + 2x_2 + x_3}{4} - \frac{f(x_1, x_2) + f(x_2, x_3)}{4f(x_1, x_2, x_3)}.$$

Use this formula to find x_0 when the following values are known:

x	3.0	3.6	3.8
f	0.13515	0.83059	0.26253

5. The function $y = x!$ has a minimum between 0 and 1. Find the abscissa from the data below.

x	$\dfrac{d}{dx}\log(x!)$	δ^2	δ^4
0.46	$-0.00158\ 05620$	-888096	-396
0.47	$+0.00806\ 64890$	-872716	-383

6. The function $y = \exp x$ is tabulated for $x = 0(0.01)1$. Find the maximal error on linear interpolation.

7. $f(x_n + ph)$ is denoted by y_{n+p}. Show that

$$y_{n+p} = y_n + p\,\Delta y_{n-1} + \binom{p+1}{2}\Delta^2 y_{n-2} + \binom{p+2}{3}\Delta^3 y_{n-3} + \cdots,$$

assuming that the series converges.

8. Find the constants A, B, C, and D in the interpolation formula

$$f(x_0 + ph) = Af_{-1} + Bf_1 + Chf'_{-1} + Dhf'_1 + R,$$

as well as the order of the error term R. Use the formula to obtain $f(2.74)$ when $f(2.6) = 0.0218502$ and $f(2.8) = 0.0168553$. The function f is defined by

$$f(x) = \int_x^\infty (e^{-t}/t)\,dt.$$

9. Determine the constants a, b, c, and d in such a way that the formula

$$y_p = ay_0 + by_1 + h^2(cy_0'' + dy_1''),$$

becomes correct to the highest possible order. Use the formula to compute $Ai(1.1)$ when $Ai(1.0) = 0.135292$ and $Ai(1.2) = 0.106126$. It is known that the function $y = Ai(x)$ satisfies the differential equation $y'' = xy$.

10. The following data are given for a certain function:

x	y	y'
0.4	1.554284	0.243031
0.5	1.561136	-0.089618

In the interval $0.4 < x < 0.5$, y has a maximum. Find its coordinates as accurately as possible.

11. A function $f(x, y)$ takes the following values in nine adjacent mesh points:

(6, 4) 8.82948	(7, 4) 11.33222	(8, 4) 14.17946
(6, 5) 9.31982	(7, 5) 11.97257	(8, 5) 14.98981
(6, 6) 9.81019	(7, 6) 12.61294	(8, 6) 15.80018

Find the function value for $x = 7.2$, $y = 5.6$.

12. The function y is given in the table below, and it is clear that it cannot be interpolated directly. Construct another function which is smooth enough to allow interpolation and find $y(0.347)$.

x	0.1	0.2	0.3	0.4	0.5
y	9.9214	6.9598	5.6372	4.8424	4.2958

13. In the table below, y is given as a function of x. It is obvious that y cannot be interpolated directly. Try to find a value p such that y can be approximately represented in the form x^p. Then find y for $x = 0.125$.

x	0.1	0.2	0.3	0.4	0.5
y	2.15204	1.70232	1.47862	1.33237	1.22327

14. A function $\varphi(x)$ is defined by $f(x) \cdot g(x)$. According to Steffensen, the following formula is valid:

$$\varphi(x_0, x_1, \ldots, x_n) = \sum_{r=0}^{n} f(x_0, x_1, \ldots, x_r) \cdot g(x_r, \ldots, x_n).$$

Prove this formula in the case $n = 2$.

15. In a table the function values have been rounded to a certain number of decimals. In order to facilitate linear interpolation one also wants to give the first differences. Determine which is best: to use the differences of the rounded values, or to give rounded values of the exact first differences.

16. One wants to compute an approximate value of π by considering regular n-sided polygons with corners on the unit circle, for such values of n that the perimeter P_n can easily be computed. Determine P_n for $n = 3, 4$, and 6 and extrapolate to ∞ taking into account that the error is proportional to $1/n^2$.

17. One wants to determine an infinite sum S_∞ by extrapolation, given $S_{15} = 0.479666$, $S_{20} = 0.482964$, $S_{25} = 0.483383$, and $S_{30} = 0.483614$. Introduce $x = a/n$ so that x becomes integer for these four values of n, and extrapolate to $x = 0$ by Lagrange's formula. What value is obtained?

18. The function $f(x) = x + x^2 + x^4 + x^8 + x^{16} + \cdots$ is defined for $-1 < x < 1$. For values close to 1 the dominating term in the expression $-\{\ln(1-x)/\ln 2 + f(x)\}$

is a constant c. Try to determine c by explicit computations for $x = 0.98, 0.99$, and 0.999 and extrapolation to $x = 1$.

19. A function is given below. Find the derivative for $x = 0.5$.

x	0.35	0.40	0.45	0.50	0.55	0.60	0.65
y	1.521525	1.505942	1.487968	1.467462	1.444243	1.418083	1.388686

20. A function $y = y(x)$ is given in the table below. Find the second derivative for $x = 3$.

x	y	x	y
2.94	0.18256 20761	3.02	0.17689 06327
2.96	0.18110 60149	3.04	0.17553 40257
2.98	0.17967 59168	3.06	0.17420 05379
3.00	0.17827 10306		

CHAPTER 15

Function Representation and Curve Fitting

A revolution is a successful effort to get rid of
a bad government and set up a worse.

§15.1   LEAST-SQUARES POLYNOMIAL APPROXIMATION

We suppose that we have a given (e.g., experimental) set of $n + 1$ points (x_0, y_0), $(x_1, y_1), \ldots, (x_n, y_n)$ with all x_k different. We now want to find a polynomial $y = y_m(x) = a_0 + a_1 x + \cdots + a_m x^m$ that provides a fit to these data. If $m = n$ we can uniquely determine an exact fit. However, if $m < n$ we can define a "best" fit in some sense. Assuming that the experimental errors are associated with the y-values only, we find it natural to minimize some suitable norm of the error vector. A popular choice is to make

$$S = \sum_{j=0}^{n} (y_m(x_j) - y_j)^2$$

a minimum by choosing the coefficients a_k conveniently. Differentiating, we obtain

$$\frac{\partial S}{\partial a_k} = 2 \sum_{j=0}^{n} (y_m(x_j) - y_j) x_j^k = 0, \qquad k = 0(1)m.$$

We introduce the notations $s_k = \sum_{j=0}^{n} x_j^k$ and $v_k = \sum_{j=0}^{n} x_j^k y_j$. Further we define the matrix C:

$$C = \begin{bmatrix} 1 & x_0 & x_0^2 & \cdots & x_0^m \\ 1 & x_1 & x_1^2 & \cdots & x_1^m \\ \vdots & & & & \\ 1 & x_n & x_n^2 & \cdots & x_n^m \end{bmatrix}$$

of type $(n + 1) \times (m + 1)$ and the vectors a, v and y:

$$a^T = (a_0, a_1, \ldots, a_m), \qquad v^T = (v_0, v_1, \ldots, v_m) \qquad \text{and} \qquad y^T = (y_0, y_1, \ldots, y_n).$$

Then $v = C^T y$, and if we write $C^T C = P = (p_{ik})$, we find $p_{ik} = s_{i+k}$ where $i, k = 0(1)m$. Hence the condition for minimizing S can be formulated as $Pa = v$. First we prove that P is nonsingular by showing that the homogeneous system $Pa = 0$ has only the trivial solution $a = 0$. This follows because we have $0 = a^T Pa = a^T C^T Ca = (Ca)^T Ca$, implying $Ca = 0$. Hence the polynomial $y_m(x)$ would vanish for $n + 1$ different x-values $x_0, x_1, \ldots, x_n$, which is possible only if all coefficients $a_0, a_1, \ldots, a_m$ vanish, i.e., $a = 0$. Since $Pa = 0$ has only the solution $a = 0$ we can draw the conclusion that P is nonsingular, and hence we have the unique solution

$$a = P^{-1}v \qquad \text{giving} \qquad S_{\min} = y^T y - v^T a.$$

This can also be written $a = (C^T C)^{-1} C^T y = C^+ y$, where C^+ is the pseudo-inverse of C (see 2.8.1). For large values of m the system becomes highly ill-conditioned and gives rise to considerable difficulties. In many cases this is an indication that the problem might not be well-posed, and perhaps other alternatives ought to be considered.

Example 1. The following data should be fitted by a quadratic function $y = a_0 + a_1 x + a_2 x^2$:

x	8	10	12	16	20	30	40	60	100
y	0.88	1.22	1.64	2.72	3.96	7.66	11.96	21.56	43.16

j	x_j	x_j^2	x_j^3	x_j^4	y_j	$x_j y_j$	$x_j^2 y_i$
0	8	64	512	4096	0.88	7.04	56.32
1	10	100	1000	10000	1.22	12.20	122.00
2	12	144	1728	20736	1.64	19.68	236.16
3	16	256	4096	65536	2.72	43.52	696.32
4	20	400	8000	160000	3.96	79.20	1584.00
5	30	900	27000	810000	7.66	229.80	6894.00
6	40	1600	64000	2560000	11.96	478.40	19136.00
7	60	3600	216000	12960000	21.56	1293.60	77616.00
8	100	10000	1000000	100000000	43.16	4316.00	431600.00
	296	17064	1322336	116590368	94.76	6479.44	537940.80

Hence we get the system of equations

$$\begin{cases} 9a_0 + 296a_1 + 17064a_2 = 94.76 \\ 296a_0 + 17064a_1 + 1322336a_2 = 6479.44 \\ 17064a_0 + 1322336a_1 + 116590368a_2 = 537940.80 \end{cases}$$

with the solution

$$\begin{cases} a_0 = -1.91915 \\ a_1 = 0.278214 \\ a_2 = 0.0017394. \end{cases}$$

With these coefficients we get the following values of $y = a_0 + a_1x + a_2x^2$; the initial data are given below for comparison.

x	8	10	12	16	20	30	40	60	100
y	0.42	1.04	1.67	2.98	4.34	7.99	11.99	21.04	43.30
	0.88	1.22	1.64	2.72	3.96	7.66	11.96	21.56	43.16

A quite common special case arises when data have to be represented by straight lines, and we shall consider two such cases separately.

1. In the first case we suppose that the errors in the x-values can be neglected compared with the errors in the y-values, and then it is natural to minimize the sum of the squares of the vertical deviations. We number the points from 1 to n:

$$S = \sum_{i=1}^{n} (kx_i + l - y_i)^2,$$

$$\frac{\partial S}{\partial l} = 2 \cdot \sum_i (kx_i + l - y_i) = 0,$$

$$\frac{\partial S}{\partial k} = 2 \cdot \sum_i (kx_i + l - y_i)x_i = 0.$$

With the notations

$$\sum_i x_i = nx_0 = s_1, \qquad\qquad \sum_i y_i = ny_0 = t_1,$$

$$\sum_i x_i^2 = s_2, \quad \sum_i x_iy_i = v_1, \quad \sum_i y_i^2 = t_2;$$

$$\begin{cases} A = s_2 - nx_0^2, \\ B = v_1 - nx_0y_0, \\ C = t_2 - ny_0^2, \end{cases}$$

we obtain $l = y_0 - kx_0$, which means that the center of gravity lies on the desired line. Further,

$$s_2k + nlx_0 - v_1 = 0;$$

that is,

$$k = \frac{v_1 - nx_0y_0}{s_2 - nx_0^2} = \frac{B}{A} \quad \text{and} \quad l = y_0 - \frac{B}{A}x_0.$$

2. Here we suppose that the x-values as well as the y-values are subject to errors of about the same order of magnitude, and then it is more natural to minimize the sum of the squares of the perpendicular distances to the line. Thus

$$S = \frac{1}{1 + k^2} \sum_{i=1}^{n} (kx_i + l - y_i)^2;$$

$$\frac{\partial S}{\partial l} = 0 \quad \text{gives, as before } l = y_0 - kx_0;$$

$$\frac{\partial S}{\partial k} = 0 \quad \text{gives } (1 + k^2) \cdot \sum_i (kx_i + l - y_i)x_i = k \sum_i (kx_i + l - y_i)^2.$$

After simplification we get:

$$k^2(v_1 - nx_0y_0) + k(s_2 - nx_0^2 - t_2 + ny_0^2) - (v_1 - nx_0y_0) = 0$$

or

$$k^2 + B^{-1}(A - C)k - 1 = 0.$$

From this equation we obtain two directions at right angles, and in practice there is no difficulty deciding which one gives a minimum.

Here we also give a simple example of how to solve a number of inconsistent equations in one variable using a maximum norm. Suppose that we have the equations:

$$\begin{cases} 2x = 4 \\ 3x = 10 \\ 5x = 14. \end{cases}$$

We then construct the functions:

$$\begin{cases} y = |2x - 4| \\ y = |3x - 10| \\ y = |5x - 14| \end{cases}$$

and form the function $z = \max y$, the maximum taken over the function values above, for different values of x. Finally we minimize $z = z(x)$ over x. If we draw the lines as in Figure 15.1, we find:

$$\begin{cases} z(x) = 14 - 5x, & x \le 2 \\ z(x) = 10 - 3x, & 2 \le x \le 2.8 \\ z(x) = 2x - 4, & 2.8 \le x \le 10/3 \\ z(x) = 5x - 14; & x \ge 10/3. \end{cases}$$

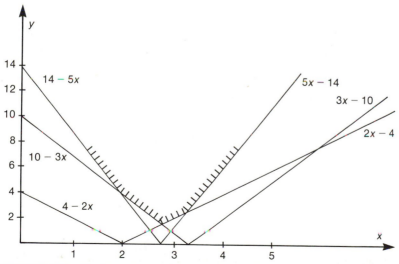

FIGURE 15.1.   Maximum norm solution of equations $2x = 4$, $3x = 10$, and $5x = 14$.

Hence z takes its minimum for $x = 2.8$, giving $y = 1.6$. It is not too difficult to devise an ascent algorithm to solve similar problems.

If instead we minimize $u = (2x - 4)^2 + (3x - 10)^2 + (5x - 14)^2$, we get $x = 54/19 = 2.84$, while the mean value of the individual solutions is $(2 + 10/3 + 14/5)/3 = 2.71$.

---

§15.2. TRIGONOMETRIC INTERPOLATION

Here we restrict ourselves to the case in which the support points are uniformly distributed, i.e., $x_k = 2\pi k/N$ where N is a positive integer. Our problem is then the following: Find coefficients $c_0, c_1, \ldots, c_{N-1}$ such that with

$$(15.2.1) \qquad p(x) = c_0 + c_1 e^{ix} + c_2 e^{2ix} + \cdots + c_{N-1} e^{(N-1)ix}$$

we have $p(x_k) = f_k$, $k = 0(1)N - 1$, where $f_0, f_1, \ldots, f_{N-1}$ are given quantities. The function $p(x)$ can be considered as a polynomial in $z = e^{ix}$ and is named the *phase polynomial*. Putting $\omega = \exp(2\pi i/N)$, we get the following system of linear equations for c_k:

$$(15.2.2) \qquad \begin{cases} c_0 + \quad c_1 + \quad c_2 + \cdots + \qquad\qquad c_{N-1} = f_0 \\ c_0 + \quad \omega c_1 + \omega^2 c_2 + \cdots + \quad \omega^{N-1} c_{N-1} = f_1 \\ c_0 + \omega^2 c_1 + \omega^4 c_2 + \cdots + \omega^{2(N-1)} c_{N-1} = f_2 \\ \qquad\qquad\qquad\qquad\qquad\qquad\qquad\qquad \vdots \end{cases}$$

Let A be the coefficient matrix with elements $a_{rs} = \omega^{rs}$, where $r, s = 0(1)N - 1$ and where c and f are vectors with components $c_r, f_r, r = 0(1)N - 1$. Then the system can be written

$$Ac = f.$$

The inverse $B = A^{-1}$ has the elements $b_{rs} = N^{-1} \omega^{-rs}$. This is seen by forming the product AB:

$$(AB)_{rs} = N^{-1} \sum_{q=0}^{N-1} \omega^{rq} \omega^{-qs} = N^{-1} \sum_{q=0}^{N-1} (\omega^{r-s})^q = N^{-1} N \delta_{rs} = \delta_{rs},$$

i.e., $AB = I$. Hence A is nonsingular and we have a unique solution, which can be written

$$(15.2.3) \qquad c_k = N^{-1} \sum_{r=0}^{N-1} \omega^{-kr} f_r, \qquad k = 0(1)N - 1.$$

To compute all coefficients we have to perform N^2 multiplications. In practical applications of Fourier analysis the number N may be very large, and moreover, the computations may have to be applied in two or three dimensions. In spite of the speed of modern computers this computation volume could become prohibitive, particularly in the 2- or 3-dimensional cases. A remedy was offered through the discovery by Cooley and Tukey (1965) of the fast Fourier transform, which requires only $O(N \log N)$ operations instead of $O(N^2)$.

§15.3. THE FAST FOURIER TRANSFORM

The method is general, but it works best when $N = 2^n$, and we limit ourselves to this case. We will demonstrate the technique on a specific example with $N = 8$ and show how the coefficients c_k are obtained from the given values f_k. Since $\omega^8 = 1$ we have $\omega^2 = i$, $\omega^3 = i\omega$, $\omega^4 = -1$, $\omega^5 = -\omega$, $\omega^6 = -i$, and $\omega^7 = -i\omega$. The matrix A then has the form:

$$A = \begin{pmatrix} 1 & 1 & 1 & 1 & 1 & 1 & 1 & 1 \\ 1 & \omega & i & i\omega & -1 & -\omega & -i & -i\omega \\ 1 & i & -1 & -i & 1 & i & -1 & -i \\ 1 & i\omega & -i & \omega & -1 & -i\omega & i & -\omega \\ 1 & -1 & 1 & -1 & 1 & -1 & 1 & -1 \\ 1 & -\omega & i & -i\omega & -1 & \omega & -i & i\omega \\ 1 & -i & -1 & i & 1 & -i & -1 & i \\ 1 & -i\omega & -i & -\omega & -1 & i\omega & i & \omega \end{pmatrix}$$

The inverse $B = A^{-1}$ becomes:

$$B = \frac{1}{8} \begin{pmatrix} 1 & 1 & 1 & 1 & 1 & 1 & 1 & 1 \\ 1 & -i\omega & -i & -\omega & -1 & i\omega & i & \omega \\ 1 & -i & -1 & i & 1 & -i & -1 & i \\ 1 & -\omega & i & -i\omega & -1 & \omega & -i & i\omega \\ 1 & -1 & 1 & -1 & 1 & -1 & 1 & -1 \\ 1 & i\omega & -i & \omega & -1 & -i\omega & i & -\omega \\ 1 & i & -1 & -i & 1 & i & -1 & -i \\ 1 & \omega & i & i\omega & -1 & -\omega & -i & -i\omega \end{pmatrix}$$

Forming $c = Bf$, we see that these eight equations can be grouped together in a suggestive way, namely equations 1 and 5, 2 and 6, 3 and 7, 4 and 8. In fact, 8 times c_0 and c_4 can be expressed as the sum and the difference of two quantities which we call g_0 and g_1:

$$\begin{cases} 8c_0 = g_0 + g_1 \\ 8c_4 = g_0 - g_1 \end{cases} \qquad \begin{cases} g_0 = f_0 + f_2 + f_4 + f_6 \\ g_1 = f_1 + f_3 + f_5 + f_7 \end{cases}$$

Similarly:

$$\begin{cases} 8c_1 = g_2 + g_3 \\ 8c_5 = g_2 - g_3 \end{cases} \quad \begin{cases} g_2 = f_0 - if_2 - f_4 + if_6 \\ g_3 = -i\omega f_1 - \omega f_3 + i\omega f_5 + \omega f_7 \end{cases}$$

$$\begin{cases} 8c_2 = g_4 + g_5 \\ 8c_6 = g_4 - g_5 \end{cases} \quad \begin{cases} g_4 = f_0 - f_2 + f_4 - f_6 \\ g_5 = -if_1 + if_3 - if_5 + if_7 \end{cases}$$

$$\begin{cases} 8c_3 = g_6 + g_7 \\ 8c_7 = g_6 - g_7 \end{cases} \quad \begin{cases} g_6 = f_0 + if_2 - f_4 - if_6 \\ g_7 = -\omega f_1 - i\omega f_3 + \omega f_5 + i\omega f_7 \end{cases}$$

We can proceed splitting in the same way:

$$\begin{cases} g_0 = h_0 + h_1 \\ g_4 = h_0 - h_1 \end{cases} \quad \begin{cases} g_1 = h_2 + h_3 \\ g_5 = -ih_2 + ih_3 \end{cases} \quad \begin{cases} g_2 = h_4 + h_5 \\ g_6 = h_4 - h_5 \end{cases} \quad \begin{cases} g_3 = ih_6 + h_7 \\ g_7 = h_6 - ih_7 \end{cases}$$

where

$$\begin{cases} h_0 = f_0 + f_4 \\ h_1 = f_2 + f_6 \end{cases} \quad \begin{cases} h_2 = f_1 + f_5 \\ h_3 = f_3 + f_7 \end{cases} \quad \begin{cases} h_4 = f_0 - f_4 \\ h_5 = -if_2 + if_6 \end{cases} \quad \begin{cases} h_6 = -\omega f_1 + \omega f_5 \\ h_7 = -\omega f_3 + \omega f_7 \end{cases}$$

The actual computations are performed in the reverse order: $f \rightarrow h \rightarrow g \rightarrow c$, i.e., roughly in three steps giving in total $3 \cdot 8 = 24$ instead of $8^2 = 64$ "operations." The procedure shown above carries over to the general case and implies a great saving in time and computational effort.

§15.4. APPROXIMATION WITH TRIGONOMETRIC FUNCTIONS

The theoretical background for the treatment of problems within this area has been given in some length in Section 3.4 on Fourier series. In particular, orthogonality formulas were presented in both the discrete and the continuous case. Orthogonality under summation forms the basis for harmonic analysis, while the same property under integration plays a decisive role in the theory of Fourier series.

First we shall try to generate a series in $\cos rj\alpha$ and $\sin rj\alpha$, which in the points $x_j = j\alpha$ takes the values y_j, and we find

$$(15.4.1) \quad \begin{cases} y_j = \dfrac{1}{2} a_0 + \displaystyle\sum_{r=1}^{(n-1)/2} (a_r \cos rj\alpha + b_r \sin rj\alpha); & n \text{ odd,} \\[4mm] y_j = \dfrac{1}{2}(a_0 + (-1)^j a_{n/2}) + \displaystyle\sum_{r=1}^{n/2-1} (a_r \cos rj\alpha + b_r \sin rj\alpha); & n \text{ even.} \end{cases}$$

The coefficients are obtained from

$$(15.4.2) \quad a_r = \frac{2}{n} \sum_{j=0}^{n-1} y_j \cos rj\alpha; \quad b_r = \frac{2}{n} \sum_{j=0}^{n-1} y_j \sin rj\alpha.$$

Similarly we can represent all periodic, absolutely integrable functions by a Fourier series:

(15.4.3)
$$f(x) = \frac{1}{2} a_0 + \sum_{r=1}^{\infty} (a_r \cos rx + b_r \sin rx).$$

In this case we get the coefficients from

(15.4.4)
$$\begin{cases} a_r = \frac{1}{\pi} \int_0^{2\pi} f(x) \cos rx \, dx; \\[2ex] b_r = \frac{1}{\pi} \int_0^{2\pi} f(x) \sin rx \, dx. \end{cases}$$

When a function is represented as a trigonometric series, this means that only certain discrete frequencies have been used. By passing to the integral form we actually use *all* frequencies. The principal point is Fourier's integral theorem, which states a complete reciprocity between the amplitude and frequency functions provided some rather mild conditions are satisfied (see (3.4.6)):

(15.4.5)
$$\begin{cases} F(t) = (2\pi)^{-1/2} \int_{-\infty}^{\infty} f(x) \exp(itx) \, dx \\[2ex] f(x) = (2\pi)^{-1/2} \int_{-\infty}^{\infty} F(t) \exp(-itx) \, dx \end{cases}$$

In practical work these integrals are evaluated numerically, and hence we must require that $f(x)$ and $F(t)$ vanish with a suitable strength at infinity. If $f(x)$ is known at a sufficient number of equidistant points, then $F(t)$ can be computed by numerical quadrature. In general, $F(t)$ will be complex, and conveniently we compute the absolute value instead.

If the computation is performed for a sufficient number of t-values, one can get a general idea of which frequencies are dominating in the data $f(x)$. Another method is autoanalysis. In this case one forms convolution integrals of the form $\int f(x)f(x-t) \, dx$; if there is a pronounced periodicity in the material, then for certain values of t the factors $f(x)$ and $f(x-t)$ will both be large, and hence the integral will also become large. It can easily be shown that if $f(x)$ contains oscillations with amplitude a_n, the convolution integral will contain the same oscillations but with amplitude a_n^2 instead.

---

§15.5. APPROXIMATION WITH EXPONENTIAL FUNCTIONS

This problem arises in several applications, e.g., in physics, chemistry, biology, and medicine, where we can have reactions with different time constants. Finding appropriate parameters from given numerical data is a difficult problem; this was demonstrated by Lanzcos, who showed that the three functions

$$f_1(x) = 0.0951 \cdot e^{-x} + 0.8607 \cdot e^{-3x} + 1.5576 \cdot e^{-5x},$$
$$f_2(x) = 0.305 \cdot e^{-1.58x} + 2.202 \cdot e^{-4.45x},$$
$$f_3(x) = 0.041 \cdot e^{-0.5x} + 0.79 \cdot e^{-2.73x} + 1.68 \cdot e^{-4.96x},$$

give the same results for $0 \leq x \leq 1.2$ to two places.

A reasonable method for solving the problem just mentioned is the following, originating with Prony. We assume that the function $y = f(x)$ is given in equidistant points with the coordinates $(x_0, y_0), (x_1, y_1), \ldots, (x_n, y_n)$, where $x_r = x_0 + rh$. We will approximate $f(x)$ by $a_1 e^{\lambda_1 x} + a_2 e^{\lambda_2 x} + \cdots + a_m e^{\lambda_m x}$. Putting $c_r = a_r e^{\lambda_r x_0}$ and $v_r = e^{h\lambda_r}$, we obtain for $r = 0, 1, 2, \ldots, m$ the equations:

(15.5.1)
$$\begin{cases} c_1 + c_2 + \cdots + c_m = y_0, \\ c_1 v_1 + c_2 v_2 + \cdots + c_m v_m = y_1, \\ \vdots \\ c_1 v_1^m + c_2 v_2^m + \cdots + c_m v_m^m = y_m. \end{cases}$$

Forming

(15.5.2)
$$(v - v_1)(v - v_2) \cdots (v - v_m) = v^m + s_1 v^{m-1} + \cdots + s_m$$
$$= \varphi(v),$$

multiplying in turn by $s_m, s_{m-1}, \ldots, s_1, s_0 = 1$, and adding, we obtain

$$\varphi(v_1)c_1 + \varphi(v_2)c_2 + \cdots + \varphi(v_m)c_m = s_m y_0 + s_{m-1} y_1 + \cdots + s_1 y_{m-1} + s_0 y_m = 0,$$

since $\varphi(v_r) = 0$, $r = 1, 2, \ldots, m$. Normally, m is, of course, substantially smaller than n, and further it is clear that we get a new equation if we shift the origin a distance h to the right. If this is repeated, we get the following system:

(15.5.3)
$$\begin{cases} y_{m-1} s_1 + y_{m-2} s_2 + \cdots + y_0 s_m = -y_m, \\ y_m s_1 + y_{m-1} s_2 + \cdots + y_1 s_m = -y_{m+1}, \\ \vdots \\ y_{n-1} s_1 + y_{n-2} s_2 + \cdots + y_{n-m} s_m = -y_n. \end{cases}$$

Thus we have $n - m + 1$ equations in m unknowns, and normally $n - m + 1 > m$; hence $s_1, s_2, \ldots, s_m$ can be determined by means of the least-squares method. Then we get $v_1, v_2, \ldots, v_m$ from (15.5.2) and $\lambda_r = \ln v_r / h$. Finally, we get $c_1, c_2, \ldots, c_m$ from the system above, and a_r from $a_r = c_r v_r^{-x_0/h}$.

---

§15.6. APPROXIMATION WITH CHEBYSHEV POLYNOMIALS

In our previous discussions we have used the least squares method, which means that the errors are measured in a Euclidean norm. However, in some applications it may be more natural to use a different norm, such as a maximum norm. This means that we have to look for approximations where the maximum deviation is minimized.

Previously we have treated the Chebyshev polynomials $T_n(x) = \cos(n \arccos x)$ as a family of orthogonal polynomials obeying the recursion rule $T_{n+1}(x) = 2xT_n(x) - T_{n-1}(x)$. However, there is another very important feature, which will form the basis for the application of Chebyshev polynomials in the theory of approximation.

Chebyshev discovered the following remarkable property of the polynomials $T_n(x)$. For $-1 \leq x \leq 1$, we consider all polynomials $p_n(x)$ of degree n and with the coefficient of x^n equal to 1 (monic polynomials). Putting $\alpha_n = \sup_{-1 \leq x \leq 1} |p_n(x)|$, we seek the polynomial $p_n(x)$ for which α_n is as small as possible. The desired polynomial is then $2^{-(n-1)}T_n(x)$, as will become clear from the following discussion.

First we observe that $T_n(x) = \cos(n \arccos x) = 0$, when

$$(15.6.1) \qquad x = \xi_r = \cos \frac{(2r+1)\pi}{2n}, \qquad r = 0, 1, 2, \ldots, n-1,$$

and that $T_n(x) = (-1)^r$ for $x = x_r = \cos(r\pi/n)$, $r = 0, 1, 2, \ldots, n$. Obviously $q_n(x) = 2^{-(n-1)}T_n(x)$ is a monic polynomial with the property $|q_n(x)| \leq 2^{-(n-1)}$. We now assume that there exists a monic polynomial $p_n(x)$ that is still smaller, i.e., $|p_n(x)| < 2^{-(n-1)}$ everywhere in the interval $-1 \leq x \leq 1$. Then we have

$$\begin{cases} 2^{-(n-1)}T_n(x_0) - p_n(x_0) > 0, \\ 2^{-(n-1)}T_n(x_1) - p_n(x_1) < 0, \\ \vdots \end{cases}$$

that is, the polynomial $f_{n-1}(x) = 2^{-(n-1)}T_n(x) - p_n(x)$ of degree $(n-1)$ would have an alternating sign in $(n+1)$ points $x_0, x_1, \ldots, x_n$. Hence $f_{n-1}(x)$ would have n roots in the interval which is possible only if $f_{n-1}(x) \equiv 0$, that is, $p_n(x) \equiv 2^{-(n-1)}T_n(x)$.

Suppose that we want to approximate a function $f(x)$ with a Chebyshev series:

$$(15.6.2) \qquad f(x) = \tfrac{1}{2}c_0 + c_1 T_1(x) + c_2 T_2(x) + \cdots + c_{n-1} T_{n-1}(x) + R_n(x).$$

If $j, k < n$ and at least one of them is different from zero, we have

$$(15.6.3) \qquad \sum_{r=0}^{n-1} T_j(\xi_r)T_k(\xi_r) = \frac{n}{2} \delta_{jk}.$$

The general formula is

$$\sum_{r=0}^{n-1} T_j(\xi_r)T_k(\xi_r) = \frac{n}{2}\left[(-1)^{(j+k)/2n}\, \delta(j+k, 2n) + (-1)^{(j-k)/2n}\, \delta(j-k, 2n)\right]$$

where as before

$$\delta(m, n) = \begin{cases} 1 & \text{if } m \text{ is divisible by } n, \\ 0 & \text{otherwise.} \end{cases}$$

The coefficients can be determined from the orthogonality relations; i.e., we get either

$$(15.6.4) \qquad c_k = \frac{2}{n}\sum_{r=0}^{n-1} f(\xi_r)T_k(\xi_r) = \frac{2}{n}\sum_{r=0}^{n-1} f(\xi_r) \cos \frac{(2r+1)k\pi}{2n}$$

or

$$(15.6.5) \qquad c_k = \frac{2}{\pi} \int_{-1}^{1} f(x) T_k(x) (1 - x^2)^{-1/2} \, dx = \frac{2}{\pi} \int_{0}^{\pi} f(\cos t) \cos(kt) \, dt.$$

The integrals are often difficult to compute, and then the first formula is preferred. It must be observed, however, that c_k depends upon n, which means that we cannot just add another term if the expansion has to be improved. With sufficiently fast convergence the remainder term is $R_n \simeq c_n T_n(x)$, and the error oscillates between $-c_n$ and $+c_n$.

Using the formulas in (4.6.8) we express the successive powers of x in Chebyshev polynomials, and in this way we express a given power series as a Chebyshev series instead. If this series is truncated after the term $c_{n-1} T_{n-1}(x)$, the error will behave essentially as $c_n T_n(x)$ and hence oscillate between c_n and $-c_n$. This truncated Chebyshev series can then be rearranged to an ordinary polynomial with the same nice properties. This procedure is called *economization* of power series expansions, and it is frequently used in applications.

If we expand e^{-x} in power series and then express the powers in Chebyshev polynomials, we find

$$e^{-x} \simeq 1.266065\,877752T_0$$
$$- 1.130318\,207984T_1 \quad + 0.271495\,339533T_2$$
$$- 0.044336\,849849T_3 \quad + 0.005474\,240442T_4$$
$$- 0.000542\,926312T_5 \quad + 0.000044\,977322T_6$$
$$- 0.000003\,198436T_7 \quad + 0.000000\,199212T_8$$
$$- 0.000000\,011037T_9 \quad + 0.000000\,000550T_{10}$$
$$- 0.000000\,000025T_{11} + 0.000000\,000001T_{12}.$$

This expansion is valid only for $-1 \leq x \leq 1$. If we truncate the series after the T_5-term and rearrange in powers of x, we obtain

$$e^{-x} \simeq 1.000045 - 1.000022x + 0.499199x^2$$
$$- 0.166488x^3 + 0.043794x^4 - 0.008687x^5.$$

The error is essentially equal to $0.000045T_6$, whose largest absolute value is 0.000045. If, instead, we truncate the usual power series after the x^5-term, we get

$$e^{-x} \simeq 1 - x + 0.5x^2 - 0.166667x^3 + 0.041667x^4 - 0.008333x^5,$$

where the error is essentially $x^6/720$; the maximum error is obtained for $x = -1$ and amounts to $1/720 + 1/5040 + \cdots = 0.001615$, that is, 36 times as large as in the former case. The error curves are represented in Fig. 15.2.* For small values of x the truncated power series is, of course, superior. On the other hand, the Chebyshev series has the power of distributing the error over the whole interval.

---

* For convenience, the latter curve is reversed.

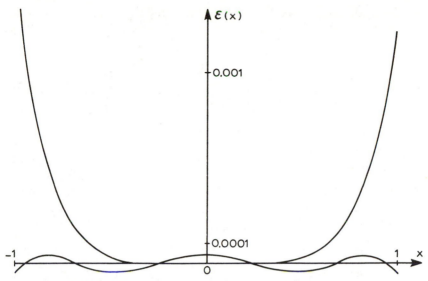

FIGURE 15.2. Approximation with Taylor and Chebyshev series.

We also give one explicit example of a direct expansion in Chebyshev series. Let $f(x) = \arcsin x$. Since $f(x)$ is an odd function, we get $c_{2k} = 0$. For the other coefficients we find

$$c_{2k+1} = \frac{2}{\pi} \int_{-1}^{+1} \arcsin x \cdot \cos((2k+1)\arccos x)(1-x^2)^{-1/2}\, dx.$$

Putting $x = \cos \varphi$, we obtain

$$c_{2k+1} = \frac{2}{\pi} \int_{0}^{\pi} \left(\frac{\pi}{2} - \varphi\right) \cdot \cos(2k+1)\varphi\, d\varphi$$

$$= \frac{4}{\pi} \frac{1}{(2k+1)^2}.$$

Thus

$$\arcsin x = \frac{4}{\pi}\left[T_1(x) + \frac{T_3(x)}{9} + \frac{T_5(x)}{25} + \cdots \right].$$

For $x = 1$ we have $T_1 = T_3 = T_5 = \cdots = 1$, and we get the formula

$$\frac{\pi^2}{8} = 1 + \frac{1}{9} + \frac{1}{25} + \cdots.$$

Expansion of other elementary functions in Chebyshev series gives rise to rather complicated integrals, and we will not treat this problem here.

When Chebyshev series are to be used on a computer, a recursive technique is desirable. Such a technique has been devised by Clenshaw for the computation of

$$f(x) = \sum_{k=0}^{n} c_k T_k(x),$$

where the coefficients c_k as well as the value x are assumed to be known. We form a sequence of numbers $a_n, a_{n-1}, \ldots, a_0$ by means of the relation

(15.6.6) $$a_k - 2xa_{k+1} + a_{k+2} = c_k,$$

where $a_{n+1} = a_{n+2} = 0$. Insertion in $f(x)$ gives

$$f(x) = \sum_{k=0}^{n-2} (a_k - 2xa_{k+1} + a_{k+2})T_k(x) + (a_{n-1} - 2xa_n)T_{n-1}(x) + a_n T_n(x)$$

$$= \sum_{k=0}^{n-2} (T_k - 2xT_{k+1} + T_{k+2})a_{k+2} + a_0 T_0 + a_1(T_1 - 2xT_0).$$

Since $T_{k+2} = 2xT_{k+1} - T_k$, we get

$$f(x) = a_0 T_0 + a_1(T_1 - 2xT_0)$$

or

(15.6.7) $$f(x) = a_0 - a_1 x.$$

§15.7. SPLINE APPROXIMATION

As demonstrated before, we can always find a polynomial of degree n satisfying the coordinates of $n + 1$ points under the assumption that all x-coordinates are different. However, if we want a smooth curve for practical purposes, this is in general a bad solution. If we abandon the condition that the curve must be defined by just one polynomial, we can instead put together several pieces of polynomials and under suitable assumptions get a much better solution. If we use an elastic rod (a "spline") and make it pass through the points (which we also call nodes), we will obtain a curve which is continuous and has continuous first and second derivatives, while the third derivative will in general be discontinuous. This suggests that we construct third degree polynomials which are then put together under the conditions just mentioned. Every third degree polynomial $y = ax^3 + bx^2 + cx + d$ is determined by four parameters. Now n curves pass through two points each, while $n - 1$ inner nodes give $2(n - 1)$ conditions for the first and second derivative. We can add two extra conditions, and we choose to make the second derivative vanish in the two extreme points at left and right. Together this will give $4n$ conditions, which is the right number and which will define a unique solution.

We shall now treat the case when the abscissas are equidistant so that $x_i = x_0 + ih$. Consider the interval $x_{i-1} \le x \le x_i$ and put $x = x_{i-1} + ph$ where $0 \le p \le 1$. Then we get

(15.7.1) $\quad S_i(x) = S_i(x_{i-1} + ph)$
$$= y_{i-1} + p(y_i - y_{i-1}) + hp(1 - p)[(k_{i-1} - d_i)(1 - p) - (k_i - d_i)p]$$

where

(15.7.2) $\qquad\qquad \begin{cases} k_i = S'_i(x_i) \\ d_i = (y_i - y_{i-1})/h. \end{cases}$

We will now verify that this formula for spline functions is in accordance with our requirements. We see that $p = 0$ gives $S_i(x_{i-1}) = y_{i-1}$ and $p = 1$ gives $S_i(x_i) = y_i$. Further

$$dS_i/dp = y_i - y_{i-1} + h(1 - 4p + 3p^2)(k_{i-1} - d_i) - h(2p - 3p^2)(k_i - d_i)$$

and hence

$$(dS_i/dp)_{p=0} = y_i - y_{i-1} + h(k_{i-1} - d_i) = hk_{i-1} = hy'_{i-1}(x);$$
$$(dS_i/dp)_{p=1} = y_i - y_{i-1} + h(k_i - d_i) = hk_i = hy'_i(x) = (dS_{i+1}/dp)_{p=0}.$$

Forming the second derivative as well, we find:

$$d^2S_i/dp^2 = h(6p - 4)(k_{i-1} - d_i) - h(2 - 6p)(k_i - d_i)$$

and hence:

$$(d^2S_i/dp^2)_{p=0} = 2h(3d_i - 2k_{i-1} - k_i)$$
$$(d^2S_i/dp^2)_{p=1} = 2h(-3d_i + k_{i-1} + 2k_i)$$
$$(d^2S_{i+1}/dp^2)_{p=0} = 2h(3d_{i+1} - 2k_i - k_{i+1}).$$

According to our rules the last two expressions must be equal for $i = 1(1)n - 1$, and hence:

(15.7.3) $\qquad\qquad k_{i-1} + 4k_i + k_{i+1} = 3(d_i + d_{i+1}).$

To this we add the two conditions:

(15.7.4) $\quad \begin{cases} (d^2S_1/dp^2)_{p=0} = 0 \\ (d^2S_n/dp^2)_{p=1} = 0 \end{cases}$ giving $\quad \begin{aligned} & 2k_0 + k_1 = 3d_1, \quad \text{and} \\ & k_{n-1} + 2k_n = 3d_n. \end{aligned}$

With $n = 4$ the full system for determination of k_i, $i = 0(1)4$, can be written:

$$\begin{pmatrix} 2 & 1 & 0 & 0 & 0 \\ 1 & 4 & 1 & 0 & 0 \\ 0 & 1 & 4 & 1 & 0 \\ 0 & 0 & 1 & 4 & 1 \\ 0 & 0 & 0 & 1 & 2 \end{pmatrix} \begin{pmatrix} k_0 \\ k_1 \\ k_2 \\ k_3 \\ k_4 \end{pmatrix} = 3 \begin{pmatrix} d_1 \\ d_1 + d_2 \\ d_2 + d_3 \\ d_3 + d_4 \\ d_4 \end{pmatrix}.$$

We now compute the integral of the spline function:

$$\int_{x_{i-1}}^{x_i} S_i(x)\, dx = h \int_0^1 \left[y_{i-1} + (y_i - y_{i-1})p + h(k_{i-1} - d_i)p(1-p)^2 \right.$$
$$\left. - h(k_i - d_i)p^2(1-p) \right] dp$$
$$= \frac{h}{2}(y_{i-1} + y_i) + (h^2/12)(k_{i-1} - k_i).$$

Summing these relations for $i = 1, 2, \ldots, n$ and denoting the union of all $S_i(x)$ by $S(x)$, we get:

$$\int_{x_0}^{x_n} S(x)\, dx = h(\tfrac{1}{2}y_0 + y_1 + y_2 + \cdots + y_{n-1} + \tfrac{1}{2}y_n) + (h^2/12)(y_0' - y_n').$$

This is the trapezoid integral formula with end correction (cf. (17.2.5)). As can be seen from Euler–McLaurin's summation formula which will be treated later, the error on integrating the spline function $S(x)$ instead of the true function $y(x)$ is $O(h^4)$.

---

§15.8.   RATIONAL APPROXIMATIONS

In order to investigate more complex approximations, we formulate the following problem. Given a function $f(x)$ and an interval (a, b), construct polynomials $P_m(x)$ and $Q_n(x)$ such that the quotient $P_m(x)/Q_n(x)$ becomes a good approximation of $f(x)$ in the given interval measured in some prescribed norm.

There are of course many ways to specialize this: how to choose m and n, how to define a suitable norm, and how to take various extra conditions into account, e.g., that f may be even or odd. We will start by discussing one case which is in fact analogous to the Taylor expansion for polynomials.

Padé Approximations

For given $m, n \geq 0$ we want to find polynomials $P_m(x)$ and $Q_n(x)$ such that $|f(x) - P_m(x)/Q_n(x)| \leq M|x^v|$ with $|x| \leq b$, and where v should be as large as possible. It turns out that typically v can be chosen equal to $m + n + 1$. If $f(x)$ is even, for example, this fact must be taken into account. We demonstrate the technique in a simple example by approximating the function $f(x) = \cos x$.

Putting

$$\cos x = 1 - \frac{x^2}{2} + \frac{x^4}{24} - \frac{x^6}{720} + \frac{x^8}{40320} - \cdots \simeq \frac{1 + a_1 x^2 + a_2 x^4}{1 + b_1 x^2 + b_2 x^4}$$

and comparing coefficients after multiplication with the denominator, we get

$$b_1 - \frac{1}{2} = a_1; \qquad b_2 - \frac{b_1}{2} + \frac{1}{24} = a_2$$

$$\frac{b_1}{24} - \frac{b_2}{2} = \frac{1}{720}; \qquad -\frac{b_1}{720} + \frac{b_2}{24} = -\frac{1}{40320}$$

giving

$$a_1 = -\frac{115}{252}, \qquad a_2 = \frac{313}{15120}, \qquad b_1 = \frac{11}{252}, \qquad \text{and} \qquad b_2 = \frac{13}{15120}.$$

Hence

$$\cos x \simeq \frac{15120 - 6900x^2 + 313x^4}{15120 + 660x^2 + 13x^4}.$$

The numerator will vanish for $x = 1.5708259$ as compared with $\pi/2 = 1.5707963$.

Rational Approximations in General

A special method is offered by continued fractions, and in many cases it is possible to deduce explicit expansions also for quite complicated functions. However, we do not intend to treat this subject in more detail, and instead we refer to some typical examples given in Section 3.6.

For rational approximations in general it is natural to use a Chebyshev norm. This means that we try to minimize the expression

$$\max |f(x) - P(x)/Q(x)|$$

taken over the given interval $a \le x \le b$. Nearest at hand are the following two representations:

$$(15.8.1) \qquad f(x) \simeq \frac{a_0 + a_1 x + \cdots + a_m x^m}{b_0 + b_1 x + \cdots + b_m x^m},$$

$$(15.8.2) \qquad f(x) \simeq \frac{\alpha_0 + \alpha_1 x^2 + \cdots + \alpha_m x^{2m}}{\beta_0 + \beta_1 x^2 + \cdots + \beta_m x^{2m}}.$$

The coefficients are determined by a semiempirical but systematic trial-and-error technique. The main idea is to adjust the zeros of the function $f(x) - P(x)/Q(x)$ until all maximal deviations are numerically equal.

Without specializing we can put $b_0 = \beta_0 = 1$. The second formula should be used if $f(x)$ is an even or an odd function; in the latter case it should first be multiplied by a suitable (positive or negative) odd power of x. It is interesting to compare the errors of the different types of approximations; naturally the type depends on the function to be approximated. However, an overall qualitative rule is the following. Let A, B, C, and D stand for power series, Chebyshev polynomial, continued fraction, and rational function, respectively, and let the signs $>$ and $\gg$ stand for "better" and "much better." Then we have, on comparing approximations with the same number of parameters,

$$D > C \gg B > A.$$

There are exceptions to this general rule, but the relation seems to be rather typical.

Last, we give a few numerical examples of rational approximation, all of them constructed in the Chebyshev sense; ε is the maximum error, and if nothing else is mentioned, the interval is $0 \leq x \leq 1$.

$$e^{-x} \simeq \frac{1.00000\,00007 - 0.47593\,58618x + 0.08849\,21370x^2 - 0.00656\,58101x^3}{1 + 0.52406\,42207x + 0.11255\,48636x^2 + 0.01063\,37905x^3},$$

$$\varepsilon = 7.34 \cdot 10^{-10}.$$

$$\ln z = \frac{-0.69314\,71773 + 0.06774\,12133x + 0.52975\,01385x^2 + 0.09565\,58162x^3}{1 + 1.34496\,44663x + 0.45477\,29177x^2 + 0.02868\,18192x^3},$$

$$x = 2z - 1; \qquad \tfrac{1}{2} \leq z \leq 1; \qquad \varepsilon = 3.29 \cdot 10^{-9}.$$

$$\frac{\arctan x}{x} \simeq \frac{0.99999\,99992 + 1.13037\,54276x^2 + 0.28700\,44785x^4 + 0.00894\,72229x^6}{1 + 1.46370\,86496x^2 + 0.57490\,98994x^4 + 0.05067\,70959x^6},$$

$$\varepsilon = 7.80 \cdot 10^{-10}.$$

$$\frac{\sin x}{x} \simeq \frac{1 - 0.13356\,39326x^2 + 0.00328\,11761x^4}{1 + 0.03310\,27317x^2 + 0.00046\,49838x^4},$$

$$\varepsilon = 4.67 \cdot 10^{-11}.$$

$$\cos x \simeq \frac{0.99999\,99992 - 0.45589\,22221x^2 + 0.02051\,21130x^4}{1 + 0.04410\,77396x^2 + 0.00089\,96261x^4},$$

$$\varepsilon = 7.55 \cdot 10^{-10}.$$

The errors were computed before the coefficients were rounded to ten places.

EXERCISES

1. The points (2, 2), (5, 4), (6, 6), (9, 9), and (11, 10) should be approximated by a straight line. Perform this, assuming
   (a) that the errors in the x-values can be neglected;
   (b) that the errors in the x- and y-values are of the same order of magnitude.

2. During the time December 1 to 28, 1981, the sun was setting between 15:30 and 15:40 (in Lund, Sweden; latitude 55°42′). The time was given in whole minutes and appears in the table below where x is the day ($1 \leq x \leq 28$), and y the number of minutes after 15:30.

x	y	x	y
1	8	19–21	3
2	7	22–23	4
3	6	24	5
4–5	5	25	6
6–7	4	26–27	7
8–9	3	28	8
10–18	2		

These data can be represented with good accuracy in the form $y = a + bx + cx^2$. Perform this computation using the least squares method. For which x (integer!) will y attain its minimum? Also state the corresponding sunset time in hours, minutes, and seconds. Auxiliary formulas:

$$\sum_{k=1}^{n} k^2 = n(n+1)(2n+1)/6; \qquad \sum_{k=1}^{n} k^3 = n^2(n+1)^2/4;$$

$$\sum_{k=1}^{n} k^4 = n(n+1)(2n+1)(3n^2 + 3n - 1)/30.$$

3. Approximate e^{-x} for $0 \leq x \leq 1$ by $a - bx$ in the Chebyshev sense.

4. For small values of x, we want to approximate the function $y = e^{2x}$, with

$$f(x) = \frac{1 + ax + bx^2 + cx^3}{1 - ax + bx^2 - cx^3},$$

Find a, b, and c.

5. The intensity of radiation I from a radioactive source varies with time t according to the formula $I = I_0 e^{-qt}$. Through measurements the following values have been obtained:

t	1	2	3	4	5	6
I	6.32	4.76	3.51	2.67	2.01	1.48

Find I_0 and q by the least squares method.

6. For $0 \leq x \leq \pi/2$, the function $\cos x$ is to be approximated by a second degree polynomial $y = ax^2 + bx + c$ such that the integral $\int_0^{\pi/2} (y - \cos x)^2 \, dx$ is minimized. Find a, b, and c.

7. The function $y(t)$ is a solution of the differential equation $y' + ay = 0$, $y(0) = b$. Using the method of least squares, find a and b if the following values are known:

t	0.1	0.2	0.3	0.4	0.5
$y(t)$	80.4	53.9	36.1	24.2	16.2

8. y is a function of x, given by the data below. It should be represented in the form $Ae^{-ax} + Be^{-bx}$. Determine the constants a, b, A, and B.

x	0.4	0.5	0.6	0.7	0.8	0.9	1.0	1.1
y	2.31604	2.02877	1.78030	1.56513	1.37854	1.21651	1.07561	0.95289

9. The function y of x given by the data below should be represented in the form $y = e^{-ax} \sin bx$. Find a and b.

x	0.2	0.4	0.6	0.8	1.0	1.2	1.4	1.6
y	0.15398	0.18417	0.16156	0.12301	0.08551	0.05537	0.03362	0.01909

10. The function e^{-x} should be approximated by $ax + b$ so that $|e^{-x} - ax - b| \leq 0.005$. The constants a and b are to be chosen so that this accuracy is attained for an interval $[0, c]$ as large as possible. Find a, b, and c to three decimals.

11. The function $y = 1/x$ should be approximated in the interval $1 \leq x \leq 2$ by two different straight lines. One line is used in $1 \leq x \leq a$ and another in

$a \le x \le 2$. Each line intersects with the curve in two points, and the two lines intersect in $x = a$. All five maximum deviations are numerically equal. Find a and the maximum deviation.

12. The parabola $y = f(x) = x^2$ is to be approximated by a straight line $y = g(x)$ in the Chebyshev sense in a certain interval I so that the deviations in the two endpoints and the maximum deviations in I all are numerically $y = \varepsilon$. Compute B/A when

$$A = \int_I (g(x) - f(x)) \, dx \qquad \text{and} \qquad B = \int_I |g(x) - f(x)| \, dx.$$

13. Determine the best solution of the equations

$$\begin{cases} 2x = 5 \\ 3x = 11 \\ 6x = 19 \end{cases}$$

using the maximum norm.

14. Approximate $x/(e^x - 1)$ by a polynomial of lowest possible degree so that the absolute value of the error is $< 5 \times 10^{-4}$ when $x \in [0, 1]$.

15. The function F is defined through

$$F(x) = \int_0^x \exp\left(-\frac{t^2}{2}\right) dt.$$

Determine a polynomial $P(x)$ of lowest possible degree so that $|F(x) - P(x)| \le 10^{-4}$ for $|x| \le 1$.

CHAPTER 16

The Monte Carlo Method

Im ächten Manne ist ein Kind versteckt; das will
spielen.

F. Nietzsche

The Monte Carlo method is an artificial sampling method that can be used in different ways. One possibility is to create a mathematical model of a certain problem and then devise a game that will simulate the model. But it is also possible that the problem itself is of a stochastic nature and hence can be treated directly by similar methods. As an example of the first kind we mention computation of an integral in many dimensions, and of the second kind scattering of a neutron against different atomic nuclei in a certain region. We can also consider traffic problems, combat models, economic models, and queuing systems. In any case, elements of randomness must be introduced, and then after a large number of trials a statistical analysis can be performed according to standard rules. Advantages of the method are that even quite complicated problems can be treated rather easily and parameters may be varied without difficulty. On the other hand, one must be prepared to accept low accuracy. As a rule, if one wants to improve the results by a factor p, then the number of trials must be increased by a factor p^2. A warning seems to be appropriate that the method might tempt one to neglect searching for analytical solutions whenever such solutions are not quite obvious.

§16.1. RANDOM NUMBERS

Random numbers play an important role in applications of the Monte Carlo method. We do not intend to go into strict definitions since this would require a rather elaborate description.

First we remind of a few fundamental concepts from the theory of probability. For an arbitrary value x the *distribution function* $F(x)$ associated with a stochastic variable X gives the probability that $X \leq x$, that is

(16.1.1)
$$F(x) = P(X \leq x).$$

A distribution function is always nondecreasing, and further $\lim_{x \to -\infty} F(x) = 0$, $\lim_{z \to +\infty} F(x) = 1$.

We define a series of random numbers from the distribution $F(x)$ as a sequence of independent observations of X. In a long series of this kind the relative amount of numbers $\leq x$ is approximately equal to $F(x)$. If, for example, the distribution is such that $F(0) = 0.2$, then in a long series of random numbers approximately 20% of the numbers should be negative or zero while the rest should be positive.

The distribution is said to be of *continuous* type if we can represent $F(x)$ in the form

(16.1.2)
$$F(x) = \int_{-\infty}^{x} f(t) \, dt$$

where $f(t)$ is continuous, except possibly in a finite number of points. The function $f(x) = F'(x)$ is called the *frequency function* and is always ≥ 0. A long series of random numbers from the corresponding distribution will have the property that the relative amount of random numbers in a small interval $(x, x + h)$ is approximately equal to $hf(x)$.

In the following we are mostly going to deal with distributions of continuous type. In this connection the uniform distribution is of special importance; it is defined by

$$f(x) = \begin{cases} 0, & x < 0, \\ 1, & 0 < x < 1, \\ 0, & x > 1. \end{cases}$$

The values for $x = 0$ and $x = 1$ can be put equal to 0 or 1 depending on the circumstances. The corresponding distribution function becomes

$$F(x) = \begin{cases} 0, & x < 0, \\ x, & 0 \leq x \leq 1, \\ 1, & x > 1. \end{cases}$$

The reason that the uniform distribution plays such an important role for the Monte Carlo method is that random numbers from other distributions can be constructed from random numbers of this simple type. If y is a random number with uniform distribution, we get a random number from the distribution $F(x)$ by solving for x from the equation

(16.1.3)
$$F(x) = y.$$

This relation means that the two numbers will cut off the same amount of the area between the frequency functions and the abscissa axes. Later on we shall provide an example of this.

We shall now briefly discuss how to construct random numbers with uniform distribution in $(0, 1)$. First we could think of some physical device, e.g., in the form of binary counters controlled by a radioactive source or by some kind of electromagnetic noise. If we lock the counters after suitable time-intervals, we can use the

contents as random numbers in binary form, provided the time-intervals are of sufficient length to guarantee randomness. However, this procedure has some serious
drawbacks. First, the production rate will not be adequate compared with the speed
of modern computers. Second, a computation using a sequence of random numbers
produced in this way cannot be repeated exactly since the numbers cannot be re-
constructed if they are not stored. For these reasons one prefers to use what we call
pseudo-random numbers formed by mathematical methods. Several procedures have
been discussed for this production, but we only mention a congruential method of
the form

$$x_{n+1} \equiv ax_n + b \bmod. P$$

where x_n, a, b, and P are integers. Here $u \equiv v$ mod. w means that $u - v$ is divisible
by w. With proper choice of the integers x_0, a, b, and P the numbers $r_n = x_n/P$ have
a uniform distribution in $(0, 1)$. It has been shown that with $P = 2^p$ the length of
the period will be 2^p provided b is odd and a is of the form $4k + 1$. Good results
have been obtained with $P = 2^{35}$, $a = 2^A + B$ where $15 \le A \le 24$, $B = 1, 5$, or 9, and
$b = 1$. However, it turns out that the digits in pseudo-random numbers generated in
this way cannot in general be considered as independent. To achieve this we must
use more elaborate methods.

   We will now discuss the problem of constructing random numbers from a few
other distributions a little more in detail.

The Normal Distribution

   First we investigate how random numbers with normal distribution can be obtained by transforming random numbers with uniform distribution in $(0, 1)$. For a
given value of s ($0 \le s \le 1$) we have to solve the following equation for x:

$$(16.1.4) \qquad (2\pi)^{-1/2} \int_{-\infty}^{x} \exp(-t^2/2)\, dt = s.$$

Writing this as $F(x) = s$, we have $F'(x) = f(x) = (2\pi)^{-1/2} \exp(-x^2/2)$; both F and f
have been tabulated extensively and could be stored in a computer with intervals
such as 0.01 [since $f(-x) = f(x)$ and $F(-x) = 1 - F(x)$, we need only consider positive values of x]. Taking a value x_0 that makes $F(x)$ as close to s as possible, we then
perform just one Newton–Raphson iteration and obtain:

$$(16.1.5) \qquad x_1 = x_0 - [F(x_0) - s]/f(x_0).$$

The result x_1 is then a good approximation of a random number from a normal
distribution with mean 0 and variance 1.

   Example 1.   Take $s = 0.6$. Then $x_0 = 0.25$ gives $F = 0.5987$ and $f = 0.3867$. Hence
$x_1 = 0.25 + 0.0013/0.3867 = 0.25336$. In fact, $F(0.25336) = 0.600005$.

   However, there has been some criticism against this method. First we choose a
number at random between 0 and 1, and then we perform a series of complicated
mathematical operations on this number. In fact, there are other, more direct ways.
One of the simplest is the following, which we quote without proof. Choose n ($\gtrsim 10$)

uniform random numbers $x_1, x_2, \ldots, x_n$ and form

(16.1.6)
$$z = \sum_{k=1}^{n} x_k - n/2.$$

Then z is asymptotically normal with mean 0 and variance $n/12$. In particular, with $n = 12$ we get the result that $z = x_1 + x_2 + \cdots + x_{12} - 6$ is asymptotically normal with mean 0 and variance 1.

A simple and elegant method is the following. Take two random numbers x and y from a uniform distribution and form:

(16.1.7)
$$\begin{cases} u = (-2 \ln x)^{1/2} \cos 2\pi y \\ v = (-2 \ln x)^{1/2} \sin 2\pi y. \end{cases}$$

Then u and v are normal random numbers with mean 0 and variance 1.

Next we assume that we want random numbers θ in the interval $(-1, 1)$ with the frequency function $f(\theta) = (1/\pi)(1 - \theta^2)^{-1/2}$. Then the distribution function is $F(\theta) = (1/\pi)(\arcsin \theta + \pi/2)$, where $\arcsin \theta$ is determined in such a way that $-\pi/2 \leq \arcsin \theta \leq \pi/2$. Hence the number θ can be determined from a random number ξ with uniform distribution in $(0, 1)$ through the equation $(1/\pi)(\arcsin \theta + \pi/2) = \xi$, that is, $\theta = -\cos \pi\xi = \cos \pi(1 - \xi)$. The same thing can be attained if we choose *two* numbers x and y from a uniform distribution. Then we form $X = 2x - 1$ and $Y = 2y - 1$ and test whether $X^2 + Y^2 < 1$; otherwise the numbers are rejected. Thus the point (X, Y) will lie inside a circle with radius 1. Then $z = X/(X^2 + Y^2)^{1/2}$ is a number from the desired distribution. Alternatively, we see that $(X^2 - Y^2)/(X^2 + Y^2)$ has the same distribution; the latter formulation should be preferred, since no square roots are present.

The Exponential Distribution

In this case we have $F(x) = 1 - e^{-\lambda x}$, and hence we have to solve the equation $F(x) = s$ where s is random with uniform distribution. However, since the same is true also for $1 - s$, we can just as well solve the equation $e^{-\lambda x} = s$, giving $x = -\lambda^{-1} \ln s$.

There are also more direct methods, for example a discrimination method devised by von Neumann. However, it is rather inefficient since it needs about six random numbers with uniform distribution to produce one number with exponential distribution.

The Poisson Distribution

Finally, we will consider a discrete frequency function that defines the Poisson distribution:
$$f(x) = m^x e^{-m}/x!; \qquad x = 0, 1, 2, \ldots; \quad m > 0.$$

We see that

$$F(x) = \sum_{t=0}^{x} e^{-m} m^t/t! \qquad \text{with } F(\infty) = e^{-m} \sum_{t=0}^{\infty} m^t/t! = 1$$

as it should be. Further, the mean value is obtained from

$$\sum_{x=0}^{\infty} xf(x) = e^{-m} \sum_{x=1}^{\infty} m^x/(x-1)! = me^{-m} \sum_{x=1}^{\infty} m^{x-1}/(x-1)! = m.$$

An integer N is constructed by taking a series of uniform random numbers u_0, $u_1, u_2, u_3, \ldots$ such that

(16.1.8) $$u_0 u_1 u_2 \cdots u_N < e^{-m}$$

where N is the smallest number making this inequality valid. It can be proved that N is a random number of Poisson type with mean value m.

§16.2. RANDOM WALKS

*"Would you tell me, please, which way I ought to
walk from here?"*
*"That depends a good deal on where you want to
go," said the Cat.*
"I don't much care where—" said Alice.
*"Then it doesn't matter which way you walk,"
said the Cat.*

Lewis Carroll

As a simple example of a random walk, we assume that a particle can move between the grid-points $x = \ldots, -2, -1, 0, 1, 2, \ldots$ on a given line. Every moment the particle moves one step right or one step left, with probability $\frac{1}{2}$ for each direction. However, it is a bit more interesting to consider a particle that can move on a square grid.

In the following example we shall assume that the probabilities for a movement to the four closest points are all equal to $\frac{1}{4}$. We denote by $P(x, y, t)$ the probability that the particle after t moves is at the point (x, y). Then we have

(16.2.1) $$P(x, y, t+1) = \tfrac{1}{4}[P(x-1, y, t) + P(x, y-1, t) \\ + P(x+1, y, t) + P(x, y+1, t)].$$

This equation can be rewritten

$$P(x, y, t+1) - P(x, y, t) = \tfrac{1}{4}[P(x+1, y, t) - 2P(x, y, t) + P(x-1, y, t) \\ + P(x, y+1, t) - 2P(x, y, t) + P(x, y-1, t)].$$

This difference equation evidently approximates the two-dimensional heat equation (cf. 20.3.1)

$$\frac{\partial \Gamma}{\partial t} = C\left(\frac{\partial^2 \Gamma}{\partial x^2} + \frac{\partial^2 \Gamma}{\partial y^2}\right).$$

Now we suppose that a random-walk process is performed in a limited domain and in such a way that the particle is absorbed when it reaches the boundary. At the same time, a certain profit V is paid out; its amount depends on the boundary point at which the particle is absorbed. Boundary points will be denoted by (x_r, y_r) and interior points by (x, y). At each interior point we have a certain probability $P(x, y, x_r, y_r)$ that the particle which starts at the point (x, y) will be absorbed at the boundary point (x_r, y_r). The expected profit is obviously

$$(16.2.2) \qquad u(x, y) = \sum_r P(x, y, x_r, y_r) V(x_r, y_r).$$

We find immediately that $u(x, y)$ satisfies the difference equation

$$(16.2.3) \quad u(x, y) = \tfrac{1}{4}[u(x + 1, y) + u(x - 1, y) + u(x, y + 1) + u(x, y - 1)],$$

with $u(x_r, y_r) = V(x_r, y_r)$. This equation is a well-known approximation of the Laplace equation (cf. 20.4.1)

$$\frac{\partial^2 u}{\partial x^2} + \frac{\partial^2 u}{\partial y^2} = 0.$$

In a simple example we will show how this equation can be solved by the random-walk technique. As our domain we choose a square with the corners $(0, 0)$, $(4, 0)$, $(4, 4)$, and $(0, 4)$, including integer grid points. As boundary values we choose $u = 0$ for $x = 0$; $u = x/(x^2 + 16)$ for $y = 4$; $u = 4/(16 + y^2)$ for $x = 4$; and $u = 1/x$ for $y = 0$. We shall restrict ourselves to considering the interior point $(1, 3)$. From this point we can reach the point $(1, 4)$ in one step with probability $\tfrac{1}{4}$. But we can also reach it in three steps in two different ways with total probability $2/4^3 = 1/32$, and in five steps in ten different ways with total probability $10/4^5 = 5/512$, and so on. Adding all these probabilities, we obtain about 0.30. In a similar way the other probabilities have been computed (see Fig. 16.1).

Insertion in (16.2.2) gives $u(1, 3) = 0.098$, in good agreement with the value 0.1 obtained from the exact solution $u = x/(x^2 + y^2)$.

0.30 0.09 0.03					$\frac{1}{17}$	$\frac{1}{10}$	$\frac{3}{25}$		

0.30 .   •Q .   .   . 0.03   0 .   . $\frac{4}{25}$

0.09 .   .   .   . 0.03   0 .   . $\frac{4}{20}$

0.03 .   .   .   . 0.02   0 .   . $\frac{4}{17}$

. . . . $\frac{4}{17}$

0.03 0.03 0.02   1   2   3   .

Probabilities   Boundary values

FIGURE 16.1

Clearly, the method can easily be programmed for a computer, and then the probabilities are obtained by the random-walk technique. It is also evident that the method can hardly compete with direct methods if such can be applied.

§16.3. COMPUTATION OF DEFINITE INTEGRALS

Computation of multidimensional integrals is an extremely complicated problem, and an acceptable conventional technique still does not exist. Here the Monte Carlo method is well suited, even if the results are far from precise. For simplicity we demonstrate the technique on one-dimensional integrals and consider

$$(16.3.1) \qquad I = \int_0^1 f(x)\,dx; \qquad 0 \le f(x) \le 1.$$

We choose N number pairs (x_i, y_i) with uniform distribution and define z_i through

$$z_i = \begin{cases} 0 & \text{if} \quad y_i > f(x_i), \\ 1 & \text{if} \quad y_i \le f(x_i). \end{cases}$$

Putting $n = \sum_i z_i$, we have $n/N \simeq I$. Somewhat more precisely, we can write

$$(16.3.2) \qquad I = \frac{n}{N} + O(N^{-1/2}).$$

Obviously, the accuracy is poor; with 100 pairs we get a precision of the order of $\pm 5\%$, and the traditional formulas, for example, Simpson's formula, are much better. In many dimensions we still have errors of this order of magnitude, and since systematic formulas (insofar as such formulas exist in higher dimensions) are extremely difficult to manage, it might well be the case that the Monte Carlo method compares favorably, at least if the number of dimensions is ≥ 6.

It is even more natural to consider the integral as the mean value of $f(\xi)$, where ξ is uniform, and then estimate the mean value from

$$(16.3.3) \qquad I \simeq \frac{1}{N} \sum_{i=1}^N f(\xi_i).$$

This formula can easily be generalized to higher dimensions.

Equation (16.3.1) can be rewritten in the following way:

$$(16.3.4) \qquad I = \int_0^1 \frac{f(x)}{g(x)} g(x)\,dx,$$

and hence I can be interpreted as the mean value of $f(\xi)/g(\xi)$, where ξ is a random number with the frequency $g(\xi)$. Thus

$$(16.3.5) \qquad I \simeq \frac{1}{N} \sum_{i=1}^N \frac{f(\xi_i)}{g(\xi_i)}.$$

where ξ_i is a random number with the frequency $g(\xi)$. Conveniently, the function $g(\xi)$ is chosen in such a way that it does not deviate too much from the function to be integrated.

§16.4. SIMULATION

The main idea in simulation is to construct a stochastic model of the real events, and then, by aid of random numbers or random walk, play a large number of games. The results are then analyzed statistically, as usual. This technique can be used for widely different purposes, for example, in social sciences, in physics, in operational analysis, and in combat problems, queuing problems, industrial problems, and so on. An example which is often quoted is the case when neutrons pass through a metallic plate. The energy and direction distributions are known, and these quantities are chosen accordingly. Further, the probabilities for reflection and absorption are known, and hence we can trace the history of each neutron. When a sufficient number of events has been obtained, the whole process can be analyzed in detail.

Many phenomena that one wants to study in this way occur with very low frequency. The probability that the neutron passes through the plate might not be more than 10^{-6}, and naturally we need a very large number of trials to get enough data. In situations like this, some kind of "importance sampling" is often used. This means that intentionally one directs the process toward the interesting but rare cases and then compensates for this by giving each trial correspondingly less weight. Exactly the same technique was used in the preceding section, where random numbers were chosen with a frequency function agreeing as far as possible with the function to be integrated. Under certain circumstances when it is difficult to state exactly whether a case is interesting, one can with probability $\frac{1}{2}$ stop a trial that seems to be unsuccessful; if it does continue, its weight is doubled. In the same way, trials that seem to be promising are split into two whose weights are halved at the same time.

The tricks that have been sketched here are of great practical importance, and in many cases they can imply great savings of computing time and, furthermore, make the use of the Monte Carlo method possible.

Apart from these indications, we cannot go into detail about the different applications. Instead we will content ourselves with a simple though important feature. Suppose that we have a choice among N different events with the probabilities $P_1, P_2, \ldots, P_N$, where $\sum_{r=1}^{N} P_r = 1$. We form so-called accumulated probabilities Q_r according to

$$(16.4.1) \qquad Q_r = \sum_{i=1}^{r} P_i \quad \text{with} \quad Q_0 = 0.$$

Choosing a uniform random number ξ, we seek a value of r such that

$$(16.4.2) \qquad Q_{r-1} < \xi < Q_r.$$

Then the event number r should be chosen. We can interpret the quantities P_r as distances with total length 1, and if the pieces are placed one after another, the point ξ will fall on the corresponding piece with probability P_r.

EXERCISES

1. Three uniform random numbers in $(0, 1)$ are ordered so that $x_k \geq y_k \geq z_k$. Using a computer, form 100 such triples and compute the mean values. Also compare with the theoretical values, which should also be computed.

2. Two uniform random numbers x_1 and x_2 in $(0, 1)$ are chosen. Then we form y_1, y_2, and y_3 through

$$y_1 = 1 - \sqrt{x_1}; \qquad y_2 = (1 - x_2)\sqrt{x_1}; \qquad y_3 = x_2\sqrt{x_1}.$$

   Finally, these numbers are sorted in decreasing order and renamed as $z_1 > z_2 > z_3$.
   (a) Find the mean values of y_1, y_2, and y_3.
   (b) Find the mean values of z_1, z_2, and z_3. (Hint: Each triple y_1, y_2, and y_3 as well as z_1, z_2, and z_3 can be represented as a point inside the triangle $x > 0$, $y > 0$, $z > 0$ in the plane $x + y + z = 1$.)
   (c) Using a computer, simulate the procedure above and compare with the theoretical results.

3. Using a computer, form 50 numbers with Poisson distribution and mean $m = 5$. Apply formula (16.1.8).

4. Construct a normal random number with mean value m and standard deviation σ from a normal random number ξ with mean value 0 and standard deviation 1 [frequency functions $\sigma^{-1}(2\pi)^{-1/2} \exp\{-(x - m)^2/2\sigma^2\}$ and $(2\pi)^{-1/2} \exp(-x^2/2)$].

5. The following method has been suggested for obtaining pseudo-random numbers. Starting from a value x_1 such that $0 < x_1 < 1$ and $x_1 \neq k/4$, $k = 0(1)4$, we form a sequence $x_1, x_2, x_3, \ldots$, where $x_{n+1} = 4x_n(1 - x_n)$. Using this method, compute

$$\int_0^1 \frac{dx}{1 + x} \simeq \frac{1}{10} \sum_{i=1}^{10} \frac{1}{1 + x_i}$$

with $x_1 = 0.120$.

Part Four

INTEGRATION and SUMMATION

CHAPTER **17**

Numerical Integration

There was an old fellow of Trinity
who solved the square root of infinity
but it gave him such fidgets
to count up the digits
that he chucked math and took up divinity.

§17.1. GENERAL RULES

Computation of integrals is an important problem in mathematics as well as in different applications. We shall here give an exposition on this subject and develop a spectrum of methods useful in different cases. If the integrand is not too complicated it may be possible to perform the integration explicitly, either directly or by partial integration or by substitution. Sometimes a recursion formula may be obtained, while in other cases series expansion may be feasible. Finally, in special situations, preferably when the integration interval is infinite, complex integration and residue calculus may lead to the goal. It is not until other possibilities have been exhausted that we should turn to purely numerical methods. It is then of great importance that the integrand is sufficiently regular. A singularity cannot be accepted, except possibly in the weight function, and even singularities in the first or second derivative may be disturbing. The same is true for singularities in the vicinity of, but outside, the interval. Below we will show in some examples how to get around this difficulty.

First we consider $I = \int_0^1 x^{-1/2} \cos x \, dx$. The integrand is singular in the origin, but the singularity is of the form x^{-p} with $p = \frac{1}{2}$, and hence the integral has a finite value. There are several possible ways to compute this value.

(a) Direct series expansion gives

$$
\begin{aligned}
I &= \int_0^1 dx \left\{ x^{-1/2} - \frac{x^{3/2}}{2!} + \frac{x^{7/2}}{4!} - \frac{x^{11/2}}{6!} + \cdots \right\} \\
&= \left[2x^{1/2} - \frac{x^{5/2}}{2! \, 5/2} + \frac{x^{9/2}}{4! \, 9/2} - \frac{x^{13/2}}{6! \, 13/2} + \cdots \right]_0^1 \\
&= 2 - \frac{1}{5} + \frac{1}{108} - \frac{1}{4680} + \frac{1}{342720} - \cdots = 1.80904\,84759.
\end{aligned}
$$

(b) The singularity is subtracted and the remaining nonsingular part is computed by aid of a suitable numerical method. In this case we can write

$$I = \int_0^1 x^{-1/2} \, dx - \int_0^1 x^{-1/2}(1 - \cos x) \, dx = 2 - \int_0^1 x^{-1/2}(1 - \cos x) \, dx.$$

In the last integral the integrand is regular in the whole interval [for $x = 0$ we get $\lim_{\alpha \to 0} ((1 - \cos x)/\sqrt{x}) = 0$], and at first glance we would not expect any numerical difficulties. However, it turns out that we obtain a very poor accuracy unless we take very small intervals close to the origin. The reason is that higher derivatives are still singular in the origin, and hence we ought to employ a more refined subtraction procedure. For example, we can write instead

$$I = \int_0^1 x^{-1/2}(1 - x^2/2) \, dx + \int_0^1 x^{-1/2}(\cos x - 1 + x^2/2) \, dx$$

$$= 1.8 + \int_0^1 x^{-1/2}(\cos x - 1 + x^2/2) \, dx.$$

We gain two advantages in this way: (1) the integrand becomes much smaller; (2) the influence of singularities in higher derivatives has been reduced strongly. Both imply that numerical accuracy is improved considerably.

(c) The substitution $x = t^2$ gives $I = 2\int_0^1 \cos t^2 \, dt$. The resulting integral has no singularity left in the integrand, and even all derivatives are regular. Hence, direct computation by use of a suitable numerical method will give no difficulties. We can also employ series expansion:

$$I = 2\int_0^1 \left[1 - \frac{t^4}{2!} + \frac{t^8}{4!} - \frac{t^{12}}{6!} + \cdots \right] dt$$

$$= 2\left[1 - \frac{1}{10} + \frac{1}{216} - \frac{1}{9360} + \cdots \right],$$

which is exactly the same series as before.

(d) Partial integration gives

$$I = [2x^{1/2} \cos x]_0^1 + \int_0^1 2x^{1/2} \sin x \, dx.$$

The last integral can be computed numerically or by series expansion; in the latter case we obtain

$$I = 2\cos 1 + 4\left(\frac{1}{5 \cdot 1!} - \frac{1}{9 \cdot 3!} + \frac{1}{13 \cdot 5!} - \frac{1}{17 \cdot 7!} + \cdots \right)$$

$$= 1.80904\,8476.$$

As our second example we take $I = \int_0^1 e^{-x} \ln x \, dx$. The integrand has a logarithmic singularity in the origin which can easily be neutralized. Expanding e^{-x} in a power series, we find

$$I = \int_0^1 e^{-x} \ln x \, dx = \int_0^1 \left(1 - x + \frac{x^2}{2!} - \frac{x^3}{3!} + \cdots \right) \ln x \, dx.$$

From

$$\int_0^1 x^k \ln x \, dx = \left[\frac{x^{k+1}}{k+1} \left(\ln x - \frac{1}{k+1} \right) \right]_0^1 = -\frac{1}{(k+1)^2},$$

we obtain

$$I = \sum_{n=1}^{\infty} (-1)^n (n \cdot n!)^{-1} = -\left(1 - \frac{1}{4} + \frac{1}{18} - \frac{1}{96} + \cdots \right).$$

This series converges very rapidly, and 8 correct decimals are obtained from only 11 terms. The result is $0.79659\,960$.

As our third example, we choose

$$I = \int_0^1 \frac{dx}{\sqrt{(x(1-x^2))}}.$$

We have a singularity of the form $1/\sqrt{x}$ in the origin and another of the form $1/\sqrt{(2(1-x))}$ when x is close to 1. We could in principle make the substitution $x = t^2$ in a suitable left subinterval, and the substitution $1 - x = t^2$ in the right subinterval. However, putting instead $x = \sin^2 t$ and replacing $\sin t$ by $\cos(\pi/2 - t)$, we find after easy reductions:

$$I = \sqrt{2} \int_0^{\pi/2} \frac{dt}{\sqrt{(1 - \frac{1}{2} \sin^2 t)}}$$

which is a standard integral of elliptic type (see (7.4.7)); it can be written $I = \sqrt{2}\,K(1/\sqrt{2}) = 2.62205\,75543$. Even without knowledge of this, one could obtain a quickly converging power series in $\sin^2 t$ and perform the integration term-wise.

Here we have indicated some methods to get rid of singularities of different kinds. For completeness we mention that perhaps the most effective method for computation of complicated definite integrals is the residue method. We refer to Section 7.1 for a few simple examples. Likewise, we just mention the technique of differentiating or integrating an integral with respect to a parameter. One example is provided by

$$\int_0^\infty e^{-px} \sin x \, dx = \frac{1}{p^2 + 1}$$

which on integration in p gives:

$$\int_0^\infty dx \sin x \int_0^\infty e^{-px} \, dp = \int_0^\infty \frac{dp}{p^2 + 1}$$

and hence the well-known formula:

$$\int_0^\infty \frac{\sin x}{x} \, dx = \frac{\pi}{2}.$$

§17.2.   METHODS BASED ON EQUIDISTANT NODE POINTS

Newton–Cotes Formulas

Assume that $y = f(x)$ is a sufficiently regular function. At once we have to distinguish between two cases: either y is known only in certain isolated points $x_1, x_2, \ldots, x_n$, or y can be computed in any point within the integration interval, the limits included. In both cases some sort of interpolation must be performed, at least in principle, and in the former case it is a matter of good judgment to choose between different alternatives. It may well be the case that spline interpolation is the best choice here.

In the more common case when y is generally known, we choose a set of abscissas where y is computed. This can be done in two ways, either with all points equidistant, or with a more sophisticated choice. We shall treat both cases, and we start with equidistant x-values.

In the section on approximation we discussed the properties of Lagrange's interpolation formula. Although the formula is exact in the nodes, it behaves rather badly in intermediate points, at least for certain functions. If we integrate this formula we obtain a whole family of integration relations called the Newton–Cotes formulas. We show briefly how the results are established and also present a short table of coefficients together with error terms.

We assume a constant interval length h and put

$$P(x) = \sum_{k=0}^{n} L_k(x) y_k,$$

where

$$L_k(x) = \frac{(x - x_0)(x - x_1) \cdots (x - x_{k-1})(x - x_{k+1}) \cdots (x - x_n)}{(x_k - x_0)(x_k - x_1) \cdots (x_k - x_{k-1})(x_k - x_{k+1}) \cdots (x_k - x_n)}.$$

Here $x_k = x_0 + kh$, and further we put $x = x_0 + sh$, obtaining $dx = h\, ds$. Hence

$$L_k = \frac{s(s-1) \cdots (s - k + 1)(s - k - 1) \cdots (s - n)}{k(k-1) \cdots (1)(-1) \cdots (k - n)},$$

and we get

$$\int_{x_0}^{x_n} P(x)\, dx = nh \frac{1}{n} \sum_{k=0}^{n} y_k \int_0^n L_k\, ds.$$

Finally, putting $(1/n) \int_0^n L_k\, ds = C_k^n$, we can write the integration formula

(17.2.1)
$$\int_{x_0}^{x_n} P(x)\, dx = nh \sum_{k=0}^{n} C_k^n y_k = (x_n - x_0) \sum_{k=0}^{n} C_k^n y_k.$$

The numbers C_k^n $(0 \le k \le n)$ are called *Cotes numbers*. As is easily proved, they satisfy $C_k^n = C_{n-k}^n$ and $\sum_{k=0}^{n} C_k^n = 1$. Later on we shall examine what can be said about the difference between the desired value, $\int_{x_0}^{x_n} f(x)\, dx$, and the numerically computed value, $\int_{x_0}^{x_n} P(x)\, dx$.

The best known in the family of Newton–Cotes formulas are the first and the second formulas, called the *trapezoidal rule* $(n = 1)$ and *Simpson's rule* $(n = 2)$. For higher order formulas some coefficients are negative, and we advise strongly against using these formulas in practice because of a considerable risk of instability.

The Trapezoidal Rule

The name of this method comes from a simple geometric interpretation: the area between the curve, the x-axis, and two ordinates is approximated by the area of a trapezoid:

$$\int_a^b f(x)\, dx \simeq \frac{b-a}{2}(f(a) + f(b)).$$

When this approximation is repeated over several intervals of length h, with $x_k = x_0 + kh$, $k = 0(1)n$, we get:

(17.2.2) $$\int_{x_0}^{x_n} f(x)\, dx \simeq \frac{h}{2}[y_0 + 2y_1 + 2y_2 + \cdots + 2y_{n-1} + y_n]$$

where we have put $y_k = f(x_k)$. We will now compute the truncation error and start by considering one interval:

$$R(h) = \frac{h}{2}\left[f\left(-\frac{h}{2}\right) + f\left(\frac{h}{2}\right) \right] - \int_{-h/2}^{h/2} f(x)\, dx.$$

Then
$$R'(h) = (h/2)(-\tfrac{1}{2}f'(-h/2) + \tfrac{1}{2}f'(h/2))$$
$$R''(h) = (h/2)(f''(-h/2) + f''(h/2))/4 + (f'(h/2) - f'(-h/2))/4$$
$$R'''(h) = (h/16)(-f'''(-h/2) + f'''(h/2)) + (f''(-h/2) + f''(h/2))/4.$$

Hence $R'(0) = 0$, $R''(0) = 0$, and $R'''(0) = \tfrac{1}{2}f''(0)$, and we see that the truncation error is essentially $h^3 f''(0)/12$. A somewhat more careful analysis gives the value $(h^3/12)f'''(\xi)$, $-h/2 < \xi < h/2$. When the trapezoidal formula is used over a number of intervals so that $b - a = nh$, we see that the truncation error is $O(h^2)$. In spite of the low accuracy the formula is quite useful in many situations due to its robustness, and further it serves as a starting solution in the Romberg scheme.

Simpson's Rule

Putting $n = 2$, we obtain the corresponding Cotes numbers:

$$\begin{cases} C_0^2 = \dfrac{1}{2}\displaystyle\int_0^2 \frac{(s-1)(s-2)}{(-1)(-2)}\, ds = \frac{1}{6}, \\[2mm] C_1^2 = \dfrac{1}{2}\displaystyle\int_0^2 \frac{s(s-2)}{1(-1)}\, ds = \frac{4}{6}, \\[2mm] C_2^2 = \dfrac{1}{2}\displaystyle\int_0^2 \frac{s(s-1)}{2\cdot 1}\, ds = \frac{1}{6}, \end{cases}$$

(17.2.3) $$\int_{x_0}^{x_2} f(x)\, dx \simeq (x_2 - x_0)\left(\frac{1}{6}y_0 + \frac{4}{6}y_1 + \frac{1}{6}y_2\right).$$

This is the famous *Simpson rule.*

We shall now perform a simplified error computation. Assume that $f(x)$ can be represented as a power series expansion in the integration interval $(-h, h)$, i.e., $f(x) = a_0 + a_1 x + a_2 x^2 + a_3 x^3 + a_4 x^4 + \cdots$. We then construct

$$R(h) = (h/3)[f(-h) + 4f(0) + f(h)] - \int_{-h}^{h} f(x)\, dx$$

and find after an easy computation:

$$R(h) = (h/3)(6a_0 + 2a_2 h^2 + 2a_4 h^4 + \cdots) - (2a_0 h + 2a_2 h^3/3 + 2a_4 h^5/5 + \cdots)$$
$$= 4a_4 h^5/15 + \cdots = (h^5/90)f^{(iv)}(0) + \cdots$$

A more careful analysis gives the result $R(h) = (h^5/90)f^{(iv)}(\xi)$ where $-h < \xi < h$. When the formula is used over n intervals, we have $2nh = b - a$ and

$$R = (b - a)(h^4/180)f^{(iv)}(\xi), \qquad a < \xi < b.$$

Similar error estimates can be made for the other Cotes formulas. The coefficients and the error terms appear in the following table.

Table of Cotes's Numbers

n	N	$NC_0^{(n)}$	$NC_1^{(n)}$	$NC_2^{(n)}$	$NC_3^{(n)}$	$NC_4^{(n)}$	$NC_5^{(n)}$	$NC_6^{(n)}$	Remainder term	
1	2	1	1						$8.3 \cdot 10^{-2}(b-a)^3 f''$	(ξ)
2	6	1	4	1					$3.5 \cdot 10^{-4}(b-a)^5 f^{(iv)}$	(ξ)
3	8	1	3	3	1				$1.6 \cdot 10^{-4}(b-a)^5 f^{(iv)}$	(ξ)
4	90	7	32	12	32	7			$5.2 \cdot 10^{-7}(b-a)^7 f^{(vi)}$	(ξ)
5	288	19	75	50	50	75	19		$3.0 \cdot 10^{-7}(b-a)^7 f^{(vi)}$	(ξ)
6	840	41	216	27	272	27	216	41	$6.4 \cdot 10^{-10}(b-a)^9 f^{(viii)}$	(ξ)

For example, we have

$$\int_0^{6h} y\, dx \simeq \frac{6h}{840}[41y_0 + 216y_1 + 27y_2 + 272y_3 + 27y_4 + 216y_5 + 41y_6].$$

Simpson's Formula with End Correction

We shall now try to improve the Simpson formula by also using derivatives in the end points. Putting

$$\int_{-h}^{+h} y\, dx \simeq h(ay_{-1} + by_0 + ay_1) + h^2(cy'_{-1} - cy'_1),$$

and expanding in series we find

$$\begin{cases} 2a + b = 2, \\ a - 2c = \tfrac{1}{3}, \\ a - 4c = \tfrac{1}{5}, \end{cases}$$

whence

$$a = \frac{7}{15}; \qquad b = \frac{16}{15}; \qquad c = \frac{1}{15}.$$

If this formula is used for several adjacent intervals, the y'-terms cancel in all interior points, and we are left with the final formula

(17.2.4) $$\int_a^b y\,dx = \frac{h}{15}(7y_0 + 16y_1 + 14y_2 + 16y_3 + \cdots + 7y_{2n})$$

$$+ \frac{h^2}{15}(y_0' - y_{2n}') + O(h^6),$$

where $b - a = 2nh$ and $y_0, y_1, \ldots, y_{2n}$ denote the ordinates in the points $a, a + h,$ $\ldots, a + 2nh = b$. Note that the error term is $O(h^6)$.
     A similar computation gives

(17.2.5) $$\int_a^b y\,dx = h\left[\frac{1}{2}y_0 + y_1 + y_2 + \cdots + y_{n-1} + \frac{1}{2}y_n\right]$$

$$+ \frac{h^2}{12}(y_0' - y_n') + O(h^4)$$

(the trapezoidal formula with end correction).

Richardson Extrapolation

     The form of the error term in Simpson's formula implies that for reasonable values of h the error will be multiplied by approximately a factor 16 when h is doubled. Hence, with obvious notations:

$$I(h) \simeq I_0 + \varepsilon$$
$$I(2h) \simeq I_0 + 16\varepsilon$$

giving $I_0 \simeq (16I(h) - I(2h))/15$. Similar extrapolations can be performed in other cases; in particular we now turn to the trapezoidal rule, which serves as a basis for Romberg's method.

Romberg's Method

     In this method we shall make use of Richardson extrapolation in a systematic way. Suppose that we start with the trapezoidal rule for evaluating $\int_a^b f(x)\,dx$. Using the interval $h_m = (b - a) \cdot 2^{-m}$, we denote the result with A_m. Since the error is proportional to h_m^2, we obtain an improved value if we form

$$B_m = A_m + \frac{A_m - A_{m-1}}{3}, \qquad m \geq 1.$$

As a matter of fact, B_m is the same result as obtained from Simpson's formula. Now we make use of the fact that the error in B_m is proportional to h_m^4 and form

$$C_m = B_m + \frac{B_m - B_{m-1}}{15}, \qquad m \geq 2.$$

This formula, incidentally, is identical with the Cotes formula for $n = 4$. Since the error is proportional to h_m^6, we form

$$D_m = C_m + \frac{C_m - C_{m-1}}{63}, \qquad m \geq 3.$$

The error is now proportional to h_m^8, but the formula is not of the Cotes type any longer. The same process can be repeated again; we stop when two successive values in the same column are sufficiently close to each other.

Example 1. $f(x) = 1/x$; $a = 1$, $b = 2$. We find the following values:

m	A_m	B_m	C_m	D_m	E_m
0	0.75				
1	0.70833 33333	0.69444 44444			
2	0.69702 38095	0.69325 39682	0.69317 46031		
3	0.69412 18504	0.69315 45307	0.69314 79015	0.69314 74777	
4	0.69339 12022	0.69314 76528	0.69314 71943	0.69314 71831	0.69314 71819

The value of the integral is $\ln 2 = 0.69314\ 71805$.

§17.3. FORMULAS OF GAUSS TYPE

In deducing the Cotes formulas, we made use of function values in equidistant points, and the weights were determined in such a way that the formula obtained an accuracy of highest possible order. We shall now choose abscissas without any special conditions and make the following attempt:

$$(17.3.1) \qquad \int_a^b w(x)f(x)\,dx = A_1 f(x_1) + A_2 f(x_2) + \cdots + A_n f(x_n) + R_n.$$

Here the weights A_i, as well as the abscissas x_i, are at our disposal. Further, $w(x)$ is a weight function, which will be specialized later on; so far, we only suppose that $w(x) \geq 0$. We shall use the same notations as in Lagrange's and Hermite's interpolation formulas, namely,

$$F(x) = \prod_{k=1}^{n} (x - x_k); \qquad F_k(x) = \frac{F(x)}{x - x_k}; \qquad L_k(x) = \frac{F_k(x)}{F_k(x_k)}.$$

Now we form the Lagrangian interpolation polynomial $P(x)$:

$$P(x) = \sum_{k=1}^{n} L_k(x)f(x_k),$$

and find $f(x) - P(x) = 0$ for $x = x_1, x_2, \ldots, x_n$.

Hence we can write

$$f(x) = P(x) + F(x)(\alpha_0 + \alpha_1 x + \alpha_2 x^2 + \cdots),$$

which inserted into (17.3.1), gives

$$\int_a^b w(x)f(x)\, dx = \sum_{k=1}^{n} \left(\int_a^b w(x)L_k(x)\, dx \right) f(x_k)$$

$$+ \int_a^b w(x)F(x)\{\alpha_0 + \alpha_1 x + \alpha_2 x^2 + \cdots\}\, dx$$

$$= \sum_{k=1}^{n} A_k f(x_k) + R_n.$$

Identifying, we obtain

(17.3.2)
$$A_k = \int_a^b w(x)L_k(x)\, dx,$$

$$R_n = \int_a^b w(x)F(x) \sum_{r=0}^{\infty} \alpha_r x^r\, dx.$$

Here the abscissas $x_1, x_2, \ldots, x_n$ are still at our disposal, but as soon as they are given, the weight constants A_k are also determined.

For determination of the abscissas we can formulate n more conditions to be fulfilled, and we choose the following:

(17.3.3)
$$\int_a^b w(x)F(x)x^r\, dx = 0; \qquad r = 0, 1, 2, \ldots, n - 1.$$

Hence $F(x)$, which is of degree n, is orthogonal to all polynomials of degree $0, 1, 2, \ldots,$ $(n - 1)$ with respect to the weight function $w(x)$; i.e., $F(x)$ belongs to the corresponding family of orthogonal polynomials. If we choose $w(x) = 1$, $a = -1$, and $b = 1$, then for $n = 3$ we get $F(x) = cP_3(x)$ where $P_n(x)$ are the Legendre polynomials (see Section 4.3). Hence we find the equation $5x^3 - 3x = 0$, which gives the abscissas $0, \pm\sqrt{0.6}$. Transforming with $x' = \frac{1}{2}(x + 1)$, we get instead $a = 0$, $b = 1$ and the abscissas 0.5, $0.5 \pm \sqrt{0.15}$.

This computation can also be done without knowledge of the theory of orthogonal polynomials. We put for $n = 3$

$$F(x) = (x - x_1)(x - x_2)(x - x_3) \equiv x^3 - s_1 x^2 + s_2 x - s_3$$

and use the conditions

$$\int_0^1 x^r(x^3 - s_1 x^2 + s_2 x - s_3)\, dx = 0$$

for $r = 0, 1, 2$. The equations become:

$$
\begin{cases}
\dfrac{1}{4} - \dfrac{s_1}{3} + \dfrac{s_2}{2} - s_3 = 0, \\[2mm]
\dfrac{1}{5} - \dfrac{s_1}{4} + \dfrac{s_2}{3} - \dfrac{s_3}{2} = 0, \\[2mm]
\dfrac{1}{6} - \dfrac{s_1}{5} + \dfrac{s_2}{4} - \dfrac{s_3}{3} = 0,
\end{cases}
$$

with the solution $s_1 = \frac{3}{2}$, $s_2 = \frac{3}{5}$, and $s_3 = \frac{1}{20}$. The abscissas are obtained from the equation

$$
x^3 - \frac{3}{2} x^2 + \frac{3}{5} x - \frac{1}{20} = 0,
$$

and we find

$$
\begin{aligned}
x_1 &= 0.5 - \sqrt{0.15} = 0.1127; \\
x_2 &= 0.5; \\
x_3 &= 0.5 + \sqrt{0.15} = 0.8873.
\end{aligned}
$$

In order to get a theoretically more complete and more general discussion we are now going to treat the problem from another point of view. Integrating Hermite's interpolation formula (14.1.8) and (14.1.9), we get

$$
(17.3.4) \qquad \int_a^b w(x)f(x)\, dx = \sum_{k=1}^n B_k f(x_k) + \sum_{k=1}^n C_k f'(x_k) + E,
$$

where

$$
(17.3.5) \qquad
\begin{cases}
B_k = \displaystyle\int_a^b w(x)[1 - 2L_k'(x_k)(x - x_k)][L_k(x)]^2 \, dx, \\[3mm]
C_k = \displaystyle\int_a^b w(x)(x - x_k)[L_k(x)]^2 \, dx, \\[3mm]
E = \displaystyle\int_a^b w(x)\frac{f^{(2n)}(\xi)}{(2n)!}[F(x)]^2 \, dx.
\end{cases}
$$

First we will consider the error term, and since $w(x) \ge 0$, we can use the first mean-value theorem of integral calculus:

$$
(17.3.6) \qquad E = \frac{f^{(2n)}(\eta)}{(2n)!} \int_a^b w(x)[F(x)]^2 \, dx.
$$

Here $a < \xi, \eta < b$ and further ξ is a function of x.

So far, the abscissas $x_1, x_2, \ldots, x_n$ are not restricted, but now we shall examine whether they can be chosen in such a way that the n constants C_k vanish. First we transform C_k [see (14.1.3)]:

$$
C_k = \int_a^b w(x)[(x - x_k)L_k(x)]L_k(x)\, dx = \frac{1}{F'(x_k)} \int_a^b w(x)F(x)L_k(x)\, dx = 0.
$$

We can say that the n polynomials $L_k(x)$ of degree $n - 1$ must be orthogonal to $F(x)$ with the weight function $w(x)$. It is also easily understood that the n powers $1, x, x^2, \ldots, x^{n-1}$ must also be orthogonal to $F(x)$ in the same way. Then (17.3.4) transforms to

$$\int_a^b w(x)f(x)\,dx = \sum_{k=1}^n B_k f(x_k) + \frac{f^{(2n)}(\eta)}{(2n)!} \int_a^b w(x)[F(x)]^2\,dx.$$

The abscissas $x_1, x_2, \ldots, x_n$ are now determined in principle through the conditions $C_k = 0$, which are equivalent to (17.3.3). Further we see that A_k in (17.3.2) can be identified with B_k in (17.3.5). Rewriting B_k we get directly

$$B_k = \int_a^b w(x)[L_k(x)]^2\,dx - 2L_k'(x_k)C_k = \int_a^b w(x)[L_k(x)]^2\,dx.$$

Thus we have the following double formula for A_k:

(17.3.7) $$A_k = \int_a^b w(x)L_k(x)\,dx = \int_a^b w(x)[L_k(x)]^2\,dx.$$

An important conclusion can be drawn at once from this relation: all weights A_k are > 0. Again, the condition for the validity of (17.3.6) is that the relation (17.3.3) is satisfied. Thus R_n vanishes if $f(x)$ is a polynomial of degree $\leq 2n - 1$.

We now choose some special cases according to the following table:

Weight function	Lower limit a	Upper limit b	Polynomial
1	-1	1	Legendre
$\exp(-x)$	0	∞	Laguerre
$\exp(-x^2)$	$-\infty$	∞	Hermite
$(1 - x^2)^{-1/2}$	-1	1	Chebyshev

Here we will treat the first case in some more detail.

The abscissas x_i, $i = 1(1)n$, are computed as zeros of the equation $P_n(x) = 0$, where P_n is the nth degree Legendre polynomial, and we will now also determine the weights A_k. Using the notation

$$F_n(x) = \prod_{k=1}^n (x - x_k)$$

we have $p_n(x) = a_n F_n(x)$ where $p_n(x)$ are orthonormalized polynomials, related to the Legendre polynomials by $P_n(x) = p_n(x)(n + \tfrac{1}{2})^{-1/2}$. We now apply Christoffel–Darboux's theorem,

$$(a_n/a_{n+1})[p_{n+1}(x)p_n(y) - p_n(x)p_{n+1}(y)] = (x - y)\sum_{k=0}^n p_k(x)p_k(y).$$

Putting $y = x_i$, multiplying by $p_0(x)/(x - x_i)$ and integrating, we get

$$-(a_n/a_{n+1})p_{n+1}(x_i)\int_{-1}^1 p_n(x)p_0(x)(x - x_i)^{-1}\,dx$$

$$= \sum_{k=0}^n p_k(x_i)\int_{-1}^1 p_k(x)p_0(x)\,dx = p_0(x_i) = 1/\sqrt{2}$$

since the last integral equals δ_{k0} and $P_n(x_i) = 0$. Hence we have

$$A_k = \int_{-1}^{1} L_k(x)\, dx = \int_{-1}^{1} (x - x_k)^{-1}(F_n(x)/F_n'(x_k))\, dx$$
$$= p_n'(x_k)^{-1} \int_{-1}^{1} (p_n(x)/(x - x_k))\, dx = -[(a_n/a_{n+1})p_n'(x_k)p_{n+1}(x_k)]^{-1}.$$

From the three-term recursion formula

$$p_{n+1} = (A_n x + B_n)p_n - C_n p_{n-1}$$

(note that here $A_n = a_{n+1}/a_n$ and has nothing to do with the weight A_k) we get for $x = x_k$:

$$p_{n+1}(x_k) = -C_n p_{n-1}(x_k)$$
$$= -(a_{n+1}a_{n-1}/a_n^2)p_{n-1}(x_k)$$

which leads to the expression

$$A_k = [(a_{n-1}/a_n)p_n'(x_k)p_{n-1}(x_k)]^{-1}.$$

But

$$a_n = \frac{(2n)!}{2^n(n!)^2}\, (n + \tfrac{1}{2})^{1/2} \qquad \text{giving} \qquad a_{n-1}/a_n = n(4n^2 - 1)^{-1/2},$$

and further

$$p_n'(x_k) = (n + \tfrac{1}{2})^{1/2} P_n'(x_k), \qquad p_{n-1}(x_k) = (n - \tfrac{1}{2})^{1/2} P_{n-1}(x_k).$$

This gives

$$A_k = \left[\frac{n}{2} P_n'(x_k)P_{n-1}(x_k)\right]^{-1}.$$

Now we have

$$(1 - x^2)P_n' = nP_{n-1} - nxP_n$$

since both sides are equal when $x = 1$ and also have the same derivative:

$$\frac{d}{dx}\left[(1 - x^2)P_n'\right] = nP_{n-1}' - nxP_n' - nP_n.$$

Using the differential equation for P_n, we can replace the left-hand side by $-n(n + 1)P_n$, and so we obtain $nP_n + P_{n-1}' - xP_n' = 0$, a formula already proved in (4.3.6). Putting $x = x_k$ we find:

$$(1 - x_k^2)P_n'(x_k) = nP_{n-1}(x_k)$$

and we finally get:

(17.3.8) $$A_k = \frac{2}{(1 - x_k^2)P_n'(x_k)^2}.$$

Similar relations for the weights can be obtained in an analogous way for the Laguerre and Hermite polynomials. These results are summarized in Tables 17.1 to 17.3.

TABLE 17.1. Abscissas and weights in the Gauss–Legendre quadrature formula

n	x_k	A_k
2	$\pm 0.57735\,02691\,89626$	$1.00000\,00000\,00000$
3	$\pm 0.77459\,66692\,41483$	$0.55555\,55555\,55556$
	$0.00000\,00000\,00000$	$0.88888\,88888\,88889$
4	$\pm 0.86113\,63115\,94053$	$0.34785\,48451\,37454$
	$\pm 0.33998\,10435\,84856$	$0.65214\,51548\,62546$
5	$\pm 0.90617\,98459\,38664$	$0.23692\,68850\,56189$
	$\pm 0.53846\,93101\,05683$	$0.47862\,86704\,99366$
	$0.00000\,00000\,00000$	$0.56888\,88888\,88889$
6	$\pm 0.93246\,95142\,03152$	$0.17132\,44923\,79170$
	$\pm 0.66120\,93864\,66265$	$0.36076\,15730\,48139$
	$\pm 0.23861\,91860\,83197$	$0.46791\,39345\,72691$

TABLE 17.2. Abscissas and weights in the Gauss–Laguerre quadrature formula

n	x_k	A_k
2	$0.58578\,64376\,27$	$0.85355\,33905\,93$
	$3.41421\,35623\,73$	$0.14644\,66094\,07$
3	$0.41577\,45567\,83$	$0.71109\,30099\,29$
	$2.29428\,03602\,79$	$0.27851\,77335\,69$
	$6.28994\,50829\,37$	$0.01038\,92565\,016$
4	$0.32254\,76896\,19$	$0.60315\,41043\,42$
	$1.74576\,11011\,58$	$0.35741\,86924\,38$
	$4.53662\,02969\,21$	$0.03888\,79085\,150$
	$9.39507\,09123\,01$	$0.00053\,92947\,05561$
5	$0.26356\,03197\,18$	$0.52175\,56105\,83$
	$1.41340\,30591\,07$	$0.39866\,68110\,83$
	$3.59642\,57710\,41$	$0.07594\,24496\,817$
	$7.08581\,00058\,59$	$0.00361\,17586\,7992$
	$12.64080\,08442\,76$	$0.00002\,33699\,723858$

TABLE 17.3. Abscissas and weights in the Gauss–Hermite quadrature formula

n	x_k	A_k
2	$\pm 0.70710\,67811\,87$	$0.88622\,69254\,53$
3	0	$1.18163\,59006\,04$
	$\pm 1.22474\,48713\,92$	$0.29540\,89751\,51$
4	$\pm 0.52464\,76232\,75$	$0.80491\,40900\,06$
	$\pm 1.65068\,01238\,86$	$0.08131\,28354\,473$
5	0	$0.94530\,87204\,83$
	$\pm 0.95857\,24646\,14$	$0.39361\,93231\,52$
	$\pm 2.02018\,28704\,56$	$0.01995\,32420\,591$
6	$\pm 0.43607\,74119\,28$	$0.72462\,95952\,24$
	$\pm 1.33584\,90740\,14$	$0.15706\,73203\,23$
	$\pm 2.35060\,49736\,74$	$0.00453\,00099\,0551$

Example 1. For $n = 3$ the Legendre polynomial is $P_3 = \frac{1}{2}(5x^3 - 3x)$ with zeros $x_1 = -\sqrt{0.6}$, $x_2 = 0$, and $x_3 = \sqrt{0.6}$. Then

$$P_3'(x) = \frac{3}{2}(5x^2 - 1), \qquad P_3'(x_1) = P_3'(x_3) = 3, \qquad P_3'(x_2) = -\frac{3}{2};$$
$$A_1 = A_3 = 2/(0.4 \cdot 9) = \frac{5}{9}; \qquad A_2 = 2/\frac{9}{4} = \frac{8}{9}.$$

The Chebyshev formulas deserve to be treated separately. The abscissas are obtained from $T_n(x) = \cos(n \arccos x) = 0$, giving $n \arccos x = (2k + 1)\pi/2$ for $k = 0(1)n - 1$, i.e.,

$$x_k = \cos((2k + 1)\pi/2n) = \cos \varphi_k.$$

The orthonormalized polynomials are $p_n(x) = cT_n$ with $c = (2/\pi)^{1/2}$, and further we have $a_n = (2/\pi)^{1/2}2^{n-1}$. Using our previous formula

$$A_k = -[(a_n/a_{n+1})p_n'(x_k)p_{n+1}(x_k)]^{-1}$$

we get:

$$A_k = -[\tfrac{1}{2}cT_n'(x_k)cT_{n+1}(x_k)]^{-1} = -\left[\pi^{-1}(-n)\frac{\sin n\varphi_k}{-\sin \varphi_k}\cos(n + 1)\varphi_k\right]^{-1}$$
$$= -(\pi/n)[\sin(k + \tfrac{1}{2})\pi \cos((k + \tfrac{1}{2})\pi + \varphi_k)/\sin \varphi_k]^{-1}.$$

But $\sin(k + \tfrac{1}{2})\pi = (-1)^k$ and $\cos((k + \tfrac{1}{2})\pi + \varphi_k) = (-1)^{k+1}\sin \varphi_k$, which gives the final result $A_k = \pi/n$. Hence we have proved that all weights are equal in this case.

Summing up, we have the following Gauss formulas:

$$\int_{-1}^{+1} f(x)\,dx = 2\sum_{k=1}^{n}\frac{f(x_k)}{(1 - x_k^2)[P_n'(x_k)]^2} + \frac{2^{2n+1}(n!)^4}{(2n + 1)[(2n)!]^3}f^{(2n)}(\xi)$$

$$x_1, x_2, \ldots, x_n \text{ roots of } P_n(x) = 0.$$

$$\int_0^\infty e^{-x}f(x)\,dx = (n!)^2\sum_{k=1}^{n}\frac{f(x_k)}{x_k[L_n'(x_k)]^2} + \frac{(n!)^2}{(2n)!}f^{(2n)}(\xi)$$

$$x_1, x_2, \ldots, x_n \text{ roots of } L_n(x) = 0.$$

$$\int_{-\infty}^{+\infty} e^{-x^2}f(x)\,dx = 2^{n+1} \cdot n!\sqrt{\pi}\sum_{k=1}^{n}\frac{f(x_k)}{[H_n'(x_k)]^2} + \frac{n!\sqrt{\pi}}{2^n(2n)!}f^{(2n)}(\xi)$$

$$x_1, x_2, \ldots, x_n \text{ roots of } H_n(x) = 0.$$

$$\int_{-1}^{+1} f(x)(1 - x^2)^{-1/2}\,dx = \frac{\pi}{n}\sum_{k=1}^{n}f\left(\cos\frac{2k - 1}{2n}\pi\right) + \frac{2\pi}{2^{2n}(2n)!}f^{(2n)}(\xi).$$

It is now natural to ask whether it is also possible to find formulas with all weights equal when $w(x) = 1$ and the interval is $(-1, 1)$. Chebyshev showed that this is possible, at least partially. He proved that the polynomial with the abscissas as roots can be written

$$F(x) = x^n \exp[-n((2 \cdot 3x^2)^{-1} + (4 \cdot 5x^4)^{-1} + (6 \cdot 7x^6)^{-1} + \cdots)]$$

where negative powers of x are discarded. It turned out that all roots are real only when $n = 1, 2, \ldots, 7$ and $n = 9$.

Summing up, we have for $w(x) = 1$:

$$\int_{-1}^{+1} f(x)\, dx = \frac{2}{n} \sum_{k=1}^{n} f(x_r) + R_n,$$

$$R_n = \frac{1}{(n+1)!} \left[\int_{-1}^{+1} x^{n+1} f^{(n+1)}(\xi)\, dx - \frac{2}{n} \sum_{r=1}^{n} x_r^{n+1} f^{(n+1)}(\xi_r) \right];$$

$$-1 < \xi < 1; \qquad -1 < \xi_r < x_r.$$

The real roots are collected in the following table:

n	x_r	n	x_r
2	$\pm 0.57735\,02691$	6	$\pm 0.26663\,54015$
			$\pm 0.42251\,86538$
3	0		$\pm 0.86624\,68181$
	$\pm 0.70710\,67812$	7	0
			$\pm 0.32391\,18105$
			$\pm 0.52965\,67753$
4	$\pm 0.18759\,24741$		$\pm 0.88386\,17008$
	$\pm 0.79465\,44723$		
		9	0
			$\pm 0.16790\,61842$
5	0		$\pm 0.52876\,17831$
	$\pm 0.37454\,14096$		$\pm 0.60101\,86554$
	$\pm 0.83249\,74870$		$\pm 0.91158\,93077$

Example 2. The mean temperature shall be determined for a 24-hour period running from midnight to midnight, by aid of four readings. Which times should be chosen?

We do not want to give one reading more weight than another, so we choose a Chebyshev formula with $n = 4$ and find the times $12 \pm 12 \cdot 0.1876$ and $12 \pm 12 \cdot 0.7947$ or, approximately, 2.30 a.m., 9.45 a.m., 2.15 p.m., and 9.30 p.m.

Lobatto Quadrature

We end this exposition by mentioning that there exists a pseudo-Gaussian method in which the endpoints -1 and $+1$ are also used as abscissas. The other abscissas x_k are obtained as zeros of the equation $P'_{n-1}(x) = 0$, while the weights are $\alpha = 2/n(n-1)$ and $\alpha_k = \alpha / P_{n-1}(x_k)^2$. Hence

$$\int_{-1}^{1} f(x)\, dx \simeq \alpha f(-1) + \alpha f(1) + \sum_{k=1}^{n-2} \alpha_k f(x_k).$$

The abscissas and weights are shown in Table 17.4.

TABLE 17.4. Table of abscissas and weights for Lobatto quadrature

n	x_k	α_k
3	$\pm 1.00000\ 00000$	$0.33333\ 33333$
	$0.00000\ 00000$	$1.33333\ 33333$
4	$\pm 1.00000\ 00000$	$0.16666\ 66667$
	$\pm 0.44721\ 35955$	$0.83333\ 33333$
5	$\pm 1.00000\ 00000$	$0.10000\ 00000$
	$\pm 0.65465\ 36707$	$0.54444\ 44444$
	$0.00000\ 00000$	$0.71111\ 11111$
6	$\pm 1.00000\ 00000$	$0.06666\ 66667$
	$\pm 0.76505\ 53239$	$0.37847\ 49563$
	$\pm 0.28523\ 15165$	$0.55485\ 83770$

By evaluating the integrand at ± 1 we lose two degrees of freedom, and a Lobatto n-point formula is exact only for polynomials of degree $2n - 3$, as compared with $2n - 1$ for a corresponding Gauss–Legendre formula. However, if $f(x)$ vanishes in the endpoints, we can use a Lobatto formula with the same number of function evaluations as a Gauss–Legendre formula with n points, while the formulas are exact for polynomials of degree $2n + 1$ and $2n - 1$ respectively.

§17.4. MULTIPLE INTEGRALS

The Good Lord said to the animals:
  Go out and multiply!
But the snake answered:
  How could I? I am an adder!

In integrating functions of several variables, one encounters essentially greater difficulties than in the case of just one variable. Already interpolation in two variables is a troublesome and laborious procedure. As in the case of one variable, one can use the operator technique and deduce interpolation coefficients for the points in a square grid. Usually one prefers to perform first a series of simple interpolations in one variable and then to interpolate the other variable by means of the new values.

An obvious method for computation of a double integral is first to perform a series of integrations in the y-direction for different values of x, followed by a final integration in the x-direction. In doing so, there are no restrictions with respect to interval lengths. However, we will now construct a family of formulas that are valid when we have a square grid. Then we conveniently use operator techniques, and we put:

$$(17.4.1) \qquad \xi = h\frac{\partial}{\partial x}; \qquad \eta = h\frac{\partial}{\partial y}.$$

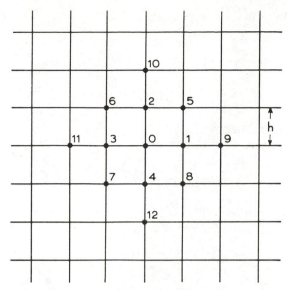

FIGURE 17.1

Further, we number a few points close to the origin (see Figure 17.1). With these notations we have, for example, $f_1 = e^{\xi}f_0$; $f_2 = e^{\eta}f_0$; $f_3 = e^{-\xi}f_0$; $f_4 = e^{-\eta}f_0$; $f_5 = e^{\xi+\eta}f_0$, and so on. We shall here use the symbols ∇ and D in the following sense:

$$(17.4.2) \quad \begin{cases} \xi^2 + \eta^2 = h^2\left(\dfrac{\partial^2}{\partial x^2} + \dfrac{\partial^2}{\partial y^2}\right) = h^2\nabla^2, \\[2mm] \xi\eta = h^2\dfrac{\partial^2}{\partial x\,\partial y} = h^2 D^2. \end{cases}$$

Further, we shall consider the following three sums:

$$S_1 = f_1 + f_2 + f_3 + f_4,$$
$$S_2 = f_5 + f_6 + f_7 + f_8,$$
$$S_3 = f_9 + f_{10} + f_{11} + f_{12}.$$

Then

$$S_1 = 2(\cosh \xi + \cosh \eta)f_0$$
$$= 4f_0 + (\xi^2 + \eta^2)f_0 + \tfrac{1}{12}(\xi^4 + \eta^4)f_0 + \cdots,$$
$$S_2 = 4 \cosh \xi \cosh \eta \cdot f_0$$
$$= 4f_0 + 2(\xi^2 + \eta^2)f_0 + \tfrac{1}{6}(\xi^4 + \eta^4)f_0 + \xi^2\eta^2 f_0 + \cdots,$$
$$S_3 = 2(\cosh 2\xi + \cosh 2\eta)f_0$$
$$= 4f_0 + 4(\xi^2 + \eta^2)f_0 + \tfrac{4}{3}(\xi^4 + \eta^4)f_0 + \cdots.$$

The most important differential expression is $\nabla^2 f$, and we find

$$(17.4.3) \quad \nabla^2 f_0 \simeq \frac{1}{h^2}(S_1 - 4f_0).$$

Further, we obtain

(17.4.4) $$4S_1 + S_2 - 20f_0 = (6h^2\nabla^2 + \tfrac{1}{2}h^4\nabla^4 + \cdots)f_0,$$

with the main disturbing term invariant under rotation, and

(17.4.5) $$\nabla^2 f_0 \simeq \frac{1}{12h^2}\,(16S_1 - S_2 - 60f_0),$$

with the error proportional to h^4.

Analogously, we find without difficulty:

(17.4.6) $$\nabla^4 f_0 \simeq \frac{1}{h^4}\,(20f_0 - 8S_1 + 2S_2 + S_3)$$

with the error proportional to h^2.

We now pass to the integration, and try to compute

$$V = \int_{-h}^{+h} \int_{-h}^{+h} f(x, y)\, dx\, dy.$$

Then

$$f(x, y) = f(hr, hs) = e^{r\xi + s\eta}f_0; \qquad \begin{cases} dx = h\, dr, \\ dy = h\, ds, \end{cases}$$

and hence

$$
\begin{aligned}
V &= h^2 \int_{-1}^{+1} \int_{-1}^{+1} e^{r\xi + s\eta} f_0\, dr\, ds \\
&= h^2 \left[\frac{e^{r\xi}}{\xi}\right]_{r=-1}^{+1} \left[\frac{e^{s\eta}}{\eta}\right]_{s=-1}^{+1} f_0 = \frac{4h^2 \cdot \sinh \xi \cdot \sinh \eta}{\xi\eta}\, f_0 \\
&= 4h^2 \left[1 + \frac{1}{3!}(\xi^2 + \eta^2) + \frac{1}{5!}\left(\xi^4 + \frac{10}{3}\xi^2\eta^2 + \eta^4\right) + \cdots \right] f_0.
\end{aligned}
$$

Setting for a moment $\xi^2 + \eta^2 = P$, $\xi^4 + \eta^4 = Q$, and $\xi^2\eta^2 = R$, and adding the condition that the error term shall be proportional to ∇^4, that is, to $(\xi^2 + \eta^2)^2 = Q + 2R$ (thus invariant under rotation), we obtain

$$V = 4h^2[\alpha f_0 + \beta S_1 + \gamma S_2 + c(Q + 2R)] + \cdots.$$

Identifying, we get

$$
\begin{cases}
\alpha + 4\beta + 4\gamma = 1, \\[2mm]
\beta + 2\gamma = \dfrac{1}{6}, \\[2mm]
\dfrac{\beta}{12} + \dfrac{\gamma}{6} + c = \dfrac{1}{120}, \\[2mm]
\gamma + 2c = \dfrac{1}{36},
\end{cases}
\quad \text{whence} \quad
\begin{cases}
\alpha = \dfrac{22}{45}, \\[2mm]
\beta = \dfrac{4}{45}, \\[2mm]
\gamma = \dfrac{7}{180}, \\[2mm]
c = -\dfrac{1}{180}.
\end{cases}
$$

Thus we find the formula

(17.4.7) $$V = \frac{h^2}{45}(88f_0 + 16S_1 + 7S_2) - \frac{h^6}{45}\nabla^4 f_0 + \cdots .$$

On the other hand, Simpson's formula in two directions gives

(17.4.8) $$V = \frac{h^2}{9}(16f_0 + 4S_1 + S_2) + \cdots ,$$

but the error term is more complicated.

All these formulas can conveniently be illustrated by so-called computation mole-cules. The examples in Figure 17.2 should need no extra explanation.

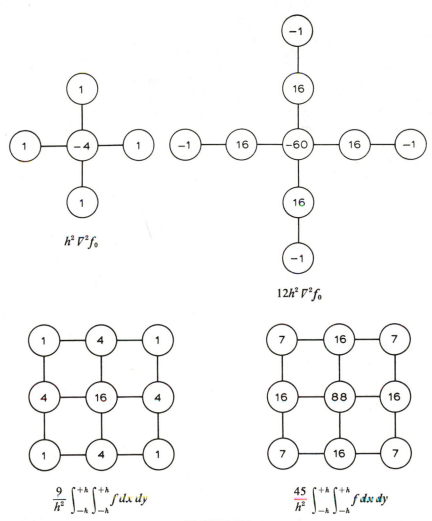

FIGURE 17.2

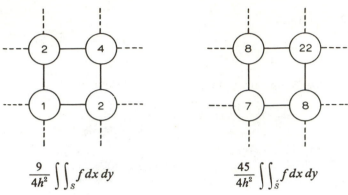

$$\frac{9}{4h^2} \iint_s f \, dx \, dy \qquad\qquad \frac{45}{4h^2} \iint_{\dot{s}} f \, dx \, dy$$

FIGURE 17.3

The two integration formulas are valid for integration over a square with the side $= 2h$. Integrating over a larger domain, we see that the coefficients at the corners receive contributions from four squares, while the coefficients in the midpoints of the sides get contributions from two squares. Taking out the factor 4, we obtain the open-type formulas shown in Figure 17.3. This standard configuration is then repeated in both directions.

Integration in higher dimensions gives rise to great difficulties. One possible way is to use the Monte Carlo method as described in a previous chapter. However, a serious drawback is the low accuracy attainable. In practical work it is recommended first to perform as many exact integrations as possible, and then to integrate each variable by itself.

EXERCISES

1. The prime number theorem states that the number of primes in the interval $a < x < b$ is approximately $\int_a^b dx/\ln x$. Use this for $a = 100$ and $b = 200$, and compare with the exact value.

2. Calculate $\int_0^1 x^{-x} \, dx$ to six decimal places.

3. Calculate $\int_0^{\pi/2} (1 - \sin^2 x/4)^{1/2} \, dx$ to five places.

4. A rocket is launched from the ground. Its acceleration during the first 80 seconds is given in the following table:

t (sec)	0	10	20	30	40	50	60	70	80
a (m/sec^2)	30.00	31.63	33.44	35.47	37.75	40.33	43.29	46.69	50.67

Find velocity and height at the time $t = 80$.

5. Find $\int_0^1 \ln x \cos x \, dx$ to five places.

6. Find $\int_0^1 (e^x + x - 1)^{-1/2} \, dx$ to four places.

7. Calculate the total arc length of the ellipse $x^2 + y^2/4 = 1$. The result should be correct to six decimal places.

8. Compute $\int_0^\infty x^{-x} \, dx$ to six decimal places.

9. Compute $\int_0^\infty dx/(e^x + x^2)$ and $\int_0^\infty dx/(e^{-x} + x^2)$.

10. Compute $\int_0^{\pi/2} \cos(\cos x) \, dx$ and $\int_0^{\pi/2} \sin(\sin x) \, dx$ to five decimal places.

11. Compute $\int_0^1 \sin \pi x [x(1 - x^2)]^{-3/2} \, dx$ to five decimal places.

12. In the inteval $0 \le x \le \pi$, y is defined as a function of x through the relation
$$y = -\int_0^x \ln(2 \sin(t/2)) \, dt.$$
Find the maximum value to five decimal places.

13. Compute $\int_0^\infty (e^x + x)^{-1/2} \, dx$ to three decimal places.

14. Compute $\int_0^\infty x \, dx/(e^x - 1)$ and $\int_0^\infty x \, dx/(e^x + 1)$.

15. Find $\int_0^\infty \exp(-x - 1/x) \, dx$ to four decimal places.

16. Compute $\int_0^1 (e^x - \cos x)^{-1/2} \, dx$ to five decimal places.

17. Find $\int_0^{\pi/2} (\sin x)^{1/3} \, dx$ to six decimal places.

18–24. Determine the weights A_i and the abscissas x_i in the following Gaussian formulas:
$$\int_a^b w(x) f(x) \, dx \simeq \sum_{i=1}^n A_i f(x_i)$$
where the limits, the weight function, and the order are as follows:

	a	b	$w(x)$	n
18.	0	1	$x^{1/2}$	2
19.	0	1	$\ln x^{-1}$	2
20.	0	1	$x^{-1/2}$	2
21.	-1	1	$2x^2 + 1$	3
22.	-1	1	$(x^2 + 1)^{-1}$	3
23.	-1	1	$\exp(-|x|)$	2
24.	0	1	$\begin{cases} 4x, & 0 \le x \le \frac{1}{2} \\ 4 - 4x, & \frac{1}{2} \le x \le 1 \end{cases}$	2

25. Find a Chebyshev quadrature formula with three abscissas for the interval $0 \le x \le 1$ and the weight function $w(x) = x^{-1/2}$.

26. Compute numerically $\iint_D (dx \, dy/(x^2 + y^2))$, where D is a square with corners $(1, 1)$, $(2, 1)$, $(2, 2)$, and $(1, 2)$.

27. Compute numerically $\int_0^\infty \int_0^\infty e^{-(x^2 + y^2)^2} \, dx \, dy$ ($h = 0.5$), and compare with the exact value.

28. For computation of $\iint_D f(x, y) \, dx \, dy$, where D is a square with the sides parallel to the axes, one wants to use the approximation $h^2(af_0 + bS)$. Here h is the side of the square, f_0 is the function value at the midpoint, and S is the sum of the function values at the corners. Find a and b, and compute

approximately

$$\int_{-1}^{+1} \int_{-1}^{+1} e^{-(x^2 + y^2)} \, dx \, dy$$

by using 13 points in all.

29. One wants to compute $\iint_c f(x, y) \, dx \, dy$, where C is a circle with radius a, by using the approximating expression $(\pi a^2 / 4)(f_1 + f_2 + f_3 + f_4)$. Here f_1, f_2, f_3, and f_4 are the function values at the corners of a square with the same center as the circle. Find the side of the square when the formula is correct to the highest possible order.

30. A function $u = u(x, y)$ sufficiently differentiable is given, together with a regular hexagon with corners $P_1, P_2, \ldots, P_6$, center P_0 and side h. The function value in the point P_v is denoted by u_v. Putting $S = u_1 + u_2 + \cdots + u_6 - 6u_0$ and $\Delta = \partial^2/\partial x^2 + \partial^2/\partial y^2$, show that

$$S = \tfrac{3}{2}h^2 \, \Delta u + \tfrac{3}{32}h^4 \, \Delta^2 u + O(h^6),$$

where Δu and $\Delta^2 u$ are taken in P_0.

CHAPTER 18

Summation

At 6 p.m. the well marked $\frac{1}{2}$ inch of water, at
nightfall $\frac{3}{4}$ and at daybreak $\frac{7}{8}$ of an inch. By noon
of the next day there was $\frac{15}{16}$ and on the next night
$\frac{31}{32}$ of an inch of water in the hold. The situation was
desperate. At this rate of increase few, if any, could
tell where it would rise to in a few days.

Stephen Leacock

§18.1. SUMS OF FACTORIALS AND POWERS

Previously (see Section 5.4) we have mentioned the fundamental properties of factorials on formation of differences and sums. We have the basic formulas:

$$p^{(n)} = p(p - 1) \cdots (p - n + 1); \qquad p^{(0)} = 1; \qquad p^{(-1)} = 1/(p + 1);$$
$$p^{(-n)} = [(p + 1)(p + 2) \cdots (p + n)]^{-1} = [(p + n)^{(n)}]^{-1};$$
$$\Delta p^{(n)} = np^{(n-1)}; \qquad \Delta^k p^{(n)} = n^{(k)} p^{(n-k)}$$

where n is an integer > 0. When n is noninteger we have to write

$$p^{(n)} = \frac{p!}{(p - n)!} = \frac{\Gamma(p + 1)}{\Gamma(p - n + 1)}.$$

Further, we have the summation formula

$$\sum_{p=P}^{Q} p^{(n)} = \frac{(Q + 1)^{(n+1)} - P^{(n+1)}}{n + 1},$$

valid for all integers n except $n = -1$. As an example we first calculate

$$\sum_{p=3}^{6} p^{(3)} = \sum_{p=3}^{6} p(p - 1)(p - 2)$$
$$= \frac{7^{(4)} - 3^{(4)}}{4} = \frac{7 \cdot 6 \cdot 5 \cdot 4 - 3 \cdot 2 \cdot 1 \cdot 0}{4} = 210,$$

which is easily verified by direct computation. We can also compute, for example,

$$\sum_{p=1}^{4} p^{(-3)} = \sum_{p=1}^{4} \frac{1}{(p + 1)(p + 2)(p + 3)}$$
$$= \frac{5^{(-2)} - 1^{(-2)}}{-2} = \frac{1}{2}\left(\frac{1}{2 \cdot 3} - \frac{1}{6 \cdot 7}\right) = \frac{1}{14},$$

which is also easily verified.

In (5.4.7) we showed how a power could be expressed as a linear combination of factorials by use of Stirling's numbers of the second kind. Thus we have directly a method for computing sums of powers, and we demonstrate the technique on an example.

$$\sum_{p=0}^{N} p^4 = \sum_{p=0}^{N} [p^{(1)} + 7p^{(2)} + 6p^{(3)} + p^{(4)}]$$

$$= \frac{(N+1)^{(2)}}{2} + \frac{7(N+1)^{(3)}}{3} + \frac{6(N+1)^{(4)}}{4} + \frac{(N+1)^{(5)}}{5}$$

$$= \frac{N(N+1)}{30}[15 + 70(N-1) + 45(N-1)(N-2) + 6(N-1)(N-2)(N-3)]$$

$$= \frac{N(N+1)(2N+1)(3N^2 + 3N - 1)}{30}.$$

For brevity we put $S_n(p) = 1^n + 2^n + \cdots + p^n$. By repeated use of the relation $\Delta B_{n+1}(x)/(n+1) = x^n$ (see (3.5.7)), we obtain

(18.1.1)
$$S_n(p) = \frac{B_{n+1}(p+1) - B_{n+1}(0)}{n+1},$$

which is an alternative formula for computation of power sums.

Here we shall also indicate how power sums with even negative exponents can be expressed by use of Bernoulli numbers. We need the following theorem which was proved in Section 8.2, Example 5:

$$\pi \cot \pi x = \frac{1}{x} + \sum_{n=1}^{\infty} \left(\frac{1}{x-n} + \frac{1}{x+n} \right)$$

We consider the series expansion for small values of x on both sides of the equation:

$$\pi x \cot \pi x = 1 + x \cdot \sum_{n=1}^{\infty} \left(\frac{1}{x-n} + \frac{1}{x+n} \right).$$

For the left-hand side, we use equation (3.5.11) and find that the coefficient of x^{2p} in $\pi x \cot \pi x$ is $-(2\pi)^{2p} B_p/(2p)!$ For the right-hand side we have, apart from the constant term,

$$x \cdot \sum_{n=1}^{\infty} \left(-\frac{1}{n}\frac{1}{1-x/n} + \frac{1}{n}\frac{1}{1+x/n} \right)$$

$$= -\sum_{n=1}^{\infty} \frac{x}{n} \left\{ 1 + \frac{x}{n} + \frac{x^2}{n^2} + \frac{x^3}{n^3} + \cdots - \left(1 - \frac{x}{n} + \frac{x^2}{n^2} - \frac{x^3}{n^3} + \cdots \right) \right\}$$

$$= -2 \cdot \sum_{n=1}^{\infty} \left(\frac{x^2}{n^2} + \frac{x^4}{n^4} + \frac{x^6}{n^6} + \cdots \right).$$

Hence the coefficient of x^{2p} is $-2 \cdot \sum_{n=1}^{\infty} n^{-2p}$, and we get

(18.1.2)
$$\sum_{n=1}^{\infty} \frac{1}{n^{2p}} = \frac{(2\pi)^{2p}}{2 \cdot (2p)!} B_p.$$

For $p = 1, 2, 3, 4,$ and 5, we obtain

$$\sum_{n=1}^{\infty} \frac{1}{n^2} = \frac{\pi^2}{6}; \quad \sum_{n=1}^{\infty} \frac{1}{n^4} = \frac{\pi^4}{90}; \quad \sum_{n=1}^{\infty} \frac{1}{n^6} = \frac{\pi^6}{945};$$

$$\sum_{n=1}^{\infty} \frac{1}{n^8} = \frac{\pi^8}{9450}; \quad \sum_{n=1}^{\infty} \frac{1}{n^{10}} = \frac{\pi^{10}}{93555}.$$

However, there are no simple expressions for the sums $\sum_{n=1}^{\infty} n^{-p}$ when p is odd.

As mentioned in Section 7.5, the relations above are special cases of the Riemann zeta function; a completely different proof was given there (see (7.5.7)).

From equation (18.1.2) several interesting conclusions can be drawn. First of all, we see that $B_p > 0$ for $p = 1, 2, 3, \ldots$ Further, we can easily obtain estimates of B_p for large values of p, since the left-hand side approaches 1 very rapidly, and we have asymptotically

(18.1.3)
$$B_p \sim \frac{2(2p)!}{(2\pi)^{2p}}.$$

This can be transformed by use of Stirling's formula:

(18.1.4)
$$B_p \simeq 4(\pi p)^{1/2} (p/\pi e)^{2p}.$$

§18.2. EULER–McLAURIN'S SUMMATION FORMULA

We will here consider the sum $y_0 + y_1 + y_2 + \cdots + y_n$ where $y = f(x)$ and $y_k = f(x_k)$ with $x_k = x_0 + kh$. We note that the sum must be a reasonably good approximation of $(h^{-1}) \int_{x_0}^{x_n} f(x)\, dx$. We form

$$f(x_0) + f(x_1) + \cdots + f(x_{n-1}) - \frac{1}{h} \int_{x_0}^{x_n} f(x)\, dx$$

$$= \frac{E^n - 1}{E - 1}\left(1 - \frac{\Delta}{U}\right) f(x_0)$$

$$= (E^n - 1)\left(\frac{1}{e^U - 1} - \frac{1}{U}\right) f(x_0)$$

$$= \left(-\frac{1}{2} + B_1 \frac{U}{2!} - B_2 \frac{U^3}{4!} + B_3 \frac{U^5}{6!} - \cdots\right)(f(x_n) - f(x_0)).$$

Adding $f(x_n)$ to both sides, we obtain the following equality:

(18.2.1) $f(x_0) + f(x_1) + \cdots + f(x_n)$

$$= \frac{1}{h} \int_{x_0}^{x_n} f(x)\, dx + \frac{1}{2}\left[f(x_0) + f(x_n)\right] + \frac{h}{12}\left[f'(x_n) - f'(x_0)\right]$$

$$- \frac{h^3}{720}\left[f'''(x_n) - f'''(x_0)\right] + \frac{h^5}{30240}\left[f^{(v)}(x_n) - f^{(v)}(x_0)\right] - \cdots$$

This is Euler–Mclaurin's well-known summation formula. If the series is truncated after the term $B_{m-1}U^{2m-3}/(2m-2)!$, the remainder term is

$$R = \frac{nB_m h^{2m}}{(2m)!} f^{(2m)}(\xi); \qquad x_0 < \xi < x_n.$$

If n tends to infinity, however, this remainder term is of little use. Instead it can be proved that the absolute value of the error is strictly less than twice the absolute value of the first neglected term, provided that $f^{(2m)}(x)$ has the same sign throughout the interval (x_0, x_n). Further, the error has the same sign as the first neglected term.

Example 1. Euler's constant is defined by

$$\gamma = \lim_{n \to \infty} \left(1 + \frac{1}{2} + \frac{1}{3} + \cdots + \frac{1}{n} - \ln n \right),$$

which can also be written

$$\gamma = 1 + \sum_{n=2}^{\infty} \left(\frac{1}{n} + \ln \frac{n-1}{n} \right).$$

We have

$$\int_{10}^{\infty} \left(\frac{1}{x} - \ln x + \ln(x-1) \right) dx = -1 + 9 \ln 10 - 9 \ln 9$$

$$\simeq -0.05175\ 53591.$$

The sum up to $n = 9$ is computed directly, and we get

$$
\begin{aligned}
S_9 &\simeq 0.63174\ 36767 \\
\int &\simeq -0.05175\ 53591 \\
\tfrac{1}{2}f(10) &\simeq -0.00268\ 02578 \\
-\tfrac{1}{12}f'(10) &\simeq -0.00009\ 25926 \\
\tfrac{1}{720}f'''(10) &\simeq 0.00000\ 01993 \\
-\tfrac{1}{30240}f^{(v)}(10) &\simeq -0.00000\ 00015 \\
\hline
S &\simeq 0.57721\ 56650
\end{aligned}
$$

The exact value of Euler's constant is $\gamma = 0.57721\ 56649\ 015\ldots$

A similar formula using differences instead of derivatives was obtained by Gauss:

$$f(x_0) + f(x_1) + \cdots + f(x_n) = \frac{1}{h} \int_{x_0}^{x_n} f(x)\, dx + \frac{1}{2} [f(x_0) + f(x_n)]$$

$$+ \frac{1}{12} [\mu\delta f(x_n) - \mu\delta f(x_0)] - \frac{11}{720} [\mu\delta^3 f(x_n) - \mu\delta^3 f(x_0)]$$

$$+ \frac{191}{60480} [\mu\delta^5 f(x_n) - \mu\delta^5 f(x_0)] - \cdots.$$

§18.3. STIRLING'S FORMULA

We are now in a position to prove the very important Stirling formula, which we stated in the section on the gamma function. To do this we simply apply Euler–McLaurin's summation formula for the function $f(x) = \ln x$ and with interval $h = 1$. Further, let m and n be large positive integers with $n > m$. Since $f^{(2r+1)}(x) = (2r)! x^{-(2r+1)}$, we obtain the following asymptotic relation:

$$\sum_{k=m}^{n} \ln k = \int_{m}^{n} \ln x \, dx + \frac{1}{2}(\ln m + \ln n) - 12^{-1}(m^{-1} - n^{-1}) + 360^{-1}(m^{-3} - n^{-3})$$

$$- 1260^{-1}(m^{-5} - n^{-5}) + \cdots + \frac{(-1)^{r+1} B_{r+1}}{(2r+1)(2r+2)} (m^{-2r-1} - n^{-2r-1}) + \cdots$$

Applying $\sum_{k=m}^{n} \ln k = \ln n! - \ln(m-1)!$ and

$$\int_{m}^{n} \ln x \, dx = n \ln n - n - m \ln m + m$$

and collecting terms in m and n, we find

$$\ln m! - (m + \tfrac{1}{2}) \ln m + m - \tfrac{1}{12}m^{-1} + \tfrac{1}{360}m^{-3} - \cdots$$
$$= \ln n! - (n + \tfrac{1}{2}) \ln n + n - \tfrac{1}{12}n^{-1} + \tfrac{1}{360}n^{-3} - \cdots,$$

where both sides must be considered as asymptotic expressions. Now let m and n tend to infinity and truncate the two series just before the terms $m^{-1}/12$ and $n^{-1}/12$; then we understand that what is left must be a constant, since the left-hand side is a function of m only while the right-hand side is a function of n only. Hence we get

$$\lim_{n \to \infty} \left[\ln n! - (n + \tfrac{1}{2}) \ln n + n \right] = K$$

where K is a constant still to be determined. Thus we have the asymptotic expression

(18.3.1) $\qquad \ln n! = K + (n + \tfrac{1}{2}) \ln n - n + \tfrac{1}{12}n^{-1} - \tfrac{1}{360}n^{-3} + \cdots$

Using this relation for the argument $2n$ instead of n, we get:

$$\ln (2n)! - 2 \ln n! = -K + (2n + \tfrac{1}{2}) \ln 2 - \tfrac{1}{2} \ln n - \tfrac{1}{8}n^{-1} + \tfrac{1}{192}n^{-3} - \cdots$$

The left-hand side can be estimated by considering the integral $I_m = \int_0^{\pi/2} \sin^m x \, dx$ for non-negative integer values of m. By partial integration we find, when $m \geq 2$,

$$I_m = \frac{m-1}{m} I_{m-2},$$

with $I_0 = \pi/2$ and $I_1 = 1$. Hence we have

$$I_{2n} = \frac{2n-1}{2n} \frac{2n-3}{2n-2} \cdots \frac{1}{2} \frac{\pi}{2} = \frac{(2n)!}{2^{2n}(n!)^2} \frac{\pi}{2}$$

and

$$I_{2n+1} = \frac{2n}{2n+1} \frac{2n-2}{2n-1} \cdots \frac{2}{3} = \frac{2^{2n}(n!)^2}{(2n)!(2n+1)}.$$

Further we have for $n \geq 1$

$$\int_0^{\pi/2} \sin^{2n+1} x \, dx < \int_0^{\pi/2} \sin^{2n} x \, dx < \int_0^{\pi/2} \sin^{2n-1} x \, dx,$$

that is, $I_{2n+1} < I_{2n} < I_{2n-1}$. From this follows

$$\frac{I_{2n+1}}{I_{2n-1}} < \frac{I_{2n}}{I_{2n-1}} < 1 \qquad \text{or} \qquad \frac{2n}{2n+1} < \frac{[(2n)!]^2 n}{2^{4n}(n!)^4} \pi < 1.$$

Hence we finally get

(18.3.2)
$$\lim_{n \to \infty} \frac{(2n)! \sqrt{n}}{2^{2n}(n!)^2} = \frac{1}{\sqrt{\pi}},$$

which is the well-known formula by Wallis. For large values of n we have

$$\ln(2n)! - 2 \ln n! \cong 2n \ln 2 - \tfrac{1}{2} \ln(n\pi)$$

and comparing with our previous expression we get when $n \to \infty$

$$K = \tfrac{1}{2} \ln 2\pi.$$

The derivation has been performed for positive integers, but the formula can be continued to arbitrary complex values z sufficiently far away from the singular points $z = 0, -1, -2, \ldots$ We now present *Stirling's formula* in the following shape ($\ln \Gamma(n) = \ln n! - \ln n$):

(18.3.3)
$$\ln \Gamma(z) = \frac{1}{2} \ln 2\pi + \left(z - \frac{1}{2}\right) \ln z - z + \frac{1}{12z} - \frac{1}{360z^3}$$
$$+ \frac{1}{1260z^5} - \cdots + \frac{(-1)^{n-1} B_n}{2n(2n-1)z^{2n-1}} + R_{n+1},$$

where

$$|R_{n+1}| \leq \frac{B_{n+1}}{(2n+1)(2n+2)|z|^{2n+1}(\cos(\varphi/2))^{2n+1}} \qquad \text{and} \qquad \varphi = \arg z.$$

For the Γ-function itself we have

(18.3.4)
$$\Gamma(z) \sim (2\pi)^{1/2} z^{z-1/2} e^{-z} \left[1 + \frac{1}{12z} + \frac{1}{288z^2} - \frac{139}{51840z^3} - \frac{71}{2488320z^4} + \cdots\right].$$

The formulas can be used only if $|\arg z| < \pi$, but it is advisable to keep to the right half-plane and apply the functional relation when necessary.

§18.4. EULER'S TRANSFORMATION AND ALTERNATING SERIES

We start from the power series $S = u_0 + u_1 x + u_2 x^2 + \cdots$, and we consider only values of x for which the series converges. Symbolically we have

$$S = (1 + Ex + E^2 x^2 + \cdots)u_0 = \frac{1}{1 - Ex} u_0 = \frac{u_0}{1 - x - x\Delta}$$

$$= \frac{1}{1 - x} \frac{u_0}{1 - (x/(1 - x))\Delta},$$

and hence

(18.4.1)
$$S = \frac{1}{1 - x} \cdot \sum_{s=0}^{\infty} \left(\frac{x}{1 - x}\right)^s \Delta^s u_0.$$

This formula is known as *Euler's transformation.*

We now turn to the alternating series $u_0 - u_1 + u_2 - u_3 + \cdots$, which is conveniently treated by putting $x = -1$ in the formula above. Hence we obtain:

(18.4.2)
$$S = u_0 - u_1 + u_2 - \cdots = \frac{1}{2} \sum_{s=0}^{\infty} \frac{(-1)^s}{2^s} \Delta^s u_0$$

$$= \frac{1}{2}\left[u_0 - \frac{1}{2}\Delta u_0 + \frac{1}{4}\Delta^2 u_0 \cdots \right].$$

If we use central differences, we get instead

(18.4.3)
$$S = \frac{1}{2} u_0 - \frac{1}{2^3}(u_1 - u_{-1}) + \frac{1}{2^5}(\delta^2 u_1 - \delta^2 u_{-1}) - \frac{1}{2^7}(\delta^4 u_1 - \delta^4 u_{-1}) + \cdots.$$

If we prefer to use derivatives, it is more convenient to make a special calculation as follows:

$$S = \frac{u_0}{1 + E} = \frac{u_0}{e^U + 1} = \frac{1}{U}\left[\frac{U}{e^U - 1} - \frac{2U}{e^{2U} - 1} \right]u_0$$

$$= \frac{1}{U}\left[1 - \frac{U}{2} + B_1 \frac{U^2}{2!} - B_2 \frac{U^4}{4!} + \cdots - 1 + \frac{2U}{2} - B_1 \frac{4U^2}{2!} + B_2 \frac{16U^4}{4!} - \cdots \right]u_0$$

$$= \left[\frac{1}{2} - B_1 \frac{2^2 - 1}{2!} U + B_2 \frac{2^4 - 1}{4!} U^3 - \cdots \right]u_0.$$

Hence we find the formula

(18.4.4)
$$S = \frac{1}{2}\left[u_0 - \frac{h}{2} u_0' + \frac{h^3}{24} u_0''' - \frac{h^5}{240} u_0^{(v)} \right.$$

$$\left. + \frac{17h^7}{40320} u_0^{(vii)} - \frac{31h^9}{725760} u_0^{(ix)} + \frac{691h^{11}}{159667200} u_0^{(xi)} - \cdots \right].$$

Example 1.   Consider

$$S = \sum_{n=0}^{\infty} \frac{(-1)^n}{2n + 1} = \frac{1}{2} \sum_{n=0}^{\infty} \frac{(-1)^n}{n + \frac{1}{2}}.$$

We use (18.4.1) for $x = -1$ and $u_n = 1/(n + \frac{1}{2}) = (n - \frac{1}{2})^{(-1)}$ [cf. (5.4.1) and (5.4.2)]. Hence

$$\Delta u_n = -(n - \tfrac{1}{2})^{(-2)}; \qquad \Delta^2 u_n = 2(n - \tfrac{1}{2})^{(-3)}; \qquad \Delta^3 u_n = -2 \cdot 3(n - \tfrac{1}{2})^{(-4)}, \text{ etc.,}$$

and from this we get

$$\Delta^s u_0 = (-1)^s \frac{(s!)^2 2^{2s+1}}{(2s + 1)!},$$

and the final result

$$S = \sum_{s=0}^{\infty} \frac{2^{s-1}(s!)^2}{(2s + 1)!}.$$

The quotient between the $(s + 1)$th and the sth term is $(s + 1)/(2s + 3)$; that is, the convergence is now slightly better than in a geometric series with quotient $\frac{1}{2}$. The remainder term is practically equal to the last term included. Already 10 terms, corrected with a corresponding remainder term, give the result 0.7855, deviating from the correct value $\pi/4$ by only one unit in the last place. In order to obtain this accuracy by use of the original series, we would have to compute about 5000 terms.

Using derivatives instead we find:

$$S = \left(1 - \frac{1}{3} + \frac{1}{5} - \cdots - \frac{1}{19}\right) + \left(\frac{1}{21} - \frac{1}{23} + \cdots\right) = S_1 + S_2.$$

By direct evaluation we get $S_1 = 0.76045\,990472\ldots$ Here

$$u = \frac{1}{2x - 1}; \qquad u' = -\frac{2}{(2x - 1)^2}; \qquad u''' = -\frac{2 \cdot 4 \cdot 6}{(2x - 1)^4} \cdots,$$

with $2x - 1 = 21$. Thus

$$u = \frac{1}{21}; \qquad u' = -\frac{2}{21^2}; \qquad u''' = -\frac{2 \cdot 4 \cdot 6}{21^4};$$

$$u^{(v)} = -\frac{2 \cdot 4 \cdot 6 \cdot 8 \cdot 10}{21^6}, \text{ etc.,}$$

and

$$S_2 = \frac{1}{2}\left[\frac{1}{21} + \frac{1}{21^2} - \frac{2}{21^4} + \frac{16}{21^6} - \frac{272}{21^8} + \frac{7936}{21^{10}} - \cdots\right];$$

that is, $S_2 = 0.02493\,825868\ldots$ Hence $S_1 + S_2 \simeq 0.78539\,81634$, compared with the exact value

$$\pi/4 = 0.78539\,81633\,974\ldots$$

§18.5. SUMMATION BY USE OF RIEMANN'S ZETA FUNCTION

Here we also mention another technique which is often quite useful. Consider

$$f(n) = \frac{\alpha_2}{n^2} + \frac{\alpha_3}{n^3} + \frac{\alpha_4}{n^4} + \cdots,$$

where the constants α_r are assumed to have such properties as to make the series convergent. For $s > 1$, Riemann's ζ-function can be defined through

$$\zeta(s) = \sum_{k=1}^{\infty} k^{-s},$$

and this function has been tabulated accurately. If one now wants to compute $S = \sum_{n=1}^{\infty} f(n)$, we immediately find the result

$$S = \alpha_2 \zeta(2) + \alpha_3 \zeta(3) + \alpha_4 \zeta(4) + \cdots.$$

In order to improve the rate of convergence, we write instead with $z_k = \zeta(k) - 1$:

$$S = f(1) + \alpha_2 z_2 + \alpha_3 z_3 + \cdots$$

For facilitating such computations, we give a 10-decimal table of z_s for $s = 2(1)33$:

s	z_s	s	z_s	s	z_s	s	z_s
2	.64493 40668	10	99 45751	18	38173	26	149
3	.20205 69032	11	49 41886	19	19082	27	75
4	8232 32337	12	24 60866	20	9540	28	37
5	3692 77551	13	12 27133	21	4769	29	19
6	1734 30620	14	6 12481	22	2385	30	9
7	834 92774	15	3 05882	23	1192	31	5
8	407 73562	16	1 52823	24	596	32	2
9	200 83928	17	76372	25	298	33	1

Example 1

$$\sum_{n=1}^{\infty} \frac{1}{n^3 + 1} = \frac{1}{2} + \sum_{n=2}^{\infty} \left(\frac{1}{n^3} - \frac{1}{n^6} + \frac{1}{n^9} - \cdots \right)$$

$$= \frac{1}{2} + z_3 - z_6 + z_9 - \cdots \simeq 0.686503 \qquad \text{(7 terms).}$$

Example 2

$$\sum_{n=1}^{\infty} \sin^2 \frac{1}{n} = \sin^2 1 + \frac{1}{2} \sum_{n=2}^{\infty} \left(1 - \cos \frac{2}{n} \right)$$

$$= \sin^2 1 + \sum_{n=2}^{\infty} \left(\frac{1}{n^2} - \frac{1}{3n^4} + \frac{2}{45n^6} - \frac{1}{315n^8} + \cdots \right)$$

$$= \sin^2 1 + z_2 - \frac{z_4}{3} + \frac{2z_6}{45} - \frac{z_8}{315} + \cdots$$

$$\simeq 1.326324 \qquad \text{(5 terms).}$$

Transformation of a Series to Alternating Form

An interesting technique has been suggested by van Wijngaarden, who transformed a series with positive terms to an alternating series. Let

$$S = u_1 + u_2 + u_3 + \cdots,$$

and put

$$
\begin{cases}
v_1 = u_1 + 2u_2 + 4u_4 + 8u_8 + \cdots, \\
v_2 = u_2 + 2u_4 + 4u_8 + 8u_{16} + \cdots, \\
v_3 = u_3 + 2u_6 + 4u_{12} + 8u_{24} + \cdots, \\
\vdots \\
v_k = u_k + 2u_{2k} + 4u_{4k} + 8u_{8k} + \cdots, \\
\vdots
\end{cases}
$$

Then it is easily shown that $S = v_1 - v_2 + v_3 - v_4 + \cdots$. The conditions for the validity of this transformation are quite mild; for example, it suffices that the terms u_k decrease as $k^{-1-\varepsilon}$, where $\varepsilon > 0$.

Example 3

$$S = \sum_{k=1}^{\infty} \frac{1}{k^3},$$

$$v_k = \frac{1}{k^3}\left(1 + \frac{2}{2^3} + \frac{4}{4^3} + \frac{8}{8^3} + \cdots\right) = \frac{4}{3}\frac{1}{k^3}.$$

Hence we have

$$S = \frac{4}{3} \cdot \sum_{k=1}^{\infty} (-1)^{k-1}\frac{1}{k^3}.$$

EXERCISES

1. Calculate $S = 1/1 \cdot 3 - 3/5 \cdot 7 + 5/9 \cdot 11 - 7/13 \cdot 15 + \cdots$ to five places.
2. Calculate $1 - 1/\sqrt{2} + 1/\sqrt{3} - 1/\sqrt{4} + 1/\sqrt{5} - \cdots$ to six places.
3. Find $S = 1/\ln 2 - 1/\ln 3 + 1/\ln 4 - 1/\ln 5 + \cdots$ to five places.
4. Compute $\arctan 1 - \arctan \frac{1}{3} + \arctan \frac{1}{5} - \arctan \frac{1}{7} + \cdots$ to six places.
5. Calculate Catalan's constant $\sum_{k=0}^{\infty} (-1)^k/(2k+1)^2$ to ten places.
6. Compute $\sum_{n=1}^{\infty} (-1)^n(n^{1/n} - 1)$.
7. Compute $1 - 1/(1+i) + 1/2 - 1/(2+i) + 1/3 - 1/(3+i) + \cdots$ to six decimal places.
8. Compute $\sum_{n=1}^{\infty} n^3/n!$ by use of factorials.

9. Show that the series
$$S = 1/1 \cdot 1 - 2/3 \cdot 3 + 1/5 \cdot 5 + 1/7 \cdot 7 - 2/9 \cdot 9$$
$$+ 1/11 \cdot 11 + 1/13 \cdot 13 - 2/15 \cdot 15 + 1/17 \cdot 17 + \cdots$$
can be rewritten as
$$S = \frac{2}{3} \sum_{n=0}^{\infty} (2n + 1)^{-2}.$$
Find the exact value.

10. The function $f(x)$ can be differentiated an infinite number of times. We put $f(x + k) = u_k$, $k = 0, 1, 2, \ldots$ The series
$$S = u_0 + u_1 - u_2 - u_3 + u_4 + u_5 - \cdots$$
is supposed to be convergent. Find S expressed in $f(x)$, $f'(x)$, $f''(x)$, $\ldots$ up to fifth order terms.

11. Find $S = \sum_{n=1}^{\infty} n^{-1} \arctan(n^{-1})$ to five decimal places.

12. Compute $\sum_{n=1}^{\infty} n^{-5/2}$ to six places.

13. Compute $\sum_{n=1}^{\infty} (1 - \cos(1/n))$.

14. Compute $\lim_{n \to \infty} (1 + 1/\sqrt{2} + 1/\sqrt{3} + \cdots + 1/\sqrt{n} - 2\sqrt{n})$.

15. The Fibonacci series is defined through $a_{n+1} = a_n + a_{n-1}$. Compute $\sum_{n=1}^{\infty} 1/a_n$ when $a_1 = a_2 = 1$ (six decimal places).

16. Determine an asymptotic expression for $\sum_{r=1}^{n} \ln r!$.

17. Compute $\prod_{n=1}^{\infty} \cos(1/n^2)$.

18. Compute $\prod_{n=1}^{\infty} n \sin(1/n)$.

19. Compute $\prod_{n=1}^{\infty} (1 + n^{-2})$.

20. Using a computer, we have obtained $\prod_{n=3}^{50000} \cos(\pi/n) = 0.11495339$. Find the corresponding infinite product, applying a suitable estimate for the remaining factor.

21. Compute $\prod_{n=1}^{\infty} (10n - 1)(10n - 9)/(10n - 5)^2$.

22. Compute $\prod_{n=1}^{\infty} (1 - q^n)^{-1}$ for $q = 0.9$.

23. Putting $z_n = \sum_{k=2}^{\infty} k^{-n}$, $n = 2, 3, 4, \ldots$, prove that
$$\frac{z_2}{2} + \frac{z_3}{3} + \frac{z_4}{4} + \cdots = 1 - \gamma$$
where γ is Euler's constant.

24. Find $\sum_{n=1}^{\infty} 1/x_n^2$ where $x_1, x_2, x_3, \ldots$ are the positive roots of the equation $\tan x = x$ taken in increasing order.

25. Compute $\prod_{n=1}^{\infty} n^{1/n^2}$.

26. Calculate explicitly A_n/B_n for $n \le 12$ when
$$\begin{cases} A_n = A_{n-1} + (n - 1)A_{n-2} \\ B_n = B_{n-1} + (n - 1)B_{n-2} \end{cases}$$
and $A_0 = 0$, $A_1 = B_0 = B_1 = 1$. Using these values, also find $\lim_{n \to \infty} A_n/B_n$ (four decimal places).

Part Five

Differential and Integral Equations

CHAPTER 19

Ordinary Differential Equations

Eppur si muove.

Galilei

§19.1. EXISTENCE OF SOLUTIONS

An ordinary differential equation is an equation containing one independent and one dependent variable and at least one of its derivatives with respect to the independent variable; no one of the two variables need enter the equation explicitly. If the equation is of such a form that the highest (*n*th) derivative can be expressed as a function of lower derivatives and the two variables, then it is possible to replace the equation by a system of *n* first-order equations by use of a simple substitution technique. The definition of linear and homogeneous equations, as well as linear equations with constant coefficients, is trivial and should need no comment.

The discussion of a system of first-order differential equations can, in essence, be reduced to an examination of the equation

(19.1.1) $$y' = f(x, y).$$

For this reason we shall pay a great deal of attention to this equation. It is obvious that the properties of the function $f(x, y)$ are of decisive importance for the character of the solution.

We will first point out a simple geometrical fact. By equation (19.1.1) every point in the domain of definition for f is assigned one or several directions according as f is single-valued or not; only the first of these cases will be treated here. In this way it is possible to give a picture of the directions associated with the equation (see Figure 19.1), and we can obtain a qualitative idea of the nature of the solutions.

We are now going to consider two important cases for $f(x, y)$: first we assume that the function is analytic, second that it satisfies the Lipschitz condition.

If $f(x, y)$ is analytic (note that x and y may be two different complex variables), then it is an easy matter to obtain a solution of (19.1.1) by aid of Taylor's formula. The differential equation can be differentiated an arbitrary number of times, and hence we can obtain as many derivatives as we want. With the initial value $y = y_0$

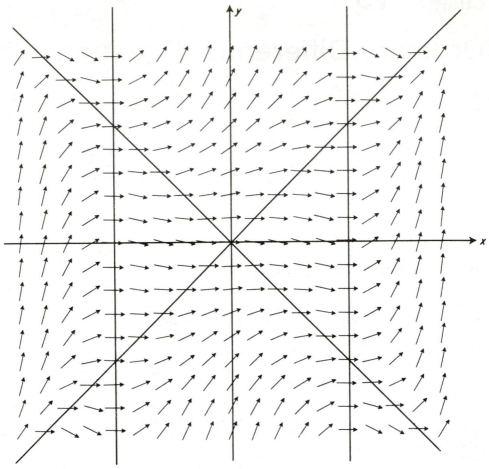

FIGURE 19.1. Direction field for the differential equation $y' = (x^2 - 1)(x^2 - y^2)$. In addition to the coordinate axes, the lines $x = \pm 1$ and $y = \pm x$ are drawn.

for $x = x_0$, we get

$$y = y_0 + (x - x_0)y'_0 + \frac{1}{2!}(x - x_0)^2 y''_0 + \cdots,$$

and provided that $|x - x_0|$ is sufficiently small, the series converges and gives a unique solution with the initial condition $y = y_0$ for $x = x_0$.

Often one does not demand analyticity from the function $f(x, y)$. However, if we require that $f(x, y)$ be bounded and continuous, this turns out not to be enough to guarantee a unique solution. A widely used extra condition is the so-called Lipschitz condition

(19.1.2) $$|f(x, y) - f(x, z)| < L|y - z|,$$

where L is a constant.

Now we suppose that this condition is fulfilled, and further that $f(x, y)$ is bounded and continuous within the domain under consideration: $|f(x, y)| < M$. Further we assume the initial condition $y = y_0$ for $x = x_0$. We integrate (19.1.1) between x_0 and x and obtain

$$(19.1.3) \qquad y = y_0 + \int_{x_0}^{x} f(\xi, y)\, d\xi.$$

Thus the differential equation has been transformed to an integral equation. Now we choose an initial approximation $y = y_1(x)$ which satisfies the conditions $|y_1'(x)| < M$ and $y_1(x_0) = y_0$. For example, we could choose $y_1(x) = y_0$, but in general it should be possible to find a better starting solution. Then we form a sequence of functions $y_i(x)$, $i = 2, 3, 4, \ldots$

$$(19.1.4) \qquad y_{i+1}(x) = y_0 + \int_{x_0}^{x} f(\xi, y_i(\xi))\, d\xi.$$

This equation defines Picard's method, which on rare occasions is used in practical work. We now obtain

$$|y_{i+1}(x) - y_i(x)| \leq \int_{x_0}^{x} |f(\xi, y_i(\xi)) - f(\xi, y_{i-1}(\xi))|\, d\xi$$
$$\leq L \int_{x_0}^{x} |y_i(\xi) - y_{i-1}(\xi)|\, d\xi.$$

We suppose $x_0 \leq x \leq X$, and putting $X - x_0 = h$, we obtain

$$|y_2(x) - y_1(x)| = \left| y_0 - y_1(x) + \int_{x_0}^{x} f(\xi, y_1(\xi))\, d\xi \right|$$
$$\leq |y_1(x) - y_0| + \int_{x_0}^{x} M\, d\xi \leq 2Mh = N.$$

Further, we find successively:

$$|y_3(x) - y_2(x)| \leq L \int_{x_0}^{x} N\, d\xi = NL(x - x_0) \leq NLh,$$

$$|y_4(x) - y_3(x)| \leq L \int_{x_0}^{x} NL(\xi - x_0)\, d\xi = N \frac{L^2(x - x_0)^2}{2!} \leq N \frac{L^2 h^2}{2!},$$

$$\vdots$$

$$|y_{n+1}(x) - y_n(x)| \leq L \int_{x_0}^{x} N \frac{L^{n-2}(\xi - x_0)^{n-2}}{(n-2)!}\, d\xi$$
$$= N \frac{L^{n-1}(x - x_0)^{n-1}}{(n-1)!} \leq N \frac{L^{n-1} h^{n-1}}{(n-1)!}.$$

But

$$(19.1.5) \qquad y_{n+1}(x) = y_1(x) + (y_2(x) - y_1(x)) + \cdots + (y_{n+1}(x) - y_n(x)),$$

and apart from $y_1(x)$, every term in this series is less in absolute value than the corresponding term in the series

$$N \left(1 + Lh + \frac{L^2 h^2}{2!} + \cdots + \frac{L^{n-1} h^{n-1}}{(n-1)!} \right),$$

which converges toward Ne^{Lh}. Thus the series (19.1.5) is absolutely and uniformly convergent toward a continuous limit function $y(x)$. The continuity follows from the fact that every partial sum has been obtained by integration of a bounded function.

Now we form

$$\left| y(x) - y_0 - \int_{x_0}^x f(\xi, y(\xi))\, d\xi \right| = \left| y(x) - y_{n+1}(x) - \int_{x_0}^x \{ f(\xi, y(\xi)) - f(\xi, y_n(\xi)) \}\, d\xi \right|$$

$$\le |y(x) - y_{n+1}(x)| + L \int_{x_0}^x |y(\xi) - y_n(\xi)|\, d\xi \to 0,$$

since $y_n(x)$ and $y_{n+1}(x)$ converge toward $y(x)$ when $n \to \infty$. Hence we find

(19.1.6)
$$y(x) = y_0 + \int_{x_0}^x f(\xi, y(\xi))\, d\xi.$$

Again we point out that $y(x) = \lim_{n \to \infty} y_n(x)$, where the formation law is given by (19.1.4). Differentiating (19.1.6), we get back (19.1.1).

It is easy to show that the solution is unique. Assuming that $z = z(x)$ is another solution, we obtain

$$z = y_0 + \int_{x_0}^x f(\xi, z(\xi))\, d\xi, \qquad y_{n+1} = y_0 + \int_{x_0}^x f(\xi, y_n(\xi))\, d\xi,$$

and hence

$$|z - y_{n+1}| \le \int_{x_0}^x |f(\xi, z) - f(\xi, y_n)|\, d\xi \le L \int_{x_0}^x |z - y_n|\, d\xi.$$

But $|z - y_0| \le Mh = N/2$, and hence $|z - y_1| \le (N/2)L(x - x_0) \le (N/2)Lh$;

$$|z - y_2| \le \frac{N}{2} \frac{L^2(x - x_0)^2}{2!} \le \frac{N}{2} \frac{L^2 h^2}{2!},$$

and finally

$$|z - y_{n+1}| \le \frac{N}{2} \frac{L^{n+1} h^{n+1}}{(n+1)!} \to 0,$$

when $n \to \infty$. Hence we have $z(x) \equiv y(x)$.

We can easily give examples of equations for which the Lipschitz condition is not fulfilled and where more than one solution exists. The equation $y' = y^{1/2}$ with $y(0) = 0$ has the two solutions $y = 0$ and $y = x^2/4$ for $x \ge 0$.

Classification of Solution Methods

In the literature one can find many examples of equations that can be solved explicitly. In spite of this fact, good arguments can be given that these equations constitute a negligible minority. We shall restrict ourselves to such methods as are of interest for equations which cannot be solved explicitly in closed form. However, the methods will often be illustrated on simple equations whose exact solution is known.

If an explicit solution cannot be constructed, one is usually satisfied by computing a solution in certain discrete points, as a rule equidistant in order to facilitate interpolation. First we shall consider the differential equation $y' = f(x, y)$, and we are going to use the notations $x_k = x_0 + kh$, and $y_k =$ the value of y obtained from the method [to be distinguished from the exact value $y(x_k)$]. The value y_{n+1} may then appear either as a function of just one y-value y_n, or as a function of several values y_n, $y_{n-1}, \ldots, y_{n-p}$; further, as a rule, the values x_n and h will also be present. In the first case we have a *single-step method*, in the second case a *multi-step method*.

The computation of y_{n+1} is sometimes performed through an *explicit*, sometimes through an *implicit*, formula. For example,

$$y_{n+1} = y_{n-1} + 2hf(x_n, y_n)$$

is an explicit multi-step formula, while

$$y_{n+1} = y_n + \frac{h}{2}\left[f(x_n, y_n) + f(x_{n+1}, y_{n+1})\right]$$

is an implicit single-step formula. In the latter case an iterative technique is often used to determine the value of y_{n+1}. If just one function value for the derivative is used, e.g., $f(\beta_k y_{n+k} + \cdots + \beta_0 y_n)$, we have a so-called *one-leg* method that can be single-step or multi-step, explicit or implicit.

Different Types of Errors

For obvious reasons error estimations play an essential role in connection with numerical methods for solution of differential equations. The errors have two different sources. First, even the simplest numerical method implies introduction of a *trunca-tion error* since the exact solutions as a rule are transcendental and need an infinite number of operations, while the numerical method only resorts to a finite number of steps. Second, all operations are performed in finite precision, and hence *rounding errors* can never be avoided.

Let $y(x_n)$ be the exact solution in the point x_n, and y_n the exact value which would result from the numerical algorithm. Then the *total truncation error* ε_n is defined through

$$\varepsilon_n = y_n - y(x_n).$$

Due to the finite precision computation we actually calculate another value $\bar{y}_n$. The rounding error ε'_n is defined through the relation

$$\varepsilon'_n = \bar{y}_n - y_n.$$

For the total error r_n we then obtain

$$|r_n| = |\bar{y}_n - y(x_n)|$$
$$= |(\bar{y}_n - y_n) + (y_n - y(x_n))| \le |\varepsilon_n| + |\varepsilon'_n|.$$

There are two main problems associated with these error estimations: (1) the form of the truncation error, and (2) the nature of the error propagation. Error propagation refers to how an error that was introduced by round-off, for example, is propagated during later stages of the process. In the first case the situation can be improved by making h smaller; in the second case such a measure will make the situation slightly worse due to an increased number of steps. When one wants to investigate the truncation errors one should first study the *local truncation error*, that is, the error resulting from use of the method in just one step.

§19.2.   EULER'S METHOD

In this section we shall discuss a method, suggested originally by Euler, but rarely used in practical work. However, the method is of great interest because of its simplicity. Applied on the differential equation $y' = f(x, y)$, $y(x_0) = \alpha$, it is defined through

$$(19.2.1) \qquad\qquad y_{n+1} = y_n + hf(x_n, y_n); \qquad y_0 = \alpha.$$

Geometrically the method has a very simple meaning: the wanted function curve is approximated by a polygon train where the direction of each part is determined as the function value $f(x, y)$ in its starting point. First we shall derive the local truncation error ε_1 assuming that the solution is twice continuously differentiable. Since ε_1 is the error after one step, we have

$$\varepsilon_1 = y_1 - y(x_1) = y_0 + hy_0' - y(x_0 + h) = -\tfrac{1}{2}h^2 y''(\xi),$$

where $x_0 < \xi < x_1$ and $y_0 = y(x_0)$. Hence the local truncation error is $O(h^2)$.

Further, we shall also estimate the total truncation error $\varepsilon_n = \varepsilon(x)$ with

$$x = x_n = x_0 + nh,$$

assuming the Lipschitz condition $\big|f(x, y_2) - f(x, y_1)\big| \le L\big|y_2 - y_1\big|$ and $\big|y''(\xi)\big| \le N$:

$$\begin{cases} y_{m+1} = y_m + hf(x_m, y_m), & m = 0, 1, 2, \ldots, n-1, \\ y(x_{m+1}) = y(x_m) + hf(x_m, y(x_m)) + \tfrac{1}{2}h^2 y''(\xi_m), & x_m < \xi_m < x_{m+1}. \end{cases}$$

Subtracting and putting $\varepsilon_m = y_m - y(x_m)$, we obtain

$$\varepsilon_{m+1} = \varepsilon_m + h[f(x_m, y_m) - f(x_m, y(x_m))] - \tfrac{1}{2}h^2 y''(\xi_m).$$

Hence

$$\big|\varepsilon_{m+1}\big| \le (1 + hL)\big|\varepsilon_m\big| + \tfrac{1}{2}h^2 N.$$

We prefer to write this equation in the form $\big|\varepsilon_{m+1}\big| \le A\big|\varepsilon_m\big| + B$ with $m = 0, 1, 2, \ldots,$ $n-1$, and by induction we easily prove for $A \ne 1$:

$$\big|\varepsilon_n\big| \le A^n\big|\varepsilon_0\big| + \frac{A^n - 1}{A - 1}\, B.$$

But $\varepsilon_0 = 0$ and $A^n = (1 + hL)^n < e^{ehL} = e^{L(x - x_0)}$, which gives

$$|\varepsilon_n| \le \frac{1}{2} hN \frac{e^{L(x - x_0)} - 1}{L}.$$

It is possible to prove the asymptotic formula: $\varepsilon_n \sim c_1 h + c_2 h^2 + c_3 h^3 + \cdots$, where the coefficients $c_1, c_2, c_3, \ldots$, depend upon x and the function $f(x, y)$. Starting from Euler's method, existence and uniqueness proofs can be constructed by showing that the sequence of functions that is obtained when $h \to 0$ converges toward a function $y(x)$ satisfying the differential equation.

Example 1. We will here demonstrate that Euler's method can actually be used in practice if it is combined with repeated Richardson extrapolation. We choose the equation $y' = e^{xy}$ with $y = 0$ for $x = 0$ and take the intervals $h = 0.05, 0.1$, and 0.2. The results are given in the table below.

x	$h = 0.05$ u	$h = 0.1$ v	$h = 0.2$ w	$U = 2u - v$	$V = 2v - w$	$y = \frac{1}{3}(4U - V)$	y exact
0	0	0	0	0	0	0	0
0.05	0.05						
0.10	0.10013	0.1		0.10025			
0.15	0.15063						
0.20	0.20177	0.20101	0.2	0.20254	0.20201	0.20271	0.20272
0.25	0.25383						
0.30	0.30711	0.30511		0.30911			
0.35	0.36193						
0.40	0.41868	0.41469	0.40816	0.42267	0.42122	0.42316	0.42321
0.45	0.47780						
0.50	0.53979	0.53274		0.54685			
0.55	0.60529						
0.60	0.67504	0.66326	0.64363	0.68682	0.68288	0.68813	0.68835

Refinement of Numerical Solutions

When solving a differential equation numerically we know, as a rule, the order of the total error, and if the computation is repeated with another interval length we can perform Richardson extrapolation, following exactly the same principles as we have just demonstrated. We shall now comment upon another idea on how to improve a numerical solution, namely *deferred correction*. For simplicity we demonstrate the technique on a specific example. Later we will give one more example in connection with boundary value problems.

Suppose that we use a trapezoid implicit one-step formula:

$$y_{n+1} = y_n + (h/2)(y'_n + y'_{n+1}).$$

It is easy to see that this formula has a local truncation error $O(h^3)$. We will now try to find an improved formula, and we make the following attempt:

$$y_{n+1} = y_n + (h/2)(y'_n + y'_{n+1}) - ah^3 y'''_{n+1/2} + \cdots$$

Using operator techniques we see that this is equivalent to

$$E = 1 + \tfrac{1}{2}(E + 1)U - aE^{1/2}U^3 + \cdots \qquad (U = hD).$$

Remember that $U = \delta - \delta^3/24 + \cdots$ (since $\delta = 2\sinh(U/2)$) and further

$$\mu = \tfrac{1}{2}(E^{1/2} + E^{-1/2}) = (1 + \delta^2/4)^{1/2} = 1 + \delta^2/8 + \cdots$$

Thus, from

$$E - 1 = \tfrac{1}{2}(E + 1)\,\delta(1 - \delta^2/24 + \cdots) - aE^{1/2}(\delta^3 - \delta^5/8 + \cdots),$$

dividing by $E^{1/2}$ and by $\delta = E^{1/2} - E^{-1/2}$ we get:

$$1 = (1 + \delta^2/8 + \cdots)(1 - \delta^2/24 + \cdots) - a\delta^2 + \cdots$$

giving $a = 1/12$. After some simplification, the correction term becomes

$$\tfrac{1}{12}(\delta^2 y_{n+1} - \delta^2 y_n) = \tfrac{1}{12}(y_{n+2} - 3y_{n+1} + 3y_n - y_{n-1}).$$

Using this correction obtained with the already computed y-values, we now solve

$$z_{n+1} = z_n + \frac{h}{2}(z'_n + z'_{n+1}) - \frac{1}{12}(y_{n+2} - 3y_{n+1} + 3y_n - y_{n-1}),$$

where we hope that z will be a better approximation of the desired solution.

Example 2. We demonstrate the technique on the equation $y' = e^{xy}$, $y(0) = 0$. Choosing $h = 0.1$ we compute y_n and $\tfrac{1}{12}\Delta^3 y_{n-1}$. After that we calculate z_n. For comparison we also determine y with $h = 0.2$ and perform Richardson extrapolation. Finally we have also computed the exact values of y using Taylor series expansion applied on the obvious relation $y'' = y'(xy' + y)$ (see the next section).

x	y	$\Delta^3 y_{n-1}/12$	z	$y(h = 0.2)$	y(extr.)	y(series)
0	0		0	0		0
0.1	0.100505	0.000191	0.100313			0.100335
0.2	0.203083	0.000234	0.202651	0.204168	0.202721	0.202721
0.3	0.310029	0.000313	0.309268			0.309427
0.4	0.424148	0.000458	0.422885	0.426959	0.423210	0.423211
0.5	0.549193					
0.6	0.690666					

We see that the values obtained by Richardson extrapolation are much better than those obtained through deferred correction. However, the comparison is not quite fair since in the former case we only get the values in the points 0.2, 0.4, ...

---

§19.3. TAYLOR SERIES EXPANSION

Assuming sufficient differentiability properties of the function $f(x, y)$, we can compute higher-order derivatives directly from the differential equation:

$$y' = f(x, y),$$

$$y'' = \frac{\partial f}{\partial x} + \frac{\partial f}{\partial y} y' = \frac{\partial f}{\partial x} + f \frac{\partial f}{\partial y},$$

$$y''' = \frac{\partial^2 f}{\partial x^2} + 2f \frac{\partial^2 f}{\partial x \, \partial y} + f^2 \frac{\partial^2 f}{\partial y^2} + \frac{\partial f}{\partial x} \frac{\partial f}{\partial y} + f \left(\frac{\partial f}{\partial y} \right)^2,$$

$$\vdots$$

After this we can compute $y(x + h) = y(x) + hy'(x) + \frac{1}{2}h^2 y''(x) + \cdots$, the series being truncated in a convenient way depending on the value of h. We demonstrate the method in an example, $y' = 1 - 2xy$. Differentiating repeatedly, we get

$$\begin{cases} y'' = -2xy' - 2y, \\ y''' = -2xy'' - 4y', \\ \vdots \\ y^{(n+2)} = -2xy^{(n+1)} - 2(n+1)y^{(n)}. \end{cases}$$

Putting $\alpha_0 = y$; $\alpha_1 = hy'$; $\alpha_2 = h^2 y''/2$; ..., we obtain

$$\alpha_{n+2} = -\frac{2h}{n+2}(x\alpha_{n+1} + h\alpha_n) \qquad \text{and} \qquad y(x+h) = \alpha_0 + \alpha_1 + \alpha_2 \cdots.$$

It is possible to compute

$$y'' = -2x(1 - 2xy) - 2y = (4x^2 - 2)y - 2x,$$
$$y''' = (-8x^3 + 4x)y + 4x^2 - 4 + 8xy = (-8x^3 + 12x)y + 4(x^2 - 1), \quad \text{etc.},$$

but it should be observed that it is better *not* to construct these explicit expressions.

Example 1

$$x = 1, \qquad h = 0.1$$
$$y = \alpha_0 = 0.53807\,95069$$

(This value comes from the solution
with initial conditions $y = 0$ for $x = 0$.)

$$\alpha_0 = 0.53807\,95069$$
$$\alpha_1 = -0.00761\,59014$$
$$\alpha_2 = -0.00461\,92049$$
$$\alpha_3 = 0.00035\,87197$$
$$\alpha_4 = 51600$$
$$\alpha_5 = -16413$$
$$\alpha_6 = 375$$
$$\alpha_7 = 36$$
$$\alpha_8 = 0$$
$$y(1.1) = 0.52620\,66801$$
$$\text{Exact value:} \quad 0.52620\,66800$$

A great advantage of this method is that it can be checked in a simple and effective manner. We have

$$y(x - h) = \alpha_0 - \alpha_1 + \alpha_2 - \alpha_3 + \cdots,$$

and this value can be compared with a previous value. In this example we find $y(0.9) = 0.54072\,43189$ compared with the exact value $0.54072\,43187$. The differences, of course, depend on round-off errors.

The error estimates in this method must, as a rule, be made empirically. Very often the values α_n vary in a highly regular way, and moreover, computational errors would be disclosed in the back-integration checking.

The exact solution can be computed in this case. With $y(0) = 0$ it is

$$y = \exp(-x^2) \int_0^x \exp(t^2)\, dt \qquad \text{(Dawson's integral)},$$

but this result is hardly of great use.

§19.4. RUNGE–KUTTA (RK) METHODS

Around 1900 there was a big search for methods utilizing a number of values of the function $f(x, y)$ taken in certain points determined recursively. The final increment of y was then obtained as a weighted mean of different f-values, multiplied with the interval length. The starting point was Euler's method, which was then improved in turn by Runge, Heun, and Kutta. Finally a method emerged which has acquired high reputation and great popularity due to its simplicity and its good accuracy. The method is known as Runge–Kutta's fourth order method, and it will be discussed in some detail below. It will become obvious that it is just one member of a whole family of methods. During the last decade there has been increased interest in this area, particularly in implicit RK methods. However, we must refrain from a more detailed discussion.

As before we consider the equation $y' = f(x, y)$ with starting point (x_0, y_0) and interval length h. Then we put

(19.4.1)
$$\begin{cases} k_1 = hf(x_0, y_0), \\ k_2 = hf(x_0 + mh, y_0 + mk_1), \\ k_3 = hf(x_0 + nh, y_0 + rk_2 + (n - r)k_1), \\ k_4 = hf(x_0 + ph, y_0 + sk_2 + tk_3 + (p - s - t)k_1), \\ k = ak_1 + bk_2 + ck_3 + dk_4. \end{cases}$$

The constants should be determined in such a way that $y_0 + k$ becomes as good an approximation of $y(x_0 + h)$ as possible. By use of series expansion, one obtains, after rather complicated calculations, the following system:

(19.4.2)
$$\begin{cases} a + b + c + d = 1, & cmr + d(nt + ms) = \tfrac{1}{6}, \\ bm + cn + dp = \tfrac{1}{2}, & cmnr + dp(nt + ms) = \tfrac{1}{8}, \\ bm^2 + cn^2 + dp^2 = \tfrac{1}{3}, & cm^2r + d(n^2t + m^2s) = \tfrac{1}{12}, \\ bm^3 + cn^3 + dp^3 = \tfrac{1}{4}, & dmrt = \tfrac{1}{24}. \end{cases}$$

We have eight equations in ten unknowns, and hence we can choose two quantities arbitrarily. If we assume that $m = n$, which seems rather natural, we find

$$m = n = \tfrac{1}{2}; \qquad p = 1; \qquad a = d = \tfrac{1}{6}; \qquad b + c = \tfrac{2}{3};$$
$$s + t = 1; \qquad cr = \tfrac{1}{6}; \qquad rt = \tfrac{1}{2}.$$

If we further choose $b = c$, we get $b = c = \tfrac{1}{3}$ and hence

$$m = n = \tfrac{1}{2}, \qquad s = 0, \qquad a = d = \tfrac{1}{6},$$
$$p = 1, \qquad t = 1, \qquad b = c = \tfrac{1}{3},$$
$$r = \tfrac{1}{2}.$$

Thus we have the final formula system:

$$(19.4.3) \quad \begin{cases} k_1 = hf(x_0, y_0), \\ k_2 = hf(x_0 + \tfrac{1}{2}h, y_0 + \tfrac{1}{2}k_1), \\ k_3 = hf(x_0 + \tfrac{1}{2}h, y_0 + \tfrac{1}{2}k_2), \\ k_4 = hf(x_0 + h, y_0 + k_3). \end{cases} \quad k = \tfrac{1}{6}(k_1 + 2k_2 + 2k_3 + k_4).$$

If f is independent of y, the formula reduces to Simpson's rule, and it can also be shown that the local truncation error is $O(h^5)$. The total error has the asymptotic form $\varepsilon(x) \sim c_4 h^4 + c_5 h^5 + c_6 h^6 + \cdots$. The explicit formula for the error term, however, is rather complicated, and in practice one keeps track of the errors by repeating the computation with $2h$ instead of h and comparing the results; we can also obtain an improvement by aid of the usual Richardson extrapolation.

Runge–Kutta's method can be applied directly to differential equations of higher order. Taking, for example, the equation $y'' = f(x, y, y')$, we put $y' = z$ and obtain the following system of first-order equations:

$$\begin{cases} y' = z, \\ z' = f(x, y, z). \end{cases}$$

This is a special case of

$$\begin{cases} y' = F(x, y, z), \\ z' = G(x, y, z), \end{cases}$$

which is integrated through:

$$\begin{cases} k_1 = hF(x, y, z), \\ k_2 = hF(x + \tfrac{1}{2}h, y + \tfrac{1}{2}k_1, z + \tfrac{1}{2}l_1), \\ k_3 = hF(x + \tfrac{1}{2}h, y + \tfrac{1}{2}k_2, z + \tfrac{1}{2}l_2), \\ k_4 = hF(x + h, y + k_3, z + l_3), \\ k = \tfrac{1}{6}(k_1 + 2k_2 + 2k_3 + k_4), \end{cases} \quad \begin{aligned} & l_1 = hG(x, y, z), \\ & l_2 = hG(x + \tfrac{1}{2}h, y + \tfrac{1}{2}k_1, z + \tfrac{1}{2}l_1), \\ & l_3 = hG(x + \tfrac{1}{2}h, y + \tfrac{1}{2}k_2, z + \tfrac{1}{2}l_2), \\ & l_4 = hG(x + h, y + k_3, z + l_3), \\ & l = \tfrac{1}{6}(l_1 + 2l_2 + 2l_3 + l_4). \end{aligned}$$

The new values are $(x + h, y + k, z + l)$.

Example 1

$$y'' = xy'^2 - y^2, \quad \begin{cases} y' = z, \\ z' = xz^2 - y^2, \end{cases} \quad \begin{array}{l} (= F(x, y, z)) \\ (= G(x, y, z)) \end{array} \quad h = 0.2.$$

Initial conditions:

$$\begin{cases} x = 0, \\ y = 1, \\ y' = 0. \end{cases}$$

x	y	$z(=F)$	$xz^2 - y^2(=G)$	k	l
0	1	0	-1	0	-0.2
0.1	1	-0.1	-0.999	-0.02	-0.1998
0.1	0.99	-0.0999	-0.979102	-0.01998	-0.1958204
0.2	0.98002	-0.1958204	-0.9527709	-0.039164	-0.1905542
0.2	0.980146	-0.196966			

The values below the line are used to initiate the next step in the integration. Note that the interval length can be changed without restrictions. On the other hand, it is somewhat more difficult to get an idea of the accuracy, as pointed out above.

We have mentioned that there is a whole family of Runge–Kutta methods of varying orders. For practical use, however, the classical formula demonstrated here has reached a dominating position by its simplicity.

A special phenomenon, which can appear when Runge–Kutta's method is applied, deserves to be mentioned. First we take an example. The equations

$$\begin{cases} y' = -12y + 9z, \\ z' = 11y - 10z \end{cases}$$

have the particular solution

$$\begin{cases} y = 9e^{-x} + 5e^{-21x}, \\ z = 11e^{-x} - 5e^{-21x}. \end{cases}$$

For $x \geq 1$, we have $e^{-21x} < 10^{-9}$, and one would hardly expect any difficulties from this component. However, starting with $x = 1$, $y = 3.3111(\simeq 9e^{-1})$, $z = 4.0469(\simeq 11e^{-1})$, and $h = 0.2$, we obtain

x	y	z	$11y/9z$
1.0	3.3111	4.0469	1.0000
1.2	2.7109	3.3133	1.0000
1.4	2.2195	2.7127	1.0000
1.6	1.8174	2.2207	1.0003
1.8	1.4892	1.8169	1.0018
2.0	1.2270	1.4798	1.0134
2.2	1.0530	1.1632	1.1064
2.4	1.1640	0.6505	2.1870
2.6	2.8360	-1.3504	-2.5668

Round-off has been performed strictly according to well-known principles. A slight change in these principles is sufficient to produce a completely different table.

The explanation is quite simple. When we use Runge–Kutta's method, we approximate e^{-ah} by $1 - ah + \frac{1}{2}a^2h^2 - \frac{1}{6}a^3h^3 + \frac{1}{24}a^4h^4$, which is acceptable when $a = 1$ and $h = 0.2$ (the error is $< 3 \cdot 10^{-6}$); however, it is a bad approximation if $a = 21$ (6.2374 compared with $e^{-4.2} \simeq 0.014996$). In spite of the fact that proper instabilities appear only in connection with multi-step methods, the phenomenon observed here is usually called *partial instability*. A reduction of the interval length (in this case to about 0.1) would give satisfactory results. If the systems of differential equations is written in the form

$$y_i' = f_i(x, y_1, y_2, \ldots, y_n); \qquad i = 1, 2, \ldots, n,$$

the stability properties are associated with a certain matrix A having the elements $a_{ik} = \partial f_i/\partial y_k$. As a matter of fact, there are several matrices A since the derivatives are considered in different points, but if we assume sufficiently small intervals we can neglect this. Denoting the characteristic values of A by λ_i, we have stability for $-2.785 < h\lambda_i < 0$ if the eigenvalues are real and for $0 < |h\lambda_i| < 2\sqrt{2}$ if the eigenvalues are purely imaginary. In the case of complex eigenvalues, $h\lambda_i$ must belong to a closed domain in the complex plane, mainly situated in the left half-plane. The value -2.785 is the real root not equal to 0 of the equation $1 + x + x^2/2! + x^3/3! + x^4/4! = 1$. In the example above, we had

$$A = \begin{pmatrix} -12 & 9 \\ 11 & -10 \end{pmatrix},$$

that is, $\lambda_1 = -1$ and $\lambda_2 = -21$. Consequently h must be chosen so that $21h < 2.785$, or $h < 0.1326$. Explicit calculations show, for example, that $h = 0.1$ gives acceptable results.

§19.5. STIFF DIFFERENTIAL EQUATIONS

The previous problem is a typical example where two widely different time scales are involved (provided x is interpreted as a time variable.) Differential equations with this property are said to be *stiff*. For small values of the independent variable it is natural that small intervals are necessary. However, when the fast components have died out one would expect that larger intervals could be used. As a matter of fact this is not true in general, and the choice of integration method turns out to be crucial.

The way one tries to master this problem is to use a suitable implicit formula. A very simple method is the implicit trapezoid rule, which can then be combined with Richardson extrapolation or deferred correction, and it has proved successful in many cases. We demonstrate the technique on the same equations as before:

$$\begin{cases} y' = -12y + 9z \\ z' = 11y - 10z \end{cases}$$

starting with $y(1) = 3.3111$, $z(1) = 4.0469$, and $h = 0.2$. Hence

$$\begin{cases} y_{n+1} = y_n + (h/2)(y'_n + y'_{n+1}) \\ z_{n+1} = z_n + (h/2)(z'_n + z'_{n+1}) \end{cases}$$

or, replacing y' and z' by the expressions above:

$$\begin{cases} 22y_{n+1} - 9z_{n+1} = -2y_n + 9z_n \\ -11y_{n+1} + 20z_{n+1} = 11y_n. \end{cases}$$

Solving for y_{n+1} and z_{n+1} we find

$$\begin{pmatrix} y_{n+1} \\ z_{n+1} \end{pmatrix} = \frac{1}{341} \begin{pmatrix} 59 & 180 \\ 220 & 99 \end{pmatrix} \begin{pmatrix} y_n \\ z_n \end{pmatrix}$$

or $U_{n+1} = AU_n$. The eigenvalues of A are $\lambda_1 = 9/11$ and $\lambda_2 = -11/31$ with the corresponding eigenvectors $\begin{pmatrix} 9 \\ 11 \end{pmatrix}$ and $\begin{pmatrix} 1 \\ -1 \end{pmatrix}$. Now the starting vector happens to be exactly the first eigenvector, and because λ_2 is absolutely much smaller than λ_1, possible rounding errors will quickly decrease since they can be viewed as a disturbance in the form of an admixture of the second eigenvector. Hence we obtain the very simple recursion

$$y_{n+1} = \frac{9}{11} y_n, \qquad z_{n+1} = \frac{9}{11} z_n$$

which should be compared with the almost exact solution

$$y_{n+1} = \exp(-0.2)y_n, \qquad z_{n+1} = \exp(-0.2)z_n.$$

We have $\exp(-0.2) = 0.818731$ while $9/11 = 0.818182$.

If we use the same method on more general differential equations, we must be prepared to solve the resulting equations, e.g., by iteration. If we use the traditional Euler method with $h = 0.2$ on the previous system of equations, we find:

$$\begin{pmatrix} y_{n+1} \\ z_{n+1} \end{pmatrix} = \frac{1}{5} \begin{pmatrix} -7 & 9 \\ 11 & -5 \end{pmatrix} \begin{pmatrix} y_n \\ z_n \end{pmatrix} = A \begin{pmatrix} y_n \\ z_n \end{pmatrix}$$

where A has the eigenvalues $\lambda_1 = 0.8$, $\lambda_2 = -3.2$ and the corresponding eigenvectors $\begin{pmatrix} 9 \\ 11 \end{pmatrix}$ and $\begin{pmatrix} 1 \\ -1 \end{pmatrix}$ (the same as before!). However, the unwanted eigenvector is associated with the absolutely largest eigenvalue and the process is unstable.

In order to get better insight, we compute the eigenvalues in the two cases for general values of h. For the implicit method the result is $\lambda_1 = (2 - h)/(2 + h)$, $\lambda_2 = (2 - 21h)/(2 + 21h)$ and hence $|\lambda_1| > |\lambda_2|$ when $0 < h < 0.4364$. We see that λ_1 is a good approximation of $\exp(-h)$ with an error $O(h^3)$. The explicit method, on the other hand, gives the eigenvalues $\lambda_1 = 1 - h$, $\lambda_2 = 1 - 21h$, and we see that we are in trouble already when h is of the order 0.1.

§19.6.   MULTI-STEP METHODS

As has been mentioned previously, a multi-step method defines the wanted value y_{n+1} as a function of several preceding values $y_n, y_{n-1}, \ldots, y_{n-k+1}$. In this case we have a k-step method, and if in particular $k = 1$, we are back with the single-step methods just discussed. The method is *explicit* if the value can be found directly, and *implicit* if the formula contains the wanted value also on the right-hand side. We shall assume equidistant abscissas throughout. Further, it is also obvious that another technique is necessary for calculation of a sufficient number of initial values. Primarily we shall treat the equation $y' = f(x, y)$, and later on also $y'' = f(x, y)$. For brevity the notation $f_k = f(x_k, y_k)$ will be used.

In order to show what kind of difficulties may arise, we will quote a classical example given by Todd as early as 1950. He considers the equation $y'' = -y$, and attempts a solution by $h^2 y'' = \delta^2 y - \frac{1}{12}\delta^4 y + \frac{1}{90}\delta^6 y - \cdots$. Truncating the series after two terms, he obtains with $h = 0.1$:

$$0.01 y_n'' = y_{n+1} - 2y_n + y_{n-1} - \tfrac{1}{12}(y_{n+2} - 4y_{n+1} + 6y_n - 4y_{n-1} + 'y_{n-2})$$
$$= -0.01 y_n,$$
$$y_{n+2} = 16 y_{n+1} - 29.88 y_n + 16 y_{n-1} - y_{n-2}.$$

As initial values, he takes 0, sin 0.1, sin 0.2, and sin 0.3, rounded to 10 and 5 decimals, respectively.

$x = x_n$	$y = \sin x$ (exact)	y_n (10 decimals)	Error	y_n (5 decimals)	Error
0	0	0		0	
0.1	0.09983 34166	0.09983 34166		0.09983	
0.2	0.19866 93308	0.19866 93308		0.19867	
0.3	0.29552 02067	0.29552 02067		0.29552	
0.4	0.38941 83423	0.38941 83685	262	0.38934	−8
0.5	0.47942 55386	0.47942 59960	4574	0.47819	−124
0.6	0.56464 24734	0.56464 90616	65882	0.54721	−1743
0.7	0.64421 76872	0.64430 99144	9 22272	0.40096	−24326
0.8	0.71735 60909	0.71864 22373	128 61464	−2.67357	−3.39093
0.9	0.78332 69096	0.80125 45441	1792 76345		
1.0	0.84147 09848	1.09133 22239	24988 12391		
1.1	0.89120 73601	4.37411 56871	3.48290 83270		

The explanation is quite simple. The characteristic equation of the difference equation is

$$r^4 - 16r^3 + 29.88r^2 - 16r + 1 = 0,$$

with two real roots $r_1 = 13.938247$ and $r_2 = 1/r_1 = 0.07174504$. The complex roots can be written $r_3 = e^{i\theta}$ and $r_4 = e^{-i\theta}$, where $\sin \theta = 0.09983347$. Thus the difference equation has the solution

$$y(n) = Ar_1^n + Br_1^{-n} + C \cos n\theta + D \sin n\theta,$$

and for large values of n the term Ar_1^n predominates. The desired solution in our case is obtained if we put $A = B = C = 0, D = 1$; it becomes $y(n) = \sin(n \cdot 0.1000000556)$ instead of $\sin(n \cdot 0.1)$. In practical computation rounding errors can never be avoided, and such errors will be interpreted as admixtures of the three suppressed solutions; of these Ar_1^n represents a rapidly increasing error. The quotient between the errors in $y(1.1)$ and $y(1.0)$ is 13.93825, which is very close to r_1.

Hence the method used is very unstable, and the reason is that the differential equation has been approximated by a difference equation of too high order.

Here it should be observed that this kind of instability cannot be defeated by making the interval length shorter, and as a rule such a step only makes the situation still worse.

The general linear k-step method is defined through the formula

$$(19.6.1) \quad \alpha_k y_{n+k} + \alpha_{k-1} y_{n+k-1} + \cdots + \alpha_0 y_n$$
$$= h[\beta_k f_{n+k} + \beta_{k-1} f_{n+k-1} + \cdots + \beta_0 f_n], \quad n = 0, 1, 2, \ldots.$$

Two polynomials will be introduced, namely,

$$\begin{cases} \rho(z) = \alpha_k z^k + \alpha_{k-1} z^{k-1} + \cdots + \alpha_0, \\ \sigma(z) = \beta_k z^k + \beta_{k-1} z^{k-1} + \cdots + \beta_0. \end{cases}$$

The polynomial $\rho(z)$ is the characteristic polynomial associated with the difference equation (19.6.1) when $h = 0$. The method is now said to be *convergent* if $\lim_{h \to 0} y = y(x_n)$, and this must be valid even for initial values that are close to the right ones and converge to these when $h \to 0$. Note that $n \to \infty$ in such a way that nh becomes finite. Then a necessary condition for convergence is that the zeros z_i of $\rho(z)$ are such that $|z_i| \leq 1$, and further that all zeros on the unit circle are simple. This condition is known as the *stability condition*. Methods for which this condition is not fulfilled are said to be *strongly unstable*.

It is easy to show that we do not get convergence if $|z_i| > 1$. To prove this we consider the equation $y' = 0$ with the initial value $y(0) = 0$ and exact solution $y(x) = 0$. Assume one root z_i such that $|z_i| = \lambda > 1$; then the solution of (19.6.1) contains a term Az_i^n (the right-hand side of the difference equation is equal to 0 since $f(x, y) = 0$). Further assume an initial value such that $A = h$ and consider y_n in the point $x = nh$. Then y_n will contain a term with the absolute value $(x/n)\lambda^n$ which tends to infinity when $h \to 0$ $(n \to \infty)$.

By discussing other simple equations we can obtain further conditions which must be satisfied in order to secure convergence of the method. Let us first consider the differential equation $y' = 0$ with the initial value $y(0) = 1$ and exact solution $y(x) = 1$. Since $f(x, y) = 0$ and all $y_r = 1$, we have

(19.6.2)
$$\alpha_k + \alpha_{k-1} + \cdots + \alpha_0 = 0.$$

Next we also consider the differential equation $y' = 1$ with the initial value $y(0) = 0$ and exact solution $y(x) = x$. Inserting this into (19.6.1) we find

$$\alpha_k y_{n+k} + \alpha_{k-1} y_{n+k-1} + \cdots + \alpha_0 y_n = h(\beta_k + \beta_{k-1} + \cdots + \beta_0).$$

Since $y_r = rh$, we obtain

$$(n + k)h\alpha_k + (n + k - 1)h\alpha_{k-1} + nh\alpha_0 = h(\beta_k + \beta_{k-1} + \cdots + \beta_0).$$

Taking (19.6.2) into account, we get

(19.6.3)
$$k\alpha_k + (k - 1)\alpha_{k-1} + \cdots + \alpha_1 = \beta_k + \beta_{k-1} + \cdots + \beta_0.$$

The conditions (19.6.2) and (19.6.3) can be written in a more compact form:

(19.6.4)
$$\begin{cases} \rho(1) = 0, \\ \rho'(1) = \sigma(1). \end{cases}$$

These relations are usually called the *consistency conditions,* and it is obvious from the derivation that they are necessary for convergence. Somewhat vaguely we can express this by saying that a method that does not satisfy the consistency conditions is mathematically wrong. On the other hand, it might well be the case that a "correct" method is strongly unstable (cf. Todd's example discussed previously).

If the local truncation error is $O(h^{p+1})$, the *order* of the method is said to be p. One is clearly interested in methods which, for a given step-number k, are of highest possible order, but at the same time stability must be maintained. Dahlquist has proved that the highest possible order that can be attained is $2k$, but if we also claim stability we cannot get more than $k + 1$ if k is odd, and $k + 2$ if k is even.

We shall now discuss a simple example in order to illustrate the effect of a multi-step method on the error propagation. We start from the equation

$$y' = ay, \qquad y(0) = 1,$$

and apply a method suggested by Milne:

$$y_{n+2} = y_n + \frac{h}{3}[y_n' + 4y_{n+1}' + y_{n+2}'],$$

that is, Simpson's formula. Also using the differential equation we obtain the following difference equation:

$$\left(1 - \frac{ah}{3}\right)y_{n+2} - \frac{4ah}{3}y_{n+1} - \left(1 + \frac{ah}{3}\right)y_n = 0.$$

The characteristic equation is

$$\left(1 - \frac{ah}{3}\right)\lambda^2 - \frac{4ah}{3}\lambda - \left(1 + \frac{ah}{3}\right) = 0$$

with the roots $(2ah/3 \pm (1 + a^2h^2/3)^{1/2})/(1 - ah/3)$ or after series expansion,

$$\lambda_1 = 1 + ah + \frac{a^2h^2}{2} + \frac{a^3h^3}{6} + \frac{a^4h^4}{24} + \frac{a^5h^5}{72} + \cdots \simeq e^{ah} + c_1 h^5,$$

$$\lambda_2 = -\left(1 - \frac{ah}{3} + \frac{a^2h^2}{18} + \frac{a^3h^3}{54} + \cdots\right) \simeq -(e^{-ah/3} + c_2 h^3),$$

where $c_1 = a^5/180$ and $c_2 = 2a^3/81$. The general solution of the difference equation is $y_n = A\lambda_1^n + B\lambda_2^n$, and since we are looking for the solution $y = e^{ax}$ of the differential equation we ought to choose $A = 1$, $B = 0$. We now compute approximate values of λ_1^n and λ_2^n, choosing n and h in such a way that $nh = x$ where x is a given value. Then we find

$$\lambda_1^n \simeq (e^{ah} + c_1 h^5)^n = e^{anh}(1 + c_1 h^5 e^{-ah})^n = e^{ax}(1 + nc_1 h^5 e^{-ah} + \cdots)$$

$$\simeq e^{ax}(1 + c_1 x h^4) = e^{ax}(1 + \alpha x),$$

where $\alpha = c_1 h^4 = a^5 h^4/180$. In a similar way we also find

$$\lambda_2^n \simeq (-1)^n e^{-ax/3}(1 + \beta x),$$

where $\beta = c_2 h^2 = 2a^3 h^2/81$. We are now interested in the solution λ_1^n, but in a numerical calculation we can never avoid errors which will then be understood as an admixture of the other solution. However, the effect of this will be quite different in the two cases $a > 0$ and $a < 0$. For simplicity, we first assume $a = 1$. Further we assume that the parasitic solution from the beginning is represented with the fraction ε and the "correct" solution with the fraction $1 - \varepsilon$. Thus, the generated solution is

$$y_n \simeq (1 - \varepsilon)e^x(1 + \alpha x) + \varepsilon(-1)^n e^{-x/3}(1 + \beta x)$$

and the error is essentially

$$E_1 = \alpha x e^x - \varepsilon e^x + \varepsilon(-1)^n e^{-x/3}.$$

Since x is kept at a fixed value, the first term depends only on the interval length ($\alpha = h^4/180$) and can be made arbitrarily small. The second term depends on ε, that is, the admixture of the parasitic solution. Even if this fraction is small from the beginning, it can grow successively because of accumulation of rounding errors. The last term finally represents an alternating and at the same time decreasing error which cannot cause any trouble.

We now give a numerical illustration and choose $h = 0.3$. Further, we treat two alternatives for the initial value y_1. We choose the first according to the *difference equation* ($y_1 = \lambda_1$), and the second according to the *differential equation* ($y_1 = \exp(0.3)$). We find

$$\lambda^2 - 4\lambda/9 = 11/9$$

TABLE 19.1   Solution of $y' = y$ by Milne's method with $y(h) = (2 + \sqrt{103})/9$ and with $y(h) = \exp(0.3)$

x	y_{I}	Error $\cdot 10^6$	$E_1^{(\mathrm{I})} \cdot 10^6$	y_{II}	Error $\cdot 10^6$	$E_1^{(\mathrm{II})} \cdot 10^6$
0	1			1	—	
0.3	1.349877	18	18	1.349859	—	0
0.6	1.822167	48	49	1.822159	40	41
0.9	2.459702	99	100	2.459676	73	74
1.2	3.320294	177	179	3.320273	156	158
1.5	4.481988	299	303	4.481948	259	262
1.8	6.050132	485	490	6.050088	441	446
2.1	8.166934	764	772	8.166864	694	703
2.4	11.024354	1178	1191	11.024270	1094	1106
2.7	14.881521	1789	1808	14.881398	1666	1686
3.0	20.088220	2683	2712	20.088062	2525	2554

and $\lambda_1 = 1.349877$; $\lambda_2 = -0.905432$. The difference equation is $9y_{n+2} = 4y_{n+1} + 11y_n$, and the results are given in Table 19.1.

For y_1 we have with good approximation $\varepsilon = 0$ and hence $E_1^{(\mathrm{I})} \simeq \alpha x e^x$ with $\alpha = h^4/180 = 0.000045$. For y_{II} we find $\varepsilon = 0.000008$, and from this, $E_1^{(\mathrm{II})}$ is computed. The difference between actual and theoretical values is due to the fact that the term αx is only an approximation of a whole series and also to some extent to rounding errors.

We now pass to the second case $a = -1$, that is, the differential equation $y' = -y$, $y(0) = 1$. In the same way as before we get

$$y_n \simeq (1 - \varepsilon)e^{-x}(1 + \alpha x) + \varepsilon(-1)^n e^{x/3}(1 + \beta x),$$

where the signs of α and β have now been changed. Hence the error is

$$E_2 \simeq (\alpha x - \varepsilon)e^{-x} + \varepsilon(-1)^n e^{x/3}(1 + \beta x).$$

Again we illustrate by a numerical example. With the same choice of interval ($h = 0.3$), the characteristic equation is $\lambda^2 + 4\lambda/11 - 9/11 = 0$ with the roots $\lambda_1 = 0.74080832$ and $\lambda_2 = -1.10444469$. Also in this case the solution is directed first according to the difference equation, that is, $y_1 = \lambda_1$, and second according to the differential equation, that is, $y_1 = \exp(-0.3)$. The difference equation has the form $11y_{n+2} = -4y_{n+1} + 9y_n$, and the results are presented in Table 19.2.

As can be seen in the table, there is good agreement between the actual and the theoretical error for the solution y_{I} up to about $x = 8$. Here the discrepancy between y_{I} and λ_1^n begins to be perceptible, which indicates that ε is now not equal to 0. We also notice the characteristic alternating error, which is fully developed and completely dominating for $x = 12$. Still the computed values y_{I} are fairly meaningful. If an error of this type appears, the method is said to be *weakly stable*. In particular it should be observed that for a given value x we can obtain any accuracy by making ε sufficiently small, and in this case, ε being equal to 0 initially, we have only to perform the calculations with sufficient accuracy. The other solution shows strong oscillations from the beginning, due to the fact that the initial value $\exp(-0.3)$

TABLE 19.2 Solution of $y' = -y$ by Milne's method with $y(h) = (-2 + \sqrt{103})/11$ and with $y(h) = \exp(-0.3)$

x	$y_I \cdot 10^8$	Error $\cdot 10^8$	$E_2^{(I)} \cdot 10^8$	$y_{II} \cdot 10^8$	Error $\cdot 10^8$	$E_2^{(II)} \cdot 10^8$
0	1 0000 0000	—		1 0000 0000	—	
0.3	7408 0832	-990	-1000	7408 1822	—	
0.6	5487 9697	-1467	-1482	5487 9337	-1826	-1842
0.9	4065 5536	-1630	-1647	4065 6277	-689	-707
1.2	3011 7812	-1609	-1626	3011 7175	-2246	-2263
1.5	2231 1525	-1491	-1506	2231 2527	-489	-505
1.8	1652 8564	-1325	-1339	1652 7679	-2210	-2223
2.1	1224 4497	-1146	-1157	1224 5638	-4	-16
2.4	907 0826	-969	-980	906 9687	-2108	-2118
2.7	671 9743	-808	-816	672 1091	$+539$	$+531$
3.0	497 8042	-665	-672	497 6620	-2087	-2093
5.4	45 1551	-107	-110	44 8346	-3313	-3314
5.7	33 4510	-87	-86	33 8055	$+3458$	$+3456$
6.0	24 7811	-64	-67	24 3899	-3976	-3976
6.3	18 3577	-53	-52	18 7900	$+4269$	$+4267$
6.6	13 5999	-38	-40	13 1227	-4810	-4808
8.1	3 0340	-14	-11	3 8183	$+7829$	$+7826$
8.4	2 2482	-5	-9	1 3820	-8667	-8664
8.7	1 6648	-11	-7	2 6215	$+9557$	$+9552$
9.0	1 2341	0	-5	1774	-10567	-10562
9.3	9133	-9	-4	2 0804	$+11661$	$+11655$
9.6	6776	$+3$	-3	-6114	-12886	-12879
12.0	628	$+14$				
12.3	439	-16				
12.6	354	$+17$				
12.9	230	-20				
13.2	206	$+21$				
13.5	113	-24				
13.8	127	$+26$				
14.1	46	-29				
14.4	83	$+27$				
14.7	7	-34				
15.0	65	$+35$				

is interpreted as an admixture of the unwanted solution λ_2^n to the desired solution λ_1^n. We find

$$\varepsilon = \frac{(\lambda_1 - e^{ah})(1 - ah/3)}{2(1 + a^2 h^2/3)^{1/2}};$$

for $a = 1$ and $h = 0.3$, we get $\varepsilon = -536 \cdot 10^{-8}$. Since $\varepsilon = O(h^5)$ the error can be brought down to a safe level by decreasing h; already a factor 4 gives $\varepsilon \sim 5 \cdot 10^{-9}$ which brings us back to the first case y_I. Again we stress the importance of steering after the *difference equation* in cases when weak stability can occur.

This case has been discussed in considerable detail because it illustrates practically all phenomena which are of interest in this connection, and the discussion of more general cases will be correspondingly facilitated. Again we choose the equation $y' = ay$, $y(0) = 1$ but now a more general method will be applied:

$$\alpha_k y_{n+k} + \alpha_{k-1} y_{n+k-1} + \cdots + \alpha_0 y_n = h(\beta_k f_{n+k} + \beta_{k-1} f_{n+k-1} + \cdots + \beta_0 f_n).$$

Putting $f = ay$ we get a difference equation with the characteristic equation

$$\rho(z) - ah\sigma(z) = 0.$$

If h is sufficiently small, the roots of this equation will be close to the roots z_i of the equation $\rho(z) = 0$. We assume the stability and consistency conditions to be fulfilled, and hence it is known that $|z_i| \leq 1$, $\rho(1) = 0$, and $\rho'(1) = \sigma(1)$. Roots whose absolute values are < 1 can never cause any difficulties, and we restrict ourselves to roots on the unit circle, all of which are known to be simple; among them is also $z_1 = 1$. Applying Newton–Raphson's method to the equation $\rho(z) - ah\sigma(z) = 0$, we find the root

$$z_r' = z_r + \frac{ah\sigma(z_r)}{\rho'(z_r)} + O(h^2).$$

We now introduce the notation $\lambda_r = \sigma(z_r)/(z_r \rho'(z_r))$; for reasons which will soon become clear these quantities are called *growth parameters*. We find

$$(z_r')^n = z_r^n(1 + a\lambda_r h + O(h^2))^n = z_r^n(\exp(a\lambda_r h + c_r h^2 + O(h^3)))^n$$
$$= z_r^n \exp(a\lambda_r nh)(1 + c_r h^2 \exp(-a\lambda r h) + O(h^3))^n$$
$$\simeq z_r^n \exp(\lambda_r ax)(1 + c_r nh^2) = z_r^n \exp(\lambda_r ax)(1 + \alpha_r x),$$

where $\alpha_r = c_r h$ and x is supposed to be a fixed value. In particular we have

$$\lambda_1 = \frac{\sigma(1)}{\rho'(1)} = 1,$$

and the desired solution is obtained in the form $e^{ax}(1 + \alpha_1 x)$. Concerning the parasitic solutions we see at once that such roots z_r', for which $|z_r| < 1$, will normally not cause any trouble (however, see the discussion below). If on the other hand $|z_r| = 1$ and $\mathrm{Re}(a\lambda_r) > 0$, we are confronted with exactly the same difficulties as have just been described (weak stability). A method that cannot give rise to weak stability is said to be *strongly stable*.

Here it should also be mentioned that the same complication as was discussed in connection with Runge–Kutta's method may appear with the present methods. For suppose that $a < 0$ and that h is chosen large enough to make $|z_r'| > 1$ in spite of the fact that $|z_r| < 1$; then the term $z_r'^n$ will initiate a fast-growing error which can be interpreted as an unwanted parasitic solution. As has been mentioned earlier, this phenomenon is called *partial instability* since it can be eliminated by choosing a smaller value of h.

Last, we shall briefly mention one more difficulty. Suppose that we are looking for a decreasing solution of a differential equation which has also an increasing solution that should consequently be suppressed. Every error introduced will be interpreted as an admixture of the unwanted solution, and it will grow quickly and independently of the difference method we are using. This phenomenon is called *mathematical* or *inherent instability*, i.e., the problem is ill-conditioned.

During the last few decades a great deal of effort has been devoted to stability properties of different methods. In many cases valuable information may be obtained from the study of how a certain method performs on the scalar test equation

$$\frac{dy}{dx} = qy.$$

It is particularly important that the method produces a bounded solution when $\text{Re}(qh) \leq 0$. The largest region in which the method is stable is called the *stability region*. As examples we only mention *A-stability* if this region contains the left half-plane, and $A(\alpha)$-*stability* if the sector $|\pi - \arg(qh)| \leq \alpha$ is included (hence $A(\pi/2)$-stability is the same as A-stability). All these notions are strongly related to stiff problems.

Milne–Simpson's Method

The general idea behind this method for solving the equation $y' = f(x, y)$ is first to perform an extrapolation by a coarse method, a "*predictor*," to obtain an approximate value of y_{n+1}, and then to improve this value by a better formula, a "*corrector*." As predictor we choose the formula:

$$y_{n+1} = y_{n-3} + \frac{4h}{3}(2y'_{n-2} - y'_{n-1} + 2y'_n)$$

with the local truncation error $\frac{14}{45}h^5 y^{(v)}(\xi_1)$, and get an approximate value of y_{n+1} with which we can compute $y'_{n+1} = f(x_{n+1}, y_{n+1})$. Then an improved value of y_{n+1} is determined from the corrector formula

$$y_{n+1} = y_{n-1} + \frac{h}{3}\left[y'_{n-1} + 4y'_n + y'_{n+1}\right],$$

with the local truncation error $-\frac{1}{90}h^5 y^{(v)}(\xi_2)$.

As before we put $y_n = y(x_n) + \varepsilon_n$, from which we get

$$y'_n = f(x_n, y_n) = f(x_n, y(x_n + \varepsilon_n)) \simeq f(x_n, y(x_n)) + \varepsilon_n \frac{\partial f}{\partial y} = y'(x_n) + \varepsilon_n \frac{\partial f}{\partial y}.$$

Assuming that the derivative $\partial f/\partial y$ varies slowly, we can approximately replace it by a constant K. The corrector formula then gives the difference equation

$$\varepsilon_{n+1} = \varepsilon_{n-1} + \frac{Kh}{3}(\varepsilon_{n-1} + 4\varepsilon_n + \varepsilon_{n+1}),$$

which describes the error propagation of the method. As has been shown previously, we have weak stability if $\partial f/\partial y < 0$. If instead we use the predictor formula

$$y_{n+1} = -4y_n + 5y_{n-1} + 2h(y'_{n-1} + 2y'_n)$$

and then the corrector formula just once, it is possible to prove that we have strong stability, in spite of the fact that the predictor formula is strongly unstable.

Methods Based on Numerical Integration

By formally integrating the differential equation $y' = f(x, y)$, we can transform it to an integral equation (cf. Chapter 21)

$$y(x) - y(a) = \int_a^x f(t, y(t))\, dt.$$

The integrand contains the unknown function $y(t)$, and, as a rule, the integration cannot be performed. However, $f(t, y(t))$ can be replaced by an interpolation polynomial $P(t)$ taking the values $f_n = f(x_n, y_n)$ for $t = x_n$ (we suppose that these values have already been computed). By choosing the limits in suitable lattice points and prescribing that the graph of $P(t)$ must pass through certain points, one can derive a whole series of interpolation formulas. It is then convenient to represent the interpolation polynomial by use of backward differences, and we consider the polynomial of degree q in s:

$$\varphi(s) = f_p + s\nabla f_p + \frac{s(s + 1)}{1 \cdot 2} \nabla^2 f_p + \cdots + \frac{s(s + 1) \cdots (s + q - 1)}{q!} \nabla^q f_p.$$

Evidently $\varphi(0) = f_p;\ \varphi(-1) = f_{p-1};\ \ldots;\ \varphi(-q) = f_{p-q}$. But $s = 0$ corresponds to $t = x_p, s = -1$ to $t = x_{p-1}$, and so on, and hence we must have $s = (t - x_p)/h$.

As an example we derive *Adams–Bashforth's* method, suggested as early as 1883. Then we must put

$$y_{p+1} - y_p = \int_{x_p}^{x_{p+1}} P(t)\, dt = h \int_0^1 \varphi(s)\, ds = h \sum_{r=0}^a c_r \nabla^r f_p,$$

where $c_r = (-1)^r \int_0^1 \binom{-s}{r}\, ds$. The first coefficients become

$$c_0 = 1; \quad c_1 = \frac{1}{2}; \quad c_2 = \int_0^1 \frac{s(s+1)}{2}\, ds = \frac{5}{12};$$

$$c_3 = \int_0^1 \frac{s(s+1)(s+2)}{6}\, ds = \frac{3}{8}; \quad c_4 = \frac{251}{720};$$

$$c_5 = \frac{95}{288}; \quad c_6 = \frac{19087}{60480}; \quad c_7 = \frac{36799}{120960}; \quad \cdots$$

It is obvious that a special starting procedure is needed in this case.

If the integration is performed between x_{p-1} and x_p, we get *Adams–Moulton's* method; the limits x_{p-1} and x_{p+1} gives *Nyström's* method, while x_{p-2} and x_p give

Milne–Simpson's method, which has already been treated in considerable detail. However, Nyström's method may give rise to weak stability.

§19.7. COWELL–NUMEROV'S METHOD

Equations of the form $y'' = f(x, y)$, that is, not containing the first derivative, can be treated by a special technique. It has been devised independently by several authors: Cowell, Crommelin, Numerov, Störmer, Milne, Manning, and Millman. We start from the operator formula (5.3.4)

$$\frac{\delta^2}{U^2} = 1 + \frac{\delta^2}{12} - \frac{\delta^4}{240} + \cdots.$$

Hence we have

$$\delta^2 y_n = \left(1 + \frac{\delta^2}{12} - \frac{\delta^4}{240} + \cdots\right) h^2 y_n'' = h^2 \left(1 + \frac{\delta^2}{12} - \frac{\delta^4}{240} + \cdots\right) f_n,$$

or

$$y_{n+1} - 2y_n + y_{n-1} = \frac{h^2}{12}(f_{n+1} + 10f_n + f_{n-1}) - \frac{h^6}{240} y_n^{\text{VI}} + O(h^8),$$

since $\delta^4 \simeq U^4 = h^4 D^4$ and $y_n'' = f_n$. Neglecting terms of order h^6 and higher, we get the formula

(19.7.1) $$y_{n+1} = 2y_n - y_{n-1} + \frac{h^2}{12}(f_{n+1} + 10f_n + f_{n-1}),$$

with a local truncation error $O(h^6)$. In general, the formula is implicit, but in the special case when $f(x, y) = yg(x)$ we get the following explicit formula:

(19.7.2) $$y_{n+1} = \frac{\beta_n y_n - \alpha_{n-1} y_{n-1}}{\alpha_{n+1}},$$

where $\alpha = 1 - (h^2/12)g$ and $\beta = 2 + (5h^2/6)g$.

In a special example we shall investigate the total error of the method. Let us discuss the equation $y'' = a^2 y$ leading to the following difference equation for y:

$$\left(1 - \frac{a^2 h^2}{12}\right) y_{n+1} - \left(2 + \frac{10a^2 h^2}{12}\right) y_n + \left(1 - \frac{a^2 h^2}{12}\right) y_{n-1} = 0,$$

with the characteristic equation

$$\lambda^2 - 2\left(\frac{1 + 5a^2 h^2/12}{1 - a^2 h^2/12}\right)\lambda + 1 = 0.$$

The roots are $\{1 + 5a^2h^2/12 \pm ah(1 + a^2h^2/6)^{1/2}\}/(1 - a^2h^2/12)$, and after series expansion we find

$$
\begin{cases}
\lambda_1 = \exp(ah) + \dfrac{a^5h^5}{480} + \dfrac{a^6h^6}{480} + \cdots, \\[2mm]
\lambda_2 = \exp(-ah) - \dfrac{a^5h^5}{480} + \dfrac{a^6h^6}{480} - \cdots.
\end{cases}
$$

Hence we have

$$
\lambda_1^n = \exp(anh)\left(1 + \frac{a^5h^5}{480}\exp(-ah) + \frac{a^6h^6}{480}\exp(-ah) + \cdots\right)^n
$$

$$
= \exp(ax)\left(1 + \frac{a^5h^5}{480} + O(h^7)\right)^n = \exp(ax)[1 + \alpha_4 h^4 + O(h^6)],
$$

where $\alpha_4 = a^5 x/480$ and $nh = x$. In a similar way we find

$$
\lambda_2^n = \exp(-ax)\left(1 - \frac{a^5h^5}{480} + O(h^7)\right)^n = \exp(-ax)[1 - \alpha_4 h^4 + O(h^6)].
$$

This means that the total truncation error can be written

$$
\varepsilon(x) \simeq c_4 h^4 + c_6 h^6 + \cdots.
$$

An error analysis in the general case gives the same result.

As is easily inferred, the method is extremely fast. The following disadvantages must, however, be taken into account: (1) The first step needs another method, for example, series expansion. (2) Interval changes are somewhat difficult. (3) The derivative is not obtained during the calculation. Just as was the case with Milne–Simpson's method for the equation $y' = f(x, y)$, Cowell–Numerov's method can be interpreted as one of a whole family of methods designed for the equation $y'' = f(x, y)$.

The following identity is easily proved by differentiation:

$$
y(x + k) - y(x) = ky'(x) + \int_x^{x+k} (x + k - z)y''(z)\, dz.
$$

The same relation with k replaced by $-k$ becomes

$$
y(x - k) - y(x) = -ky'(x) + \int_x^{x-k} (x - k - z)y''(z)\, dz.
$$

Replacing z by $2x - z$, putting $f(x, y(x)) = f(x)$, and adding the relations for k and $-k$ we obtain

$$
(19.7.3) \qquad y(x + k) - 2y(x) + y(x - k) = \int_x^{x+k} (x + k - z)[f(z) + f(2x - z)]\, dz.
$$

Instead of $f(z)$ we then introduce a suitable interpolation polynomial of degree q through the points $x_p, x_{p-1}, \ldots, x_{p-q}$, and by different choices of x, k, and q we are now able to derive a whole family of methods. If we choose $x = x_p$ and $x + k = x_{p+1}$, we get *Störmer's* method:

$$y_{p+1} - 2y_p + y_{p-1} = h^2 \sum_{m=0}^{q} a_m \nabla^m f_p,$$

where

$$a_m = (-1)^m \int_0^1 (1-t) \left\{ \binom{-t}{m} + \binom{t}{m} \right\} dt.$$

One finds $a_0 = 1$, $a_1 = 0$, $a_2 = \frac{1}{12}$, $a_3 = \frac{1}{12}$, $a_4 = \frac{19}{240}$, $a_5 = \frac{3}{40}, \ldots$

If instead we choose $x = x_{p-1}$ and $x + k = x_p$ with $q \geq 2$, we again find Cowell–Numerov's method:

$$y_p - 2y_{p-1} + y_{p-2} = h^2 \sum_{m=0}^{q} b_m \nabla^m f_p,$$

where

$$b_m = (-1)^m \int_0^1 t \left\{ \binom{t}{m} + \binom{2-t}{m} \right\} dt.$$

The following values are obtained: $b_0 = 1$, $b_1 = -1$, $b_2 = \frac{1}{12}$, $b_3 = 0$, $b_4 = -\frac{1}{240}$, $b_5 = -\frac{1}{240}, \ldots$

Finally, we shall also briefly discuss the stability problems for equations of the form $y'' = f(x, y)$. First, we define the *order* of the method: if the local truncation error is $O(h^{p+2})$, then the order of the method is said to be p. This means, for example, that Cowell–Numerov's method as just discussed is of order 4. Again, we use the same notations $\rho(z)$ and $\sigma(z)$ as for the equation $y' = f(x, y)$. Our method can now be written

$$(19.7.4) \qquad\qquad \rho(E)y_n = h^2 \sigma(E)f_n.$$

The *stability condition* can then be formulated as follows. A necessary condition for convergence of the multi-step method defined by (19.7.4) is that all zeros z_i of $\rho(z)$ are such that $|z_i| \leq 1$, and further for the roots on the unit circle the multiplicity is not greater than 2.

The proof is conducted by discussing the problem $y'' = 0$, $y(0) = y'(0) = 0$, with the exact solution $y(x) = 0$.

In a similar way we derive the *consistency conditions*

$$(19.7.5) \qquad\qquad \rho(1) = 0; \qquad \rho'(1) = 0; \qquad \rho''(1) = 2\sigma(1),$$

with the simple meaning that if the integration formula is developed in powers of h we must have identity in terms of the orders h^0, h^1, and h^2. For Cowell–Numerov's method we have

$$\begin{cases} \rho(z) = z^2 - 2z + 1, \\ \sigma(z) = \frac{1}{12}(z^2 + 10z + 1), \end{cases}$$

and hence both the stability and the consistency conditions are satisfied. In Todd's example we have $\rho(z) = z^4 - 16z^3 + 30z^2 - 16z + 1$ and $\sigma(z) = -12$. The roots of

$\rho(z) = 0$ are 1, 1, $7 + 4\sqrt{3}$, and $7 - 4\sqrt{3}$, and since $7 + 4\sqrt{3} > 1$ the stability condition is not fulfilled. On the other hand, the consistency conditions are satisfied since

$$\rho(1) = \rho'(1) = 0 \quad \text{and} \quad \rho''(1) = -24 = 2\sigma(1).$$

§19.8. SYSTEMS OF FIRST-ORDER LINEAR DIFFERENTIAL EQUATIONS

We denote the independent variable by t, and the n dependent variables by $x_1, x_2, \ldots, x_n$. They will be considered as components of a vector x, and hence the whole system can be written in the compact form

$$(19.8.1) \qquad \frac{dx}{dt} = A(t)x.$$

Here $A(t)$ is a square matrix whose elements are functions of t. Now we assume that $A(t)$ is continuous for $0 \leq t \leq T$. Then it can be proved that the system above with the initial condition $x(0) = c$ has a unique solution in the interval $0 \leq t \leq T$; c is, of course, a constant vector. The proof is essentially the same as that given in Section 19.1.

The uniqueness is proved in complete analogy to the one-dimensional case.

In the remainder of this section we shall concentrate on a very important special case, namely, when A is constant. Then the system has the simple form

$$(19.8.2) \qquad \frac{dx}{dt} = Ax; \qquad x(0) = x_0.$$

We can write down the solution directly:

$$(19.8.3) \qquad x = e^{tA}x_0.$$

Surprisingly enough, this form is well suited for numerical work. First we choose a value of t, which is so small that e^{tA} can easily be computed. Then we use the properties of the exponential function to compute $e^{mtA} = e^{TA}$, where $T = mt$ is the desired interval length. After this has been done, one step in the computation consists of multiplying the solution vector by the matrix e^{TA}.

It is of special interest to examine under what conditions the solution is bounded also for $t \to \infty$. We then restrict ourselves to the case when all eigenvalues are different. In this case there exists a regular matrix S such that $S^{-1}AS = D$, where D is a diagonal matrix. Hence $e^{tA} = Se^{tD}S^{-1}$, and we have directly one sufficient condition: for all eigenvalues we must have

$$\text{Re}(\lambda_i) \leq 0.$$

Problems of this kind are common in biology, where $x_1, x_2, \ldots, x_n$ (the components of the vector x) represent, for example, the amounts of a certain substance at different places in the body. This implies that $\sum_{i=1}^{n} x_i = \text{constant}$ and hence

344 ORDINARY DIFFERENTIAL EQUATIONS

$\sum_{i=1}^{n} (dx_i/dt) = 0$, which means that the sum of the elements in each column of A is zero. Thus the matrix A is *singular*, and at least one eigenvalue is zero. Often the matrix has the following form:

$$A = \begin{pmatrix} -a_{11} & a_{12} & \cdots & a_{1n} \\ a_{21} & -a_{22} & \cdots & a_{2n} \\ \vdots & & & \\ a_{n1} & a_{n2} & \cdots & -a_{nn} \end{pmatrix}$$

with $a_{ij} \geq 0$ and

$$a_{jj} = \sum_{\substack{i=1 \\ i \neq j}}^{n} a_{ij}.$$

From the estimate (2.6.2), we have

$$|\lambda + a_{jj}| \leq \sum_{\substack{i=1 \\ i \neq j}}^{n} a_{ij},$$

or

(19.8.4) $$|\lambda + a_{jj}| \leq a_{jj}.$$

Hence for every eigenvalue, either $\lambda = 0$ or $\text{Re}(\lambda) < 0$.

When $t \to \infty$, the contribution from those eigenvalues which have a negative real part will vanish, and the limiting value will be the eigenvector belonging to the eigenvalue $\lambda = 0$. This can also be inferred from the fact that the final state must be stationary, that is, $dx/dt = 0$ or $Ax = 0$.

If the problem is changed in such a way that A is singular, as before, but at least some eigenvalue has a positive real part, we have still a stationary solution which is not 0, but it is unstable. Every disturbance can be understood as an admixture of the components which correspond to eigenvalues with positive real parts, and very soon they will predominate completely.

---

§19.9.   BOUNDARY VALUE PROBLEMS

I eat my peas with honey,
I've done it all my life.
It makes the peas taste funny,
But it keeps them on my knife.

The solution of $y' = f(x, y)$ is, in general, a function of the form $F(x, y, C) = 0$, where C is an arbitrary constant. The integral curves form a one-parameter family, where a special curve corresponds to a special choice of the constant C. Usually we specialize by assigning a function value $y = y_0$ which should correspond to the abscissa $x = x_0$. Then the integration can begin at this point, and we are solving what is called an *initial-value problem*. For equations of second and higher order, we can specialize by conditions at *several* points, and in this way we obtain a *boundary value problem*. Later we shall consider only equations of the second order; the methods can be generalized directly to equations of higher order.

Often the boundary conditions have the form

(19.9.1)
$$\begin{cases} a_0 y_0 + b_0 y_0' = C_0 \\ a_1 y_1 + b_1 y_1' = C_1. \end{cases}$$

In nth-order equations such conditions can be imposed in n points:

(19.9.2)
$$\sum_{k=0}^{n-1} a_{ik} y_i^{(k)} = b_i; \qquad i = 0, 1, \ldots, n-1.$$

There are at least five methods that are widely used for boundary value problems, namely the *shooting method*, the *collocation method*, the *repeated integration method*, the *band matrix method*, and the *minimization method*. We will consider the *finite element method* as a special case of the minimization method. In the following we are going to treat examples on simple problems of the type $y'' = f(x, y, y')$ with boundary conditions $y(a) = c$, $y(b) = d$, where $a \leq x \leq b$.

The shooting method is a trial-and-error technique where we guess an initial value for the first derivative, perform the integration using some standard method, and record the y-value obtained for $x = b$. After a few trials it should be possible to carry out an approximate interpolation, which will then make further improvements possible.

The collocation method works in the following way. We define a function $P(x)$ through

$$P(x) = p(x) + \sum_{i=1}^{n} a_i q_i(x)$$

where $p(a) = c$, $p(b) = d$, and $q_i(a) = q_i(b) = 0$. The coefficients a_i should be chosen in such a way that in some arbitrarily selected *collocation points* $x_1, x_2, \ldots, x_n$ the relation

$$P''(x_i) = f(x_i, P(x_i), P'(x_i))$$

is satisfied. Often $p(x)$ and $q_i(x)$ are polynomials, but this is by no means necessary.

The repeated integration method starts with a suitable function y_0 satisfying the boundary conditions. Then we construct a series of successively improved approximations from $y_{n+1}'' = f(x_n, y_n, y_n')$ where the integration constants are determined to fulfill the boundary conditions.

The band matrix method should preferably be used on linear equations; it will then produce a system of linear equations with the coefficient matrix of band form. It can also be used on some "weakly" nonlinear equations. The obtained solution can be improved, either by deferred correction or by Richardson extrapolation.

In the minimization method suggested by Ritz and Galerkin,* we consider the given differential equation as the Euler equation of a certain definite integral. This can then be approximated by expressing the wanted solution as a linear combination of certain basis functions, often orthogonal polynomials. After that the minimum will be taken over the corresponding coefficients, which of course must be chosen in such a way that the boundary conditions are satisfied.

Example 1.   Consider the differential equation $y'' = \frac{1}{2}y^2$ with the boundary conditions $y(0) = 0$, $y(1) = 2$. First we guess a value $a = y'(0)$ and then we perform a numerical integration by, e.g., Runge–Kutta's method. Using $h = 0.5$ and $h = 1$ we find after Richardson extrapolation:

$$y(1) = 1.9405 \qquad \text{for } a = 1.8,$$
$$y(1) = 2.0569 \qquad \text{for } a = 1.9.$$

A linear interpolation then gives $a = 1.8511$, which implies $y(1) = 2.0004$.

In this case it is also possible to construct a series expansion, and with a straight-forward technique we find, with $y'(0) = a$:

$$y^{(4)}(0) = a^2; \qquad y^{(7)}(0) = 5a^3; \qquad y^{(10)}(0) = 75a^4; \qquad y^{(13)}(0) = 2475a^5; \qquad \ldots,$$

all other derivatives being zero. Putting $x = 1$ we have to solve the equation

$$a + \frac{a^2}{4!} + \frac{5a^3}{7!} + \frac{75a^4}{10!} + \frac{2475a^5}{13!} + \cdots = 2$$

which gives $a = 1.8507$, in good agreement with the previous value.

Example 2.   Turning to the collocation method, we now consider the differential equation $y'' + x^2 y = 0$ with boundary conditions $y(0) = 0$, $y(1) = 1$. We put simply $y = ax + bx^2 + cx^3 + dx^4$ with the obvious condition $y(1) = a + b + c + d = 1$. Further we assume $x = 1/4, 1/2$, and $3/4$ as collocation points and get three equations of the form:

$$2b + 6cx + 12dx^2 + x^2(ax + bx^2 + cx^3 + dx^4) = 0$$

which should be satisfied for these three values. This leads to the system $Au = b$ where

$$A = \begin{pmatrix} 1 & 1 & 1 & 1 \\ 8 & 132 & 194 & 193 \\ 64 & 8208 & 6148 & 3073 \\ 1728 & 9488 & 19404 & 28377 \end{pmatrix}, \quad u = \begin{pmatrix} a \\ b \\ c \\ d \end{pmatrix}, \quad b = \begin{pmatrix} 1 \\ 0 \\ 0 \\ 0 \end{pmatrix}$$

with the solution

$$\begin{cases} a = & 1.0591563 \\ b = & -0.0463835 \\ c = & 0.1145303 \\ d = & -0.1273032. \end{cases}$$

---

* Pronounced Galyórkin.

Moreover, $y(\frac{1}{2}) = a/2 + b/4 + c/8 + d/16 = 0.524342$, while the correct value, obtained from the series expansion

$$y = k\left[x - \frac{x^5}{4\cdot 5} + \frac{x^9}{4\cdot 5\cdot 8\cdot 9} - \frac{x^{13}}{4\cdot 5\cdot 8\cdot 9\cdot 12\cdot 13} + \cdots\right]$$

with $k = 1.051867581$ (obtained from $y(1) = 1$) is $y(\frac{1}{2}) = 0.52429167$.

We now solve the same problem with the repeated integration method, choosing $y_0 = x$. The following results were computed:

n	y_n	$y_n(\frac{1}{2})$
0	x	0.5
1	$(21x - x^5)/20$	0.52344
2	$(7573x - 378x^5 + 5x^9)/7200$	0.5242635
3	$(5907277x - 295347x^5 + 4095x^9 - 25x^{13})/5616000$	0.5242908

Example 3. The use of the band matrix method is demonstrated on the equation

$$y'' - (1 + 2/x)y + x + 2 = 0,$$

with the boundary conditions $y = 0$ for $x = 0$ and $y = 2$ for $x = 1$. Approximating y'' by $\delta^2 y/h^2$ and choosing $h = 0.2$, we obtain

$$y_{n+1} - (2.04 + 0.08/x)y_n + y_{n-1} + 0.04(x + 2) = 0.$$

This results in a linear system of equations with four unknowns, which are denoted as follows: $y(0.2) = y_1$, $y(0.4) = y_2$, $y(0.6) = y_3$, and $y(0.8) = y_4$. The system becomes

$$\begin{cases} 2.44y_1 - y_2 & = 0.088, \\ -y_1 + 2.24y_2 - y_3 & = 0.096, \\ -y_2 + 2.1733y_3 - y_4 = 0.104, \\ -y_3 + 2.14y_4 = 2.112. \end{cases}$$

We find

$$\begin{cases} y_1 = 0.2902 & \text{(exact 0.2899)}, \\ y_2 = 0.6202 & (\;\text{''}\quad 0.6195), \\ y_3 = 1.0030 & (\;\text{''}\quad 1.0022), \\ y_4 = 1.4556 & (\;\text{''}\quad 1.4550), \end{cases}$$

in good agreement with the exact solution $y = x(e^{x-1} + 1)$.

Even if the differential equation is not linear, it may be possible to use the band matrix method. Then the resulting equations must be solved by iteration. Applied for $h = 1/4$ on the equation $y'' = \frac{1}{2}y^2$ with $y(0) = 0$ and $y(1) = 2$, we find after a few iterations $y(1/4) = 0.4651$, $y(1/2) = 0.9370$, and $y(3/4) = 1.4362$ as compared with the correct values 0.4632, 0.9343, and 1.4340 obtained from the series expansion.

We will also demonstrate briefly how a deferred correction technique can be applied. We choose the equation $u'' + u = f(x)$ with the boundary conditions $u(0) = u(1) = 0$. Using the operators $hD = U = \delta - \delta^3/24 + \cdots$ we find $U^2 = \delta^2 - (1/12)\delta^4 + \cdots$ and from $U^2u + h^2u = h^2f(x)$ we get

$$\delta^2u + h^2u = h^2f(x) + (1/12)\delta^4u + \cdots$$

This is then replaced by

$$\begin{cases} \delta^2u + h^2u = h^2f(x) \\ \delta^2z + h^2z = h^2f(x) + (1/12)\delta^4u \end{cases}$$

or

$$\begin{cases} h^{-2}(u_{i-1} - (2 - h^2)u_i + u_{i+1}) = f(x_i) \\ h^{-2}(z_{i-1} - (2 - h^2)z_i + z_{i+1}) = f(x_i) + (1/12)h^{-2}\delta^4u_i. \end{cases}$$

Let us take $h = 1/5$ and $f(x) = x^2 + 3x$. In order to compute δ^4u we also need $u(-1/5) = u_{-1}$ and $u(6/5) = u_6$. The first value is obtained from $u_{-1} - 2u_0 + u_1 + h^2u_0 = h^2f(0)$ where $u_0 = 0$ and hence $u_{-1} = -u_1$. Similarly, $u_6 = h^2f(1) - u_4 = 0.35292236$.

The system becomes:

$$\begin{cases} -49u_1 + 25u_2 & = 0.64 \\ 25u_1 - 49u_2 + 25u_3 & = 1.36 \\ 25u_2 - 49u_3 + 25u_4 = 2.16 \\ 25u_3 - 49u_4 = 3.04 \end{cases}$$

After solving this system, we construct a new system where the right-hand sides have been slightly modified, and in this way the following table is constructed. For comparison we also give the correct values computed from the exact solution

$$u = 2 \cos x - 2 \cot \tfrac{1}{2} \sin x + x^2 + 3x - 2.$$

x	u	δ^4u	z	u (exact)
-0.2	0.12707764			
0	0		0	0
0.2	-0.12707764	0.0019727	-0.12719057	-0.12719039
0.4	-0.22347218	0.0006664	-0.22352911	-0.22352900
0.6	-0.25652782	-0.0006664	-0.25647089	-0.25647100
0.8	-0.19292236	-0.0019727	-0.19280943	-0.19280961
1	0		0	0
1.2	0.35292236			

We now turn to the last method depending on reformulation to a minimum problem. We demonstrate the basic ideas on some examples and start with the equation $y'' - y = 0$ together with the boundary conditions $y(0) = 0$, $y(1) = 1$. The exact solution is obviously

$$y = \frac{\sinh x}{\sinh 1}.$$

Forming

$$G = \int_0^1 (y'^2 + y^2)\, dx$$

we see that the Euler equation (see Section 9.1) is exactly $y'' - y = 0$. Hence we want to find a function $y = y(x)$ that makes G a minimum. Taking the exact solution, we find $G_{\min} = \coth 1 = 1.3130352855$. Already $y = x$ is a reasonable approximation, giving $G = 4/3 = 1.3333\ldots$, remarkably close to the correct value.

Putting $y = ax^2 + bx$, we must have $y(1) = a + b = 1$, and we find after simple computations

$$G = \tfrac{23}{15}a^2 + \tfrac{5}{2}ab + \tfrac{4}{3}b^2$$

or, since $b = 1 - a$,

$$G = (11a^2 - 5a + 40)/30.$$

This expression is minimized for $a = 5/22$, $b = 17/22$ giving $G = 1.314394$. A similar computation with three parameters gives

$$y = (77x^3 - 8x^2 + 404x)/473 \qquad \text{and} \qquad G = 1.31303735$$

with an error of only $2 \cdot 10^{-6}$.

Our previous example $y'' = \tfrac{1}{2}y^2$ can also be treated by this method. We then have to minimize

$$G = \int_0^1 \left(y'^2 + \frac{1}{3}y^3 \right) dx$$

under the conditions $y(0) = 0$, $y(1) = 2$. With three parameters we find after somewhat lengthy computations $y = ax + bx^2 + cx^3$ where $a = 1.8852$, $b = -0.1985$, and $c = 0.3133$.

Finally, we will briefly discuss a method that has attracted much attention during the last decades and which in a way represents a combination of the minimum and the band matrix methods. It is known as the *finite element method*, and it has proved to be very powerful and flexible, particularly when applied to partial differential equations. We demonstrate the technique on a simple differential equation: $y'' - y = 0$ with boundary conditions $y(0) = 0$, $y(1) = 1$ which has been discussed in considerable detail before. We will use a set of simple basis functions which do not form an orthogonal system but which possess other nice properties. They are defined as follows. Let $x_i = ih$, $i = 0(1)N$ and put

$$\varphi_i(x) = \begin{cases} x/h - (i-1) & (i-1)h \le x \le ih \\ i+1 - x/h & ih \le x \le (i+1)h \\ 0 & \text{otherwise.} \end{cases}$$

Hence, in the interval $x_i < x < x_{i+1}$ only φ_i and φ_{i+1} are $\neq 0$, and further the function $f(x) = a\varphi_i(x) + b\varphi_{i+1}(x)$ is linear with $f(x_i) = a$, $f(x_{i+1}) = b$. Using this technique we can cut a curve in pieces replacing each piece by the corresponding chord.

Let us now try to minimize

$$G = \int_0^1 (y'^2 + y^2)\, dx$$

under the conditions $y(0) = 0$, $y(1) = 1$. We choose $N = 4$, $h = 1/N = 1/4$ and put $y = a\varphi_0 + b\varphi_1 + c\varphi_2 + d\varphi_3 + e\varphi_4$. From the boundary conditions we obtain immediately $a = 0$ and $e = 1$. After standard computations and simplification we find

$$G = (1/12)[98(b^2 + c^2 + d^2 + \tfrac{1}{2}) - 95(bc + cd + d)].$$

From the minimization conditions $\partial G/\partial b = \partial G/\partial c = \partial G/\partial d = 0$ we get

$$\begin{cases} 196b - 95c & = 0 \\ -95b + 196c - 95d = 0 \\ -95c + 196d = 95 \end{cases}$$

giving $b = 0.2147875$, $c = 0.4431405$, $d = 0.6994814$. For a general value N we find the matrix elements $-(6N^2 - 1)$ and $12N^2 + 4$. The parameters are obtained as solutions to a difference equation with characteristic equation

$$\lambda^2 - [(12N^2 + 4)/(6N^2 - 1)]\lambda + 1 = 0.$$

Denoting the roots of this equation by λ_1 and λ_2, we have the general solution $a_r = A\lambda_1^r + B\lambda_2^r$ where A and B can be determined from $a_0 = 0$, $a_N = 1$. Hence $a_r = (\lambda_1^r - \lambda_2^r)/(\lambda_1^N - \lambda_2^N)$. The value of G becomes:

N	G	
4	1.31455 2878	
5	1.31401 0273	
6	1.31371 3772	
(∞)	1.31303 5317	(extrapolated)
	1.31303 52855	(exact)

This technique can be generalized by constructing more general basis functions, e.g., consisting of piecewise cubic polynomials in three parts. These are chosen in such a way that the function and its derivative are continuous, and further:

$$\varphi_i(x_{i-1}) = \varphi_i(x_{i+2}) = 0; \qquad \varphi_i'(x_{i-1}) = \varphi_i'(x_{i+2}) = 0;$$
$$\varphi_i''(x_{i-1}) = \varphi_i''(x_{i+2}) = 0; \qquad \varphi_i(x_i) = \varphi_i(x_{i+1}) = 1 \qquad (x_i = ih).$$

For $h = 1$ and $i = 0$ we get:

$$\begin{aligned} \varphi_0 &= (x + 1)^3 & -1 \le x \le 0 \\ \varphi_0 &= 1 + 3x - 3x^2 & 0 \le x \le 1 \\ \varphi_0 &= (2 - x)^3 & 1 \le x \le 2. \end{aligned}$$

§19.10. EIGENVALUE PROBLEMS

Consider the following boundary value problem:

$$y'' + a^2 y = 0, \qquad y(0) = y(1) = 0.$$

The differential equation has the solution $y = A \cos ax + B \sin ax$. From $y(0) = 0$ we get $A = 0$, while $y(1) = 0$ gives $B \sin a = 0$. If $\sin a \neq 0$ we have $B = 0$, that is, the only possible solution is the trivial one $y(x) = 0$. If, on the other hand, $\sin a = 0$, that is, if $a = n\pi$ where n is an integer, then B can be chosen arbitrarily. These special values $a^2 = n^2\pi^2$ are called *eigenvalues* and the corresponding solutions *eigenfunctions*.

A linear differential equation with boundary values corresponds to a linear system of equations, and the situation discussed here corresponds to the case when the right-hand sides are equal to 0. If the coefficient matrix is nonsingular, we have only the trivial solution zero, but if the matrix is singular, that is, if the determinant is equal to 0, we have an infinity of nontrivial solutions.

The eigenvalue problems play an important role in modern physics, and as a rule, the eigenvalues represent quantities that can be measured experimentally (for example, energies). Usually the differential equation in question is written

(19.10.1)
$$\frac{d}{dx}\left(p\frac{dy}{dx}\right) - qy + \lambda\rho y = 0,$$

where p, q, and ρ are real functions of x.

The problem of solving this equation with regard to the boundary conditions

$$\begin{cases} \alpha_0 y(a) + \alpha_1 y'(a) = 0, \\ \beta_0 y(b) + \beta_1 y'(b) = 0, \end{cases}$$

or $y(a) = y(b)$; $p(a)y'(a) = p(b)y'(b)$ is called Sturm–Liouville's problem.

If the interval between a and b is divided into equal parts, and the derivatives are approximated by difference expressions, we obtain

$$\frac{p_r}{h^2}(y_{r-1} - 2y_r + y_{r+1}) + \frac{p'_r}{2h}(y_{r+1} - y_{r-1}) - qy_r + \lambda\rho_r y_r = 0.$$

This can obviously be written in the form

$$(A - \lambda I)y = 0,$$

where A is a band matrix and y is a column vector. Nontrivial solutions exist if $\det(A - \lambda I) = 0$, and hence our eigenvalue problem has been transformed into an algebraic eigenvalue problem. However, only the lowest eigenvalues of the original problem can be obtained in this way.

Here we mention an interesting method by which we can advance via functions fulfilling *both* boundary conditions and simultaneously try to improve the fit of the differential equation.

In the Sturm–Liouville case we compute y_n iteratively from

(19.10.2)
$$\frac{d}{dx}\left(p(x)\frac{dy_n}{dx}\right) = q(x)y_{n-1} - \lambda_n \rho(x)y_{n-1}.$$

When integrating, we obtain two integration constants, which can be determined from the boundary conditions; further, λ_n is obtained from the condition $\int_a^b y_n^2\,dx = 1$, or from some similar condition. However, in general the usual discretization method leading to an algebraic eigenvalue problem is computationally much simpler and should be preferred.

Example 1.   Determine the lowest eigenvalue of the problem:

$$y'' + \lambda y/(x+1)^2 = 0 \qquad \text{with } y(0) = y(1) = 0.$$

We replace y'' by $[y(x-h) - 2y(x) + y(x+h)]/h^2$ where $h = N^{-1}$ ($N = 2, 3, 4, 5$). As an example, for $N = 4$ we obtain

$$\begin{cases} -2y_1 + y_2 & + \lambda y_1/25 = 0 \\ y_1 - 2y_2 + y_3 + \lambda y_2/36 = 0 \\ y_2 - 2y_3 + \lambda y_3/49 = 0 \end{cases}$$

and hence

$$\begin{vmatrix} \lambda - 50 & 25 & 0 \\ 36 & \lambda - 72 & 36 \\ 0 & 49 & \lambda - 98 \end{vmatrix} = 0.$$

This gives $\lambda^3 - 220\lambda^2 + 12892\lambda - 176400 = 0$ with the roots 19.7301, 67.1751, and 133.0948. The roots in the other cases were found in a similar way and are presented below.

N	λ_1	λ_2	λ_3	λ_4
2	18			
3	19.0683	62.9317		
4	19.7301	67.1751	133.0948	
5	20.0876	71.5331	138.2993	230.0800
extrap.	20.8035	83.02	147.55	
exact	20.7923	82.42	185.13	328.93

The extrapolation is performed using the error expression $\varepsilon = ah^2 + bh^4 + \cdots$. It can be shown that the exact eigenvalues are $\pi^2 n^2/(\ln 2)^2 + 1/4$. The results show clearly that only the lowest eigenvalues can be obtained numerically with reasonable accuracy.

EXERCISES

1. Solve the differential equation $y' = x - y^2$ by series expansion for $x = 0.2(0.2)1$. Sketch the function graphically and read off the minimum point. Initial value: $x = 0$, $y = 1$.

2. Solve the differential equation $y' = 1/(x + y)$ for $x = 0.5(0.5)2$ by using Runge–Kutta's method. Initial value: $x = 0$, $y = 1$.

3. Solve the differential equation $y'' = xy$ for $x = 0.5$ and $x = 1$ by use of series expansion and Runge–Kutta's method. Initial values: $x = 0$, $y = 0$, $y' = 1$.

4. Find y for $x = 0.2$, 0.4, and 0.6 when $y' = \ln(x + y + 1)$ and $y(0) = 0$ (four decimals).

5. Perform two integration steps with Runge–Kutta's method for the equation $y' = \cos x + \sin y$ with $y(0) = 0$ and $h = 0.2$. Also determine a series expansion and compare the results.

6. Apply Euler's method to compute $x(0.2)$ and $y(0.2)$ from
$$\begin{cases} dx/dt = (1 + t)x + y \\ dy/dt = (1 - t)x - y \end{cases}$$
with $x(0) = 1$, $y(0) = 0$. Use $h = 0.1$ and $h = 0.05$ and Richardson extrapolation.

7. The differential equation $y' = e^y$ is given with $y(0) = 0$ and $y'(0) = -0.4636$. Perform two integration steps with Runge–Kutta's method and $h = 0.5$.

8. Solve the differential equation $xy'' - yy' = 0$ by Runge–Kutta's method for $x = 0.5$ and $x = 1$ ($h = 0.5$) when $y(0) = 1$ and $y''(0) = 2$ (four decimals). Then compare with the exact solution, which is of the form
$$y = \frac{ax^2 + b}{cx^2 + d}.$$

9. The differential equation $y' = 1 + x^2y^2$ with $y = 0$ for $x = 0$ is given. When x increases from 0 to ξ, y increases from 0 to ∞. The quantity ξ is to be determined to three places in the following way. First we compute $y(0.8)$, using Picard's method. Then we introduce $z = 1/y$, and the differential equation for z is integrated with $h = 0.3$ by use of, for example, Runge–Kutta's method. When z is sufficiently small, we extrapolate to $z = 0$, using a series expansion.

10. The differential equation $y' = axy + b$ is given. Then we can write $y^{(n)} = P_ny + Q_n$, where P_n and Q_n are polynomials in x. Show that $z_n = P_nQ_{n+1} - P_{n+1}Q_n$ is a constant which depends on n, and compute z_n.

11. The differential equation $y'' + x^3y = 0$ is given. For large values of x the solution y behaves like a damped oscillation with decreasing wavelength. The distance between the Nth and the $(N + 1)$st zero is denoted by z_N. Show that $\lim_{N \to \infty} z_N^5 \cdot N^3 = 8\pi^2/125$.

12. Given the system of differential equations
$$\begin{cases} xy' + z' + \frac{1}{2}y = 0, \\ xz' - y' + \frac{1}{2}z = 0, \end{cases}$$

with the initial conditions $x = 0$, $y = 0$, $z = 1$. Show that $y^2 + z^2 = (1 + x^2)^{-1/2}$ and find $y(\frac{1}{2})$ and $z(\frac{1}{2})$ to five decimals.

13. The differential equation $y' = (x^2 - y^2)/(x^2 + y^2)$ is given. It has solutions that asymptotically approach a certain line $y = px$, which is also a solution of the equation. Another line $y = qx$ is perpendicular to all integral curves. Show that $pq = 1$ and find p and q to five decimals.

14. The differential equation $y'' = -y$ with initial conditions $y(0) = 0$ and $y(h) = k$ is solved by Numerov's method. Find the explicit solution in the simplest possible form. Then compute y_6 when $h = \pi/6$ and $k = \frac{1}{2}$.

15. Solve the differential equation $y'' = (x - y)/(1 + y^2)$ for $x = 2.4(0.2)3$ by use of Cowell–Numerov's method. Initial values: $(2, 1)$ and $(2.2, 0.8)$. Also find the coordinates of the minimum point graphically.

16. What is the largest value of the interval length h such that all solutions of the Cowell–Numerov difference equation corresponding to $y'' + y = 0$ are finite when $x \to \infty$?

17. The differential equation $y'' + ay' + by = 0$ is given with $0 < a < 2\sqrt{b}$. Find the largest possible interval length h such that the numerical solutions stay finite when the equation is solved by use of the formulas $y_{n+1} = y_n + hy'_n$; $y'_{n+1} = y'_n + hy''_n$.

18. The differential equation $y'' + x^2(y + 1) = 0$ is given together with the boundary values $y(1) = y(-1) = 0$. Find approximate values of $y(0)$ and $y(\frac{1}{2})$ by assuming $y = (1 - x^2)(a + bx^2 + cx^4)$, which satisfies the boundary conditions. Use the points $x = 0$, $x = \frac{1}{2}$, and $x = 1$ (three significant figures).

19. Determine $y(0)$ of Exercise 18 approximating by a difference equation of second order. Use $h = \frac{1}{5}$ and solve the linear system $Az = a + Dz$ iteratively by the formula

$$z_{n+1} = A^{-1}a + A^{-1}Dz_n.$$

The inverse of an $n \times n$-matrix A with elements $a_{11} = 1$, $a_{ik} = 2\delta_{ik} - \delta_{i-1, k} - \delta_{i+1, k}$ otherwise, is a matrix B with elements $b_{ik} = n + 1 - \max(i, k)$.

20. Investigate the stability properties of the method

$$y_{n+3} = y_n + \frac{3h}{8}(y'_n + 3y'_{n+1} + 3y'_{n+2} + y'_{n+3}),$$

when applied to the equation $y' = -y$.

21. The following system is given

$$\begin{cases} y'' - 2xy' + z = 0, \\ z'' + 2xz' + y = 0, \end{cases}$$

with boundary values $y(0) = 0$, $y'(0) = 1$, $z(\frac{1}{2}) = 0$, and $z'(\frac{1}{2}) = -1$. Find $z(0)$ and $y(\frac{1}{2})$ to 3 decimals. Also show that $y^2 + z^2 + 2y'z'$ is independent of x.

22. The solution of the differential equation $y'' = y$, $0 \le x \le 1$, with the boundary

conditions $y(0) = y(1) = 1$, can be associated with minimization of the integral

$$I = \int_0^1 (y'^2 + y^2)\, dx.$$

Compute I using the attempt $y = 1 + ax(1 - x)$ and compare with the exact value.

23. One wants to determine a solution y of the equation $y'' = xy$ with $y(0) = 0$, $y(1) = 1$, by trying $y'(0) = p$. Find p and the series expansion for y, and further

$$I = \int_0^1 (y'^2 + xy^2)\, dx$$

which is minimized by the solution.

Alternatively, solve the problem by trying the approximation $y = ax + bx^2 + cx^3$ with the same boundary conditions, and solve the minimum equations for a, b, and c. Also find the value of I.

24. One wants to solve the boundary value problem $y'' + y^2 = 0$, $y(0) = 0$, $y(1) = 1$ by minimizing a suitable integral. Find this and try $y = ax + bx^2$. Compute a and b and the minimum.

25. Find the solution of the differential equation $y'' + x^2 y = 0$ with the boundary conditions $y(0) = 0$, $y(1) = 1$ at the points 0.25, 0.50, and 0.75:
    (a) by approximating the differential equation with a second-order difference equation;
    (b) by performing two steps in the iteration $y''_{n+1} = -x^2 y_n$, $y_0 = x$.

26. A certain eigenvalue λ of the differential equation

$$y'' - 2xy' + 2\lambda y = 0, \qquad y(0) = y(1) = 0,$$

is associated with the eigenfunction $y(x) = \sum_{n=0}^{\infty} a_n(\lambda) x^n$. Determine $a_n(\lambda)$ and also give the lowest eigenvalue with two correct decimals.

27. The differential equation $y'' + \lambda xy = 0$ is given, together with the boundary conditions $y(0) = 0$, $y(1) = 0$. Find the smallest eigenvalue λ by approximating the second derivative with the second difference for $x = \frac{1}{4}, \frac{1}{2}$, and $\frac{3}{4}$. Also use $h = \frac{1}{5}$ and extrapolate.

28. Find the smallest eigenvalue of the differential equation $xy'' + y' + \lambda xy = 0$, with the boundary conditions $y(0) = y(1) = 0$. The equation is approximated by a system of difference equations, first with $h = \frac{1}{2}$, then with $h = \frac{1}{4}$, and finally Richardson extrapolation is performed to $h = 0$ (the error is proportional to h^2).

29. Find the smallest positive eigenvalue of the differential equation $y^{(iv)}(x) = \lambda y(x)$ with the boundary conditions $y(0) = y'(0) = y''(1) = y'''(1) = 0$.

30. The differential equation $y'' + (1/x)y' + \lambda^2 y = 0$ is given together with the boundary conditions $y'(0) = 0$, $y(1) = 0$. Determine the lowest eigenvalue by assuming

$$y = a(x^2 - 1) + b(x^3 - 1) + c(x^4 - 1)$$

which satisfies the boundary conditions. Use the points $x = 0$, $x = \frac{1}{2}$, and $x = 1$.

31. The differential equation $y'' + \lambda y/(x^2 + 1) = 0$ is given with $y(0) = y(1) = 0$. Find the lowest eigenvalue λ by discretizing with $h = \frac{1}{2}, \frac{1}{3}$, and $\frac{1}{4}$ and using Richardson extrapolation.

32. The differential equation $(1 + x)y'' + y' + \lambda(1 + x)y = 0$ is given together with the boundary conditions $y'(0) = 0$, $y(1) = 0$. Show that by a suitable transformation $x = \alpha t + \beta$ of the independent variable, the equation can be brought to a Bessel equation of order 0 (cf. Chapter 7). Then show that the eigenvalues λ can be obtained from $\lambda = \xi^2$ where ξ is a root of the equation

$$J_1(\xi)Y_0(2\xi) - Y_1(\xi)J_0(2\xi) = 0.$$

It is known that $J_0'(\xi) = -J_1(\xi)$ and $Y_0'(\xi) = -Y_1(\xi)$.

CHAPTER 20

Partial Differential Equations

Le secret d'ennuyer est celui de tout dire.

Voltaire

§20.1. CLASSIFICATION

Partial differential equations and systems of such equations appear in the description of physical processes, for example, in hydrodynamics, the theory of elasticity, the theory of electromagnetism (Maxwell's equations), and quantum mechanics. The solutions of the equations describe possible physical reactions that have to be fixed through boundary conditions, which may be of quite a different character. Here we will mainly restrict ourselves to second-order partial differential equations, which dominate in the applications. Such equations are often obtained when systems of the kind described above are specialized and simplified in different ways.

We shall assume that the equations are linear in the second derivatives, that is, of the form

$$(20.1.1) \qquad a\frac{\partial^2 u}{\partial x^2} + 2b\frac{\partial^2 u}{\partial x\,\partial y} + c\frac{\partial^2 u}{\partial y^2} = e,$$

where a, b, c, and e are functions of x, y, u, $\partial u/\partial x$, and $\partial u/\partial y$. We introduce the conventional notations:

$$p = \frac{\partial u}{\partial x}; \qquad q = \frac{\partial u}{\partial y}; \qquad r = \frac{\partial^2 u}{\partial x^2}; \qquad s = \frac{\partial^2 u}{\partial x\,\partial y}; \qquad t = \frac{\partial^2 u}{\partial y^2}.$$

Then we have the relations:

$$du = \frac{\partial u}{\partial x}\,dx + \frac{\partial u}{\partial y}\,dy = p\,dx + q\,dy,$$

$$dp = \frac{\partial}{\partial x}\left(\frac{\partial u}{\partial x}\right)dx + \frac{\partial}{\partial y}\left(\frac{\partial u}{\partial x}\right)dy = r\,dx + s\,dy,$$

$$dq = \frac{\partial}{\partial x}\left(\frac{\partial u}{\partial y}\right)dx + \frac{\partial}{\partial y}\left(\frac{\partial u}{\partial y}\right)dy = s\,dx + t\,dy.$$

With the notations introduced, the differential equation itself can be written

$$ar + 2bs + ct = e.$$

When treating ordinary differential equations numerically, it is customary to start in a certain point where also a number of derivatives are known. For a second-order equation it is usually sufficient to know $y_0 = y(x_0)$ and y_0'; higher derivatives are then obtained from the differential equation by successive differentiation. For a second-order partial differential equation it would be natural to conceive a similar procedure, at least for "open" problems for which the boundary conditions do not prevent this. Then it would be reasonable to replace the starting point by an initial curve along which we assume the values of u, p, and q given. With this background we first encounter the problem of determining r, s, and t in an arbitrary point of the curve. We suppose that the equation of the curve is given in parameter form:

$$x = x(\tau), \qquad y = y(\tau).$$

Since u, p, and q are known along the curve, we can simply write $u = u(\tau)$, $p = p(\tau)$, and $q = q(\tau)$, and introducing the notations $x' = dx/d\tau$ and so on, we find

(20.1.2)
$$\begin{cases} x'r + y's \quad\;\; = p', \\ \qquad x's + y't = q', \\ ar + 2bs + \;\; ct = e. \end{cases}$$

From this system we can determine r, s, and t as functions of τ, provided that the coefficient determinant $D \neq 0$. We find directly

(20.1.3)
$$D = ay'^2 - 2bx'y' + cx'^2.$$

It is rather surprising that the most interesting situation appears in the exceptional case $D = 0$. This equation has a solution consisting of two directions $y'/x' = dy/dx$, or rather a field of directions assigning two directions to every point in the plane. These directions then define two families of curves which are called *characteristics*. They are real if $b^2 - ac > 0$ (*hyperbolic* equation) or if $b^2 - ac = 0$ (*parabolic* equation) but imaginary if $b^2 - ac < 0$ (*elliptic* equation).

If $D = 0$ there is no solution of (20.1.2) unless one of the following three relations is fulfilled:

$$\begin{vmatrix} p' & x' & y' \\ q' & 0 & x' \\ e & a & 2b \end{vmatrix} = 0; \qquad \begin{vmatrix} p' & x' & 0 \\ q' & 0 & y' \\ e & a & c \end{vmatrix} = 0; \qquad \begin{vmatrix} p' & y' & 0 \\ q' & x' & y' \\ e & 2b & c \end{vmatrix} = 0.$$

(Incidentally, they are equivalent if $D = 0$.) This means that one cannot prescribe arbitrary initial values $(x, y, u, p,$ and $q)$ on a characteristic.

As is easily understood, an equation can be elliptic in one domain and hyperbolic in another. A well-known example is gas flow at high velocities; the flow can be subsonic at some places, supersonic at others.

The following description gives the typical features of equations belonging to these three kinds, as well as additional conditions. In the *hyperbolic* case we have an open domain bounded by the x-axis between $x = 0$ and $x = a$, and the lines $x = 0$ and $x = a$ for $y \geq 0$. On the portion of the x-axis between 0 and a, the functions $u(x, 0)$ and $\partial u / \partial y$ are given as initial conditions. On each of the vertical lines a boundary condition of the form $\alpha u + \beta(\partial u / \partial x) = \gamma$ is given. In the *parabolic* case we have the same open domain, but on the portion of the x-axis only $u(x, 0)$ is given as an initial condition. Normally, $u(0, y)$ and $u(a, y)$ are also given as boundary conditions. Finally, in the *elliptic* case we have a closed curve on which u or the normal derivative $\partial u / \partial n$ (or a linear combination of both) is given; together with the equation, these conditions determine u in all interior points.

Hyperbolic equations, as a rule, are connected with oscillating systems (example: the wave equation), and parabolic equations are connected with some kind of diffusion. An interesting special case is the Schrödinger equation, which appears in quantum mechanics. Elliptic equations usually are associated with equilibrium states and especially with potential problems of all kinds.

§20.2. HYPERBOLIC EQUATIONS

We shall first give an explicit example, and we choose the wave equation in one dimension, setting the propagation velocity equal to c:

$$(20.2.1) \qquad \frac{\partial^2 u}{\partial x^2} = \frac{1}{c^2} \frac{\partial^2 u}{\partial t^2}.$$

In this case we can easily write down the general solution:

$$(20.2.2) \qquad u = f(x + ct) + g(x - ct).$$

The solution contains two arbitrary functions f and g. Its physical meaning is quite simple: u can be interpreted as the superposition of two waves traveling in opposite directions. If, in particular, we choose $f(z) = -\frac{1}{2} \cos \pi z$ and $g(z) = \frac{1}{2} \cos \pi z$, we get $u = \sin \pi x \sin \pi ct$, which describes a standing wave with $u(0, t) = u(1, t) = 0$ and $u(x, 0) = 0$. Physically, the phenomenon can be realized by a vibrating string stretched between the points $x = 0$ and $x = 1$.

Often the initial conditions have the form (Cauchy's initial value problem)

$$\begin{cases} u(x, 0) = f(x), \\ \dfrac{\partial u}{\partial t}(x, 0) = g(x), \end{cases}$$

and then we have the general solution

$$(20.2.3) \qquad u(x, t) = \frac{f(x + ct) + f(x - ct)}{2} + \frac{1}{2c} \int_{x-ct}^{x+ct} g(\xi)\, d\xi.$$

Explicit solutions can be given for the two- and three-dimensional wave equations as well.

Setting $y = ct$ in (20.2.1), we get the equation $\partial^2 u/\partial x^2 - \partial^2 u/\partial y^2 = 0$ with the explicit solution $u = f(x + y) + g(x - y)$. It is interesting to observe that the partial difference equation

$$(20.2.4) \qquad u(x + 1, y) + u(x - 1, y) = u(x, y + 1) + u(x, y - 1)$$

has exactly the same solution. This equation is obtained if we approximate the differential equation by second-order differences. However, if the coefficients of the equation do not have this simple form, it is, in general, impossible to give explicit solutions, and instead we must turn to numerical methods. A very general method is to replace the differential equation with a difference equation. This procedure will be exemplified later, and then we shall also treat the corresponding stability problems. At present we shall consider a method which is special for hyperbolic equations and which makes use of the properties of the *characteristics*.

First, we take the simple hyperbolic equation $\partial^2 u/\partial x\, \partial y = 0$ with the equation for the characteristics $dx\, dy = 0$, that is, $x = $ constant, $y = $ constant. The characteristics have the property that if we know u along two intersecting characteristics, then we also know u in the whole domain where the equation is hyperbolic. In this special case we have the general solution $u = \varphi(x) + \psi(y)$. Now suppose that u is known on the line $x = x_0$, as well as on the line $y = y_0$. Thus we have

$$\begin{cases} \varphi(x) + \psi(y_0) = F(x), \\ \varphi(x_0) + \psi(y) = G(y), \end{cases}$$

where both $F(x)$ and $G(y)$ are known functions. Adding and putting $x = x_0$ and $y = y_0$, we obtain

$$\varphi(x_0) + \psi(y_0) = \tfrac{1}{2}[F(x_0) + G(y_0)],$$

where $F(x_0) = G(y_0)$, and

$$u = \varphi(x) + \psi(y) = F(x) + G(y) - \tfrac{1}{2}[F(x_0) + G(y_0)].$$

In general, the characteristics are obtained from the equation

$$ay'^2 - 2bx'y' + cx'^2 = 0$$

or

$$\frac{dy}{dx} = \frac{b \pm (b^2 - ac)^{1/2}}{a};$$

these two values will be denoted by f and g (they are functions of x, y, u, p, and q). Further, consider an arc AB which is not characteristic and on which u, p, and q are given (see Figure 20.1). Let PR be a characteristic of f-type, QR one of g-type. Further, let $P = (x_1, y_1)$; $Q = (x_2, y_2)$; $R = (x_3, y_3)$. In the first approximation we have

$$\begin{cases} y_3 - y_1 = (x_3 - x_1)f_1 \\ y_3 - y_2 = (x_3 - x_2)g_2. \end{cases}$$

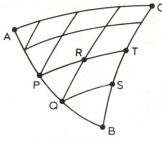

FIGURE 20.1

From this system x_3 and y_3 can be computed. The second of the three equivalent determinantal equations can be written

$$ex'y' - ap'y' - cq'x' = 0$$

or

$$e\frac{dy}{dx}dx - a\frac{dy}{dx}dp - c\,dq = 0.$$

Along PR this relation is approximated by

$$e_1(x_3 - x_1)f_1 - a_1(p_3 - p_1)f_1 - c_1(q_3 - q_1) = 0,$$

and along QR by

$$e_2(x_3 - x_2)g_2 - a_2(p_3 - p_2)g_2 - c_2(q_3 - q_2) = 0.$$

Since x_3 and y_3 are now approximately known, p_3 and q_3 can be solved from this system. Then it is possible to compute u_3 from

$$u_3 = u_1 + \frac{\partial u}{\partial x}dx + \frac{\partial u}{\partial y}dy,$$

where we approximate $\partial u/\partial x$ by $\frac{1}{2}(p_1 + p_3)$ and $\partial u/\partial y$ by $\frac{1}{2}(q_1 + q_3)$; further, dx is replaced by $x_3 - x_1$ and dy by $y_3 - y_1$. Hence we get

$$u_3 = u_1 + \tfrac{1}{2}(x_3 - x_1)(p_1 + p_3) + \tfrac{1}{2}(y_3 - y_1)(q_1 + q_3).$$

From the known approximations of u_3, p_3, and q_3, we can compute f_3 and g_3. Then improved values for x_3, y_3, p_3, q_3, and u_3 can be obtained in the same way as before, but with f_1 replaced by $\frac{1}{2}(f_1 + f_3)$, g_2 by $\frac{1}{2}(g_2 + g_3)$, a_1 by $\frac{1}{2}(a_1 + a_3)$, and so on. When the value in R has been obtained, we can proceed to S, and from R and S we can reach T. We observe that we can only obtain values within a domain ABC of triangular shape and with AB as "base."

We shall now demonstrate in an example how it is possible to apply an ordinary series expansion technique. Again we start from the wave equation

$$\frac{\partial^2 u}{\partial x^2} - \frac{\partial^2 u}{\partial y^2} = 0$$

with the initial conditions

$$u(x, 0) = x^2; \qquad \frac{\partial u}{\partial y}(x, 0) = e^{-x}.$$

(Note that the straight line $y = 0$, where the initial conditions are given, is not a characteristic.) We are then going to compute $u(0.2, 0.1)$ starting from the origin. Differentiating the initial conditions, we obtain

$$\frac{\partial u}{\partial x}(x, 0) = 2x, \qquad \frac{\partial^2 u}{\partial x \, \partial y}(x, 0) = -e^{-x},$$

$$\frac{\partial^2 u}{\partial x^2}(x, 0) = 2, \qquad \frac{\partial^3 u}{\partial x^2 \, \partial y}(x, 0) = e^{-x},$$

$$\frac{\partial^3 u}{\partial x^3}(x, 0) = 0.$$

From the differential equation we get

$$\frac{\partial^2 u}{\partial y^2}(x, 0) = 2$$

and then successively:

$$\frac{\partial^3 u}{\partial x^3} - \frac{\partial^3 u}{\partial x \, \partial y^2} = 0; \qquad \frac{\partial^3 u}{\partial x^2 \, \partial y} - \frac{\partial^3 u}{\partial y^3} = 0,$$

and hence

$$\frac{\partial^3 u}{\partial x \, \partial y^2}(x, 0) = 0; \qquad \frac{\partial^3 u}{\partial y^3}(x, 0) = e^{-x},$$

and so on. We then finally obtain the desired value through

$$u(x, y) = \exp(xD_x + yD_y)u(0, 0)$$

$$= 0.1 + \frac{1}{2}(2 \cdot 0.04 - 0.04 + 0.02)$$

$$+ \frac{1}{6}(3 \cdot 0.004 + 0.001) + \frac{1}{24}(-4 \cdot 0.0008 - 4 \cdot 0.0002) + \cdots$$

$$= 0.13200.$$

The exact solution is

$$u = \frac{(x + y)^2 + (x - y)^2}{2} + \frac{1}{2}\int_{x-y}^{x+y} e^{-t} \, dt = x^2 + y^2 + e^{-x} \sinh y,$$

which for $x = 0.2$, $y = 0.1$ gives the value $u = 0.13201$.

As mentioned before, it is of course also possible to approximate a hyperbolic differential equation with a difference equation. From the wave equation

$$\frac{\partial^2 u}{\partial x^2} - \frac{\partial^2 u}{\partial y^2} = 0,$$

we get the difference equation

(20.2.5)
$$\frac{u_{r,\,s+1} - 2u_{r,s} + u_{r,\,s-1}}{k^2} = \frac{u_{r+1,\,s} - 2u_{r,s} + u_{r-1,\,s}}{h^2}.$$

As before, the indices r and s occur symmetrically except for h and k, but a difference is introduced through the initial conditions which define an integration direction, viz., toward increasing values of s. In 1928, Courant, Friedrichs, and Lewy proved that the difference equation is stable only if $k \le h$. In the following sections we shall give much attention to stability problems.

Example 1

$$\frac{\partial^2 u}{\partial x^2} - \frac{\partial^2 u}{\partial y^2} = 0; \qquad u(x, 0) = x^2; \qquad \frac{\partial u}{\partial y}(x, 0) = e^{-x}.$$

Since $a = 1$ and $c = -1$, the equation for the characteristics becomes $y'^2 - x'^2 = 0$ or, after integration, $y = \pm x + \alpha$ where α is constant.

First we compute the values on levels -1 and 1 (i.e., for $y = -0.1$ and $y = 0.1$); for example, we have

$$\frac{\partial u}{\partial y}(0.3, 0) \simeq \frac{d_1 - d_{-1}}{0.2} \simeq e^{-0.3} = 0.7408.$$

Using the difference equation (20.2.5) with $h = k = 0.1$, we get $d_1 + d_{-1} = 0.04 + 0.16$. Hence we have the system:

$$\begin{cases} d_1 - d_{-1} = 0.14816 \\ d_1 + d_{-1} = 0.20 \end{cases}$$

which gives $d_1 = 0.17408$. The exact value is $0.09 + 0.01 + e^{-0.3} \sinh 0.1 = 0.174205$. For points on higher levels we get for example:

$$c_2 + 0.04 = b_1 + d_1$$

and we see that we need values inside a right-angled equilateral triangle as indicated in Figure 20.2. This triangle is, of course, formed by a pair of characteristics. In this way the table below was constructed (exact values within parenthesis).

n	a_n	b_n	c_n	d_n	e_n
1	0.11	0.11048	0.13187	0.17408	0.23703
	(0.11017)	(0.11063)	(0.13201)	(0.17421)	(0.23714)
2		0.23187	0.24456	0.27890	
		(0.23218)	(0.24484)	(0.27915)	
			0.37890		
3			(0.37932)		

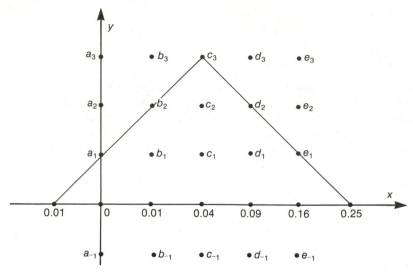

FIGURE 20.2. Values of u in grid points ($h = k = 0.1$).

§20.3. PARABOLIC EQUATIONS

The simplest nontrivial parabolic equation has the form

(20.3.1)
$$\frac{\partial^2 u}{\partial x^2} = \frac{\partial u}{\partial t},$$

with the initial values $u(x, 0) = f(x)$, $0 \le x \le 1$, and the boundary conditions $u(0, t) = \varphi(t)$; $u(1, t) = \psi(t)$. This is the *heat equation* in one dimension.

We shall first solve the problem by approximating $\partial^2 u / \partial x^2$, but not $\partial u / \partial t$, by a difference expression. The portion between $x = 0$ and $x = 1$ is divided into n equal parts in such a way that $x_r = rh$ and $nh = 1$. Then we have approximately

$$\frac{\partial u_r}{\partial t} = \frac{du_r}{dt} = \frac{1}{h^2} (u_{r-1} - 2u_r + u_{r+1}).$$

Writing out the equations for $r = 1, 2, \ldots, n - 1$, we obtain

(20.3.2)
$$\begin{cases} h^2 \dfrac{du_1}{dt} = u_0 - 2u_1 + u_2, \\[2mm] h^2 \dfrac{du_2}{dt} = u_1 - 2u_2 + u_3, \\[2mm] \quad\vdots \\[2mm] h^2 \dfrac{du_{n-1}}{dt} = u_{n-2} - 2u_{n-1} + u_n. \end{cases}$$

Hence we get a system of $n - 1$ ordinary differential equations in $u_1, u_2, \ldots, u_{n-1}$; the quantities u_0 and u_n are, of course, known functions of t:

$$\begin{cases} u_0 = \varphi(t), \\ u_n = \psi(t). \end{cases}$$

The values of $u_1, u_2, \ldots, u_{n-1}$ for $t = 0$ are $f(h), f(2h), \ldots, f((n-1)h)$. Putting

$$U = \begin{pmatrix} u_1 \\ u_2 \\ \vdots \\ u_{n-1} \end{pmatrix}; \quad V = \begin{pmatrix} u_0 \\ 0 \\ \vdots \\ 0 \\ u_n \end{pmatrix}; \quad A = \begin{pmatrix} -2 & 1 & 0 & \cdots & 0 & 0 \\ 1 & -2 & 1 & \cdots & 0 & 0 \\ \vdots & & & & & \vdots \\ 0 & 0 & 0 & \cdots & 1 & -2 \end{pmatrix},$$

we can write

$$h^2 \frac{dU}{dt} = AU + V.$$

As an example, we solve the heat equation for a thin, homogeneous rod of uniform thickness, which at time $t = 0$ has temperature $u = 0$. Further we assume $u(0, t) = 0$ and $u(1, t) = t$. We will then consider the temperature u_1 for $x = \frac{1}{4}$, u_2 for $x = \frac{1}{2}$, and u_3 for $x = \frac{3}{4}$, and we find with $h = \frac{1}{4}$:

$$\begin{cases} \dfrac{du_1}{dt} = 16(-2u_1 + u_2), \\[2mm] \dfrac{du_2}{dt} = 16(u_1 - 2u_2 + u_3), \\[2mm] \dfrac{du_3}{dt} = 16(u_2 - 2u_3 + t). \end{cases}$$

Putting $u_k = a_k + b_k t$, we obtain the particular solution

$$\begin{pmatrix} u_1 \\ u_2 \\ u_3 \end{pmatrix} = \frac{1}{128} \begin{pmatrix} -5 + 32t \\ -8 + 64t \\ -7 + 96t \end{pmatrix}.$$

The solution of the homogeneous system is best obtained by putting $u_k = e^{\lambda t} = e^{16\mu t}$. This gives a homogeneous system of equations whose coefficient determinant must be zero:

$$\begin{vmatrix} -2 - \mu & 1 & 0 \\ 1 & -2 - \mu & 1 \\ 0 & 1 & -2 - \mu \end{vmatrix} = 0.$$

The roots are

$$\mu_1 = -2; \quad \mu_2 = -2 + \sqrt{2}; \quad \mu_3 = -2 - \sqrt{2},$$

and the corresponding eigenvectors are

$$\begin{pmatrix} 1 \\ 0 \\ -1 \end{pmatrix}; \quad \begin{pmatrix} 1 \\ \sqrt{2} \\ 1 \end{pmatrix}; \quad \begin{pmatrix} 1 \\ -\sqrt{2} \\ 1 \end{pmatrix}.$$

Hence the solution is

$$\begin{cases} u_1 = A \cdot e^{-32t} + B \cdot e^{-16(2-\sqrt{2})t} + C \cdot e^{-16(2+\sqrt{2})t} + \dfrac{1}{128}(-5 + 32t), \\[2mm] u_2 = B\sqrt{2} \cdot e^{-16(2-\sqrt{2})t} - C\sqrt{2} \cdot e^{-16(2+\sqrt{2})t} + \dfrac{1}{16}(-1 + 8t), \\[2mm] u_3 = -A \cdot e^{-32t} + B \cdot e^{-16(2-\sqrt{2})t} + C \cdot e^{-16(2+\sqrt{2})t} + \dfrac{1}{128}(-7 + 96t); \end{cases}$$

A, B, and C can be obtained from $u_1(0) = u_2(0) = u_3(0) = 0$:

$$A = -\frac{1}{128}; \quad B = \frac{3 + 2\sqrt{2}}{128}; \quad C = \frac{3 - 2\sqrt{2}}{128}.$$

For large values of t we have

$$u_1(t) \simeq \frac{t}{4}; \quad u_2(t) \simeq \frac{t}{2}; \quad u_3(t) \simeq \frac{3t}{4}.$$

More generally, taking $h = 1/n$ we get a matrix $-2I + A$ with $(A)_{ik} = 1$ if $i - k = \pm 1$ and 0 otherwise. The eigenvalues of A can be computed as in (12.1.9), and become $2\cos(\pi r/n)$, $r = 1, 2, \ldots, n - 1$; hence

$$\mu_r = 2(\cos(\pi r/n) - 1).$$

The quotient between the absolutely largest and smallest eigenvalues is approximately $4n^2/\pi^2$. This means that the problem becomes stiffer the larger n is. Our conclusion must be that the method just described cannot be recommended. However, we can also approximate $\partial u/\partial t$ by a difference expression. Then it often turns out to be necessary to use smaller intervals in the t-direction. As before, we take $x_r = rh$, but now we also choose $t_s = sk$. Then we obtain the following partial difference equation:

(20.3.3)
$$\frac{u_{r-1,s} - 2u_{r,s} + u_{r+1,s}}{h^2} = \frac{u_{r,s+1} - u_{r,s}}{k}.$$

With $\alpha = k/h^2$, we get

(20.3.4)
$$u_{r,s+1} = \alpha u_{r-1,s} + (1 - 2\alpha)u_{r,s} + \alpha u_{r+1,s}.$$

Later on we shall examine for what values of α the method is stable; here we restrict ourselves to computing the truncation error. We get

$$\frac{u(x, t + k) - u(x, t)}{k} = \frac{\partial u}{\partial t} + \frac{k}{2!}\frac{\partial^2 u}{\partial t^2} + \frac{k^2}{3!}\frac{\partial^3 u}{\partial t^3} + \cdots,$$

$$\frac{u(x - h, t) - 2u(x, t) + u(x + h, t)}{h^2} = \frac{\partial^2 u}{\partial x^2} + \frac{h^2}{12}\frac{\partial^4 u}{\partial x^4} + \frac{h^4}{360}\frac{\partial^6 u}{\partial x^6} + \cdots.$$

Since u must satisfy $\partial u/\partial t = \partial^2 u/\partial x^2$, we also have, under the assumption of sufficient differentiability,

$$\frac{\partial^2 u}{\partial t^2} = \frac{\partial^4 u}{\partial x^4}; \qquad \frac{\partial^3 u}{\partial t^3} = \frac{\partial^6 u}{\partial x^6}; \cdots$$

Hence we obtain the truncation error

$$F = \frac{h^2}{2}\left(\alpha\frac{\partial^2 u}{\partial t^2} - \frac{1}{6}\frac{\partial^4 u}{\partial x^4}\right) + \frac{h^4}{6}\left(\alpha^2\frac{\partial^3 u}{\partial t^3} - \frac{1}{60}\frac{\partial^6 u}{\partial x^6}\right) + \cdots$$

$$= \frac{h^2}{2}\left(\alpha - \frac{1}{6}\right)\frac{\partial^2 u}{\partial t^2} + \frac{h^4}{6}\left(\alpha^2 - \frac{1}{60}\right)\frac{\partial^3 u}{\partial t^3} + \cdots.$$

Obviously, we get a truncation error of highest order when we choose $\alpha = \frac{1}{6}$, that is, $k = h^2/6$. Then we find

$$F = \frac{h^4}{540}\frac{\partial^3 u}{\partial t^3} + \cdots.$$

Now we solve the heat equation again with the initial condition $u(x, 0) = 0$ and the boundary conditions $u(0, t) = 0$, $u(1, t) = t$, choosing $h = \frac{1}{4}$ and $k = \frac{1}{96}$ corresponding to $\alpha = \frac{1}{6}$. The recursion formula takes the following simple form:

$$u(x, t + k) = \frac{1}{6}(u(x - h, t) + 4u(x, t) + u(x + h, t)).$$

After 12 steps we have reached $t = 0.125$ and the u-values, $u_1 = 0.00540$, $u_2 = 0.01877$, $u_3 = 0.05240$, and, of course, $u_4 = 0.125$. From our previous solution, we obtain instead

$$u_1 = 0.00615, \qquad u_2 = 0.01994, \qquad \text{and} \qquad u_3 = 0.05331.$$

The exact solution of the problem is given by

$$u(x, t) = \frac{1}{6}(x^3 - x + 6xt) + \frac{2}{\pi^3}\sum_{n=1}^{\infty}\frac{(-1)^{n-1}}{n^3}e^{-n^2\pi^2 t}\sin n\pi x,$$

which gives $u_1 = 0.00541$, $u_2 = 0.01878$, and $u_3 = 0.05240$. Hence the method with $\alpha = \frac{1}{6}$ gives an error which is essentially less than what is obtained if we use ordinary derivatives in the t-direction, corresponding to $\alpha = 0$.

The method defined by (20.3.3) or (20.3.4) gives the function values at time $t + k$ as a linear combination of three function values at time t; these are then supposed to be known. For this reason the method is said to be *explicit*. Since the time step k has the form αh^2, it is interesting to know how large values of α we can choose and still have convergence. Let us again consider the same equation

$$\frac{\partial^2 u}{\partial x^2} = \frac{\partial u}{\partial t}$$

but now with the boundary values $u(0, t) = u(1, t) = 0$ when $t \geq 0$, and with the initial condition $u(x, 0) = f(x), 0 \leq x \leq 1$. If $f(x)$ is continuous we can write down an analytic solution

$$u = \sum_{m=1}^{\infty} a_m \sin m\pi x \, e^{-m^2\pi^2 t},$$

where the coefficients are determined from the initial condition which requires

$$f(x) = u(x, 0) = \sum_{m=1}^{\infty} a_m \sin m\pi x.$$

Thus we obtain:

$$a_m = 2 \int_0^1 f(x) \sin m\pi x \, dx.$$

We shall now try to produce an analytic solution also for the difference equation, which we write in the form

$$u_{r, s+1} - u_{r,s} = \alpha(u_{r-1, s} - 2u_{r,s} + u_{r+1, s}).$$

In order to separate the variables we try with the expression $u_{r,s} = X(rh)T(sk)$ and obtain

$$\frac{T((s + 1)k) - T(sk)}{T(sk)} = \alpha\frac{X((r + 1)h) - 2X(rh) + X((r - 1)h)}{X(rh)} = -p^2.$$

(Since the left-hand side is a function of s and the right-hand side a function of r, both must be constant; $-p^2$ is called the *separation constant*.) We notice that the function $X(rh) = \sin m\pi rh$ satisfies the equation

$$X((r + 1)h) - 2X(rh) + X((r - 1)h) = -\frac{p^2}{\alpha} X(rh),$$

if we choose $p^2 = 4\alpha \sin^2(m\pi h/2)$. Further we assume $T(sk) = e^{-qsk}$ and obtain $e^{-qk} = 1 - 4\alpha \sin^2(m\pi h/2)$. In this way we finally get

$$u_{r,s} = \sum_{m=1}^{M-1} b_m \sin m\pi rh \left(1 - 4\alpha \sin^2 \frac{m\pi h}{2}\right)^s.$$

Here $Mh = 1$, and further the coefficients b_m should be determined in such a way that $u_{r,0} = f(rh)$, which gives

$$b_m = \frac{2}{M} \sum_{n=1}^{M-1} f(nh) \sin m\pi nh.$$

The time dependence of the solution is expressed through the factor $\{1 - 4\alpha \sin^2(m\pi h/2)\}^s$, and for small values of α we have

$$1 - 4\alpha \sin^2 \frac{m\pi h}{2} \simeq \exp(-\alpha m^2 \pi^2 h^2).$$

If this is raised to the power s we get

$$\exp(-\alpha m^2 \pi^2 h^2 s) = \exp(-m^2 \pi^2 ks) = \exp(-m^2 \pi^2 t),$$

that is, the same factor as in the solution of the differential equation. If α increases, the factor will deviate more and more from the right value, and we observe that the situation becomes dangerous if $\alpha > \frac{1}{2}$ since the factor then may become absolutely > 1. Therefore, it is clear that h and k must be chosen such that $k < h^2/2$. Even if $h \to 0$ one cannot guarantee that the solution of the difference equation is close to the solution of the differential equation unless k is chosen such that $\alpha \leq \frac{1}{2}$.

A numerical method is usually said to be *stable* if an error that has been introduced remains finite all the time. Now it is obvious that the errors satisfy the same difference equation, and hence we have proved that the stability condition is exactly $\alpha \leq \frac{1}{2}$. It should be pointed out, however, that the dividing line between *convergence* on one side and *stability* on the other is rather vague. In general, convergence means that the solution of the difference equation when $h \to 0$ and $k \to 0$ approaches the solution of the differential equation, while stability implies that introduced rounding errors do not grow during the procedure. Normally these two properties go together, but examples have been given of convergence with an unstable method.

Before we enter upon a more systematic error investigation, we shall demonstrate a technique which has been used before to illustrate the error propagation in a difference scheme. We choose the heat conduction equation, which we approximate according to (20.3.4) with $\alpha = \frac{1}{2}$:

$$u_{r, s+1} = \tfrac{1}{2}u_{r-1, s} + \tfrac{1}{2}u_{r+1, s}.$$

The following scheme is obtained:

	$s = 0$	$s = 1$	$s = 2$	$s = 3$	$s = 4$
$r = -4$	0	0	0	0	$\frac{1}{16}$
$r = -3$	0	0	0	$\frac{1}{8}$	0
$r = -2$	0	0	$\frac{1}{4}$	0	$\frac{4}{16}$
$r = -1$	0	$\frac{1}{2}$	0	$\frac{3}{8}$	0
$r = 0$	1	0	$\frac{2}{4}$	0	$\frac{6}{16}$
$r = 1$	0	$\frac{1}{2}$	0	$\frac{3}{8}$	0
$r = 2$	0	0	$\frac{1}{4}$	0	$\frac{4}{16}$
$r = 3$	0	0	0	$\frac{1}{8}$	0
$r = 4$	0	0	0	0	$\frac{1}{16}$

The numerators contain the binomial coefficient $\binom{s}{n}$ and the denominators 2^s, and the method seems to be stable. If the same equation is approximated with the difference equation

$$\frac{u_{r-1,s} - 2u_{r,s} + u_{r+1,s}}{h^2} = \frac{u_{r,s+1} - u_{r,s-1}}{2k},$$

which has considerably smaller truncation error than (20.3.3), we get with $\alpha = k/h^2 = \frac{1}{2}$:

$$u_{r,s+1} = u_{r,s-1} + u_{r-1,s} + u_{r+1,s} - 2u_{r,s}$$

	$s = -1$	$s = 0$	$s = 1$	$s = 2$	$s = 3$	$s = 4$
$r = -4$	0	0	0	0	0	1
$r = -3$	0	0	0	0	1	-8
$r = -2$	0	0	0	1	-6	31
$r = -1$	0	0	1	-4	17	-68
$r = 0$	0	1	-2	7	-24	89
$r = 1$	0	0	1	-4	17	-68
$r = 2$	0	0	0	1	-6	31
$r = 3$	0	0	0	0	1	-8
$r = 4$	0	0	0	0	0	1

Evidently, the method is very unstable and useless in practical computation. It is easy to give a sufficient condition for stability. Suppose that the term to be computed is given as a linear combination of known terms with coefficients $c_1, c_2, c_3, \ldots, c_n$. Then the method is stable if

$$\sum_{i=1}^{n} |c_i| \le 1.$$

We see, for example, that (20.3.4) is stable if $\alpha \le \frac{1}{2}$.

We shall now treat the problem in a way that shows more clearly how the error propagation occurs. We start again from the same system of equations:

$$u_{r,s+1} = \alpha u_{r-1,s} + (1 - 2\alpha)u_{r,s} + \alpha u_{r+1,s}.$$

Introducing the notation u_s for a vector with the components $u_{1,s}, \ldots, u_{M-1,s}$, we find $u_{s+1} = Cu_s$; where

$$C = \begin{bmatrix} 1 - 2\alpha & \alpha & 0 & 0 & \cdots & 0 \\ \alpha & 1 - 2\alpha & \alpha & 0 & \cdots & 0 \\ 0 & \alpha & 1 - 2\alpha & \alpha & \cdots & 0 \\ \vdots & & & & & \vdots \\ 0 & 0 & 0 & \cdots & \alpha & 1 - 2\alpha \end{bmatrix}.$$

If we also introduce an error vector e, we get trivially

$$e_s = C^s e_0$$

and we have stability if e_s stays finite when $s \to \infty$. Now $C = (1 - 2\alpha)I + \alpha A$ where A has the eigenvalues $2 \cos(m\pi/M)$, $m = 1(1)M - 1$ (see (12.1.9)), so we obtain $\lambda(C) = 1 - 4\alpha \sin^2(m\pi/2M)$. Hence we have the stability condition

$$-1 \le 1 - 4\alpha \sin^2 \frac{m\pi}{2M} \le 1,$$

that is, $\alpha \le [2 \sin^2(m\pi/2M)]^{-1}$. The worst case occurs when $m = M - 1$ and the expression within brackets is very close to 2, and we conclude that the method is stable if we choose $\alpha \le \frac{1}{2}$. The desire to avoid such small time-steps has led to the development of other methods which allow greater steps without giving rise to instability. The best known of these methods is that of *Crank–Nicolson*. It coincides with the forward-difference method, except that $\partial^2 u/\partial x^2$ is approximated by the *mean value* of the second difference quotients taken at times t and $t + k$. Thus the method can be described through

$$(20.3.5) \qquad \frac{u_{r,s+1} - u_{r,s}}{k}$$

$$= \frac{1}{2h^2}(u_{r+1,s} - 2u_{r,s} + u_{r-1,s} + u_{r+1,s+1} - 2u_{r,s+1} + u_{r-1,s+1}).$$

We have now three unknowns, $u_{r-1,s+1}$, $u_{r,s+1}$, and $u_{r+1,s+1}$, and hence the method is *implicit*. In practical computation the difficulties are not very large since the coefficient matrix is tridiagonal.

We are now going to investigate the convergence properties of Crank–Nicolson's method. As before we put

$$u_{r,s} = X(rh)T(sk)$$

and obtain

$$\frac{T[(s+1)k] - T(sk)}{T[(s+1)k] + T(sk)} = \frac{\alpha}{2} \frac{X[(r+1)h] - 2X(rh) + X[(r-1)h]}{X(rh)} = -p^2.$$

Trying with $X(rh) = \sin m\pi rh$ we get

$$p^2 = 2\alpha \sin^2\left(\frac{m\pi h}{2}\right),$$

and putting $T(sk) = e^{-\beta sk}$ we find

$$\frac{1 - e^{-\beta k}}{1 + e^{-\beta k}} = 2\alpha \sin^2\left(\frac{m\pi h}{2}\right)$$

or finally

$$e^{-\beta k} = \frac{1 - 2\alpha \sin^2(m\pi h/2)}{1 + 2\alpha \sin^2(m\pi h/2)}.$$

We can now write down the solution of the Crank–Nicolson difference equation:

$$(20.3.6) \qquad u_{r,s} = \sum_{m=1}^{M-1} b_m \sin m\pi rh \left(\frac{1 - 2\alpha \sin^2(m\pi h/2)}{1 + 2\alpha \sin^2(m\pi h/2)}\right)^s,$$

where b_m has the same form as before. The form of the solution immediately tells us that it can never grow with time, regardless of what positive values we assign to α. It is also easy to see that when $h, k \to 0$, then (20.3.6) converges toward the analytical solution.

In order to examine the stability properties we again turn to matrix technique. We note that the system can be written

$$(2I + \alpha T)u_{s+1} = (2I - \alpha T)u_s,$$

that is,

$$u_{s+1} = \left(I + \frac{\alpha}{2} T\right)^{-1} \left(I - \frac{\alpha}{2} T\right) u_s = Cu_s.$$

A similar relation holds for the error vector: $e_s = C^s e_0$. The eigenvalues of C determine the stability properties of the method. We have already computed the eigenvalues of T:

$$\lambda_m = 4 \sin^2 \frac{m\pi}{2M}, \qquad m = 1, 2, 3, \ldots, M - 1,$$

and from this we get the eigenvalues of C (note that $Mh = 1$):

$$\frac{1 - 2\alpha \sin^2(m\pi/2M)}{1 + 2\alpha \sin^2(m\pi/2M)}.$$

Hence Crank–Nicolson's method is stable for all positive values of α.

We shall now also compute the truncation error of the method, which is best done by operator technique. Let L' be the difference operator, and $L = D_t - D_x^2$ (observe that $D_x^2 u = \partial^2 u/\partial x^2$, not $(\partial u/\partial x)^2$). We get

$$L' - L = \frac{e^{kD_t} - 1}{k} - \frac{1}{2h^2} (e^{hD_x} - 2 + e^{-hD_x})(1 + e^{kD_t}) - D_t + D_x^2$$

$$= D_t + \frac{k}{2} D_t^2 + \frac{k^2}{6} D_t^3 + \cdots$$

$$- \left(\frac{1}{2} D_x^2 + \frac{h^2}{24} D_x^4 + \cdots\right)\left(2 + kD_t + \frac{k^2}{2} D_t^2 + \cdots\right) - D_t + D_x^2$$

$$= \frac{k}{2} D_t(D_t - D_x^2) + \frac{k^2}{6} D_t^3 - \frac{k^2}{4} D_x^2 D_t^2 - \frac{h^2}{12} D_x^4 + \cdots$$

$$= O(h^2) + O(k^2),$$

since $D_t u = D_x^2 u$ if u is a solution of the differential equation. This result shows that without any risk we can choose k of the same order of magnitude as h, that is, $k = ch$, which is a substantial improvement compared with $k = \alpha h^2$.

Crank–Nicolson's method can be generalized to equations of the form

$$\frac{\partial u}{\partial t} = \frac{\partial^2 u}{\partial x^2} + \frac{\partial^2 u}{\partial y^2},$$

but usually one prefers another still more economic method, originally suggested by Peaceman and Rachford. The same technique can be used in the elliptic case, and it will be briefly described in the next section.

§20.4. ELLIPTIC EQUATIONS

When one is working with ordinary differential equations, the main problem is to find a solution depending on the same number of parameters as the order of the equation. If such a solution has been found, it is usually a simple matter to adapt the parameter values to the given boundary conditions. For partial differential equations, the situation is completely different. In many cases it is relatively simple to find the general solution, but usually it is a difficult problem to adapt it to the boundary conditions. As examples we shall take the Laplace equation

$$(20.4.1) \qquad \Delta u = \frac{\partial^2 u}{\partial x^2} + \frac{\partial^2 u}{\partial y^2} = 0,$$

and the Poisson equation

$$(20.4.2) \qquad \Delta u = \frac{\partial^2 u}{\partial x^2} + \frac{\partial^2 u}{\partial y^2} = F(x, y).$$

Here Δ is the Laplace operator in two dimensions defined by $\Delta = \partial^2/\partial x^2 + \partial^2/\partial y^2$. In the remainder of this section, we will give much attention to these two most important equations.

The Laplace equation has the general solution

$$u = f(x + iy) + g(x - iy).$$

The equation is considered inside a given region R; the boundary is often denoted ∂R. In order to determine a unique solution we must prescribe certain conditions that must be fulfilled on the boundary. Usually three different problems are defined:

1. Dirichlet's problem when $u = f(x, y)$ on ∂R;
2. Neumann's problem when the normal derivative $\partial u/\partial n = g(x, y)$ is given on ∂R;
3. Robin–Churchill's problem with mixed boundary conditions $a(x, y)u + b(x, y)\dfrac{\partial u}{\partial n} = c(x, y)$ on ∂R.

Here we also mention a very important principle, the so-called *maximum principle*, valid for the linear elliptic equation

$$a\frac{\partial^2 u}{\partial x^2} + 2b\frac{\partial^2 u}{\partial x\, \partial y} + c\frac{\partial^2 u}{\partial y^2} + d\frac{\partial u}{\partial x} + e\frac{\partial u}{\partial y} = 0$$

where a, b, c, d, and e are functions of x and y. The principle states that all non-constant solutions assume their maximum and minimum values on the boundary ∂R of R.

We shall now consider the numerical treatment of the Poisson equation; the Laplace equation is obtained as a special case if we put $F(x, y) = 0$. According to

(17.4.3) we have approximately

$$(20.4.3) \qquad \nabla^2 u(x, y) = \frac{1}{h^2} \left[u(x + h, y) + u(x, y + h) + u(x - h, y) \right.$$

$$\left. + u(x, y - h) - 4u(x, y) \right].$$

The equation $\nabla^2 u = F$ is then combined with boundary conditions in such a way that u is given, for example, on the sides of a rectangle or a triangle. If the boundary is of a more complex shape, we can take this into account by modifying (20.4.3). Now we get a partial difference equation, which can be solved approximately by Liebmann's iteration method (20.4.4). Here index n indicates a certain iteration:

$$(20.4.4) \qquad u_{n+1}(x, y) = \frac{1}{4} \left[u_{n+1}(x, y - h) + u_{n+1}(x - h, y) + u_n(x, y + h) \right.$$

$$\left. + u_n(x + h, y) \right] - \frac{h^2}{4} F(x, y).$$

We start with guessed values and iterate row by row, moving upward, and repeat the process until no further changes occur. As an example we treat the equation

$$\frac{\partial^2 u}{\partial x^2} + \frac{\partial^2 u}{\partial y^2} = \frac{1}{x^2} + \frac{1}{y^2}$$

in the interior of a triangle with vertices at the points $(1, 1)$, $(2, 1)$, and $(2, 2)$, and with the boundary values $u = x^2 - \ln x - 1$ on the horizontal side, $u = 4 - \ln 2y - y^2$ on the vertical side, and $u = -2 \ln x$ on the oblique side. With $h = 0.2$, we obtain six interior grid points: $A(1.4, 1.2)$; $B(1.6, 1.2)$, $C(1.8, 1.2)$, $D(1.6, 1.4)$; $E(1.8, 1.4)$; and $F(1.8, 1.6)$. Choosing the starting value 0 at all these points, we obtain

A	B	C	D	E	F
-0.1155	0.2327	0.8824	-0.3541	0.3765	-0.3726
-0.0573	0.3793	1.0131	-0.2233	0.3487	-0.3795
-0.0207	0.4539	1.0248	-0.2116	0.3528	-0.3785
-0.0020	0.4644	1.0285	-0.2079	0.3549	-0.3779
$+0.0006$	0.4669	1.0296	-0.2068	0.3556	-0.3778
$+0.0012$	0.4676	1.0300	-0.2064	0.3558	-0.3777
$+0.0014$	0.4679	1.0301	-0.2063	0.3559	-0.3777
$+0.0015$	0.4679	1.0301	-0.2063	0.3559	-0.3777
$+0.0012$	0.4677	1.0299	-0.2065	0.3557	-0.3778

The exact solution is $u = x^2 - y^2 - \ln xy$, from which the values in the last line have been computed.

The variant of Liebmann's method that has been demonstrated here is closely related to Gauss–Seidel's method. In general, when a discretization of the differential equation has been performed, we have an ordinary linear system of equations of high but finite order. For this reason we can apply the discussion of methods for such systems given in Section 10.4; in particular, this is the case for the overrelaxation method.

All these methods (Jacobi, Gauss–Seidel, SOR) are *point-iterative*, that is, of the form

(20.4.5)
$$u^{(n+1)} = Mu^{(n)} + c.$$

Obviously, this equation describes an *explicit* method. A natural generalization is obtained if a whole group of interrelated values (for example, in the same row) are computed simultaneously. This is called *block iteration*, and it evidently defines an *implicit* method. We give an example with the Laplace equation in a rectangular domain 4 × 3:

·9	·10	·11	·12
·5	·6	·7	·8
·1	·2	·3	·4

The coefficient matrix becomes

4 −1				−1							
−1 4 −1				−1							
−1 4 −1				−1							
−1 4				−1							
−1				4 −1				−1			
−1				−1 4 −1				−1			
−1				−1 4 −1				−1			
−1				−1 4				−1			
				−1				4 −1			
				−1				−1 4 −1			
				−1				−1 4 −1			
				−1				−1 4			

where all empty spaces should be filled with zeros. Putting

$$D = \begin{pmatrix} 4 & -1 & 0 & 0 \\ -1 & 4 & -1 & 0 \\ 0 & -1 & 4 & -1 \\ 0 & 0 & -1 & 4 \end{pmatrix} \qquad \text{we obtain} \qquad \begin{pmatrix} D & -I & 0 \\ -I & D & -I \\ 0 & -I & D \end{pmatrix}\begin{pmatrix} U \\ V \\ W \end{pmatrix} = \begin{pmatrix} A \\ B \\ C \end{pmatrix}.$$

Here U is a 4-dimensional vector giving the values in the bottom row, and similarly V and W are related to the rows above. Further, A, B, and C are computed from the boundary values. We get the system:

$$
\begin{cases}
DU - \ \ V \qquad\quad = A \\
-U + DV - \ \ W = B \\
\qquad - \ \ V + DW = C
\end{cases}
$$

which is solved e.g. by:

$$
\begin{cases}
U_{n+1} = R(A + V_n) \\
V_{n+1} = R(B + U_{n+1} + W_n) \\
W_{n+1} = R(C + V_{n+1})
\end{cases}
$$

where

$$
R = D^{-1} = \frac{1}{209}
\begin{pmatrix}
56 & 15 & 4 & 1 \\
15 & 60 & 16 & 4 \\
4 & 16 & 60 & 15 \\
1 & 4 & 15 & 56
\end{pmatrix}.
$$

During the last decade, new iterative methods for solving large linear systems have appeared, depending on so-called *preconditioning*. For instance, given the system $Ax = b$ we can instead solve $B^{-1}Ax = B^{-1}b$ where both A and B^{-1} are symmetric and positive definite, by iteration. Then a modification of the conjugate gradient method (see section 10.5) can be applied, where in each step, a linear system with the matrix B is solved. As can be understood from simple geometrical considerations, convergence is improved if the eigenvalues get closer together. Frequently, B is a product of two *sparse* matrices, lower and upper triangular, respectively. The calculation of such sparse factors is numerically stable for so-called *M-matrices* characterized by $a_{ik} \leq 0$, $i \neq k$, A nonsingular, $A^{-1} \geq 0$. Further, preconditioning can be generalized to block matrices.

Finally, we shall also treat alternating direction implicit ($=$ADI) methods, and we then restrict ourselves to the Peaceman–Rachford method. First we discuss application of the method on an equation of the form

$$
(20.4.6) \qquad \frac{\partial}{\partial x}\left[A(x, y)\frac{\partial u}{\partial x}\right] + \frac{\partial}{\partial y}\left[C(x, y)\frac{\partial u}{\partial y}\right] - Fu + G = 0,
$$

where A, C, and F are ≥ 0.

The derivatives are approximated as follows:

$$
\frac{\partial}{\partial x}\left[A(x, y)\frac{\partial u}{\partial x}\right] \simeq A\left(x + \frac{h}{2}, y\right)\frac{u(x + h, y) - u(x, y)}{h^2}
$$
$$
- A\left(x - \frac{h}{2}, y\right)\frac{u(x, y) - u(x - h, y)}{h^2}
$$

and analogously for the y-derivative. Further, we introduce the notations

$$a = A\left(x + \frac{h}{2}, y\right), \qquad c = A\left(x - \frac{h}{2}, y\right), \qquad 2b = a + c,$$

$$\alpha = C\left(x, y + \frac{h}{2}\right), \qquad \gamma = C\left(x, y - \frac{h}{2}\right), \qquad 2\beta = \alpha + \gamma,$$

and define

$$(20.4.7) \quad \begin{cases} H_0 u(x, y) = a(x, y)u(x + h, y) - 2b(x, y)u(x, y) + c(x, y)u(x - h, y), \\ V_0 u(x, y) = \alpha(x, y)u(x, y + h) - 2\beta(x, y)u(x, y) + \gamma(x, y)u(x, y - h). \end{cases}$$

The differential equation will then be approximated by the difference equation

$$(H_0 + V_0 - Fh^2)u + Gh^2 = 0.$$

Supposing boundary conditions according to Dirichlet, we construct a vector k from the term Gh^2 and the boundary values, and further we put $Fh^2 = S$. The equation is then identically rewritten in the following two forms by use of matrix notations:

$$\begin{cases} \left(\rho I - H_0 + \frac{S}{2}\right)u = \left(\rho I + V_0 - \frac{S}{2}\right)u + k, \\ \left(\rho I - V_0 + \frac{S}{2}\right)u = \left(\rho I + H_0 - \frac{S}{2}\right)u + k, \end{cases}$$

where ρ is an arbitrary parameter. Putting $H = -H_0 + S/2$, $V = -V_0 + S/2$, we now define Peaceman–Rachford's ADI-method through

$$\begin{cases} (\rho_n I + H)u^{(n+1/2)} = (\rho_n I - V)u^{(n)} + k, \\ (\rho_n I + V)u^{(n+1)} = (\rho_n I - H)u^{(n+1/2)} + k, \end{cases}$$

where ρ_n are so-called iteration parameters, so far at our disposal.

We now assume that H and V are symmetric and positive definite. First we restrict ourselves to the case when all ρ_n are equal $(= \rho)$. The intermediate result $u^{(n+1/2)}$ can then be eliminated and we find:

$$u^{(n+1)} = Tu^{(n)} + g,$$

where

$$T = (\rho I + V)^{-1}(\rho I - H)(\rho I + H)^{-1}(\rho I - V),$$
$$g = (\rho I + V)^{-1}(\rho I - H)(\rho I + H)^{-1}k + (\rho I + V)^{-1}k.$$

The convergence speed depends on the spectral radius $\lambda(T)$, but when we try to estimate this it is suitable to construct another matrix W by a similarity transformation:

$$W = (\rho I + V)T(\rho I + V)^{-1},$$

that is,

$$W = \{(\rho I - H)(\rho I + H)^{-1}\} \cdot \{(\rho I - V)(\rho I + V)^{-1}\}.$$

If the eigenvalues of H and V are λ_i and μ_i we find

$$\lambda(T) = \lambda(W) \leq \|(\rho I - H)(\rho I + H)^{-1}\| \cdot \|(\rho I - V)(\rho I + V)^{-1}\|$$
$$= \lambda\{(\rho I - H)(\rho I + H)^{-1}\} \cdot \lambda\{(\rho I - V)(\rho I + V)^{-1}\}$$
$$= \max_{1 \leq i \leq M} \left|\frac{\rho - \lambda_i}{\rho + \lambda_i}\right| \max_{1 \leq i \leq N} \left|\frac{\rho - \mu_i}{\rho + \mu_i}\right|.$$

As H is positive definite, all $\lambda_i > 0$ and we can assume $0 < a \leq \lambda_i \leq b$. Let us first try to make

$$\max_{1 \leq i \leq M} \left|\frac{\rho - \lambda_i}{\rho + \lambda_i}\right|$$

as small as possible. It is easily understood that large values of $|(\rho - \lambda_i)/(\rho + \lambda_i)|$ are obtained if ρ is far away from λ_i, and hence our critical choice of ρ should be such that $(\rho - a)/(\rho + a) = (b - \rho)/(b + \rho)$, that is, $\rho = \rho_1 = (ab)^{1/2}$. In the same way, for the other factor we ought to choose $\rho = \rho_2 = (\alpha\beta)^{1/2}$ where $0 < \alpha \leq \mu_i \leq \beta$. If these two values are not too far apart, we can expect that the "best" of them will give a fair solution to the whole problem. From the beginning we want to choose ρ such that $\lambda(T)$ is minimized. Instead we have split a majorant for $\lambda(T)$ into two factors which have been minimized separately since the general problem is considerably more difficult.

In the special case when we are treating the usual Laplace-equation inside a rectangular domain $A \times B$, where $A = Mh$ and $B = Nh$, H is operating only horizontally and V only vertically. The $M - 1$ points along a given horizontal line form a closed system with the same matrix R as was denoted by T in the section on parabolic equations. Hence the eigenvalues of R have the form

$$\lambda_m = 4 \sin^2 \frac{m\pi}{2M}, \qquad m = 1, 2, \ldots, M - 1.$$

$$H = \begin{pmatrix} R & & \\ & R & \\ & & R \end{pmatrix}, \qquad V = \begin{pmatrix} S & & & \\ & S & & \\ & & S & \\ & & & S \\ & & & & S \end{pmatrix},$$

where

$$R = \begin{pmatrix} 2 & -1 & & & \\ -1 & 2 & -1 & & \\ & -1 & 2 & -1 & \\ & & -1 & 2 & -1 \\ & & & -1 & 2 \end{pmatrix} \qquad \text{and} \qquad S = \begin{pmatrix} 2 & -1 & \\ -1 & 2 & -1 \\ & -1 & 2 \end{pmatrix}.$$

In a similar way we obtain the eigenvalues of S:

$$\mu_n = 4 \sin^2 \frac{n\pi}{2N}.$$

Hence we get

$$a = 4 \sin^2 \frac{\pi}{2M}; \qquad b = 4 \sin^2 \frac{\pi[M-1]}{2M} = 4 \cos^2 \frac{\pi}{2M};$$

$$\alpha = 4 \sin^2 \frac{\pi}{2N}; \qquad \beta = 4 \cos^2 \frac{\pi}{2N}$$

and consequently

$$\rho_1 = 2 \sin \frac{\pi}{M}, \qquad \rho_2 = 2 \sin \frac{\pi}{N}.$$

Now we can compute

$$\frac{\rho_1 - a}{\rho_1 + a} \frac{\rho_1 - \alpha}{\rho_1 + \alpha} \simeq 1 - \frac{1}{N} \frac{\pi(M^2 + N^2)}{MN},$$

and analogously

$$\frac{\rho_2 - a}{\rho_2 + a} \frac{\rho_2 - \alpha}{\rho_2 + \alpha} \simeq 1 - \frac{1}{M} \frac{\pi(M^2 + N^2)}{MN}.$$

Finally we choose the value of ρ that gives the smallest limit, that is,

$$\rho = 2 \sin \frac{\pi}{P},$$

where $P = \max(M, N)$. Note that $P = M$ produces the result containing $1/N$.

For the general case when different parameters are used in cyclical order, there is no complete theory, but nevertheless there are a few results that are useful in practical computation. We put $c = \min(a, \alpha)$ and $d = \max(b, \beta)$. In the case when one wants to use n parameters, Peaceman–Rachford suggest the following choice:

$$\rho_j = d \left(\frac{c}{d} \right)^{(2j-1)/2n}, \qquad j = 1, 2, \ldots, n.$$

Wachspress suggests instead

$$\rho_j = d \left(\frac{c}{d} \right)^{(j-1)/(n-1)}, \qquad n \geq 2, \quad j = 1, 2, \ldots, n.$$

More careful investigations have shown that convergence with Wachspress parameters is about twice as fast as with Peaceman–Rachford parameters.

When the iterations are performed in practical work, this means that one solves a system of equations with tridiagonal coefficient matrices, which can be done by usual Gaussian elimination. Since all methods for numerical solution of elliptic equations are based upon the solution of finite (but large) linear systems of equations, one need not, as a rule, be worried about stability. Almost exclusively an iterative technique is used, and then it is about enough that the method converges. Since the domain is closed the boundary values, so to speak, by force will keep the inner function values under control, while we do not have a similar situation for hyperbolic or parabolic equations.

§20.5.   THE FINITE ELEMENT METHOD (FEM)

We have already discussed this general technique in connection with boundary value problems for ordinary differential equations. When we use conventional discretization (e.g., in two dimensions), we introduce a uniform grid, and if possible we make the mesh sizes equal in both directions. However, in practical problems it may well be the case that the boundary is irregular, and further that a large mesh size is satisfactory in some parts of the region while a much smaller mesh size may be necessary elsewhere. We now decide to divide the region by straight lines into a finite number of parts, which we call *elements*. Among possible finite elements the triangle is most widely used. In this way the boundary will be approximated by a polygon train. In principle we are going to use an expansion of the wanted solution in terms of spline functions, with a simultaneous reformulation to a minimum problem.

We will restrict ourselves to treating the Poisson equation

$$\frac{\partial^2 u}{\partial x^2} + \frac{\partial^2 u}{\partial y^2} + f(x, y) = 0$$

or in abbreviated form $\Delta u + f = 0$. We choose an appropriate triangular mesh and define suitable basis functions as shown in Figure 20.3. For simplicity we consider only piecewise linear functions, but functions of higher degree can also occur. We

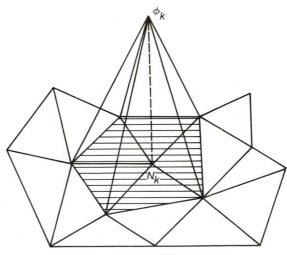

FIGURE 20.3

associate one basis function φ_j with each node N_j and claim that $\varphi_j(N_k) = \delta_{jk}$. This means that we approximate the wanted function u by a collection of planes, one for each triangle. Suppose that we try the u-values $a_1, a_2, \ldots, a_m$ in the nodes $N_1, N_2, \ldots, N_m$, where N_k has coordinates (x_k, y_k). For simplicity, let a triangle have the corners $(x_1, y_1), (x_2, y_2),$ and (x_3, y_3); then the corresponding plane has the equation

$$
\begin{vmatrix}
x & y & u & 1 \\
x_1 & y_1 & a_1 & 1 \\
x_2 & y_2 & a_2 & 1 \\
x_3 & y_3 & a_3 & 1
\end{vmatrix} = 0.
$$

We see that $u(x_1, y_1) = a_1$, $u(x_2, y_2) = a_2$, and $u(x_3, y_3) = a_3$.

We now return to the Poisson equation, which we want to solve in a finite region S with $u = 0$ on the boundary ∂S. We notice that a solution u to our problem is also a solution to the variational problem

$$(\nabla u, \nabla v) = (f, v)$$

or written out in full:

$$
\iint_S \left[\frac{\partial u}{\partial x} \frac{\partial v}{\partial x} + \frac{\partial u}{\partial y} \frac{\partial v}{\partial y} \right] dx\, dy = \iint_S fv\, dx\, dy
$$

for all $v(x, y)$ that are continuous on S, zero on the boundary, and with piecewise continuous and bounded derivatives. This relation is easy to prove by partial integration, and we have for example

$$
\int \frac{\partial u}{\partial x} \frac{\partial v}{\partial x}\, dx = \left[v \frac{\partial u}{\partial x} \right] - \int v \frac{\partial^2 u}{\partial x^2}\, dx
$$

where the integrated part vanishes because $v = 0$ on the boundary ∂S. A similar manipulation can be made for the y-derivatives, and hence we get

$$
\iint_S v[\Delta u + f]\, dx\, dy = 0.
$$

But here v is an arbitrary function, and we conclude that we must have $\Delta u + f = 0$. With only minor changes the reasoning can be made in the reverse order, and in fact the two problems are equivalent.

Our problem is now to determine the coefficients a_j in the expansion

$$
U = \sum_{j=1}^{m} a_j \varphi_j
$$

where φ_j are the pyramid-shaped basis functions. We are then going to use the formulation $(\nabla u, \nabla v) = (f, v)$ valid for all functions v. In particular, let v be one of the basis functions φ_k. Hence we get

$$
\left(\sum_j a_j \nabla \varphi_j, \nabla \varphi_k \right) = (f, \varphi_k)
$$

or in matrix form $\mathbf{Ca} = \mathbf{g}$ where

$$C_{jk} = \iint\limits_{S} (\nabla\varphi_j, \nabla\varphi_k) \, dx \, dy; \qquad g_k = \iint\limits_{S} f\varphi_k \, dx \, dy.$$

The symmetric (3×3)-matrix

$$\begin{pmatrix} C_{ii} & C_{ij} & C_{ik} \\ C_{ji} & C_{jj} & C_{jk} \\ C_{ki} & C_{kj} & C_{kk} \end{pmatrix}$$

associated with a triangle with corners N_i, N_j, N_k (and hence computed by integration over the triangle) is called the *stiffness matrix* for this triangle. If a similar computation is done for every triangle and the results are added, we get the global stiffness matrix $\mathbf{C}$. The right-hand side $\mathbf{g}$ is computed in an analogous way. The process just described is called *assemblage*. Obviously an element C_{jk} is nonzero only if the nodes N_j and N_k are corners in the same triangle. This implies that the matrix $\mathbf{C}$ is sparse. In some special cases we get band matrices, and if a sufficiently regular mesh is used we may recover the results from the usual difference methods.

The finite element method has been developed very quickly during recent decades and is now widely used, particularly on technical problems. As described above, it applies to linear elliptic problems. However, it can also be used on parabolic and hyperbolic equations, and it can even be generalized to nonlinear equations. There are many program packages available, but we must refrain from going into more detail.

---

§20.6. EIGENVALUE PROBLEMS

As we have seen already in connection with ordinary differential equations, some problems containing a parameter can be solved only for special values of this parameter. Such eigenvalue problems often appear also in applications leading to partial differential equations. We will demonstrate in a few examples different methods for computing at least the lowest eigenvalue, which in many cases has a very clear physical significance.

We start by considering the equation

$$\frac{\partial^2 u}{\partial x^2} + \frac{\partial^2 u}{\partial y^2} + \lambda u = 0$$

inside a unit square with $u = 0$ on the sides; the equation then describes a vibrating membrane. We take three mesh sizes: $h = \frac{1}{3}, \frac{1}{4}$, and $\frac{1}{5}$, and we require symmetric eigenfunctions. We then get Figures 20.4(a) and 20.4(b) in the first two cases, while the third is left to the reader.

Approximating Δu by second order differences, we get for $h = \frac{1}{3}$:

$$9(U + U + 0 + 0 - 4U) + \lambda U = 0 \qquad \text{giving} \qquad \lambda = 18.$$

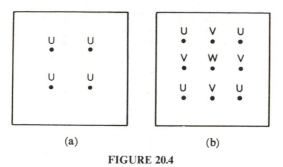

(a) (b)

FIGURE 20.4

For $h = \frac{1}{4}$, putting $\mu = \lambda/16$ we find:

$$\begin{cases} 2V - 4U + \mu U = 0, \\ 2U + W - 4V + \mu V = 0, \\ 4V - 4W + \mu W = 0. \end{cases}$$

The condition for a nontrivial solution is

$$\begin{vmatrix} 4 - \mu & -2 & 0 \\ -2 & 4 - \mu & -1 \\ 0 & -4 & 4 - \mu \end{vmatrix} = 0.$$

The result is $\mu^3 - 12\mu^2 + 40\mu - 32 = 0$, with the smallest root

$$\mu = 4 - \sqrt{8} = 1.1716,$$

and hence $\lambda = 18.7456$. Similarly, for $h = \frac{1}{5}$ we get $\lambda = 19.0983$. Richardson extrapolation with second and fourth order terms taken into account gives $\lambda = 19.7372$. The eigenfunctions are

$$u = \sin(m\pi x)\sin(n\pi y), \qquad m \text{ and } n \text{ positive integers,}$$

which gives $\lambda = (m^2 + n^2)\pi^2$. The lowest eigenvalue, obtained from $m = n = 1$, is $2\pi^2 = 19.7392$; hence the relative error is only about 10^{-4}

We now consider a slightly more difficult problem, namely to find the smallest eigenvalue of the equation $\Delta\,\Delta u - \lambda u = 0$, which describes the vibrations of a clamped thin plate. More precisely, we choose a square domain and discretize our equation written in a more explicit form

$$\frac{\partial^4 u}{\partial x^4} + \frac{\partial^4 u}{\partial y^4} + 2\frac{\partial^4 u}{\partial x^2\,\partial y^2} - \lambda u = 0$$

with the boundary conditions $u = 0$ and $\partial u/\partial n = 0$ (the normal derivative). This last condition is implemented by claiming that the image value is equal to the initial value (also see Figure 20.5). We want a solution with complete symmetry, i.e.,

$$u(x, y) = u(1 - x, y) = u(x, 1 - y) = u(1 - x, 1 - y)$$

if the square is defined by the corners $(0, 0)$, $(1, 0)$, $(1, 1)$, and $(0, 1)$. From (17.4.6)

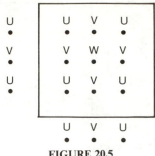

FIGURE 20.5

we have:

$$\nabla^4 f_0 \simeq h^{-4}(20 f_0 - 8 S_1 + 2 S_2 + S_3).$$

With $h = \frac{1}{4}$ and putting $\mu = \lambda h^4 = \lambda/256$, we get for $f_0 = U$:

$$S_1 = 0 + 0 + V + V = 2V; \quad S_2 = 0 + 0 + W + 0 = W; \quad S_3 = U + U + U + U = 4U.$$

Hence the first equation takes the form:

$$(24 - \mu)U - 16V + 2W = 0.$$

In a similar way we find V and W:

$$-16U + (26 - \mu)V - 8W = 0 \quad \text{and} \quad 8U - 32V + (20 - \mu)W = 0.$$

The condition for nontrivial solutions is

$$\begin{vmatrix} 24 - \mu & -16 & 2 \\ -16 & 26 - \mu & -8 \\ 8 & -32 & 20 - \mu \end{vmatrix} = 0$$

or

$$\mu^3 - 70\mu^2 + 1096\mu - 2848 = 0.$$

The smallest positive root is 3.236706, giving $\lambda = 828.597$.

In the case when $h = \frac{1}{8}$ we obtain the following matrix:

$$\begin{pmatrix}
22 & -16 & 2 & 0 & 2 & 0 & 0 & 0 & 0 & 0 \\
-8 & 23 & -8 & 1 & -8 & 3 & 0 & 0 & 0 & 0 \\
1 & -8 & 22 & -8 & 2 & -8 & 2 & 1 & 0 & 0 \\
0 & 2 & -16 & 21 & 0 & 4 & -8 & 0 & 1 & 0 \\
2 & -16 & 4 & 0 & 20 & -16 & 2 & 2 & 0 & 0 \\
0 & 3 & -8 & 2 & -8 & 23 & -8 & -8 & 3 & 0 \\
0 & 0 & 4 & -8 & 2 & -16 & 20 & 4 & -8 & 1 \\
0 & 0 & 2 & 0 & 2 & -16 & 4 & 22 & -16 & 2 \\
0 & 0 & 0 & 1 & 0 & 6 & -8 & -16 & 25 & -8 \\
0 & 0 & 0 & 0 & 0 & 0 & 4 & 8 & -32 & 20
\end{pmatrix}$$

The smallest positive eigenvalue is $\mu = 0.27599639$, giving $\lambda = 1130.481$. In the table below we display the results for $h = 1/n$, $n = 2(1)8$, together with the order of the corresponding determinant (N). Further, five extrapolated values are also given; they were obtained with the Lagrange formula as shown in an example in Section 14.1.

n	N	λ	Extrapolated from	λ
2	1	384	First 5	1289.32
3	1	648	Last 5	1292.72
4	3	828.597	First 6	1290.18
5	3	949.628	Last 6	1293.24
6	5	1031.914	All 7	1293.43
7	6	1089.315		
8	10	1130.481		

The extrapolated values show a good consistency, and the results indicate that the correct eigenvalue is about 1293.5.

Finally we will give one example showing how we can use the Ritz principle on an eigenvalue problem. We consider an elliptic plate that is kept fixed along the boundary of the ellipse $x^2/4 + y^2 = 1$. Denoting the displacement by u, this means that $u = 0$ along the boundary while the vibrations of the plate are described by the equation

$$\Delta u + \lambda u = 0$$

where as usual Δ is the Laplace operator $\partial^2/\partial x^2 + \partial^2/\partial y^2$. We now multiply by u and integrate

$$\iint_S u\, \Delta u\, dx\, dy + \lambda \iint_S u^2\, dx\, dy = 0$$

where S is the quarter ellipse defined by $0 \le x \le 2$, $0 \le y \le 1$. We introduce the integral

$$I = \iint_S [(\partial u/\partial x)^2 + (\partial u/\partial y)^2]\, dx\, dy$$

and conclude that $I > 0$. By partial integration we obtain

$$\int (\partial u/\partial x)^2\, dx = [u(\partial u/\partial x)] - \int u(\partial^2 u/\partial x^2)\, dx$$

where the first term vanishes because of the boundary conditions. A similar relation holds for the second term, and altogether we see that

$$I = -\iint_S u\, \Delta u\, dx\, dy.$$

Since

$$J = \iint_S u^2\, dx\, dy > 0$$

we find that $\lambda = I/J$ must be > 0.

For an exact solution u (an eigenfunction) and an exact eigenvalue λ, we have $G = I - \lambda J = 0$. If we make the following attempt:

$$u = (4 - x^2 - 4y^2)(a + bx^2 + cy^2)$$

the boundary condition is fulfilled for all a, b, and c. Then $G = I - \lambda J$ becomes an ordinary function of a, b, and c while λ can be considered as a parameter. Since we cannot in general hope for an exact solution $G = 0$, we do the next best thing by demanding that a, b, and c be chosen in such a way that G^2 is minimized. This leads to the conditions $\partial G/\partial a = \partial G/\partial b = \partial G/\partial c = 0$, a linear homogeneous system in a, b, and c. In order to obtain a nontrivial solution, we must claim that the determinant vanishes. From this we get a third degree equation in λ.

The computations are somewhat lengthy and we will only discuss some of the basic ideas. First we observe that integrals of the following form appear (for convenience we carry an extra factor π):

$$\pi p_{2m,2n} = \iint_S x^{2m} y^{2n} \, dx \, dy.$$

We find

$$\int_0^2 x^{2m} \, dx \int_0^{(1 - x^2/4)^{1/2}} y^{2n} \, dy = \frac{1}{2n + 1} \int_0^2 x^{2m} (1 - x^2/4)^{n + 1/2} \, dx,$$

and putting $x^2 = 4z$ we get:

$$p_{2m,2n} = \pi^{-1} 2^{2m} (2n + 1)^{-1} \int_0^1 z^{m - 1/2} (1 - z)^{n + 1/2} \, dz.$$

This integral is a beta integral (see Section 7.3):

$$B(m + \tfrac{1}{2}, n + \tfrac{3}{2}) = \frac{\Gamma(m + \tfrac{1}{2})\Gamma(n + \tfrac{3}{2})}{\Gamma(m + n + 2)}$$

and after standard simplifications by use of the functional relation for the gamma function, we obtain:

$$p_{2m,2n} = \frac{(2m)!(2n)!}{2^{2n+1} m! n! (m + n + 1)!}.$$

Some values of this expression are displayed in the following table.

n m	0	1	2	3	4
0	1/2	1/8	1/16	5/128	7/256
1	1/2	1/12	1/32	1/64	
2	1	1/8	3/80		
3	5/2	1/4			
4	7				

Using

$$\Delta u = -10a + (8 - 20x^2 - 8y^2)b + (8 - 2x^2 - 50y^2)c$$

we evaluate the integrals and find:

$$I/\pi = 10a^2 + \tfrac{40}{3}ab + \tfrac{10}{3}ac + \tfrac{46}{3}b^2 + \tfrac{5}{3}bc + \tfrac{47}{24}c^2$$
$$J/\pi = \tfrac{8}{3}a^2 + \tfrac{8}{3}ab + \tfrac{2}{3}ac + \tfrac{8}{5}b^2 + \tfrac{4}{15}bc + \tfrac{1}{10}c^2.$$

The last coefficient $\tfrac{1}{10}$ is obtained as

$$16p_{04} - 8p_{24} + p_{44} - 32p_{06} + 8p_{26} + 16p_{08}.$$

After computing the partial derivatives of $G = I - \lambda J$, we get the determinant (after multiplication of the columns by 3, 30, and 120 followed by division of the rows with 2, 4, and 1):

$$\begin{vmatrix} 15 - 4\lambda & 100 - 20\lambda & 100 - 20\lambda \\ 5 - \lambda & 115 - 12\lambda & 25 - 4\lambda \\ 5 - \lambda & 25 - 4\lambda & 235 - 12\lambda \end{vmatrix} = 0.$$

This leads to the equation $12\lambda^3 - 545\lambda^2 + 6100\lambda - 15375 = 0$ with the roots 3.56924, 12.04497, and 29.80245. With $c = 0$ we get with only two parameters $\lambda = \tfrac{5}{14}(22 \pm \sqrt{141})$ or 3.6163 and 12.098, while just one parameter yields $30/8 = 3.75$.

EXERCISES

1. The function $f(x, y)$ satisfies the Laplace equation

$$\frac{\partial^2 f}{\partial x^2} + \frac{\partial^2 f}{\partial y^2} = 0$$

inside a square with corners $(0, 0)$, $(1, 0)$, $(1, 1)$, and $(0, 1)$. We introduce a grid with the mesh size $h = \tfrac{1}{4}$ in both directions. The boundary values in the points $(0, 0)$, $(\tfrac{1}{4}, 0)$, ... taken around the square in this order are: -1, -0.6, -0.3333, -0.1429, 0, 0.0154, 0.0588, 0.1233, 0.2, 0.1385, 0.0769, 0.0244, 0, -0.28, -0.6, and -0.8824. Using Liebmann's method, compute the values in the nine inner mesh points (to 3 decimals).

2. Using Liebmann's method, solve the differential equation

$$\frac{\partial^2 u}{\partial x^2} + \frac{\partial^2 u}{\partial y^2} = \frac{y}{10}$$

in a square with corners $(1, 0)$, $(3, 0)$, $(3, 2)$, and $(1, 2)$ ($h = \tfrac{1}{2}$). The boundary values are:

x	1.5	2	2.5	3	3	3
y	0	0	0	0.5	1	1.5
u	-0.3292	0.3000	1.1625	2.3757	1.9500	1.0833

x	2.5	2	1.5	1	1	1
y	2	2	2	1.5	1	0.5
u	-1.0564	-1.4500	-1.4775	-0.8077	-0.6500	-0.7500

3. Find an approximate value of u in the origin when u satisfies the equation

$$\frac{\partial^2 u}{\partial x^2} + 2\frac{\partial^2 u}{\partial y^2} = 1$$

and further $u = 0$ on the sides of a square with corners $(\pm 1, \pm 1)$.

4. The function $u(x, y)$ satisfies the differential equation

$$\frac{\partial^2 u}{\partial x^2} + (1 + x^2 + y^2)\frac{\partial^2 u}{\partial y^2} + 2y\frac{\partial u}{\partial y} = 0.$$

On the sides of a square in the xy-plane with corners $(0, 0)$, $(0, 1)$, $(1, 0)$, and $(1, 1)$ we have $u(x, y) = x + y$. The square is split into smaller parts with side equal to $\frac{1}{3}$. Find the value of $u(x, y)$ in the four inner lattice points by approximating the differential equation with a difference equation so that the error becomes $O(h^2)$.

5. The equation

$$\frac{\partial^2 u}{\partial x^2} + \frac{1}{x}\frac{\partial u}{\partial x} = \frac{\partial u}{\partial t}$$

is given. Using separation, determine a solution satisfying $u(0, t) = 1$, $u(1, t) = 0$, and explain how a more general solution fulfilling the same conditions can be constructed.

6. In the wave equation

$$\frac{\partial^2 u}{\partial x^2} - \frac{\partial^2 u}{\partial t^2} = 0$$

the left-hand side is approximated by the expression $h^{-2}[\delta_t^2 - (1 + \alpha\delta_t^2)\delta_z^2]u$. Compute the truncation error.

7. Let $u(x, t)$ denote a sufficiently differentiable function of x and t and put $u(mh, nk) = u_{m,n}$ where h and k are the mesh sizes of a rectangular grid. Show that the equation

$$(2k)^{-1}(u_{m,n+1} - u_{m,n-1}) - h^{-2}(u_{m+1,n} - u_{m,n+1} - u_{m,n-1} + u_{m-1,n}) = 0$$

gives a difference approximation of the heat equation

$$\frac{\partial u}{\partial t} - \frac{\partial^2 u}{\partial x^2} = 0$$

if $k = h^2$ and $h \to 0$, but a difference approximation of the equation

$$\frac{\partial u}{\partial t} + \frac{\partial^2 u}{\partial t^2} - \frac{\partial^2 u}{\partial x^2} = 0$$

if $k = h$ and $h \to 0$.

8. Compute the smallest eigenvalue of the equation

$$\frac{\partial^2 u}{\partial x^2} + \frac{\partial^2 u}{\partial y^2} + \lambda u = 0$$

by replacing it with a difference equation. Boundary values are $u = 0$ on two horizontal and one vertical side of a square with side 1, (with two sides along the positive x- and y-axes), and $\partial u/\partial x = 0$ on the remaining side along the y-axis. This latter condition is conveniently replaced by $u(-h, y) = u(h, y)$. Choose $h = \frac{1}{2}$ and $h = \frac{1}{3}$, extrapolating to $h^2 = 0$. Then find an analytic solution by trying $u = \cos \alpha x \sin \beta y$, and compute an exact value of λ.

9. Find the smallest eigenvalue of the equation

$$\frac{\partial^2 u}{\partial x^2} + \frac{\partial^2 u}{\partial y^2} + \lambda(x^2 + y^2)u = 0$$

for a triangular domain with corners $(-1, 0)$, $(0, 1)$, and $(1, 0)$, where we have $u = 0$ on the boundary. Use $h = \frac{1}{2}$ and $h = \frac{1}{3}$, and extrapolate to $h^2 = 0$.

10. The equation

$$\frac{\partial^2 u}{\partial x^2} + \frac{\partial^2 u}{\partial y^2} + \lambda u = 0$$

is considered in a triangular domain defined by the lines $x = 0$, $y = 0$, and $x + y = 1$ with $u = 0$ on the boundary. Compute the smallest eigenvalue when we discretize with $h = \frac{1}{4}$ and $h = \frac{1}{3}$, and perform Richardson extrapolation Compare with the exact eigenvalue obtained from a solution of the form

$$u = \sin m\pi x \sin n\pi y + \sin n\pi x \sin m\pi y.$$

11. Find the smallest eigenvalue of the differential equation

$$\frac{\partial^2 u}{\partial x^2} + \frac{\partial^2 u}{\partial y^2} + \lambda u = 0$$

in a square with corners $(1, 0)$, $(-1, 0)$, $(0, 1)$, and $(0, -1)$ and with $u = 0$ on the boundary. Use $h = \frac{1}{2}$ and $h = \frac{1}{3}$ and perform Richardson extrapolation.

12. Compute the smallest eigenvalue of the differential equation

$$\frac{\partial^4 u}{\partial x^4} + \frac{\partial^4 u}{\partial y^4} - \lambda u = 0$$

in a square with corners $(0, 0)$, $(1, 0)$, $(1, 1)$, and $(0, 1)$. Boundary values are $u = 0$ and $\partial u/\partial n = 0$ where n is the normal. Take $h = \frac{1}{3}, \frac{1}{4}$, and $\frac{1}{5}$, and extrapolate to $h = 0$ under the assumption that the error term is of the form $ah^2 + bh^4$.

13. Perform one integration step for the equation

$$\frac{\partial u}{\partial t} = \frac{\partial^2 u}{\partial x^2}$$

using Crank–Nicolson's method when $0 \le x \le 1$ and $t \ge 0$. Choose $\Delta x = 0.1$ and $\Delta t = 0.01$. Boundary conditions are $u(x + 1, t) = u(x, t)$; initial conditions are

$$u(x, 0) = \begin{cases} 1 - 2x, & 0 \le x \le \frac{1}{2} \\ 2x - 1, & \frac{1}{2} \le x \le 1. \end{cases}$$

For symmetry reasons we have

$$u(x, t) = u(1 - x, t) \quad \text{and} \quad u(x, t) + u(\tfrac{1}{2} - x, t) = 1.$$

14. The equation

$$\frac{\partial u}{\partial t} + \frac{\partial u}{\partial x} = \frac{\partial^2 u}{\partial x^2}$$

is given with the initial values $u(x, 0) = \cos 2\pi x$ and the boundary condition $u(x, t) = u(x + 1, t)$. Perform one integration step in the variable t using Crank–Nicolson's method with $\Delta x = \frac{1}{6}$ and $\Delta t = \frac{1}{96}$. Note that the resulting linear system of equations can be simplified if the equations are added pairwise in a suitable manner.

CHAPTER 21

Integral Equations

*In some countries the drain pipes are placed outside
the houses in order to simplify repair service.
Repairs are necessary because the pipes have
been placed outside the houses.*

§21.1. CLASSIFICATION

An equation containing an integral where an unknown function to be determined enters under the integral sign is called an *integral equation*. If the desired function is present only in the integral, the equation is said to be of the *first kind*; if the function is also present outside the integral, the equation is said to be of the *second kind*. If the function to be determined only occurs linearly, the equation is said to be *linear*, and we shall restrict ourselves to such equations. If the limits in the integral are constant, the equation is said to be of *Fredholm's* type; if the upper limit is variable we have an equation of *Volterra's* type.

In the following the unknown function will be denoted by $y(x)$. We then suppose that $y(t)$ is integrated with another function $K(x, t)$ (with t as integration variable); the function $K(x, t)$ is called the *kernel* of the equation. While the function $y(x)$ as a rule has good regularity properties (continuity or differentiability), it is rather common that the kernel $K(x, t)$ has some discontinuity or at least discontinuous derivatives. The equation is still said to be of Fredholm's type if the limits are constant and the kernel is quadratically integrable over the domain. If $K(x, t)$ is singular in the considered domain, or if one or both of the limits is infinite, the integral equation is said to be *singular*. If the equation has a term which does not contain the wanted function $y(x)$, the equation is said to be *inhomogeneous*, otherwise *homogeneous*. Often a parameter λ is placed as a factor in front of the integral, and it may then happen that the equation can be solved only for certain values of λ (eigenvalues), or that there is no solution at all. In the examples below, the terminology just described is illustrated.

$$f(x) = \int_a^b K(x, t)y(t)\, dt, \qquad \text{Fredholm, 1. kind, inhomogeneous,}$$

$$y(x) = f(x) + \lambda \int_a^b K(x, t)y(t)\, dt, \qquad \text{Fredholm, 2. kind, inhomogeneous,}$$

$$y(x) = \lambda \int_a^b K(x, t)y(t)\, dt, \qquad \text{Fredholm, 2. kind, homogeneous,}$$

$$f(x) = \int_a^x K(x, t)y(t)\, dt, \qquad \text{Volterra, 1. kind, inhomogeneous,}$$

$$y(x) = f(x) + \lambda \int_a^x K(x, t)y(t)\, dt, \qquad \text{Volterra, 2. kind, inhomogeneous.}$$

Again it is pointed out that the types referred to here do not in any way cover all possibilities. The reason that just these have been chosen is their strong dominance in the applications.

The fundamental description of the integral equations was performed by Fredholm and Volterra, but important contributions to the theory have been made by Hilbert and Schmidt. An extensive literature exists within this field, as well as on numerical methods. For obvious reasons, it is only possible to give a very incomplete presentation. Primarily we shall discuss direct methods based on linear systems of equations, but iterative methods will also be treated. Further, special kernels (symmetric, degenerate) and methods related to them will be mentioned briefly. In order not to make the description unnecessarily heavy we shall assume that all functions used have the properties needed to make the performed operations legitimate.

§21.2. FREDHOLM'S INHOMOGENEOUS EQUATION OF THE SECOND KIND

Here we shall primarily study Fredholm's inhomogeneous equation of the second kind:

$$(21.2.1) \qquad y(x) = f(x) + \lambda \int_a^b K(x, t)y(t)\, dt,$$

since the treatment of other types in many respects can be related and traced back to this one. The interval between a and b is divided into n equal parts, each of length h. We introduce the notations

$$x_r = a + rh, \qquad y(x_r) = y_r, \qquad f(x_r) = f_r, \qquad K(x_r, t_s) = K_{rs}.$$

The integral is then replaced by a finite sum essentially according to the trapezoidal rule, and then the following system of equations appears:

$$Ay = f,$$

where

$$(21.2.2) \quad \begin{cases} A = \begin{pmatrix} 1 - \lambda h K_{11} & -\lambda h K_{12} & \cdots & -\lambda h K_{1n} \\ -\lambda h K_{21} & 1 - \lambda h K_{22} & \cdots & -\lambda h K_{2n} \\ \vdots & & & \\ -\lambda h K_{n1} & -\lambda h K_{n2} & \cdots & 1 - \lambda h K_{nn} \end{pmatrix}, \\ \\ y = \begin{pmatrix} y_1 \\ y_2 \\ \vdots \\ y_n \end{pmatrix}, \quad f = \begin{pmatrix} f_1 \\ f_2 \\ \vdots \\ f_n \end{pmatrix}. \end{cases}$$

Fredholm himself gave this system careful study. Of special interest is the limiting value of $\Delta = \det A$ when $h \to 0$ (and $n \to \infty$). This limiting value is called *Fredholm's determinant* and is usually denoted by $D(\lambda)$. Applying Cramer's rule we can write down the solution in explicit form according to

$$y_s = \frac{1}{\Delta} \sum_{r=1}^{n} f_r \Delta_{rs},$$

where, as usual, Δ_{rs} is the algebraic complement of the (r, s)-element in the matrix A. Expanding Δ in powers of λ we find

$$\Delta = 1 - \lambda h \sum_i K_{ii} + \frac{\lambda^2 h^2}{2!} \sum_{i,j} \begin{vmatrix} K_{ii} & K_{ij} \\ K_{ji} & K_{jj} \end{vmatrix} - \cdots$$

and after passage to the limit $h \to 0$:

$$D(\lambda) = 1 - \lambda \int_a^b K(t, t)\, dt + \frac{\lambda^2}{2!} \iint_a^b \begin{vmatrix} K(t_1, t_1) & K(t_1, t_2) \\ K(t_2, t_1) & K(t_2, t_2) \end{vmatrix} dt_1\, dt_2$$

$$- \frac{\lambda^3}{3!} \iiint_a^b \begin{vmatrix} K(t_1, t_1) & K(t_1, t_2) & K(t_1, t_3) \\ K(t_2, t_1) & K(t_2, t_2) & K(t_2, t_3) \\ K(t_3, t_1) & K(t_3, t_2) & K(t_3, t_3) \end{vmatrix} dt_1\, dt_2\, dt_3 + \cdots.$$

A similar expansion can be made for Δ_{rs}, and after division by h and passage to the limit we get

$$D(x, t; \lambda) = \lambda K(x, t) - \lambda^2 \int_a^b \begin{vmatrix} K(x, t) & K(x, t_1) \\ K(t_1, t) & K(t_1, t_1) \end{vmatrix} dt_1$$

$$+ \frac{\lambda^3}{2!} \iint_a^b \begin{vmatrix} K(x, t) & K(x, t_1) & K(x, t_2) \\ K(t_1, t) & K(t_1, t_1) & K(t_1, t_2) \\ K(t_2, t) & K(t_2, t_1) & K(t_2, t_2) \end{vmatrix} dt_1\, dt_2 - \cdots,$$

where $x = a + sh$, $t = a + rh$; however, $D(x, x; \lambda) = D(\lambda)$. Introducing the notation $H(x, t; \lambda) = D(x, t; \lambda)/D(\lambda)$ we can write the solution in the form

$$(21.2.3) \quad y(x) = f(x) + \int_a^b H(x, t; \lambda) f(t)\, dt.$$

The function $H(x, t; \lambda)$ is usually called the *resolvent* of the integral equation. Obviously, as a rule λ must have such a value that $D(\lambda) \neq 0$. On the other hand, if we are considering the homogeneous Fredholm equation of second kind, obtained by making $f(x) = 0$ in (21.2.1), then λ must be a root of the equation $D(\lambda) = 0$ in order to produce a nontrivial solution. In this case we have an eigenvalue problem. The technique which led to the solution (21.2.3) in some simple cases can be used also in practice, and we give an example of this. Consider the equation

$$y(x) = f(x) + \lambda \int_0^1 (x + t)y(t)\, dt.$$

We compute $D(\lambda)$ and $D(x, t; \lambda)$, noticing that all determinants of higher order than the second vanish:

$$D(\lambda) = 1 - \lambda - \frac{\lambda^2}{12}; \qquad D(x, t; \lambda) = \lambda\left[\left(1 - \frac{\lambda}{2}\right)(x + t) + \lambda\left(xt + \frac{1}{3}\right)\right].$$

Thus we find the solution:

$$y(x) = f(x) + \frac{\lambda}{1 - \lambda - \lambda^2/12} \int_0^1 \left[\left(1 - \frac{\lambda}{2}\right)(x + t) + \lambda\left(xt + \frac{1}{3}\right)\right] f(t)\, dt,$$

which can obviously be written $y(x) = f(x) + Ax + B$, where A and B are constants depending on λ and the function $f(t)$. The solution exists unless $1 - \lambda - \lambda^2/12 = 0$, that is, $\lambda = -6 \pm 4\sqrt{3}$.

We shall now study an iterative technique. We again start from (21.2.1), constructing a series of functions $y_0, y_1, y_2, \ldots$, according to

$$y_{n+1}(x) = f(x) + \lambda \int_a^b K(x, t)y_n(t)\, dt; \qquad y_0(x) = 0$$

(cf. Picard's method for ordinary differential equations). If this sequence of functions converges toward a limiting function, this is also a solution of the integral equation. As is easily understood, the method is equivalent to a series expansion in λ (Neumann series):

$$y = f(x) + \lambda\varphi_1(x) + \lambda^2\varphi_2(x) + \cdots.$$

Inserting this and identifying coefficients for different powers of λ, we get:

$$\varphi_1(x) = \int_a^b K(x, t)f(t)\, dt,$$

$$\varphi_2(x) = \int_a^b K(x, t)\varphi_1(t)\, dt = \int_a^b K_2(x, t)f(t)\, dt,$$

$$\varphi_3(x) = \int_a^b K(x, t)\varphi_2(t)\, dt = \int_a^b K_3(x, t)f(t)\, dt,$$

$$\vdots$$

Here

$$K_2(x, t) = \int_a^b K(x, z)K(z, t)\, dz$$

and more generally

$$K_n(x, t) = \int_a^b K_r(x, z) K_{n-r}(z, t)\, dz, \qquad K_1(x, t) = K(x, t).$$

The functions K_2, K_3, ... are called *iterated kernels*. It can be shown that the series converges for $\lambda < \|K\|^{-1}$, where

$$\|K\| = \int\int_a^b K^2(x, y)\, dx\, dy.$$

Thus we obtain the final result:

$$y(x) = f(x) + \int_a^b \left[\lambda K_1(x, t) + \lambda^2 K_2(x, t) + \lambda^3 K_3(x, t) + \cdots \right] f(t)\, dt$$

or

$$y(x) = f(x) + \int_a^b H(x, t; \lambda) f(t)\, dt,$$

where

$$H(x, t; \lambda) = \lambda K_1(x, t) + \lambda^2 K_2(x, t) + \lambda^3 K_3(x, t) + \cdots$$

as before stands for the resolvent.

For small λ, the described method is well suited for numerical treatment on a computer. We shall work an example analytically with this technique. Consider the equation

$$y(x) = f(x) + \lambda \int_0^1 \exp(x - t) y(t)\, dt.$$

Thus we have $K(x, t) = \exp(x - t)$ and consequently

$$K_2(x, t) = \int_0^1 e^{x-z} e^{z-t}\, dz = e^{x-t} = K(x, t),$$

and all iterated kernels are equal to $K(x, t)$. Hence we obtain

$$y = f(x) + \int_0^1 e^{x-t} f(t)(\lambda + \lambda^2 + \lambda^3 + \cdots)\, dt$$

or, if $|\lambda| < 1$,

$$y = f(x) + \frac{\lambda}{1 - \lambda} e^x \int_0^1 e^{-t} f(t)\, dt.$$

In this special case we can attain a solution in a still simpler way. From the equation follows

$$y(x) = f(x) + \lambda e^x \int_0^1 e^{-t} y(t)\, dt = f(x) + C\lambda e^x.$$

Inserting this solution again for determining C, we find

$$f(x) + C\lambda e^x = f(x) + \lambda \int_0^1 e^{x-t} [f(t) + C\lambda e^t]\, dt,$$

giving

$$C = \int_0^1 e^{-t} f(t)\, dt + C\lambda$$

and we get back the same solution as before. However, it is now sufficient that $\lambda \neq 1$. If $\lambda = 1$ no solution exists unless $\int_0^1 e^{-t} f(t)\, dt = 0$; if so, there are an infinity of solutions $y = f(x) + Ce^x$ with arbitrary C.

This last method obviously depends on the fact that the kernel e^{x-t} could be written $e^x e^{-t}$ making it possible to move the factor e^x outside the integral sign. If the kernel decomposes in this way so that $K(x, t) = F(x)G(t)$ or, more generally, so that

$$K(x, t) = \sum_{i=1}^{n} F_i(x)G_i(t),$$

the kernel is said to be *degenerate*. In this case, the equation can be treated in an elementary way. Let us start from

$$y(x) = f(x) + \lambda \int_a^b \left[\sum_{i=1}^{n} F_i(x)G_i(t) \right] y(t)\, dt = f(x) + \sum_{i=1}^{n} F_i(x) \int_a^b G_i(t)y(t)\, dt.$$

These last integrals are all constants:

$$\int_a^b G_i(t)y(t)\, dt = c_i$$

and so we get

$$y(x) = f(x) + \lambda \sum_{i=1}^{n} c_i F_i(x).$$

The constants c_i are determined by insertion into the equation:

$$c_i = \int_a^b G_i(t) \left[f(t) + \lambda \sum_{k=1}^{n} c_k F_k(t) \right] dt.$$

With the notations

$$\int_a^b G_i(t)F_k(t)\, dt = a_{ik} \qquad \text{and} \qquad \int_a^b G_i(t)f(t)\, dt = b_i,$$

we get a linear system of equations for the c_i:

$$(I - \lambda A)c = b,$$

where $A = (a_{ik})$ and b and c are column vectors. This system has a unique solution if $D(\lambda) = \det(I - \lambda A) \neq 0$. Further, we define $D(x, t; \lambda)$ as a determinant of order $(n + 1)$:

$$D(x, t; \lambda) = \begin{vmatrix} 0 & -F_1(x) & -F_2(x) & \cdots & -F_n(x) \\ G_1(t) & 1 - \lambda a_{11} & -\lambda a_{12} & \cdots & -\lambda a_{1n} \\ G_2(t) & -\lambda a_{21} & 1 - \lambda a_{22} & \cdots & -\lambda a_{2n} \\ \vdots & & & & \\ G_n(t) & -\lambda a_{n1} & -\lambda a_{n2} & \cdots & 1 - \lambda a_{nn} \end{vmatrix}.$$

Then we can also write

$$\dot{y}(x) = f(x) + \frac{\lambda}{D(\lambda)} \int_a^b D(x, t; \lambda) f(t) \, dt = f(x) + \lambda \int_a^b H(x, t; \lambda) f(t) \, dt,$$

with the resolvent as before represented in the form

$$H(x, t; \lambda) = \frac{D(x, t; \lambda)}{D(\lambda)}.$$

We now treat a previous example by this technique:

$$y(x) = f(x) + \lambda \int_0^1 (x + t) y(t) \, dt.$$

Putting

$$\int_0^1 y(t) \, dt = A, \qquad \int_0^1 t y(t) \, dt = B, \qquad \int_0^1 f(t) \, dt = C, \qquad \int_0^1 t f(t) \, dt = D,$$

the constants C and D are known in principle, while A and B have to be computed. Now, $y(x) = f(x) + A\lambda x + B\lambda$ and $y(t) = f(t) + A\lambda t + B\lambda$; inserting these expressions into the equation gives

$$A = C + \frac{1}{2} A\lambda + B\lambda,$$

$$B = D + \frac{1}{3} A\lambda + \frac{1}{2} B\lambda.$$

Solving for A and B we get

$$A = \frac{C(1 - \frac{1}{2}\lambda) + D\lambda}{\Delta}, \qquad B = \frac{C\lambda/3 + D(1 - \frac{1}{2}\lambda)}{\Delta},$$

where we have supposed $\Delta = 1 - \lambda - \lambda^2/12 \neq 0$. Hence we get as before

$$y(x) = f(x) + \frac{\lambda}{1 - \lambda - \lambda^2/12} \int_0^1 \left[x\left(1 - \frac{\lambda}{2}\right) + \lambda tx + \frac{\lambda}{3} + t\left(1 - \frac{\lambda}{2}\right) \right] f(t) \, dt.$$

§21.3. FREDHOLM'S HOMOGENEOUS EQUATION OF THE SECOND KIND

As mentioned earlier, the case $f(x) \equiv 0$, that is, when we have a homogeneous equation, gives rise to an eigenvalue problem. The eigenvalues are determined, for example, from the equation $D(\lambda) = 0$, or, if we are approximating by a finite linear system of equations, in the usual way from a determinantal condition. There exists an extensive theory for these eigenvalue problems, but here we shall only touch upon a few details that have exact counterparts in the theory of Hermitian or real symmetric matrices. First we prove that if the kernel is real and symmetric, that is, $K(x, t) = K(t, x)$, then the eigenvalues are real. For we have $y(x) = \lambda \int_a^b K(x, t)y(t)\, dt$. Multiplying by $y^*(x)$ and integrating from a to b we get

$$\int_a^b y^*(x)y(x)\, dx = \lambda \int\!\!\int_a^b y^*(x)K(x, t)y(t)\, dx\, dt.$$

The left-hand integral is obviously real and so is the right-hand integral, since its conjugate value is

$$\int\!\!\int_a^b y(x)K(x, t)y^*(t)\, dx\, dt = \int\!\!\int_a^b y(x)K(t, x)y^*(t)\, dx\, dt,$$

which goes over into the original integral if the integration variables change places. This implies that λ also must be real.

Next we shall prove that two eigenfunctions $y_i(x)$ and $y_k(x)$ corresponding to two different eigenvalues λ_i and λ_k are orthogonal. For

$$y_i(x) = \lambda_i \int_a^b K(x, t)y_i(t)\, dt,$$

$$y_k(x) = \lambda_k \int_a^b K(x, t)y_k(t)\, dt.$$

Now multiply the first equation by $\lambda_k y_k(x)$ and the second by $\lambda_i y_i(x)$, subtract and integrate:

$$(\lambda_k - \lambda_i) \int_a^b y_i(x)y_k(x)\, dx = \lambda_i\lambda_k \left[\int\!\!\int_a^b y_k(x)K(x, t)y_i(t)\, dx\, dt - \int\!\!\int_a^b y_i(x)K(x, t)y_k(t)\, dx\, dt \right].$$

Since $K(x, t) = K(t, x)$ and the variables x and t can change places, the two double integrals are equal, and as $\lambda_i \neq \lambda_k$ we get

(21.3.1) $$\int_a^b y_i(x)y_k(x)\, dx = 0.$$

We shall also try to construct a solution of the homogeneous equation

$$y(x) = \lambda \int_a^b K(x, t)y(t)\, dt.$$

To do this we start from the matrix A in (21.2.2). If we choose the elements of column j and multiply by the algebraic complements of column k, the sum of these products, as is well known, equals zero. Hence we have

$$(1 - \lambda h K_{jj})\Delta_{jk} - \lambda h K_{kj}\Delta_{kk} - \sum_{i \neq j,k} \lambda h K_{ij}\Delta_{ik} = 0,$$

where the products (j) and (k) have been written out separately. Dividing by h (note that Δ_{jk} and Δ_{ik} as distinguished from Δ_{kk} contain an extra factor h) we get after passage to the limit $h \to 0$:

$$D(x_k, x_j; \lambda) - \lambda K(x_k, x_j)D(\lambda) - \lambda \int_a^b K(t, x_j)D(x_k, t; \lambda)\, dt = 0.$$

Replacing x_k by x and x_j by z, we get finally

(21.3.2) $$D(x, z; \lambda) - \lambda K(x, z)D(\lambda) = \lambda \int_a^b K(t, z)D(x, t; \lambda) \, dt.$$

Expanding by rows instead of columns we find in a similar way

(21.3.3) $$D(x, z; \lambda) - \lambda K(x, z)D(\lambda) = \lambda \int_a^b K(x, t)D(t, z; \lambda) \, dt.$$

These two relations are known as *Fredholm's first* and *second* relations.

Now suppose that $\lambda = \lambda_0$ is an eigenvalue, that is, $D(\lambda_0) = 0$. Fredholm's second relation then takes the form:

$$D(x, z; \lambda_0) = \lambda_0 \int_a^b K(x, t)D(t, z; \lambda_0) \, dt$$

independently of z. Hence we can choose $z = z_0$ arbitrarily, and under the assumption that $D(x, z; \lambda_0)$ is not identically zero, we have the desired solution apart from a trivial constant factor:

(21.3.4) $$y(x) = D(x, z_0; \lambda_0).$$

Example 1

$$y(x) = \lambda \int_0^1 (x + t)y(t) \, dt.$$

The eigenvalues are obtained from $D(\lambda) = 0$, that is,

$$\lambda^2 + 12\lambda = 12, \qquad \lambda = -6 \pm 4\sqrt{3}.$$

As before we have $D(x, t; \lambda) = \lambda[(1 - \frac{1}{2}\lambda)(x + t) + \lambda(xt + \frac{1}{3})]$, and can choose for example, $t = 0$. One solution is consequently

$$y(x) = \lambda\left[\left(1 - \frac{\lambda}{2}\right)x + \frac{\lambda}{3}\right].$$

Taking $\lambda = 4\sqrt{3} - 6$ we have, apart from an irrelevant constant factor,

$$y = x\sqrt{3} + 1.$$

If we choose other values for t we get, as is easily understood, the same result.

Another method can also be used. The equation shows that the solution must have the form $y(x) = Ax + B$, and this implies

$$Ax + B = \lambda \int_0^1 (x + t)(At + B) \, dt$$

which gives the same result.

Turning to purely numerical methods, we can discretize using Simpson's rule in the points 0, $\frac{1}{2}$, and 1. Putting $\alpha = 6/\lambda$, we find the condition

$$\begin{vmatrix} \alpha & -2 & -1 \\ -\frac{1}{2} & \alpha - 4 & -\frac{3}{2} \\ -1 & -6 & \alpha - 2 \end{vmatrix} = 0$$

or $\alpha^3 - 6\alpha^2 - 3\alpha = 0$, from which $\alpha = 0$ and $\alpha = 3 \pm \sqrt{12}$. If $\alpha = 0$ we get

$$\int_0^1 (x + t)y(t)\, dt = 0$$

or

$$\int_0^1 y(t)\, dt = 0; \qquad \int_0^1 ty(t)\, dt = 0.$$

This can be interpreted as orthogonality relations, and hence $y(x)$ must be a linear combination of shifted Legendre polynomials of degree ≥ 2. Since $P_2(x) = \frac{1}{2}(3x^2 - 1)$, and the transformation $x := 2x - 1$ maps $x = 0$ into -1 and $x = 1$ into $+1$, we get the simplest solution as

$$y(x) = \frac{1}{2}[3(2x - 1)^2 - 1] = 6x^2 - 6x + 1.$$

If instead $\alpha = 3 + \sqrt{12}$, we get $\lambda = 4\sqrt{3} - 6$ and $y(x) = x\sqrt{3} + 1$; then $\alpha = 3 - \sqrt{12}$ gives $\lambda = -6 - 4\sqrt{3}$ and $y(x) = -x\sqrt{3} + 1$. The reason that we obtain exact results is obviously that Simpson's rule is exact for polynomials of degree ≤ 3.

§21.4. FREDHOLM'S EQUATION OF THE FIRST KIND

The equation which we shall study very superficially in this section has the form

$$(21.4.1) \qquad f(x) = \int_a^b K(x, t)y(t)\, dt.$$

A parameter λ can of course be brought together with the kernel. In general, this equation has no solution, as can be realized if we choose $f(x) = \sin x$ and $K(x, t) = \exp(x - t)$:

$$\sin x = \int_0^1 e^{x-t}y(t)\, dt,$$

where the right-hand side is proportional to e^x.

However, let us suppose that $f(x)$, $K(x, t)$, and $y(t)$ can be expanded in a series of orthonormalized functions u_i:

$$(21.4.2) \qquad \begin{cases} f(x) = \sum_{i=1}^{\infty} b_i u_i(x), \\[2mm] K(x, t) = \sum_{i,k=1}^{\infty} a_{ik} u_i(x) u_k(t), \\[2mm] y(t) = \sum_{i=1}^{\infty} c_i u_i(t). \end{cases}$$

The functions $u_i(x)$ being orthonormalized over the interval (a, b) means that

$$(21.4.3) \qquad \int_a^b u_i(x)u_k(x)\, dx = \delta_{ik}.$$

After insertion into (21.4.1) we obtain:

$$\sum_{i=1}^{\infty} b_i u_i(x) = \int_a^b \sum_{i,k=1}^{\infty} a_{ik} u_i(x) u_k(t) \sum_{r=1}^{\infty} c_r u_r(t)\, dt = \sum_{i,k=1}^{\infty} a_{ik} c_k u_i(x).$$

Identifying, we get an infinite linear system of equations

$$\sum_{k=1}^{\infty} a_{ik} c_k = b_i,$$

where the coefficients c_k should be determined from the known quantities a_{ik} and b_i. If the summation is truncated at the value n we have the system

(21.4.4) $$\mathbf{Ac} = \mathbf{b}$$

with well-known properties regarding existence of solutions.

Another possibility is to approximate directly by a suitable quadrature formula, giving rise to a linear system of equations:

(21.4.5) $$\sum_{j=0}^{n} a_j K_{ij} y_j = f_i.$$

However, if the integral equation has no solution, there is an apparent risk that one might not observe this fact because the discretized problem may possess a solution.

---

§21.5. VOLTERRA'S EQUATIONS

First we are going to discuss Volterra's equation of the second kind

(21.5.1) $$y(x) = f(x) + \lambda \int_a^x K(x, t) y(t)\, dt.$$

It can be considered as a special case of Fredholm's equation if we define the kernel so that $K(x, t) = 0$ when $x < t < b$. However, there are some advantages in discussing the case separately, in about the same way as for triangular systems of equations compared with the general case. For simplicity we assume $a = 0$, we use the interval length h, and we decide to use the trapezoidal rule (more sophisticated methods can also easily be applied). Then we obtain

(21.5.2)
$$
\begin{aligned}
y_0 &= f_0, \\
y_1 &= f_1 + \lambda h\big[\tfrac{1}{2}K_{10}y_0 + \tfrac{1}{2}K_{11}y_1\big], \\
y_2 &= f_2 + \lambda h\big[\tfrac{1}{2}K_{20}y_0 + K_{21}y_1 + \tfrac{1}{2}K_{22}y_2\big], \\
&\vdots
\end{aligned}
$$

and we can determine the desired values $y_0, y_1, y_2, \ldots$ successively. This method is very well adapted for use on computers.

As is the case for Fredholm's equation, we can also construct a Neumann series with iterated kernels. The technique is exactly the same, and there is no reason to discuss it again. However, there is one important difference: if $K(x, t)$ and $f(x)$ are real and continuous, then the series converges for all values of λ.

Example 1

$$y(x) = x + \int_0^x (t - x)y(t)\, dt.$$

1. The trapezoidal rule with interval length $= h$ gives

$$y_0 = 0$$

$$y_1 = h \qquad\qquad\qquad y_1 - \sin h = \frac{h^3}{6} + \cdots$$

$$y_2 = 2h - h^3 \qquad\qquad y_2 - \sin 2h = \frac{2h^3}{6} + \cdots$$

$$y_3 = 3h - 4h^3 + h^5 \qquad y_3 - \sin 3h = \frac{3h^3}{6} + \cdots$$

$$\vdots$$

Already Simpson's formula supplemented by the $\frac{3}{8}$-rule gives a substantial improvement.

2. Using successive approximations we get

$$y_0 = x,$$

$$y_1 = x + \int_0^x (t - x)t\, dt = x - \frac{x^3}{6},$$

$$y_2 = x + \int_0^x (t - x)\left(t - \frac{t^3}{6}\right) dt = x - \frac{x^3}{6} + \frac{x^5}{120},$$

$$\vdots$$

and it is easy to show that we obtain the series expansion for $\sin x$.

3. Differentiating the equation, we find

$$y'(x) = 1 - \int_0^x y(t)\, dt; \qquad y'(0) = 1.$$

$$y''(x) = -y(x) \qquad \text{with the solution} \qquad y = A\cos x + B\sin x.$$

But since $y(0) = 0$ and $y'(0) = 1$, the result is $y = \sin x$.

Very briefly we shall also comment upon Volterra's equation of the first kind,

$$(21.5.3) \qquad\qquad f(x) = \int_a^x K(x, t)y(t)\, dt.$$

Differentiation with respect to x gives

$$f'(x) = K(x, x)y(x) + \int_a^x \frac{\partial K(x, t)}{\partial x} y(t)\, dt,$$

and assuming $K(x, x) \neq 0$ in the considered interval, we can divide by $K(x, x)$, obtaining a Volterra equation of the second kind. If necessary, the procedure can be repeated. It is also possible to use the same difference method as before, the first function value being determined from the relation above:

$$y(a) = \frac{f'(a)}{K(a, a)}.$$

A special equation of Volterra type is Abel's integral equation

(21.5.4) $$f(x) = \int_0^x (x - t)^{-\alpha} y(t)\, dt \qquad (0 < \alpha < 1).$$

Writing instead $f(z) = \int_0^z (z - t)^{-\alpha} y(t)\, dt$, dividing by $(x - z)^{1-\alpha}$, and integrating in z between the limits 0 and x, we obtain

$$\int_0^x \frac{f(z)}{(x - z)^{1-\alpha}}\, dz = \int_0^x \frac{dz}{(x - z)^{1-\alpha}} \int_0^z \frac{y(t)\, dt}{(z - t)^{\alpha}}.$$

If z is a vertical and t a horizontal axis, the integration is performed over a triangle with corners in the origin, in $(0, x)$, and in (x, x), and reversing the order of integration we find

$$\int_0^x \frac{f(z)}{(x - z)^{1-\alpha}}\, dz = \int_0^x y(t)\, dt \int_t^x \frac{dz}{(z - t)^{\alpha}(x - z)^{1-\alpha}}.$$

The last integral is transformed through $z - t = (x - t)\xi$ and goes over into

$$\int_0^1 \xi^{-\alpha}(1 - \xi)^{\alpha - 1}\, d\xi = \Gamma(\alpha)\Gamma(1 - \alpha) = \frac{\pi}{\sin \pi\alpha},$$

that is,

$$\int_0^x y(t)\, dt = \frac{\sin \pi\alpha}{\pi} \int_0^x \frac{f(z)}{(x - z)^{1-\alpha}}\, dz.$$

Differentiating this relation, we obtain the desired solution

(21.5.5) $$y(x) = \frac{\sin \alpha\pi}{\pi} \frac{d}{dx} \int_0^x \frac{f(z)}{(x - z)^{1-\alpha}}\, dz.$$

Abel's initial problem was concerned with the case $\alpha = \frac{1}{2}$ and arose when he tried to determine tautochronous curves, that is, curves having the property that the time required for a particle to slide along the curve to a given end point is a given function of the position of the starting point.

§21.6. NUMERICAL TREATMENT OF INTEGRAL EQUATIONS

Most methods depend upon some kind of discretization, and it is in fact striking how many features in the treatment of differential equations carry over to integral equations. It is hardly surprising that most of the difficulties which do appear have obvious counterparts in the differential equations case.

We have already discussed one simple eigenvalue problem, which we also solved by discretization. We now take a slightly more complicated example:

$$\int_0^1 K(x, t)y(t)\, dt = \lambda y(x), \qquad 0 \le x \le 1$$

with

$$K(x, t) = \min(x, t) - \tfrac{1}{2}xt.$$

Putting $\lambda = \alpha^2$, we can write:

$$\int_0^x (t - \tfrac{1}{2}xt)y(t)\, dt + \int_x^1 (x - \tfrac{1}{2}xt)y(t)\, dt = \alpha^2 y(x).$$

Differentiating twice we get:

$$-\tfrac{1}{2}\int_0^1 ty(t)\, dt + \int_x^1 y(t)\, dt = \alpha^2 y'(x) \qquad \text{and} \qquad \alpha^2 y'' + y = 0.$$

Since $y(0) = 0$, we get apart from a numerical factor $y(x) = \sin(x/\alpha)$. Introducing this into the equation and performing some partial integrations, we obtain the equation

$$\tan\left(\frac{1}{\alpha}\right) + \frac{1}{\alpha} = 0.$$

Supposing $\alpha > 0$, we get a series of solutions with $1/\alpha$ slightly more than $\pi/2$, $3\pi/2$, $5\pi/2$, and so forth. We find

$$\begin{aligned}
1/\alpha_1 &= 2.0287578381 & \lambda_1 &= 0.242962685 \\
1/\alpha_2 &= 4.9131804394 & \lambda_2 &= 0.0414261498 \\
1/\alpha_3 &= 7.978665712 & \lambda_3 &= 0.0157086716 \\
1/\alpha_4 &= 11.08553841 & \lambda_4 &= 0.0081374141.
\end{aligned}$$

We will now try to solve the problem by discretizing, and we choose $h = \tfrac{1}{4}$. For convenience we give a short table of $32K(x, t)$:

x \ t	$\frac{1}{4}$	$\frac{1}{2}$	$\frac{3}{4}$	1
$\frac{1}{4}$	7	6	5	4
$\frac{1}{2}$	6	12	10	8
$\frac{3}{4}$	5	10	15	12
1	4	8	12	16

All integrals are then approximated according to Simpson's rule, and putting $\mu = 96\lambda$ we find the condition:

$$\begin{vmatrix}
7 - \mu & 3 & 5 & 1 \\
6 & 6 - \mu & 10 & 2 \\
5 & 5 & 15 - \mu & 3 \\
4 & 4 & 12 & 4 - \mu
\end{vmatrix} = 0$$

or $\mu^4 - 32\mu^3 + 208\mu^2 - 416\mu + 256 = 0$. The roots are, in descending order: 24.053110, 5.004228, 1.666243, and 1.276419, giving the approximate eigenvalues $\lambda = 0.250553$, 0.052127, 0.0173567, and 0.013296.

In our second example we will demonstrate the numerical treatment of a Fredholm equation of the second kind. We consider

$$y(x) - \frac{1}{\pi} \int_{-1}^{1} \frac{y(t)}{1 + (x - t)^2} \, dt = 1$$

and take 2^N intervals with $N = 2, 3, 4$, and 5. Obviously we have $y(-x) = y(x)$. Again using Simpson's rule, we get a system of linear equations and obtain the solutions:

x	$N = 2$	$N = 3$	$N = 4$	$N = 5$
0	1.917649	1.919059	1.919034	1.919032115
1/8			1.914167	1.914165293
1/4		1.899665	1.899618	1.899615338
3/8			1.875564	1.875560542
1/2	1.847130	1.842461	1.842390	1.842385096
5/8			1.800803	1.800797221
3/4		1.752072	1.751961	1.751955077
7/8			1.697544	1.697539076
1	1.640348	1.639754	1.639698	1.639695416

Richardson extrapolation in h^4 can be performed, and the error will probably be at most a few units in the 8th decimal. We also mention that deferred correction, either straightforward or iterated, is widely used in a similar way as for differential equations.

Among other useful numerical methods are the collocation method and methods of Galerkin type where the wanted solution is expressed as an expansion in suitable, not necessarily orthogonal, functions.

Finally, we point out that the stability theory for ordinary differential equations in many cases carries over to integral equations. Hence it is possible to analyze multiple-step methods and also to apply them in practical computation.

§21.7. CONNECTIONS BETWEEN DIFFERENTIAL AND INTEGRAL EQUATIONS

There exists a fundamental relationship between linear differential equations and integral equations of Volterra's type. Let us start from the equation

$$(21.7.1) \qquad \frac{d^n u}{dx^n} + a_1(x) \frac{d^{n-1} u}{dx^{n-1}} + \cdots + a_n(x)u = F(x),$$

where the coefficients are continuous. Further we have the initial conditions

$$u(0) = c_0, \qquad u'(0) = c_1, \ldots, \qquad u^{(n-1)}(0) = c_{n-1}.$$

Putting $d^n u/dx^n = y$ we find

$$D^{-1}y = \int_0^x y(t)\, dt \qquad \text{(defining the operator } D^{-1}\text{)},$$

$$D^{-2}y = D^{-1}(D^{-1}y) = \int_0^x (x - t)y(t)\, dt \qquad \text{(by partial integration)},$$

$$\vdots$$

$$D^{-n}y = D^{-1}(D^{-n+1}y) = \frac{1}{(n-1)!} \int_0^x (x - t)^{n-1} y(t)\, dt.$$

By integrating the equation $d^n u/dx^n = y$ successively and taking the initial conditions into account, we get

$$\frac{d^{n-1}u}{dx^{n-1}} = c_{n-1} + D^{-1}y,$$

$$\frac{d^{n-2}u}{dx^{n-2}} = c_{n-1}x + c_{n-2} + D^{-2}y,$$

$$\vdots$$

$$u = c_{n-1}\frac{x^{n-1}}{(n-1)!} + c_{n-2}\frac{x^{n-2}}{(n-2)!} + \cdots + c_1 x + c_0 + D^{-n}y.$$

These values are now inserted into (21.7.1) and the following equation results:

$$(21.7.2) \qquad y(x) = f(x) + \int_0^x K(x, t)y(t)\, dt,$$

where

$$(21.7.3) \quad f(x) = F(x) - a_1(x)c_{n-1} - a_2(x)(c_{n-1}x + c_{n-2}) - \cdots$$

$$- a_n(x)\left(c_{n-1}\frac{x^{n-1}}{(n-1)!} + c_{n-2}\frac{x^{n-2}}{(n-2)!} + \cdots + c_1 x + c_0 \right)$$

and

$$(21.7.4) \quad K(x, t) = -\left[a_1(x) + a_2(x)(x - t) + a_3(x)\frac{(x - t)^2}{2!} + \cdots + a_n(x)\frac{(x - t)^{n-1}}{(n-1)!} \right].$$

Thus we can say that the integral equation describes the differential equation together with its initial conditions.

Example 1. $u'' + u = 0;\ u(0) = 0,\ u'(0) = 1.$ One finds

$$y(x) = -x + \int_0^x (t - x)y(t)\, dt,$$

with the solution $y(x) = -\sin x$ (note that $y = u''$).

Example 2. $u'' - 2xu' + u = 0; u(0) = 1, u'(0) = 0.$ Hence

$$y(x) = -1 + \int_0^x (x + t)y(t) \, dt.$$

EXERCISES

1. Compute $D(\lambda)$ and $D(x, t; \lambda)$ when $K(x, t) = xt, a = 0, b = 1$.

2. Use the result in Exercise 1 to solve the integral equation

$$y(x) = \frac{1}{\sqrt{x}} + \int_0^1 xty(t) \, dt.$$

3. Find the solution of the Fredholm integral equation

$$y(x) = x^2 + \lambda \int_0^1 (x^2 + t^2)y(t) \, dt.$$

4. Determine the eigenvalues of the homogeneous integral equation

$$y(x) = \lambda \int_0^1 (x + t + xt)y(t) \, dt.$$

5. Find an approximate solution of the integral equation

$$y(x) = x + \int_0^1 \sin(xt)y(t) \, dt,$$

by replacing $\sin(xt)$ by a degenerate kernel consisting of the first three terms in the Maclaurin expansion.

6. Show that the integral equation $y(x) = 1 - x + \int_0^1 K(x, t)y(t) \, dt$ where

$$K(x, t) = \begin{cases} t(1 - x) & \text{for } t \leq x \\ x(1 - t) & \text{for } t > x \end{cases}$$

is equivalent to the boundary value problem $d^2y/dx^2 + y = 0; \; y(0) = 1;$ $y(1) = 0$.

7. Solve the integral equation

$$g(s) = s + \int_0^1 (s^2 + t^2 - st)g(t) \, dt.$$

8. Solve the integral equation

$$g(s) = s + \int_0^1 (s + t + 1)^{-1}g(t) \, dt.$$

9. Find the eigenvalues of the integral equation $y(x) = \lambda \int_0^{\pi/2} \cos(x + t)y(t) \, dt$.

10. Solve the integral equation $y(x) = x + \int_0^x xt^2 y(t) \, dt$.

11. Solve the integral equation $y(x) = x + \int_0^x ty(t) \, dt$.

12. Determine eigenvalues and eigenfunctions of the problem

$$y(x) = \lambda \int_0^1 K(x, t)y(t) \, dt$$

where $K(x, t) = \min(x, t)$.

13. Find the solution of the integral equation $3x^2 + 5x^3 = \int_0^x (x + t)y(t) \, dt$.

14. Using a numerical technique, solve the integral equation

$$y(x) = 1 + \int_0^x \frac{y(t)\, dt}{x + t + 1}$$

for $x = 0(0.2)1$. Apply the trapezoidal rule.

15. Solve the Volterra equation

$$g(s) = 1 + \int_0^s (s - t)g(t)\, dt.$$

16. Find an approximation of the eigenvalue of the integral equation

$$g(s) = \lambda \int_0^1 |s - t|g(t)\, dt$$

using Fredholm's determinant. Compare with the value obtained from the differential equation.

17. Find an integral equation corresponding to $u' - u = 0$; $u(0) = u'(0) = 1$.

18. Find an integral equation corresponding to $u''' - 3u'' - 6u' + 8u = 0$; $u(0) = u'(0) = u''(0) = 1$.

19. Transform the following boundary value problem to a Fredholm integral equation:

$$u'' + xu = 1; \qquad u(0) = u(1) = 0.$$

20. Solve the integral equation

$$y(x) = 1 + \int_0^x (x + y(t))(1 + y(t)^2)^{-1}\, dt$$

numerically for $x = 0(0.2)1$ using the trapezoidal rule.

Answers

CHAPTER 1

1. 11259375   **2.** 0.2406631213 . . .   **3.** 0.3%

4. The error in the latter case is $14(2 + \sqrt{3})^5 = 10136$ times larger than in the former case.

5. The errors are $4.355 \cdot 10^{-6}$ and $2.726 \cdot 10^{-9}$, the former being about 1600 times larger.

6.

n	2	3	4	5	6	7
$10^6 J_n$	13958	10196	8045	6411	7170	-7609

The process breaks down for $n \simeq 6$.

CHAPTER 2

1. $(4, -11, 6)$ **2.** $a = 2$; $b = 1$; $c = -1$

3. Eigenvalues: $0((n-1)\text{-fold})$, and $x^T y$ (simple). Any vector u orthogonal to y is an eigenvector when $\lambda = 0$, while x is the eigenvector when $\lambda = x^T y$. Characteristic equation: $\lambda^{n-1}(\lambda - x^T y) = 0$. Minimal equation: $\lambda^2 - \lambda(x^T y) = 0$.

7. $\begin{pmatrix} \cos a & \sin a \\ -\sin a & \cos a \end{pmatrix}$ **8.** $\begin{cases} a = 0.469182 \\ b = -2.53037 \end{cases}$ **9.** $\dfrac{1}{7}\begin{pmatrix} 6 & 2 \\ 3 & 1 \end{pmatrix}$

11. $\|A\|_1 = 5$; $\|A\|_2 = (5 + \sqrt{5})/\sqrt{2} = 5.11667$; $\|A\|_\infty = 6$; $\|A\|_E = \sqrt{30} = 5.4772$; $\rho(A) = 5$

12. $\begin{pmatrix} -430 & 512 & 332 \\ -516 & 614 & 396 \\ 234 & -278 & -177 \end{pmatrix}$

14. For instance, $U = 2^{-1/2}\begin{pmatrix} 1 & i \\ i & 1 \end{pmatrix}$; $D = \begin{pmatrix} a + ib & 0 \\ 0 & a - ib \end{pmatrix}$

15. $\alpha = (d - a \pm R)/2b$ where $R^2 = (d - a)^2 + 4bc$; $p = \pm(\alpha\alpha^* + 1)^{-1/2}$;
$a' = 7 + 4i$, $d' = 7 - 4i$.

16. P_r has at least one characteristic value $= 0$; hence det $P_r = 0$.

$$P_2 = \frac{1}{20}\begin{pmatrix} 7 & -3 & -9 & -1 \\ -3 & 7 & 1 & 9 \\ -9 & 1 & 13 & -3 \\ -1 & 9 & -3 & 13 \end{pmatrix}$$

17. $A^n = \dfrac{1}{10}\begin{pmatrix} 10 + 2n & -n \\ 4n & 10 - 2n \end{pmatrix} = nA - (n-1)I$

18. Eigenvalues: $7, 3, 1, 1$. Eigenvectors:

$$\begin{pmatrix} 1 \\ 1 \\ 1 \\ 1 \end{pmatrix}, \quad \begin{pmatrix} 1 \\ -1 \\ -1 \\ 1 \end{pmatrix}, \quad \begin{pmatrix} 1 \\ 0 \\ 0 \\ -1 \end{pmatrix}, \quad \begin{pmatrix} 0 \\ 1 \\ -1 \\ 0 \end{pmatrix}.$$

Characteristic equation: $\lambda^4 - 12\lambda^3 + 42\lambda^2 - 52\lambda + 21 = 0$.
Minimal equation: $\lambda^3 - 11\lambda^2 + 31\lambda - 21 = 0$.
A is derogatory but not defective.

19. Eigenvalues: 2 (3-fold).

Eigenvectors: $\begin{pmatrix} 1 \\ 3 \\ 0 \end{pmatrix}$ and $\begin{pmatrix} 1 \\ 0 \\ 3 \end{pmatrix}$.

Characteristic equation: $(\lambda - 2)^3 = 0$.
Minimal equation: $(\lambda - 2)^2 = 0$.
A is defective and derogatory.

20. C can be transformed to block diagonal form using $S = \begin{pmatrix} I & I \\ I & -I \end{pmatrix}$ with $S^{-1} = \tfrac{1}{2}S$.

21. $a < 11$; $a^2 < 72$; $a^3 < 510$; $a^4 < 3624$
$(a < 11; a < 8.49; a < 7.99; a < 7.76)$
Characteristic equation: $\lambda^4 - 6\lambda^3 - 9\lambda^2 + 10\lambda - 6 = 0$ giving $a = 7.0876$.

23. $\dfrac{1}{3}\begin{pmatrix} \ln 50 & \ln(25/4) \\ \ln(5/2) & \ln 20 \end{pmatrix}$ **25.** $\dfrac{1}{7}\begin{pmatrix} 7 & 0 & -7 & 7 \\ 6 & 1 & -9 & 13 \\ 8 & -1 & -12 & 15 \end{pmatrix}$ **26.** $\dfrac{1}{294}\begin{pmatrix} 1 & 2 & 4 \\ 2 & 4 & 8 \\ 3 & 6 & 12 \end{pmatrix}$

CHAPTER 3

1. $1 + \dfrac{x}{3} - \dfrac{x^2}{9} + \dfrac{5x^3}{81} - \dfrac{10x^4}{243} + \dfrac{22x^5}{729} - \cdots$ **2.** $y = \dfrac{3x^4}{512} - \dfrac{9x^5}{1024} + \dfrac{159x^6}{16384} - \cdots$

$1 - \dfrac{x}{3} + \dfrac{2x^2}{9} - \dfrac{14x^3}{81} + \dfrac{35x^4}{243} - \dfrac{91x^5}{729} + \cdots$

3. $A_n = a_n + n^{-1} \sum_{k=1}^{n-1} k a_k A_{n-k}$

$e^{\sin x} = 1 + x + \dfrac{x^2}{2} - \dfrac{x^4}{8} - \dfrac{x^5}{15} + \cdots$

4. $f(x) = a_0 + a_1 x + a_2 x^2 + \cdots = [(1-x)^2(1-x^2)]^{-1}$

$\quad a_n = 2a_{n-1} - 2a_{n-3} + a_{n-4}$

n	0	1	2	3	4	5	6	7	8
a_n	1	2	4	6	9	12	16	20	25

Using the theory of difference equations (Ch. 6), we find $a_n = \lfloor (n+2)^2/4 \rfloor$.

5. $y \simeq \dfrac{1}{2x} + \dfrac{1}{2^2 x^3} + \dfrac{1\cdot 3}{2^3 x^5} + \dfrac{1\cdot 3\cdot 5}{2^2 x^7} + \cdots$

6. $a_{2k} = 0;\ a_{2k+1} = (-1)^k(2k)!;\ b_{2k} = (-1)^{k+1}(2k-1)!;\ b_{2k+1} = 0$

7. $A = 1,\ B = \frac{2}{3},\ C = \frac{13}{15};\ x = 20.371303$ **8.** $A = -\frac{5}{6},\ B = \frac{169}{120}$

9. $\alpha_1 = \frac{1}{20},\ \alpha_2 = \frac{2}{525},\ \alpha_3 = \frac{13}{37800};\ x = \pm 0.551909$

10. $a_1 = a_2 = 1;\ a_3 = 2;\ a_4 = 5;\ a_5 = 11;\ a_6 = 28$

11. $y = 1 - \dfrac{p^2}{2!}x^2 - \dfrac{p^2(1-p^2)}{4!}x^4 - \dfrac{p^2(1-p^2)(4-p^2)}{6!}x^6 - \cdots$

12. $|\sin x| = \dfrac{2}{\pi} - \dfrac{4}{\pi}\sum_{k=1}^{\infty} \dfrac{\cos 2kx}{4k^2 - 1}$

$\quad |\cos x| = \dfrac{2}{\pi} - \dfrac{4}{\pi}\sum_{k=1}^{\infty} \dfrac{(-1)^k \cos 2kx}{4k^2 - 1}$

13. $f(x) = \dfrac{3\pi}{8} + \dfrac{2}{\pi}\left[\dfrac{\cos x}{1^2} - \dfrac{2\cos 2x}{2^2} + \dfrac{\cos 3x}{3^2} + \dfrac{\cos 5x}{5^2} - \dfrac{2\cos 6x}{6^2} + \dfrac{\cos 7x}{7^2} \right.$

$\quad\quad \left. + \dfrac{\cos 9x}{9^2} - \dfrac{2\cos 10x}{10^2} + \cdots \right]$

$\quad\quad = \dfrac{3\pi}{8} + \dfrac{2}{\pi}\sum_{k=0}^{\infty} \dfrac{\cos(2k+1)x}{(2k+1)^2} - \dfrac{4}{\pi}\sum_{k=0}^{\infty} \dfrac{\cos(4k+2)x}{(4k+2)^2}$

14. $a \simeq 1.2525;\ b \simeq 0.2449.$ The exact values are $\begin{cases} a = (\pi/2)^{1/2} = 1.2533 \\ b = 1/4 \end{cases}$

15. First case: $\quad\quad\quad\quad y = \sum_{k=0}^{\infty} a_k x^k / k!;$

Second case: $\quad\quad\quad\quad y = \sum_{k=0}^{\infty} (-1)^k b_k x^k / k!$

n	0	1	2	3	4	5	6	7	8	9	10
a_n	1	1	2	5	15	52	203	877	4140	21147	115975
b_n	1	1	0	-1	1	2	-9	9	50	-267	413

Alternatively we have:

$$a_n = e^{-1}\sum_{k=0}^{\infty} k^n / k!$$

$$b_n = e\sum_{k=0}^{\infty} (-1)^k k^n / k!$$

and

$$a_n = \sum_{k=1}^{n} \beta_k^{(n)}; \qquad b_n = \sum_{k=1}^{n} (-1)^k \beta_k^{(n)}$$

where $\beta_k^{(n)}$ are Stirling's numbers of the second kind (see equations (5.4.7) and (5.4.8)).

From the simple differential equations $y' = e^x y$ and $y' = e^{-x} y$ we obtain the recursion formulas:

$$a_{n+1} = \sum_{k=0}^{n} \binom{n}{k} a_k \qquad \text{and} \qquad b_{n+1} = \sum_{k=0}^{n} (-1)^{n+k} \binom{n}{k} b_k.$$

16. $a_1 = -1 \quad a_3 = \frac{1}{3} \quad a_5 = \frac{1}{5} \quad a_7 = \frac{1}{7} \qquad a_9 = \frac{8}{81}$
$a_2 = \frac{1}{2} \qquad a_4 = \frac{3}{8} \quad a_6 = \frac{13}{72} \quad a_8 = \frac{27}{128} \quad a_{10} = \frac{91}{800}$

17. $\dfrac{16}{\pi^2} \displaystyle\sum_{r,\,s=0}^{\infty} \dfrac{\sin(2r+1)\pi x \, \sin(2s+1)\pi y}{(2r+1)(2s+1)}$

CHAPTER 4

1. $Q_1 = x - \frac{1}{2}$; $Q_2 = x^2 - x + \frac{1}{6}$; $Q_3 = x^3 - \frac{3}{2}x^2 + \frac{3}{5}x - \frac{1}{20}$;
  $a_1 = a_2 = a_3 = \frac{1}{2}$; $b_2 = \frac{1}{12}$; $b_3 = \frac{1}{15}$

2. Discuss $P_k'(x)$.

3. $P_0 = 1$; $P_1 = 3.834553x - 1.505825$; $P_2 = 15.148233x^2 - 13.363231x + 1.8814545$

7. $P_0 = 1$; $P_1 = x\sqrt{2}$; $P_2 = \sqrt{3}(2x^2 - 1)$; $P_3 = \sqrt{4}(3x^3 - 2x)$; $P_4 = \sqrt{5}(6x^4 - 6x^2 + 1)$;
  $P_5 = \sqrt{6}(10x^5 - 12x^3 + 3x)$; $P_6 = \sqrt{7}(20x^6 - 30x^4 + 12x^2 - 1)$.
  The highest coefficient is $a_n = (n+1)^{1/2}\binom{n}{m}$; $m = \lfloor n/2 \rfloor$.

8. $f(x) = 2[T_1(x) + \frac{1}{3}T_3(x) + \frac{1}{5}T_5(x) + \cdots]$

CHAPTER 5

5. $\delta^2(x^2) = 2$; $\delta^2(x^3) = 6x$; $\delta^2(x^4) = 12x^2 + 2$; $f(x) = 2x^4 + 4x^3$. (To this we can add an arbitrary solution of the equation $\mu f(x) = 0$.)

7. $\delta y_0 = \frac{1}{2}(y_1 - y_{-1}) - \frac{1}{16}(\delta^2 y_1 - \delta^2 y_{-1}) + \frac{3}{256}(\delta^4 y_1 - \delta^4 y_{-1}) - \cdots$

8. $c_0 = 0$; $c_1 = 1$; $c_2 = 30$; $c_3 = 150$; $c_4 = 240$; $c_5 = 120$; $N = 63$

9. $J_0 y_0 = h(y_0 + \frac{1}{24}\delta^2 y_0 - \frac{17}{5760}\delta^4 y_0 + \cdots)$

10. $\delta'^2 = n^2\delta^2 + \dfrac{n^2(n^2-1)}{12}\delta^4 + \dfrac{n^2(n^2-1)(n^2-4)}{360}\delta^4 + \cdots$

CHAPTER 6

2. $x_{n+2} = 2x_{n+1} + 2x_n$

n	0	1	2	3	4	5	6
x_n	1	2	6	16	44	120	328

3. $y_n = ap^n + q(p^n - 1)/(p - 1)^2 + r(p^n - 1)/(p - 1) - qn/(p - 1);\ p \ne 1.$
$y_n = n(n - 1)q/2 + rn + a;\ p = 1.$

4. $y_n = 12^{-1}(n^4 - 4n^3 + 5n^2 + 10n)$ **5.** e **6.** $A = \frac{1}{6},\ \ B = \frac{1}{3},\ \ C = \frac{1}{2}$

7. (a) $\cos nx$ (b) $\sin nx/\sin x$

9. $k = N \tan(p\pi/n),\ \ p = 1, 2, 3, \dots;$ $p < N/2.$
$y_n = \left|\cos(p\pi/N)\right|^{-n} \sin(pn\pi/N)$

10. $\tan(n \arctan x)$ **11.** $\begin{cases} x_n = 5 \cdot 9^n - 2 \cdot 2^n \\ y_n = 9^n + 2^n \end{cases}$ **12.** $2, 1, 3, 4, 7, \dots$

CHAPTER 7

1. $\pi/4$ **2.** $\pi/60$ **3.** 3.625610 **4.** $\displaystyle\sum_{k=0}^{\infty} (-1)^k(4k + 1)^{-2} = 0.968532$ **5.** $\pi^2/4$

6. 0.78343051 **7.** $2\pi\sqrt{3}/9$ **8.** $(\pi/n)/\sin(\pi/n)$ **9.** $4/3$

10. $\pi/2 \cos(n\pi/2);\ -1 < n < 1$ **11.** $\dfrac{\pi}{2} \displaystyle\sum_{n=0}^{\infty} (-1)^n/(2^n n!)^2 = 1.2019697$

12. $\pi \displaystyle\sum_{n=0}^{\infty} (n!)^{-2} = 7.1615283$ **13.** $27\pi/1729$ **14.** $\pi/4$

16. $\sin(\pi/10) = (\sqrt{5} - 1)/4 = 0.309017$ **17.** $27\pi^2/1024$ **18.** $\pi/\cosh(\pi\sqrt{3}/2) = 0.411829$

19. $[\Gamma(1 + 1/p)]^2/\Gamma(1 + 2/p);\ \pi^2/6$ **20.** $ze^{z/2} \displaystyle\prod_{n=1}^{\infty} (1 + z^2/4\pi^2 n^2)$

23. $\cos \alpha x = \displaystyle\sum_{n=-\infty}^{\infty} (-1)^n J_{2n}(\alpha) T_{2n}(x);\ \cosh \alpha x = \displaystyle\sum_{n=-\infty}^{\infty} I_{2n}(\alpha) T_{2n}(x)$

25. $\cos(x \sin \theta) = J_0(x) + 2J_2(x) \cos 2\theta + 2J_4(x) \cos 4\theta + \cdots$
$\sin(x \sin \theta) = 2J_1(x) \sin \theta + 2J_3(x) \sin 3\theta + 2J_5(x) \sin 5\theta + \cdots$
In addition to those in Ex. 21, we also get $\sin x = 2J_1(x) - 2J_3(x) + 2J_5(x) - \cdots$

26. $2^{n+1}(n!)^2/(2n + 1)!$ **28.** $\dfrac{(-1)^k(2k)!}{2^{2k+1}k!(k + 1)!}$

29. $P'_n = (2n - 1)P_{n-1} + (2n - 5)P_{n-3} + \cdots + 3P_1$ (n even).
The value of the integral is $n(n + 1)$. Similar results are obtained if n is odd.

31. (a) $2^{n+1}r!\left(\dfrac{r + n}{2}\right)!\bigg/(r + n + 1)!\left(\dfrac{r - n}{2}\right)!$ (b) $(-1)^n(r!)^2/(n - r)!$ (c) $\pi^{1/2}r!/2^{r-n}\left(\dfrac{r - n}{2}\right)!$

CHAPTER 8

1. $p^{-1}\tanh(p/2)$ **2.** $\ln(p+1)$ **3.** $\frac{1}{2}\ln(p^2+1)$ **4.** $(p+1)^{-1/2}$ **5.** $\tanh(\pi p/2)$

6. $\arctan(1/p)$ **7.** $(e^p-1)\ln(1+e^{-p})$ **8.** $U(x)(e^x-1)^n/n!$ and $U(x)(1-e^{-x})^n/n!$

9. $U(x)\sum_{k=1}^{n}\dfrac{(-1)^{k-1}}{(n-k)!(n+k)!}\,4\sin^2\dfrac{kx}{2}$ **10.** $U(x)\left(1-e^{-x}\sum_{k=0}^{n-1}x^k/k!\right)$

11. $\pi/4-\tanh(\pi p/2)/2p=(2/\pi)p^2\sum_{k=0}^{\infty}(2k+1)^{-2}(p^2+(2k+1)^2)^{-1}$

12. $1/p-\pi/\sinh(\pi p)=2p[1/(p^2+1)-1/(p^2+4)+1/(p^2+9)-\cdots]$ or putting $p=iq$:
$$\pi/\sin(\pi q)=1/q+1/(1-q)-1/(2-q)+1/(3-q)-\cdots$$
$$-1/(1+q)+1/(2+q)-1/(3+q)+\cdots$$

13. $\dfrac{(\pi p/2)\cot(\pi p/2)}{1-p^2}=1+2p^2\sum_{n=1}^{\infty}\dfrac{1}{(4n^2-1)(4n^2-p^2)}$ (obtained when p is replaced by ip)

14. $xJ_0(x)$ and $xJ_1(x)$; $x(J_0(x)\cos x+J_1(x)\sin x)$ and $x(J_0(x)\sin x-J_1(x)\cos x)$.

15. $(m^{-1}+n^{-1})x^{-1}J_{m+n}(x)$ **16.** $(2/\pi)^{1/2}\sin((2x)^{1/2})$ **17.** $\pm 2\sin x$

18. x^2-6x+6 **19.** $(x^4+5x^3)e^x$ **20.** $\begin{cases}y=5e^{-x}+3e^{4x}\\z=5e^{-x}-2e^{4x}\end{cases}$

21. $\begin{cases}u=2e^{2x}\\v=10e^x-5e^{2x}\\w=10e^x-6e^{2x}\end{cases}$ **22.** e^x **23.** $y=e^x+2e^{-x/2}\cos(x\sqrt{3}/2)-3$

25. $y(x)=xU(x)+2(x-1)U(x-1)+2(x-2)U(x-2)+\cdots$

i.e.
$$
\begin{array}{llll}
y=x & \text{when} & 0\le x<1 \\
y=3x-2 & & 1\le x<2 \\
y=5x-6 & & 2\le x<3 \\
y=7x-12 & & 3\le x<4 \\
y=9x-20 & & 4\le x<5 \\
y=11x-30 & & 5\le x<6 \\
\vdots
\end{array}
$$

26. $y=3\cos x+\sin x$ **27.** $U(x)(\sinh x+\sin x)$ **28.** $\sin 2nx/2n$

29. $y=\displaystyle\int_0^{\infty}\exp(-t-x/t)\,dt;\ x>0.$

30. $y=2\sqrt{x}[AI_1(2\sqrt{x})+BK_1(2\sqrt{x})]$. The integral is $2\sqrt{x}\,K_1(2\sqrt{x})$.

CHAPTER 9

1. $\begin{cases}c=0.164178\\a=1.409319\end{cases}$ or $\begin{cases}c=0.777306\\a=0.424822\end{cases}$ **2.** $(2.81119,\ 1.78966)$

3. Straight lines   **4.** Circles

CHAPTER 10

1. $L = \begin{pmatrix} 1 & 0 & 0 & 0 \\ 3 & 1 & 0 & 0 \\ 5 & -2 & 1 & 0 \\ -4 & 2 & 1 & 1 \end{pmatrix}$ $R = \begin{pmatrix} 3 & 7 & -6 & 2 \\ 0 & 2 & 5 & 4 \\ 0 & 0 & 4 & 8 \\ 0 & 0 & 0 & 1 \end{pmatrix}$ $c = \begin{pmatrix} 4 \\ -3 \\ 4 \\ 1 \end{pmatrix}$ $x = \begin{pmatrix} 1 \\ -1 \\ -1 \\ 1 \end{pmatrix}$.

2. $L = \begin{pmatrix} 1 & 0 & 0 & 0 \\ 2 & 1 & 0 & 0 \\ -3 & 0 & 1 & 0 \\ 1 & 4 & -2 & 1 \end{pmatrix}$ $D = \begin{pmatrix} 4 & & & \\ & 3 & & \\ & & 2 & \\ & & & 1 \end{pmatrix}$ $c = \begin{pmatrix} 4 \\ -2 \\ -3 \\ 0 \end{pmatrix}$ $x = \begin{pmatrix} -1 \\ -2 \\ -3 \\ 0 \end{pmatrix}$.

3. $x = 8;\ y = -2;\ z = 3$

4. $L = \begin{pmatrix} 1 & 0 & 0 & 0 \\ 3 & 1 & 0 & 0 \\ -2 & 5 & 1 & 0 \\ 1 & -3 & 4 & 1 \end{pmatrix}$ $D = \begin{pmatrix} 2 & & & \\ & -1 & & \\ & & -3 & \\ & & & 4 \end{pmatrix}$ $x = \begin{pmatrix} 2 \\ 1 \\ 0 \\ -1 \end{pmatrix}$

5. $L = \begin{pmatrix} \sqrt{10} & 0 & 0 & 0 \\ 7/\sqrt{10} & 1/\sqrt{10} & 0 & 0 \\ 8/\sqrt{10} & 4/\sqrt{10} & \sqrt{2} & 0 \\ 7/\sqrt{10} & 1/\sqrt{10} & 3/\sqrt{2} & 1/\sqrt{2} \end{pmatrix}$ **6.** $L = \begin{pmatrix} 1 & 0 & 0 & 0 & 0 \\ 1 & 1 & 0 & 0 & 0 \\ 1 & 2 & 1 & 0 & 0 \\ 1 & 3 & 3 & 1 & 0 \\ 1 & 4 & 6 & 4 & 1 \end{pmatrix}$

7. $\begin{pmatrix} 5 & -10 & 10 & -5 & 1 \\ -10 & 30 & -35 & 19 & -4 \\ 10 & -35 & 46 & -27 & 6 \\ -5 & 19 & -27 & 17 & -4 \\ 1 & -4 & 6 & -4 & 1 \end{pmatrix}$ **8.** $p = 1/(1 - n);\ n \geq 2.$

10. (b) 6930   (c) $m_1 = -148;\ m_2 = 888;\ m_3 = 1750;\ m_4 = 2472;\ m_5 = 4208$

11. $c_2 = 1, c_3 = 1, c_4 = 1/2, c_5 = 2/7;\ d_1 = 1, d_2 = 1, d_3 = 2, d_4 = 7/2, d_5 = 33/7$

12. $\begin{cases} x = 1.49789 \\ y = 1.15190 \\ z = 1.13502 \end{cases}$ (exact values: $\frac{1}{711}(1065, 819, 807)$)   **13.** $\begin{cases} x = 0.41816 \\ y = 0.69964 \\ z = 0.69030 \end{cases}$

14. $\omega = 4 - 2\sqrt{2} = 1.1716 \simeq 1.2$
$\begin{cases} x_{n+1} = -0.2x_n + 0.6y_n + 4.2 \\ y_{n+1} = 0.6x_{n+1} - 0.2y_n + 0.6z_n + 0.6 \\ z_{n+1} = 0.6y_{n+1} - 0.2z_n + 0.6 \end{cases}$

x_k	3	4.8	5.448	5.929	5.9878
y_k	2	3.68	4.698	4.956	4.9979
z_k	1	2.608	2.897	2.994	2.9999

15. $z_1 = \begin{pmatrix} 1.93112 \\ 1.28029 \\ 1.86223 \end{pmatrix}$ $z_2 = \begin{pmatrix} 2.59767 \\ 1.29634 \\ 1.60808 \end{pmatrix}$

$z_3 = \begin{pmatrix} 2.999997 \\ 2.000004 \\ 1.000001 \end{pmatrix}$

CHAPTER 11

1. 1.403602   **2.** 1.7684 and 2.2410   **3.** 2.36502   **4.** 0.947747   **5.** $a = 0.85124$

6. $C = 1.50888$ **7.** $x = 1.4458$ **8.** 0.267949   **9.** $ab < 3/4$

10. $a = 0.3619$, $b = 1.3093$; 0.8858   **11.** $p = 0.804743$

12. 0.278465. This value is a root of the equation $\ln z + z + 1 = 0$.

13. 6   **14.** $2y^4 + y^3 + 3y^2 - 4y + 2$ where $y = x - 1$ **15.** $\begin{cases} x = 0.3181315 \\ y = 1.3372357 \end{cases}$

16. $(-0.325199, 0.785257)$ **17.** Max. $= 100$ for $(0, 5)$; Min. $= 7.2$ for $(2.4, 0.6)$

    (3.653496, 5.026068)

    (5.099901, 11.747201)

    (5.905204, 18.222801)

18. $f_{min} = 0.827329$ for $\begin{cases} x = 0.281213 \\ y = 0.083158 \end{cases}$

19. Local max. $x = 0$ Boundary min. $x = 2.0176887$

                  $y = 0.356272$ $y = 0$

                  $u = 34.703498$ $u = 22.28427$

  Boundary max. $x = 3$ Saddle point: $x = 1.51436$

                  $y = 3$ $y = 4/3$

                  $u = 37.76$ $u = 20\sqrt{2} + 80/9 = 37.17316$

  Boundary min. $x = 0$

                  $y = 3$

                  $u = 23.3333$

20. (a) 1.032801536; (b) 163.0000080; (c) -0.9701381291 **21.** 48.960664   **22.** 6.17417

CHAPTER 12

1. $7, -1, 2;$ $\begin{pmatrix} 1 \\ 1 \\ 0 \end{pmatrix}$, $\begin{pmatrix} 1 \\ -1 \\ 0 \end{pmatrix}$, $\begin{pmatrix} 0 \\ 0 \\ 1 \end{pmatrix}$,   **2.** 180.77772   **3.** 98.522

4. If u and v are eigenvectors of λ and $-\lambda$, then $A(u + v) = \lambda(u - v)$, $A^2(u + v) = \lambda^2(u + v)$.
Perform 2 steps and compute λ^2. Or: form $A + 2I$. Exact value: $\lambda \simeq 2.7461575$.

5. $\lambda_1 = \dfrac{R_1 R_3 - R_2^2}{R_1 - 2R_2 + R_3} = R_3 - \dfrac{(R_3 - R_2)^2}{R_1 - 2R_2 + R_3}$ or with $R = R_1 R_3 - R_2^2$, $S = R_1 - 2R_2 + R_3$,

$$\lambda_1 = R/S = R_3 - (R_3 - R_2)^2/S$$

6. $\mu_1 = 1.5002142$; $\lambda_1 = 10341.01524$; $\lambda_4 = \mu_1^{-1}$.
  $\mathrm{Tr}(H_4^{-1}) = 10496 = \lambda_1 + \lambda_2 + \lambda_3 + \lambda_4$; $\lambda_2 = 148.40597$; $\lambda_3 = 5.912219$.

7. 12.054; $(1, 0.5522i, 0.0995(3 + 2i))^T$ **8.** 19.286; -7.077 **9.** 4.040129   **10.** 70.21

11. $p = 2a + (n - 2)b$; $q = (a - b)[a + (n - 1)b]$; $\lambda_1 = a + (n - 1)b$ (simple);
  $\lambda_2 = a - b$ ($(n - 1)$-fold)

12. $A_1 = \begin{pmatrix} 9.5714 & 6 \\ 0.6122 & 1.4286 \end{pmatrix}$ $A_2 = \begin{pmatrix} 9.9552 & 6 \\ 0.0668 & 1.0448 \end{pmatrix}$ $A_3 = \begin{pmatrix} 9.9955 & 6 \\ 0.0067 & 1.0045 \end{pmatrix}$

Exact eigenvalues: 10, 1.

13. An arbitrary vector is a linear combination of the eigenvectors.

CHAPTER 13

1. $y_{max} = 4.4$ for $x^T = (4.4, 0, 0, 0.6)$ **2.** $y_{min} = -2$ for $x^T = (6, 8, 0)$

3. $\lambda \le -1$: $f_{min} = 6\lambda$; $-1 \le \lambda \le -\frac{1}{2}$: $f_{min} = 2\lambda - 4$; $-\frac{1}{2} \le \lambda$: $f_{min} = -5$

4. $f_{max} = 1$ for $x^T = (2, 2, 0, 0, 0)$ **5.** $f_{min} = 0$ for $x^T = (0, 3 - c, c, 0)$; $0 \le c \le 3$

6. $y_{min} = \frac{67}{12}$ for $x^T = (0, \frac{11}{12}, \frac{5}{4})$ **7.** $0 \le \alpha \le 2$; $f_{max} = 2\alpha$
                                                                          $2 \le \alpha \le 8$; $f_{max} = 2 + \alpha$
                                                                          $8 \le \alpha \le 12$; $f_{max} = 10$

8. $f_{min} = 3$ for $x = 3$, $y = 3$, $z = -3$.   **9.** $z_{max} = \frac{857}{16}$ for $x^T = (\frac{43}{16}, 1, 3, \frac{11}{2})$
  z can be written $z = u - v$ where $u \ge 0$, $v \ge 0$.

10. 0 lb of A, 17.5 lb of B, and 15 lb of C give a maximum profit of $1,375.

11. Mine A should be running 2 days, and B 6 days. Minimum cost: $18,000 per week.

12.

3	1	
	2	4
		5

Min = 186

13.

	75		25
25		75	
50		50	

Min = 5000

14.

5		7	6
5	5		

or

10		7	1
			5
	5		

or any combination of these two solutions.
Min = 135.

15. Several optimal solutions exist, for example,

4		4
3		
	5	1

or

4	4	
3		
	1	5

Min = 38

16. There are several optimal solutions, for example,

20	10			
	20	10	20	
40		10	25	
		20		

or

10	20			
	10	10	30	
10	40			25
			20	

Min = 430

17.

70	
60	10
30	20

18.

10	
	60
5	15
	10

Min = 730

CHAPTER 14

1. $f(3.63) = 0.136482$ **2.** 3.625   **3.** 0.55247 22945   **4.** 3.4159

5. 0.46163 21441 (correct value 0.46163 21450)   **6.** 0.000034

8. $A = \dfrac{1}{4}(p-1)^2(2+p)$; $B = \dfrac{1}{4}(p+1)^2(2-p)$; $C = \dfrac{1}{4}(p-1)^2(p+1)$;

$D = \dfrac{1}{4}(p+1)^2(p-1)$; $|R| = \dfrac{h^4}{24}(p^2-1)^2|f^{(iv)}(\xi)|$, $f(2.74) = 0.0182128$

9. $a = q$; $b = p$; $c = -\frac{1}{6}pq(q+1)$; $d = -\frac{1}{6}pq(p+1)$; Ai(1.1) = 0.120052.

10. (0.46996, 1.56250)   **11.** 12.95052

12. 5.2216. The function $z = x^{1/2}y$ can be interpolated.

13. $z = x^{1/3}y$; 1.99652   **15.** Differences of the rounded function values.   **16.** 3.1415

17. 0.484150   **18.** 0.33275   **19.** −0.43658   **20.** 0.061215

CHAPTER 15

1. (a) $y = 0.9431x - 0.0244$ (b) $y = 0.9537x - 0.0944$

2. $\begin{cases} a = 8.729243 \\ b = -0.9662646 \\ c = 0.03381963 \end{cases}$ $x = 14$, $y = 1.830176$ giving the time 15:31:50.

3. $a = 0.96103$, $b = 0.63212$ **4.** $a = 1$, $b = 2/5$, $c = 1/15$ **5.** $I_0 = 8.44$; $q = 0.289$

6. $a = -0.33824$, $b = -0.13313$, $c = 1.01937$ **7.** $a = 4.00$, $b = 120$

8. $a = 1.619$, $b = 0.729$, $A = 3.003$, $B = 0.996$. Use the fact that y satisfies a difference equation of the form $\Delta^2 y + (\alpha + \beta)\,\Delta y + \alpha\beta y = 0$ and determine $\alpha + \beta$ and $\alpha\beta$ by the least squares method; $\alpha = 1 - e^{-ah}$, $\beta = 1 - e^{-bh}$.

9. $a = 3.21$, $\alpha = 0.736$, $b = 1.47$. Use the fact that $\Delta^2 y/y = -(1 - e^{-\alpha h})$.

10. $a = 0.862$, $b = 0.995$, $c = 0.305$

11. $a = 8(3 - \sqrt{8}) = 1.3726$; $\varepsilon = (3 - \sqrt{8})/8 = 0.02145$ 12. $2\sqrt{2} - 1$

13. $x = 3.2$ $(11 - 3x = 2x - 5)$ 14. $0.99992 - 0.49878x + 0.08090x^2$

15. $P(x) = 0.999736x - 0.164497x^3 + 0.020443x^5$

CHAPTER 16

1. Theoretical values $\frac{3}{4}, \frac{1}{2}, \frac{1}{4}$ 2. (a) $\frac{1}{3}$ (b) $\frac{11}{18}, \frac{5}{18}, \frac{2}{18}$ 4. $m + \sigma\xi$ 5. 0.688

CHAPTER 17

1. 20.066 (exact value 21)   2. $\sum\limits_{n=1}^{\infty} n^{-n} = 1.291286$ 3. 1.46746

4. 3087 m/sec; 112.75 km   5. -0.94608 6. 1.3503 7. 9.688448 8. 1.995456

9. $\sum\limits_{k=0}^{\infty} (-1)^k (2k)!/(k+1)^{2k+1} = 0.818759$; 1.985522 10. 1.20197; 0.89324

11. 11.18866 12. $y_{\max} = 1.01494$ for $x = \pi/3$ 13. 1.816 14. $\pi^2/6$ and $\pi^2/12$

15. 0.27973 16. 1.74747 17. 1.293555 18. $A_1 = 0.389111$ $x_1 = 0.821162$
$A_2 = 0.277556$ $x_2 = 0.289949$

19. $A_1 = 0.718539$ $x_1 = 0.1120088$ 20. $A_1 = 1.30429$ $x_1 = 0.1155871$
$A_2 = 0.281461$ $x_2 = 0.6022769$ $A_2 = 0.69571$ $x_2 = 0.7415557$

21. $A_1 = A_3 = 1.10719$ $x_1 = -x_3 = 0.8138413$
$A_2 = 1.11895$ $x_2 = 0$

22. $A_1 = A_3 = 0.387883$ $x_1 = -x_3 = 0.743817$
$A_2 = \quad\ 0.795030$ $x_2 = 0$

23. $A_1 = A_2 = 1 - e^{-1}$ $x_1 = -x_2 = [(2e - 5)/(e - 1)]^{1/2} = 0.504053$

24. $A_1 = A_2 = \frac{1}{2}$ $x_1 = (6 - \sqrt{6})/12$; $x_2 = (6 + \sqrt{6})/12$

25. $k = \frac{2}{3}$; $x_1 = 0.0711$; $x_2 = 0.1785$; $x_3 = 0.7504$ 26. 0.2313

27. 0.722 (exact value $\pi^{3/2}/8 = 0.696$)   28. $a = \frac{2}{3}$; $b = \frac{1}{12}$; 2.241

29. The side of the square is a.

CHAPTER 18

1. 0.27768 **2.** 0.6048986 **3.** 0.92430 **4.** 0.580438 **5.** 0.91596 55942

6. 0.18785 964 **7.** 0.671866 + 1.076674i **8.** 5e **9.** $\pi^2/12$

10. $S = f(x) - \frac{1}{2}f'(x) - \frac{1}{4}f''(x) + \frac{1}{6}f'''(x) + \frac{5}{48}f^{(iv)}(x) - \frac{1}{15}f^{(v)}(x) + \cdots$ **11.** 1.405871

12. 1.341487 **13.** 0.77875 858 **14.** -1.4603545 (exact answer: $\zeta(\frac{1}{2}) = -1.46035\ 45088$)

15. 3.35988 56662

16. $\frac{1}{2}(n + 1)^2 \ln n - \frac{3}{4}n^2 - n + \frac{1}{2}n \ln(2\pi) - \frac{1}{12} \ln n + C + O(n^{-1})$, $C = 0.75351\ 73895$

17. 0.518335 **18.** 0.75536339 **19.** 3.676078 (exact value: $(\sinh \pi)/\pi = 3.67607\ 79103\ 76$)

20. 0.11494204 **21.** 0.309017 (exact value $\sin(\pi/10)$) **22.** 777564

24. 0.1 (incidentally, the answer is exact!) **25.** 2.55371 2683 **26.** 0.6557

CHAPTER 19

1.

x	0.2	0.4	0.6	0.8	1
y	0.8512	0.7798	0.7620	0.7834	0.8334

Minimum: (0.58, 0.76)

2.

x	0.5	1.0	1.5	2.0
y	1.3571	1.5837	1.7555	1.8956

3.

x	0.5	1.0
y(R.-K.)	0.50521	1.08508
y(ser.)	0.50522	1.08533

4. 0.01994; 0.07910; 0.17582

5. $y(0.2) = 0.2199$; $y(0.4) = 0.4783$
The series expansion $y = x + x^2/2 - x^4/24 - x^5/20 - x^6/36 + \cdots$ gives the same result.

6. $x(0.2) = 1.2650$, $y(0.2) = 0.1813$ (exact values: 1.2664, 0.1822)

7. $y(0.5) = -0.1135$; $y(1) = 0.0002$
Exact solution: $y = \ln(2\alpha^2/\cos^2(\alpha x - \alpha/2))$ where $\cos(\alpha/2) = \alpha\sqrt{2}$, i.e., $\alpha = 0.66802785$

8.

x	0.5	1.0
y	1.2604	2.2799

9. $\xi = 1.455$ **10.** $z_n = (-1)^n n! a^n b$

Exact solution: $y = (3x^2 + 4)/(4 - x^2)$

11. Try $y = \sin(\alpha x^n)$, $n > 1$ **12.** $y(1/2) = 0.21729$; $z(1/2) = 0.92044$

13. $p = 0.54369$; $q = 1.83928$

14. $y = k \sin n\varphi/\sin \varphi$ where $\cos \varphi = (12 - 5h^2)/(12 + h^2)$; $y_6 = 0.00049754$

15.

	2.4	2.6	2.8	3.0
y	0.6343	0.5189	0.4687	0.4944

16. $h \le \sqrt{6}$ **17.** $h \le a/b$

Minimum: (2.84, 0.47)

18. $y(0) = 0.0883$; $y(\frac{1}{2}) = 0.0828$ **19.** $y(0) = 0.0856$

20. Weak stability determined by the values $(1 + h/8)(-1 \pm i\sqrt{3})/2$ and $1 - h$

21. $y(\frac{1}{2}) = 0.496$; $z(0) = 0.547$

22. $I_{min.} = 61/66 = 0.924242$ when $a = -5/11$.
Exact value: $I_{min.} = 2\tanh(\frac{1}{2}) = 0.924234$ obtained from the exact solution
$y = \cosh(x - \frac{1}{2})/\cosh(\frac{1}{2})$.

23. $p = 0.92137056$; $y = p\left[x + \dfrac{2x^4}{4!} + \dfrac{2 \cdot 5x^7}{7!} + \dfrac{2 \cdot 5 \cdot 8x^{10}}{10!} + \cdots\right]$;
$I = 1.241495318$; $a = 0.9383704$; $b = -0.09972787$; $c = 0.16135747$; $I = 1.241539939$

24. $a = 1.1673$; $b = -0.1673$; $I_{min.} = 2.47493$ (correct value 2.47014)

25.

x	0.25	0.50	0.75
y_a	0.2617	0.5223	0.7748
y_b	0.2629	0.5243	0.7764

26. $a_1 = 1$; $a_3 = \dfrac{2(1 - \lambda)}{3!}$; $a_5 = \dfrac{2^2(1 - \lambda)(3 - \lambda)}{5!}$; $a_{2n+1} = \dfrac{2^n(1 - \lambda)(3 - \lambda) \cdots (2n - 1 - \lambda)}{(2n + 1)!}$
$\lambda = 4.57558$

27. $h = 1/4$: $\lambda = 17.87$; $h = 1/5$: $\lambda = 18.25$; $\lambda_{extr.} = 18.93$
(exact value through series expansion: $\lambda = 18.9562655$)

28. 8.60   **29.** $\lambda = \alpha^4 = 12.36236338$ where α is the smallest root of $\cosh x \cos x + 1 = 0$

30. $\lambda = 2.40$ (exact value: $\lambda = 2.4048256$)

31. 10, 11.21, 11.70; extrapolated to 12.28; $h = 1/5$ gives $\lambda = 11.93$; extrapolated to 12.336

CHAPTER 20

1.

y \ x	1/4	1/2	3/4
3/4	−0.176	−0.067	0.035
1/2	−0.379	−0.200	−0.057
1/4	−0.538	−0.297	−0.120

2.

y \ x	1.5	2	2.5
1.5	−0.84	−0.57	0.05
1	−0.46	0.00	0.78
0.5	−0.32	0.27	1.14

3.

h	u_0
1	−0.166667
1/2	−0.186275
1/3	−0.190797
1/4	−0.192474
extr.	−0.1947

Theoretical value:

$$-\frac{64}{\pi^4} \sum_{r,s=0}^{\infty} \frac{(-1)^{r+s}}{(2r + 1)(2s + 1)[(2r + 1)^2 + 2(2s + 1)^2]} = -0.19471072$$

4. $u(1/3, 1/3) = 0.71$; $u(2/3, 1/3) = 1.04$; $u(1/3, 2/3) = 1.05$; $u(2/3, 2/3) = 1.38$

5. $u = J_0(px) \exp(-p^2t)$ where $J_0(p) = 0$ (smallest value of p is $\simeq 2.4048$). More generally:

$$u = \sum_{r=1}^{\infty} a_r J_0(p_r x) \exp(-p_r^2 t) \quad \text{where} \quad J_0(p_r) = 0$$

6. $-\alpha h^2 \, \partial^4 u/\partial x^2 \, \partial t^2 + O(h^4)$ **8.** $\lambda = 12.27$ (exact value $5\pi^2/4 \simeq 12.34$)

9. $h = 1/2$ gives $\lambda_1 = 64$
$h = 1/3$ gives $\lambda_1 = 71.85$; extrapolated: 78.13
$h = 1/4$ gives $\lambda_1 = 75.328$
Extrapolated from all three: 80.356.

10. $h = 1/4$ gives $\lambda_1 = 41.3726$; extrapolated: $\lambda_1 = 48.94$ **11.** 9.8 (exact value $\pi^2 = 9.8696$)
$h = 1/5$ gives $\lambda_1 = 44.0983$; exact: $5\pi^2 = 49.35$
$h = 1/6$ gives $\lambda_1 = 45.6462$.
Extrapolated from all three $\lambda_1 = 49.348$.

12. $h = 1/3, \lambda = 486$ The condition $\partial u/\partial n = 0$ is taken care of through relations of the
$h = 1/4, \lambda = 642.784$ form $u(x + h, y) = u(x - h, y)$.
$h = 1/5, \lambda = 743.594$
$h = 1/6, \lambda = 809.455$
$h = 1/7, \lambda = 853.923$
$h = 1/8, \lambda = 884.999$
$h = 1/9, \lambda = 907.416$
Extrapolation: $h = 1/N$

N	3	4	5	6	7	8	9	λ
	•	•	•					966.939
		•	•	•				999.391
			•	•	•			998.668
				•	•	•		1000.547
			•	•	•	•		1000.932
			•	•	•	•	•	1001.109

Required value: 967;   correct value: 1001.1

13.
$u(0, 0.01) = 0.7697$ (0.7744) $u(0.3, 0.01) = 0.4121$ (0.4167)
$u(0.1, 0.01) = 0.7394$ (0.7205) $u(0.4, 0.01) = 0.2606$ (0.2795)
$u(0.2, 0.01) = 0.5879$ (0.5833) $u(0.5, 0.01) = 0.2303$ (0.2256)

The values within parentheses have been computed from the exact solution:

$$u = \frac{1}{2} + \frac{4}{\pi^2} \sum_{n=0}^{\infty} \exp(-4\pi^2(2n + 1)^2 t) \cos 2\pi(2n + 1)x/(2n + 1)^2.$$

14.
$u(0, 1/96) = 0.68334 = -u(1/2, 1/96)$ (0.66141)
$u(1/6, 1/96) = 0.37489 = -u(2/3, 1/96)$ (0.36825)
$u(1/3, 1/96) = -0.30845 = -u(5/6, 1/96)$ (−0.29316)

The values within parentheses have been computed from the exact solution:

$$u = \exp(-4\pi^2 t) \cos 2\pi(x - t).$$

CHAPTER 21

1. $D(\lambda) = 1 - \lambda/3$; $D(x, t; \lambda) = \lambda x t$ **2.** $y = x^{-1/2} + x$
3. $y = [(45 - 15\lambda)x^2 + 9\lambda]/(45 - 30\lambda - 4\lambda^2)$ **4.** $\lambda = -8 \pm \sqrt{76}$
5. $y = 1.486x - 0.483x^3 + 0.001725x^5$

7. $g(s) \simeq 0.964s^2 + 0.404s + 0.441$; $g(s) = (12/1307)(105s^2 + 44s + 48)$

8. $g(0) \simeq 0.67346$; $g(1/4) \simeq 0.82473$; $g(1/2) \simeq 1.00181$; $g(3/4) \simeq 1.19559$; $g(1) \simeq 1.40086$

9. $\lambda = \pm 4(\pi^2 - 4)^{1/2}$. Note that $\cos(x + t)$ can be written as a degenerate kernel.

10. $y = x \exp(x^2/4)$

11. $y = \exp(x^2/2) \int_0^x \exp(-t^2/2)\, dt = x + x^3/1 \cdot 3 + x^5/1 \cdot 3 \cdot 5 + x^7/1 \cdot 3 \cdot 5 \cdot 7 + \cdots$

12. $\lambda_n = (2n + 1)^2 \pi^2/4$; $y_n = \sin(n + \tfrac{1}{2})\pi x$.   **13.** $y(x) = 6x + 2$

14.

x	0	0.2	0.4	0.6	0.8	1
y	1	1.167	1.289	1.384	1.461	1.525

15. $g(s) = \cosh s$

16. We find $D(\lambda) = 1 - \dfrac{\lambda^2}{12} - \dfrac{\lambda^3}{90} - \dfrac{\lambda^4}{1680} - \cdots$ and $\lambda \simeq 2.88298$.

Exact value: $\lambda = 2\mu^2$ where $\mu \tanh \mu = 1$.

17. $y(x) = 1 + \int_0^x y(t)\, dt$ **18.** $y(x) = 1 - 2x - 4x^2 + \int_0^x [3 + 6(x - t) - 4(x - t)^2] y(t)\, dt$

19. $y = 1 + \int_0^1 K(x, t) y(t)\, dt$ where $K(x, t) = \begin{cases} tx(1 - x), & t < x \\ x^2(1 - t), & t > x \end{cases}$

20. $y(0) = 1$; $y(0.2) = 1.1087$; $y(0.4) = 1.2322$; $y(0.6) = 1.3656$; $y(0.8) = 1.5048$; $y(1) = 1.6467$

References

M. Abramowitz and I. A. Stegun (eds.): Handbook of Mathematical Functions with Formulas, Graphs, and Mathematical Tables. Dover, New York, 1964.

N. I. Achieser: Theory of Approximations. Translated from the Russian by C. Hyman. Frederick Ungar, New York, 1956.

F. S. Acton: Numerical Methods that Work. Harper & Row, New York, 1970.

J. Ahlberg, E. Nilson, and J. Walsh: The Theory of Splines and Their Application. Academic Press, New York, 1967.

L. V. Ahlfors: Complex Analysis (2nd ed.) McGraw-Hill, New York, 1966.

A. V. Aho, J. E. Hopcroft, and J. D. Ullman: The Design and Analysis of Computer Algorithms. Addison-Wesley, Reading, Mass., 1974.

W. F. Ames: Numerical Methods for Partial Differential Equations (2nd ed.). Academic Press, New York, 1977.

J. Babuska, M. Práger, and E. Vitásek: Numerical Processes in Differential Equations. Prag, 1965.

C. T. H. Baker: The Numerical Treatment of Integral Equations. Clarendon Press, Oxford, 1977.

R. Bellman: Dynamic Programming. Princeton University Press, Princeton, 1957.

E. K. Blum: Numerical Analysis and Computation: Theory and Practice. Addison-Wesley, Reading, Mass., 1972.

C. de Boor: A Practical Guide to Splines. Springer-Verlag, New York, 1978.

E. O. Brigham: The Fast Fourier Transform. Prentice-Hall, Englewood Cliffs, N.J., 1974.

R. L. Burden, J. D. Faires, and A. C. Reynolds: Numerical Analysis. Prindle, Weber and Schmidt, Boston, 1981.

E. W. Cheney: Introduction to Approximation Theory. McGraw-Hill, New York, 1966.

E. W. Cheney and D. Kincaid: Numerical Mathematics and Computing. Brooks-Cole Publishing Company, Monterey, 1980.

L. Collatz: The Numerical Treatment of Differential Equations. Springer-Verlag, Berlin, 1960.

L. Collatz: Functional Analysis and Numerical Mathematics. Academic Press, New York, 1966.

S. Conte and C. de Boor: Elementary Numerical Analysis (2nd ed.). McGraw-Hill, New York, 1972.

G. Dahlquist and Å. Björck: Numerical Methods. Prentice-Hall, Englewood Cliffs, N.J., 1974.

G. B. Dantzig: Linear Programming and Extensions. Princeton University Press, Princeton, 1963.

P. Davis: Interpolation and Approximation. Blaisdell, New York, 1963.

P. Davis and P. Rabinowitz: Methods of Numerical Integration. Academic Press, New York, 1975.

J. J. Dongarra, J. R. Bunch, C. B. Moler, and G. W. Stewart: LINPACK User's Guide. SIAM, Philadelphia, 1979.

D. K. Faddeev and V. N. Faddeeva: Computational Methods of Linear Algebra. Freeman, San Francisco, 1963.

G. Forsythe, M. Malcolm, and C. Moler: Computer Methods for Mathematical Computations. Prentice-Hall, Englewood Cliffs, N.J., 1977.

G. Forsythe and C. Moler: Computational Solution of Linear Algebraic Systems. Prentice-Hall, Englewood Cliffs, N.J., 1967.

G. E. Forsythe and W. R. Wasow: Finite Difference Methods for Partial Differential Equations. John Wiley, New York, 1960.

F. R. Gantmacher: Matrizenrechnung, I–II. Translated from the Russian. Berlin, 1958.

S. T. Gass: Linear Programming, McGraw-Hill, New York, 1958.

C. W. Gear: Numerical Initial Value Problems in Ordinary Differential Equations. Prentice-Hall, Englewood Cliffs, N.J., 1971.

C. F. Gerald: Applied Numerical Analysis. Addison-Wesley, Reading, Mass., 1978.

G. H. Golub and C. Reinsch: Singular Value Decomposition and Least Squares Solutions, Handbook for Automatic Computation, Vol. II: Linear Algebra. Springer, Heidelberg, 1971.

A. R. Gourlay and G. A. Watson: Computational Methods for Matrix Eigen-problems. John Wiley, New York, 1973.

R. T. Gregory and D. L. Karney: A Collection of Matrices for Testing Computational Algorithms. Wiley Interscience, New York, 1969.

W. Gräbner and N. Hofreiter: Integraltafel. Springer, Berlin, 1961.

P. G. Guest: Numerical Methods of Curve Fitting. Cambridge, 1961.

G. Hadley: Nonlinear and Dynamic Programming. Addison-Wesley, Reading, Mass., 1964.

R. W. Hamming: Numerical Methods for Scientists and Engineers. McGraw-Hill, New York, 1973.

Harvard Computational Laboratory: Table of Bessel Functions, Vol. I–XII. Cambridge, Mass., 1947–1951.

P. Henrici: Elements of Numerical Analysis. John Wiley, New York, 1964.

P. Henrici: Discrete Variable Methods in Ordinary Differential Equations. John Wiley, New York, 1962.

P. Henrici: Applied and Computational Complex Analysis, Vol. I. John Wiley, New York, 1974.

F. B. Hildebrand: Introduction to Numerical Analysis. McGraw-Hill, New York, 1974.

A. S. Householder: Principles of Numerical Analysis. McGraw-Hill, New York, 1953.

A. S. Householder: The Theory of Matrices in Numerical Analysis. Blaisdell, New York, 1964.

A. Hurwitz and R. Courant: Funktionentheorie. Berlin, 1929.

E. Isaacson and H. B. Keller: Analysis of Numerical Methods. John Wiley, New York, 1966.

D. Jacobs (ed.): The State of The Art in Numerical Analysis. Academic Press, New York, 1977.

H. Keller: Numerical Methods for Two-Point Boundary-Value Problems. Blaisdell, Waltham, Mass., 1968.

D. E. Knuth: The Art of Computer Programming, Vol. 2: Seminumerical Algorithms. Addison-Wesley, Reading, Mass., 1969.

V. Krylov: Approximate Calculation of Integrals. Macmillan, New York, 1962.

J. D. Lambert: Computational Methods in Ordinary Differential Equations. John Wiley, New York, 1973.

C. Lanczos: Applied Analysis. London, 1957.

L. Lapidus and J. H. Sinfeld: Numerical Solution of Ordinary Differential Equations. Academic Press, New York, 1971.

C. L. Lawson and R. J. Hanson: Solving Least-Squares Problems. Prentice-Hall, Englewood Cliffs, N.J., 1974.

G. I. Marchuk: Methods of Numerical Mathematics. Springer-Verlag, New York, 1975.

L. M. Milne–Thomson: The Calculus of Finite Differences. Macmillan, London, 1933.

A. R. Mitchell: Computational Methods for Partial Differential Equations. John Wiley & Sons, London, 1969.

A. R. Mitchell and R. Wait: The Finite Element Method in Partial Differential Equations. John Wiley, New York, 1977.

R. E. Moore: Interval Analysis. Prentice-Hall, Englewood Cliffs, N.J., 1966.

I. P. Natanson: Konstruktive Funktionentheorie. Akademie-Verlag, Berlin, 1955.

National Bureau of Standards (NBS): Basic Theorems in Matrix Theory, Appl. Math. Series, No. 57. Washington, 1960.

National Bureau of Standards (NBS): Simultaneous Linear Equations and the Determination of Eigenvalues, Appl. Math. Series, No. 29. Washington, 1953.

National Bureau of Standards (NBS): Contributions to the Solution of Linear Equations and the Determination of Eigenvalues, Appl. Math. Series, No. 39. Washington, 1954. Further Contributions ..., No. 49, 1956.

National Bureau of Standards (NBS): Handbook of Mathematical Functions, AMS 55, Washington.

National Physical Laboratory: Modern Computing Methods. London, 1961.

J. Oden and J. Reddy: An Introduction to the Mathematical Theory of Finite Elements. John Wiley, New York, 1976.

J. Ortega: Numerical Analysis, A Second Course. Academic Press, New York, 1972.

J. Ortega and W. Rheinboldt: Iterative Solution of Equations in Several Variables. Academic Press, New York, 1970.

E. D. Rainville: Special Functions. Macmillan, New York, 1960.

A. Ralston and P. Rabinowitz: A First Course in Numerical Analysis (2nd ed.). McGraw-Hill, New York, 1978.

A. Ralston and H. S. Wilf (eds.): Numerical Methods for Digital Computers, Vols. 1 and 2. John Wiley, New York, 1960 and 1967.

H. Ratschek and J. Rokne: Computer methods for the Range of Functions, Ellis Horwood Ltd., Chichester, England.

G. D. Smith: Solution of Partial Differential Equations. Oxford University Press, London, 1965.

B. T. Smith et al.: Matrix Eigensystems Routines—EISPACK Guide, Lecture Notes in Computer Science 6 (2nd ed.). Springer-Verlag, New York, 1976.

G. W. Stewart: Introduction to Matrix Computations. Academic Press, New York, 1973.

J. Stoer and R. Bulirsch: Introduction to Numerical Analysis. Springer-Verlag, New York, 1980.

G. Strang: Linear Algebra and Its Applications. Academic Press, New York, 1976.

A. Stroud: Approximate Calculation of Multiple Integrals. Prentice-Hall, Englewood Cliffs, N.J., 1971.

A. H. Stroud and D. Secrest: Gaussian Quadrature Formulas. Prentice-Hall, Englewood Cliffs, N.J., 1966.

G. Szegö: Orthogonal Polynomials. Amer. Math. Soc., New York, 1959.

J. F. Traub: Iterative Methods for the Solution of Equations. Prentice-Hall, Englewood Cliffs, N.J., 1964.

S. Vajda: Mathematical Programming. Addison-Wesley, Reading, Mass., 1961.

R. Varga: Matrix Iterative Analysis. Prentice-Hall, Englewood Cliffs, N.J., 1962.

G. A. Watson: Approximation Theory and Numerical Methods. John Wiley, New York, 1980.

G. N. Watson: A Treatise on the Theory of Bessel Functions. Cambridge University Press, 1966.

E. T. Whittaker and G. N. Watson: A Course of Modern Analysis. Cambridge University Press, 1946.

J. H. Wilkinson: Rounding Errors in Algebraic Processes. Prentice-Hall, Englewood Cliffs, N.J., 1964.

J. H. Wilkinson: The Algebraic Eigenvalue Problem. Oxford University Press, 1965.

J. Wilkinson and C. Reinsch (eds.): Handbook for Automatic Computation, Vol. II: Linear Algebra. Springer-Verlag, New York, 1971.

S. Yakowitz: Computational Probability and Simulation. Addison-Wesley, Reading, Mass., 1977.

D. Young: Iterative Methods for the Solution of Large Linear Systems. Academic Press, New York, 1971.

D. Young and R. T. Gregory: A Survey of Numerical Mathematics, Vols. I–II. Addison-Wesley, Reading, Mass., 1972.

Index